3rd International Symposium on Sustainable Landscape Development (ISSLD 2017)

IOP Conference Series: Earth and Environmental Science
Volume 179

Bogor, Indonesia
14 - 15 November 2017

Editors:

Regan Leonardus Kaswanto
Yulius Budi Prastiyo
Nike Dyah Permata
Merisa Asrina
Riska Nabilah
Keni Ramandhani
Jennie Jesica
Istiqamah

ISBN: 978-1-5108-7331-5
ISSN: 1755-1307

Printed from e-media with permission by:

Curran Associates, Inc.
57 Morehouse Lane
Red Hook, NY 12571

Some format issues inherent in the e-media version may also appear in this print version.

Printed by Curran Associates, Inc. (2018)

For permission requests, please contact the Institute of Physics
at the address below.

Institute of Physics
Dirac House, Temple Back
Bristol BS1 6BE UK

Phone: 44 1 17 929 7481
Fax: 44 1 17 920 0979

techtracking@iop.org

Additional copies of this publication are available from:

Curran Associates, Inc.
57 Morehouse Lane
Red Hook, NY 12571 USA
Phone: 845-758-0400
Fax: 845-758-2633
Email: curran@proceedings.com
Web: www.proceedings.com

Table of contents

Volume 179

3rd International Symposium for Sustainable Landscape Development (ISSLD 2017)

14–15 November 2017, Bogor, Indonesia

Accepted papers received: 23 July 2018
Published online: 17 October 2018

Preface

Papers

Mitigation

Resilience

IOP Conf. Series: Earth and Environmental Science **179** (2018) 011001 doi:10.1088/1755-1315/179/1/011001

FOREWORD

The world's awareness of sustainable landscape development has been proclaimed by the United Nations (UN) as the Sustainable Development Goals (SDGs), particularly at Goal 11th: *to make cities and human settlements inclusive, safe, resilient and sustainable* (source: https://sustainabledevelopment.un.org/sdg11). The process to achieve entire SDGs have to involve communities for implementing a sustainable landscape for living. The Indonesian government is also accomplishing those SDGs by a working plan: **nawacita**. The **nawacita** is required to be bridged and facilitated by the community which can be obtained by scientific activities such as a symposium. The 3rd International Symposium for Sustainable Landscape Development (**The 3rd ISSLD**) is a continuation of the annual symposium. The sustainability of this activity is very important to enhance the **nawacita**. The topic for this annual symposium has been adapted to specific conditions for each year.

The 3rd ISSLD was held on November 14-15, 2017 at the IPB International Convention Centre (IICC), Bogor, Indonesia. The 3rd ISSLD is jointly organized by the Department of Landscape Architecture, Faculty of Agriculture, Bogor Agriculture University (IPB), and the Directorate of Research and Innovation IPB (DRI IPB). This symposium produced 43 articles which are published in this issue of IOP Conference Series: Earth and Environmental Science (EES). Those articles came from subtopics including mitigation (9 articles), sustainability (22 articles), and resilience (12 articles). The authors came from various countries and institutions, as well as professional and academicians. More than 200 participants from five countries joined the symposium, such as Japan, Netherland, Malaysia, China, Korea, Philippines, and others.

The 3rd ISSLD aims to assess the landscape issue through a variety of relevant case studies for achieving sustainable landscape development, both from developed and developing countries. Various experiences towards a sustainable landscape are currently an intense research topic in the international level such as research on urban resilience, green technology, sustainable city, until mitigation and adaptation of climate change. At the 3rd ISSLD, the researchers, students, businessman, practitioners, stakeholders, and governments can discuss and exchange experiences to formulate an awareness to create a sustainable and comfortable tropical landscape. The 3rd ISSLD has successfully combined those participants from academicians (A), businessman (B), government (G), community (C) and mass media (M) to reach an ABGCM pentahelix collaboration for achieving sustainable landscape development.

We are really thankful to all authors who have contributed high scientific articles to become this proceeding of the 3rd ISSLD. We are also grateful to IOP Publishing, especially Anete Ashton (IOP Conference Commissioning Editor), for allowing us to arrange some of the symposium results to be published in the IOP Conference Series: Earth and Environmental Science (EES). We are trully thank to all dedicated 11 reviewers, 6 editors, 7 assitant editors, and 18 scientific committes who have reviewed and encouraged the process to publish 43 arcticles in excellent ways. We also would like to thank the Ministry of Research, Technology and Higher Education (*Kementerian Riset, Teknologi dan Perguruan Tinggi* - RISTEKDIKTI) for the opportunity to join the World Class University (WCU) agenda.

We also give our gratitude to our sponsor: Ministry of Agriculture for providing excellent support. And also to Sentul City, EPSON and Bank BNI for supporting this event which has succesffully been held over two days. Thank you.

Regan Leonardus Kaswanto
Editor-in-Chief
Proceeding the 3rd ISSLD 2017

IOP Conf. Series: Earth and Environmental Science **179** (2018) 011001 doi:10.1088/1755-1315/179/1/011001

Editorial Committees

Regan Leonardus Kaswanto, *Department of Landscape Architecture, Faculty of Agriculture, Bogor Agricultural University IPB, Jl. Meranti Kampus IPB Darmaga 16680, Bogor, West Java, Indonesia.*

Nobukazu Nakagoshi, *International Development and Environmental Cooperation (IDEC), Hiroshima University, 1–5–1 Kagamiyama, Higashi - Hiroshima shi, 739-8529 Hiroshima, Japan.*

Theo Spek, *Faculty of Arts Landscape History - Research Centre Arts in Society, University of Groningen, Oude Boteringestraat 34, 9712 GK Groningen, Netherlands.*

Hadi Susilo Arifin, *Department of Landscape Architecture, Faculty of Agriculture, Bogor Agricultural University IPB, Jl. Meranti Kampus IPB Darmaga 16680, Bogor, West Java, Indonesia.*

Nizar Nasrullah, *Department of Landscape Architecture, Faculty of Agriculture, Bogor Agricultural University IPB, Jl. Meranti Kampus IPB Darmaga 16680, Bogor, West Java, Indonesia.*

Akhmad Arifin Hadi, *Department of Landscape Architecture, Faculty of Agriculture, Bogor Agricultural University IPB, Jl. Meranti Kampus IPB Darmaga 16680, Bogor, West Java, Indonesia.*

Scientific Committees

Afra DN Makalew, *Bogor Agricultural University, Indonesia*
Akhmad Arifin Hadi, *Bogor Agricultural University, Indonesia*
Andi Gunawan, *Bogor Agricultural University, Indonesia*
Aris Munandar, *Bogor Agricultural University, Indonesia*
Damian Tang, *International Federation of Landscape Architects – Asia Pasific Region (IFLA-APR), Singapore*
Hadi Susilo Arifin, *Bogor Agricultural University, Indonesia*
Indung Sitti Fatimah, *Bogor Agricultural University, Indonesia*
Nobukazu Nakagoshi, *Hiroshima University, Japan*
Nizar Nasrullah, *Bogor Agricultural University, Indonesia*
Nurhayati Hadi Susilo Arifin, *Bogor Agricultural University, Indonesia*
Qodarian Pramukanto, *Bogor Agricultural University, Indonesia*
Regan Leonardus Kaswanto, *Bogor Agricultural University, Indonesia*
Siti Nurisjah, *Indonesian Society of Landscape Architects (ISLA / IALI), Indonesia*
Saiful Arif Abdullah, *the National University of Malaysia, Malaysia*
Syartinilia, *Bogor Agricultural University, Indonesia*
Tati Budiarti, *Bogor Agricultural University, Indonesia*
Theo Spek, *Groningen University, Netherlands*
Wahju Qamara Mugnisjah, *Bogor Agricultural University, Indonesia*

Assistant Editors

Yulius Budi Prastiyo, Nike Dyah Permata, Merisa Asrina, Riska Nabilah, Keni Ramandhani, Jennie Jesica, and Istiqamah

Department of Landscape Architecture, Faculty of Agriculture, Bogor Agricultural University IPB, Jl. Meranti Kampus IPB Darmaga 16680, Bogor, West Java, Indonesia.

Opening Speech from Vice Rector of Bogor Agricultural University (IPB):
Prof. Dr. Ir. Hermanto Siregar, MEc

A Reporting Address from Chairman of Organizing Committee:
Dr. Regan Leonardus Kaswanto, SP, MSi

Keynote Speech from Food Security Agency, Ministry of Agriculture:
Ir. Tri Agustin Satriani, MM

Keynote Speech from IDEC, Hiroshima University, Japan:
Prof. Nobukazu Nakagoshi

Keynote Speech from Fac. Arts Landscape History, Groningen University, Netherland:
Prof. Theo Spek

Picture together of keynote speaker, invited speaker, and some delegates from various institutions

Picture together with all participant of the 3rd ISSLD 2017 (First Day)

First Day the 3rd ISSLD 2017

Invited Speaker from IFLA-APR, Singapore:
Mr. Damian Tang

Invited Speaker from ISLA, Indonesia:
Mr. Rahman Andra Wijaya

Invited Speaker from Sentul City, Urban Developer:
Mr. Ricky Kinanto Teh

IOP Conf. Series: Earth and Environmental Science **179** (2018) 011001 doi:10.1088/1755-1315/179/1/011001

Excursion to Informal Settlement Landscape: Pulo Geulis.

Excursion to natural and man-made landscape: Bogor Botanical Garden (BBG).

Excursion inside Bogor Botanical Garden: Sri Gunting Lake.

IOP Conf. Series: Earth and Environmental Science **179** (2018) 011001 doi:10.1088/1755-1315/179/1/011001

Excursion inside Bogor Botanical Garden: Sri Gunting Lake.

Excursion to Danau Northridge, Sentul City, Bogor.

SPONSORS

THANK YOU TO OUR SPONSORS

FOUNDATION SPONSORS

INSTITUT PERTANIAN BOGOR

ASOSIASI PENDIDIKAN ARSITEKTUR LANSKAP INDONESIA APALI

iali IOP

GOLD SPONSORS

Food Security Agency
Minostry of Agriculture

SILVER SPONSORS

Sentul City
My First Choice

BNI

EPSON
EXCEED YOUR VISION

IOP Conf. Series: Earth and Environmental Science **179** (2018) 011002 doi:10.1088/1755-1315/179/1/011002

Peer review statement

All papers published in this volume of *IOP Conference Series: Earth and Environmental Science* have been peer reviewed through processes administered by the proceedings Editors. Reviews were conducted by expert referees to the professional and scientific standards expected of a proceedings journal published by IOP Publishing.

Correlation of microclimates dynamic to plants survivorship and space amenity of Cibodas Botanical Garden

I W Hidayat[1*] and A Darmawan[2]

[1] Researcher, Cibodas Botanical Garden-Indonesian Institute of Sciences, PO Box 19 SDL, Cipanas-Cianjur, West Java Province, Indonesia 43253
[2] IT and Database Technician, Cibodas Botanical Garden-Indonesian Institute of Sciences, PO Box 19 SDL, Cipanas-Cianjur, West Java Province, Indonesia 43253

[*] Email: imawan.wahyu.hidayat@lipi.go.id

Abstract.The microclimate extreme amplitudes decrease plant survival chance in the field. In order to find out the impact of microclimatesto plants survivorship, an action research was conducted in Cibodas Botanical Garden by inventoried microclimate data series of temperature, humidity, rainfall, wind velocity, air pressure, and solar radiation, from 2012 to 2016, and thecorrelation to plants loss of the garden. The study hasalso conducted the analysis of the dynamics of space amenity. Space amenity was assumed comparable with temperature-humidity index (THI) value. The quality of space amenity was measured bythe correlation between THItothe number of tourists. The results indicated that an increase in plants loss caused by an extreme microclimate, especially heavy rainfall that occurred in early and end of the year, along with the decrease of the visits. Total plants loss were 790 specimens and the lost which occurred at early-end of the years contributed 64.2% of total lost, bigger than the middle of the year.Despite THInever exceeded comfort value (< 26), still occurred decreased of the visits.Therefore, to minimize the impact of uncertain future microclimate conditions, operators should give better plants management toanticipate the greater losses.

Keywords: microclimate impact, plants management, Temperature Humidity Index (THI).

1. Introduction

The microclimate, which can be defined as a small scale pattern of climate that is influenced bysite topography as well as local built forms and materials, can be heavily affected by vegetation [1].Favorable microclimates in otherwise inhospitable landscapes allowed the survival of climate relict tree populations and associated biodiversity in the past, and are predicted to buffer populations against ongoing environmental change [2, 3]. When discussing individual sites, we thus use the term 'micro-climate' to refer to vegetation and environmental variability on a sub-landscape scale, at which topographic factors can create suitable conditions for localized tree populations and woodland communities to survive potentially unfavorable regional climatic regimes [4].

Microclimates have a direct effect to plant growth and survivorship in the field. An extreme amplitude can cause a decrease in plant survival and development. The rate of plant growth and development is dependent upon the temperature surrounding the plant and each species has a specific temperature range represented by a minimum, maximum, and optimum [5]. These values were summarized by Hatfield et al. [5] for a number of different species typical of grain and fruit production. Heat waves or extreme temperature events are projected to become more intense, more frequent, and last longer than what is being currently been observed in recent years [6]. The extreme temperature events may have short-term durations of a few days with temperature increases of over 5°C above the normal temperatures. Extreme events occurring during the summer period would have the most dramatic impact on plant productivity; however, there has been little research conducted to document these effects as found by Kumudini et al. [7].

Space amenity is not straightforward because this amenity is especially difficult to characterize [8]. Space amenity is not only perceived the experience of a high value of aesthetical view of the site but also provide a cohesive environment and a more identifiable character. A cohesive environment comprises man-made infrastructures, natural qualities, such as climate, topography, water resources and cultivated landscapes. In this paper, microclimates elements i.e. temperature and humidity, are also forming space amenity, in the context of the results of the complex interaction between built and green spaces. A comfort temperature-humidity will invite visitors when associated with a global climate change and the impacts to discomforts of surrounding environment. Recently, urban residents prefer to visit tourist area which offers a 'natural sense' as their attraction. Related to both, mountain area which provides a comfort temperature-humidity frequently become a tourist destination. These are interconnected with comfort felt, balmy temperature. Space microclimates comfort which formed by temperature and humidity can be represented by temperature-humidity index (THI). THI describes the degree measure of comfort that is experienced by the human from the combination of air temperature and humidity [9]. Generally, the value THI can describe initial microclimates amenity surrounding the area. This is important to support for tourism, especially for the visits of tourist.

Cibodas Botanical Garden (CBG) as a conservation area for *ex-situ* plant, research and education and also ecotourism destination, is closely related to microclimates condition. Fluctuation and dynamic condition of microclimates are influencing of plants growth and space amenity for visitors. Therefore, this study purposes were to analysis the impact of microclimates factors (i.e. temperature, humidity, rainfall, wind velocity, air pressure, and solar radiation) to plants survivorship, which represented by the number of lost plants and their correlation, and to assess space amenity, which represented by the correlation between THI and the number of visits along last five years. The results were expected to be the main consideration for site planning and management action, in order to minimize the impact of uncertain future microclimate conditions, which operators should give better attention to plants and management to anticipate greater losses.

2. Methods

The study was conducted at CBG which administratively located in Cipanas District, Cianjur Regency, West Java Province. CBG is located at mountainside of Mount Gede and Mount Pangrango at altitude of approximately 1,300-1,425 meter above sea level, with area of 84.99 hectares. In accordance with its functions, CBG had an obligation to perform of *ex situ* conservation efforts, especially for the types of plants from tropical wet highlands zone, for research and education, and also for ecotourism. CBG is a comfortable place to rest while enjoying the beauty of the various types of plants that originated from Indonesia, especially from tropical wet highland, and other foreign countries. CBG is located± 100 km from Jakarta and ± 80 km from Bandung.

2.1. Data acquisition

Collected data were consisted three main data, microclimates series data (i.e. temperature, humidity, rainfall, average wind velocity and solar radiation), lost plants data of each month and the number of tourists who visits CBG. All the data were recorded and collected from 2012 to 2016. Microclimates data were collected from the weather station which established at CBG. Weather equipment was obtained from *Precision Weather Station Davis Instruments Vantage Pro2 Plus*™. The equipment was automatic constantly sent current data to the server regarding microclimates condition of CBG. The observation equipment was located in front of management office of CBG. The collected data were accurate to describe weather condition surrounding equipment minimum in a radius of the one-kilometer square.

The second data was the plants loss, which defines as the reduced of garden plants collection. Plants loss were assumption closely related with survivorship, increasingly of the plants loss was indicated that the survivorship decreased, and vice versa. The dynamic of plants richness of CBG were routinely recorded by 'Registration Unit' of CBG [10].

The third data was the number of CBG visitors. CBG is a government institution which also conducting ecotourism. Tourists who visit CBG will be charged in form of tickets for entrance based on accordance with applicable regulations.

2.2. Data analysis

The differences between three databases, both in value and unit, were transferred in natural logarithm, $ln(x_i)$. A standard deviation (σ) was used, in order to quantify the amount of variation of microclimates series data. A low σ indicates that the data points tend to be close to the mean or less variety, while a high σ indicates that the data are spread out over a wider range of values or much variety.

In addition, in order to analyze space amenity, the study was used temperature-humidity index (THI).THI is representing an index to specify the comfort by quantitatively which combined with temperature and relative humidity, with formula [9]:

$$THI = 0.8 * T_a + \frac{(RH * T_a)}{500} \quad (1)$$

The equation 1 resulted in THI value is representing an index to specify the comfort of the space which combined with temperature (T_a)and relative humidity (RH).Based on tropical condition, the THI value which defined as a'comfort condition' is the value between 20 to 26. If the value less than 20 the condition concludes as 'too cold' condition and if exceed than 26 concludes as 'too hot' condition [9].

In order to analyze the correlation between two factors, microclimates unit($\boldsymbol{x_1}$)and plants loss ($\boldsymbol{y_1}$), and between THI($\boldsymbol{x_2}$) and the number of tourist visits ($\boldsymbol{y_2}$) were tested with Pearson correlation (equation 2). The microclimates unit are including temperature, humidity, rainfall, average wind velocity and solar radiation. The Pearson correlation coefficient was tested using the formula:

$$r = \frac{n\sum x_i y_i - \sum x_i \sum y_i}{[(n\sum x_i{}^2 (\sum x_i)^2]^{1/2} \; [(n\sum y_i{}^2 (\sum y_i)^2]^{1/2}} \quad (2)$$

In order to statistically test, then arranged the hypothetic that H_0: $r = 0$; H_1: $r \neq 0$, $\alpha = 0.01$. The value of $r = 0$ means that there is no any correlation between two factors, in reverse, the value of $r \neq 0$ means that there is a correlation. The r value will not exceed from $|1|$ when approaching to the value of -1 means that there is a negative linear relationship, and when approaching to the value of 1 means that there is a positive linear relationship. In order to assess the degree of significance, a t-test in *ANOVA* with $\alpha = 0.01$ was also conducted. If $r \neq 0$, p-value $<\alpha$ means that the correlation between two factors is significant, and if p-value $>\alpha$ means that the correlation between two factors is not significant.

3. Results and Discussions

Based on data series from 2012 to 2016, the microclimates data, the number of plants loss and the number of the tourist of CBG were described in the below. There was a dynamic value trend both in month and year.

3.1. Data inventory

Microclimates series data of CBG from 2012 to 2016 were details presented in Table 1. Nevertheless, several times occurred errors either in the weather equipment or the server, it cause these data can not be recorded.

Table 1. Microclimates dynamic series data CBG.

Year	Microclimates unit	Months											
		Jan	Feb	Mar	Apr	May	Jun	Jul	Aug	Sep	Oct	Nov	Dec
2012	Temperature (°C)	19.73	19.74	19.70	19.60	19.50	19.20	18.90	18.80	19.40	19.90	19.80	19.50
	Humidity (%)	85.60	85.20	84.30	91.50	89.10	86.00	83.80	80.90	82.00	86.40	91.20	92.80
	Rainfall (mm)	162.00	366.00	113.80	371.60	179.80	36.60	24.60	54.20	74.00	235.60	323.90	250.00
	Avg. wind velocity (kph)	5.21	17.26	6.00	3.00	3.30	3.60	3.60	4.20	3.80	3.10	3.00	2.80
	Air pressure (mbar)	999.00	998.96	998.73	999.17	998.56	999.42	999.41	1000.40	1000.20	999.15	998.61	998.50
	Solar rad. (W m^{-2})	87.10	110.00	91.20	109.00	115.00	119.00	130.00	151.00	140.00	137.00	107.00	90.20
2013	Temperature (°C)	19.70	19.70	19.80	20.20	20.00	20.10	19.10	19.10	19.80	19.90	20.00	n.d.[a]
	Humidity (%)	91.00	90.80	91.30	92.80	92.60	88.50	90.90	85.50	87.10	85.00	88.50	n.d.[a]
	Rainfall (mm)	319.20	291.40	n.d.[a]	444.40	n.d.[a]	n.d.[a]	n.d.[a]	n.d.[a]	n.d.[a]	n.d.[a]	n.d.[a]	n.d.[a]
	Avg. wind velocity (kph)	4.30	5.40	3.80	2.90	2.80	3.50	2.60	3.40	3.10	3.60	3.10	n.d.[a]
	Air pressure (mbar)	998.42	997.12	998.71	997.72	998.23	997.32	998.70	999.54	999.14	999.83	997.52	n.d.[a]
	Solar rad. (W m^{-2})	97.60	89.49	97.60	100.00	92.20	106.00	97.90	146.00	131.00	142.00	134.00	n.d.[a]
2014	Temperature (°C)	19.80	20.40	n.d.[a]	n.d.[a]	n.d.[a]	n.d.[a]	n.d.[a]	n.d.[a]	n.d.[a]	n.d.[a]	n.d.[a]	19.10
	Humidity (%)	91.10	86.30	n.d.[a]	n.d.[a]	n.d.[a]	n.d.[a]	n.d.[a]	n.d.[a]	n.d.[a]	n.d.[a]	n.d.[a]	92.80
	Rainfall (mm)	n.d.[a]	2.20	n.d.[a]	n.d.[a]	n.d.[a]	n.d.[a]	n.d.[a]	n.d.[a]	n.d.[a]	n.d.[a]	n.d.[a]	291.80
	Avg. wind velocity (kph)	3.50	3.70	n.d.[a]	n.d.[a]	n.d.[a]	n.d.[a]	n.d.[a]	n.d.[a]	n.d.[a]	n.d.[a]	n.d.[a]	1.50
	Air pressure (mbar)	997.57	998.95	n.d.[a]	n.d.[a]	n.d.[a]	n.d.[a]	n.d.[a]	n.d.[a]	n.d.[a]	n.d.[a]	n.d.[a]	867.70
	Solar rad. (W m^{-2})	100.00	169.00	n.d.[a]	n.d.[a]	n.d.[a]	n.d.[a]	n.d.[a]	n.d.[a]	n.d.[a]	n.d.[a]	n.d.[a]	108.00
2015	Temperature (°C)	18.30	18.50	18.90	18.90	18.20	18.10	17.80	18.10	18.50	19.00	19.50	19.10
	Humidity (%)	92.40	92.70	91.00	92.90	91.40	88.10	85.20	83.97	80.70	80.30	90.20	92.60
	Rainfall (mm)	242.80	254.20	376.80	328.20	180.80	31.90	32.40	5.40	6.40	25.60	418.20	252.90
	Avg. wind velocity (kph)	1.90	1.40	2.60	1.60	1.80	2.30	2.40	2.40	2.50	2.40	1.50	1.40
	Air pressure (mbar)	868.70	868.90	869.30	868.50	869.50	869.30	869.70	869.60	869.90	870.30	868.70	869.20
	Solar rad. (W m^{-2})	106.00	110.00	111.00	112.00	126.00	142.00	155.00	173.00	184.00	155.00	132.00	97.20
2016	Temperature (°C)	19.50	19.50	19.69	19.69	19.65	19.24	18.76	18.23	18.99	20.29	n.d.[a]	n.d.[a]
	Humidity (%)	91.90	93.80	94.80	94.50	94.10	91.60	92.00	91.90	93.10	89.20	n.d.[a]	n.d.[a]
	Rainfall (mm)	278.00	305.40	315.00	423.00	244.00	201.00	218.00	138.00	311.00	62.20	n.d.[a]	n.d.[a]
	Avg. wind velocity (kph)	1.20	1.50	1.32	1.23	1.25	1.40	1.71	1.38	1.37	0.66	n.d.[a]	n.d.[a]
	Air pressure (mbar)	869.80	869.30	869.50	869.10	868.80	869.40	868.70	868.60	868.60	868.20	n.d.[a]	n.d.[a]
	Solar rad. (W m^{-2})	114.00	98.50	90.80	105.00	94.70	117.00	127.00	108.00	109.00	223.00	n.d.[a]	n.d.[a]

[a)] n.d.: no data, there wereno recorded data to be collected, because of error or malfunction of the equipment. *Courtesy of* Agus Darmawan.

Based on microclimates series data, the unit which has the widest amplitude was rainfall (Table 2), and the others have a minor variety. This was supposed because of the location of CBG in the elevation of the mountain with wet tropic zone characteristics [11], it means temperature, humidity, wind velocity ,air pressure and solar radiation less diverse. Heavy rainfall ($\geq$ 100 mm per-month) which occurred in early and end of the year (wet season) given a significant variant of microclimates condition surrounding CBG. The number of wet months more than dry months, between seven to eight months per year, and accumulatively above 1,000 mm annual rainfall [12] [13]. Based on Schmidt-Ferguson classification, CBG was the type of C to B, or less wet to wet, and based on Koppen, it can be included to tropical rainy climates or wet tropic 'Am'.

Table 2. Standard deviation(σ) value of the normalized microclimates unit

Microclimates unit	σvalue of the normalized microclimates unit, ln (x_i)
Temperature	0.033
Humidity	0.045
Rainfall	1.316
Average wind velocity	0.556
Air pressure	0.070
Solar rad.	0.212

The next data inventory was the plants loss. The loss can be caused by several reasons, e.g. fell or hit by other fallen bigger plant(s) or tree(s), pests and diseases causes, dried, and withered. The effect of microclimates was dominant in these cases. The others are old ages, disappear from its location (eaten by wild animals, erosion, flood, and revoked by humans) or intended to be cutting off for human safety importance or infrastructures development (very rare). The details number of plants loss of CBG were presented in Table 3 [10].

Table 3. The number of plants loss (*n*-specimens) of CBG.

Year	Months											
	Jan	Feb	Mar	Apr	May	Jun	Jul	Aug	Sep	Oct	Nov	Dec
2012	92	6	7	23	7	7	6	8	33	4	32	0
2013	14	17	10	7	7	5	7	3	5	20	12	7
2014	4	4	11	6	6	2	4	6	12	11	42	49
2015	19	38	16	13	3	21	15	15	9	14	25	0
2016	7	11	15	6	6	2	0	33	6	1	8	11

The plants loss in CBG tended to occur in early-end of the year (Figure 1). At this time, accumulatively the number of plants loss was 507 (totally 790) specimens or 64.2%, then in the middle of the year. The cause of loss was dominated by fell or hit by other fallen tree(s), float off and increasing diseases attack [10]. The biggest loss occurred in January 2012, 92 specimens and secondly, in December 2014, 49 specimens were lost, both have occurred in early-end of the year.

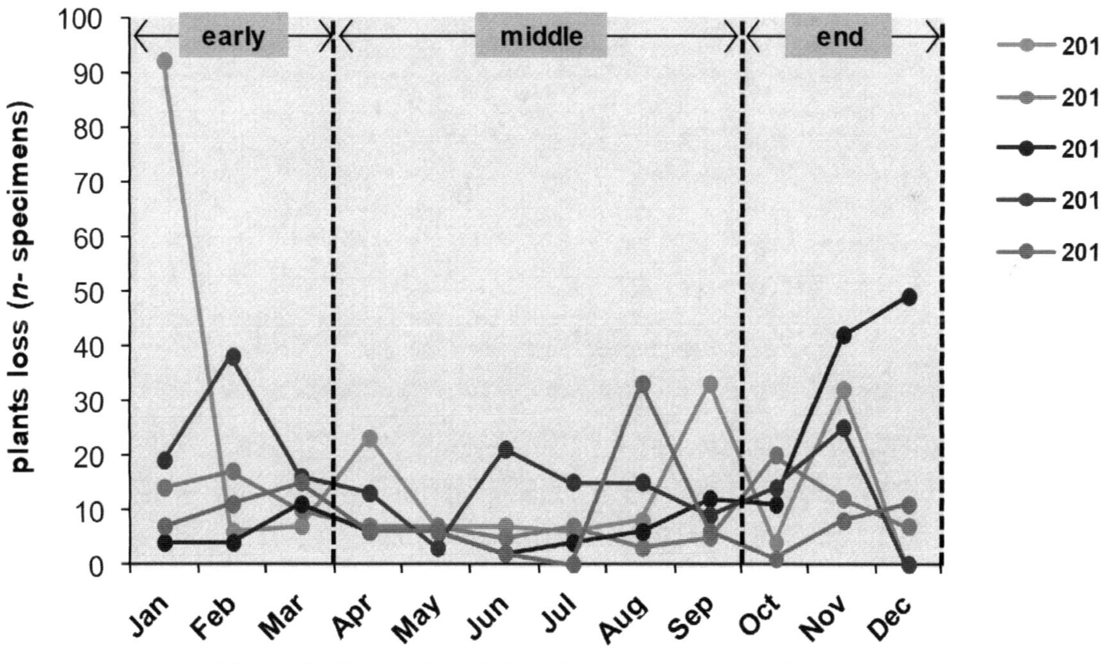

Figure 1. The number of plants loss (*n*-specimens) of CBG.

Furthermore, as an eco-tourism destination, CBG has also organized tourist visits, both in regular tourism and for research and education interest. The number of tourists based on sold entrance tickets were presented in Table 4.

Table 4. The number of tourists of CBG.

Year	Months											
	Jan	Feb	Mar	Apr	May	Jun	Jul	Aug	Sep	Oct	Nov	Dec
2012	58,402	35,524	29,303	39,021	63,254	78,206	56,512	91,190	62,237	49,551	19,030	21,049
2013	29,802	19,555	37,125	33,759	68,911	59,359	13,757	88,745	41,936	32,708	24,195	24,875
2014	37,128	21,770	42,674	32,879	47,299	50,756	32,126	88,021	41,497	38,952	26,176	28,914
2015	35,290	23,801	38,285	35,502	74,582	57,612	77,727	72,210	45,844	40,446	26,503	30,625
2016	47,367	27,960	35,095	33,737	83,551	18,982	114,759	52,945	35,664	36,105	37,197	46,164

The peak visiting occurred in June to August each year (Figure 2), which coincide with 'school long holidays'. Nevertheless, even at the end of December to early January also has long holidays, the visits to CBG still lower than the middle of the year. In 2012 to 2016, accumulatively the number of tourist in the middle of the year, April to September, was 1,692,580 persons or contributed 62.8%, then in early-end of the year, those were 19.3% and 17.9%.

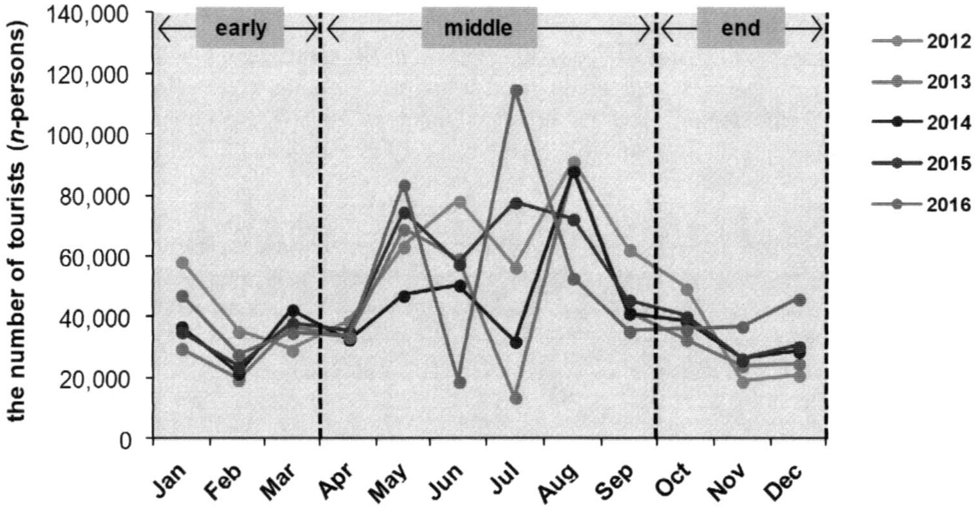

Figure 2. The number of tourists (*n*-persons) of CBG.

3.2. THI value

There were occurred microclimates dynamic in CBG between early-end of the year condition than in the middle. These data can be generated THI value of local CBG along 2012 to 2016 (Table 5).

Table 5. The THI value of CBG.

Year	Months											
	Jan	Feb	Mar	Apr	May	Jun	Jul	Aug	Sep	Oct	Nov	Dec
2012	19.2	19.2	19.1	19.3	19.1	18.7	18.3	18.1	18.7	19.4	19.5	19.2
2013	19.3	19.3	19.5	19.9	19.7	19.6	18.8	18.5	19.3	19.3	19.5	n.a[a].
2014	19.4	19.8	n.a[a]	n.a[a]	n.a[a]	n.a[a]	n.a[a]	n.a[a]	n.a[a]	n.a[a]	n.a[a]	18.8
2015	18.0	18.2	18.6	18.6	17.9	17.7	17.3	17.5	17.8	18.3	19.1	18.8
2016	19.2	19.3	19.5	19.5	19.4	18.9	18.5	17.9	18.7	n.a[a]	n.a[a]	n.a[a]

[a)] n.a.: not available (because there weren't any data can be further analyzed).

THI value of CBG was range about 17.3 to 19.9 (Table 5) and never exceeded to value 26, a boundary value of discomfort [9]. These conditions included 'too cold' condition [9], which THI below 20 to 21, although it was understandable because of the altitude of CBG located on the mountainous side. Temperatures were tended to drop in July to September or in dry months, but humidity tended to constantly in high values (> 80%) or very humid condition.

3.3. Correlation between microclimates and plants survivorship

Rainfall in CBG as microclimates unit has most diverse and biggest deviation amplitude. Heavy rainfall has occurred in early-end of the year, and along with solar radiation decreased. This can be understood when heavy rainfall occurred, cloud very thick that prevent solar radiation penetrations. Furthermore, rainfall occurred closely related to air pressure differences, bigger differences will give a bigger chance for rain occurrence.

Data showed that the biggest plants lost was also occurred in the early-end year. The lost in time range was contributed more than 64%of the total. The biggest loss, such as in January 2012 and December 2014, the level of rainfall was also very high (> 150 mm per month). Along with the high-level volume of rainfall at previous months, and this influenced to reduce soil aggregates and surface runoff which can cause surface erosion and flood [14],and reducethe capability of plants to adsorb the soil. The secaused the plants will befell because they may not be able any longer to hold the standing. These conditions were caused decreasing of plants survivorship [15][16] in the gardenand vulnerable have occurred of plants loss. Fallen plant(s), especially tree(s), is not only harm human safety in the garden, but also hurt other plants collection because of stroked of this coverage. A tall tree(s) with widespread canopy and stem diameter can cause massif damages to other plants collection up to tens.

The disadvantages of a massif rainfall condition with plants loss [15] can be analyzed based on the correlation between them in Table 6. Each microclimates unit has a correlation to plants loss ($r \neq 0$), then reject H_0.Based on Pearson test, the correlation between rainfall and plants loss has the highest value than others. The correlation value was positive, which mean that the higher value of rainfall can cause a higher risk of plants loss. Although the index value was not close with value '1' (0.496), or correlation was moderate or less strong, but based on *t*-test value found that a highly significant correlation between them ($p < \alpha = 0.01$). So this can be described that rainfall was linear significantly influence to plants loss at a moderate level.

Table 6. Average correlation index (r) between microclimates unit and plants loss[a)].

Microclimates unit	Plants loss		
	r	α	p-value
Temperature	0.433	0.01	1.44×10^{-4}
Humidity	0.156	0.01	1.08×10^{-9}
Rainfall	0.496	0.01	1.18×10^{-7}
Avg. wind velocity	0.031	0.01	3.09×10^{-7}
Air pressure	0.236	0.01	4.06×10^{-13}
Solar rad.	-0.139	0.01	1.15×10^{-9}

[a)] because the lack of data for several months (n.d.: no data) in microclimates, then the data were averaged for available data in the same month.

The occurrences of heavy rainfall in the early-end of the year were not only directly reduced the ability of plants standing but also increased of diseases stroked [17]. Rotten roots, stem and/ or other parts of plants and fungus stroked were also increased in this time range. Strokes of fallen plants and diseases have dominated the causes of plants loss, more than three-quarter, than other causes [10].

Several actions have been conducted by CBG operator in order to anticipate of this disadvantages, such as regular pruning, pests and diseases controlling, strengthening of hills side of the garden by concrete. And for the results, in 2016 the number of plants loss was the lowest than previously, nevertheless lost which caused by extreme microclimates, especially severe rainfall still dominated [10]. This was difficult to anticipate especially the fallen risk of tall and big tree(s).

3.4. Correlation between THI and number of visits

The diverse of THI values along five years were also relatively constant with low amplitude. This condition should be encouraged tourist to keep visiting CBG or at least constantly. But in fact, there was a big difference of visits between early-end compared to the middle of the year. Data was showed that 'school long holiday' (June to August) which occurred in the middle of the year, not always contributed to the leap of the number of visits than in the early-end year. These can be analyzed on data, such as in July 2013 was the lowest visits along time range observation, or in July 2014 was lower than March, June, or September in the same year. So that in June 2016 was the lowest visits of the year.

Based on Pearson test, it was obtained the value of r was -0.71 with p-value $<\alpha = 0.01$, so reject H_0or there is a correlation between THI and the number of visits. The negative value was meant that bigger THI value (being more discomfort) then the number of visits became lower. Although the result was shown a significant correlation between THI and the number of visits, this still has not explained the dynamics of the number of visits which occurred. This because in fact, THI value for five years was relatively constant, but in the other part, the number of visits was dynamically raised up and downed, especially between the early-end which compared to the middle of the year (Figure 3).

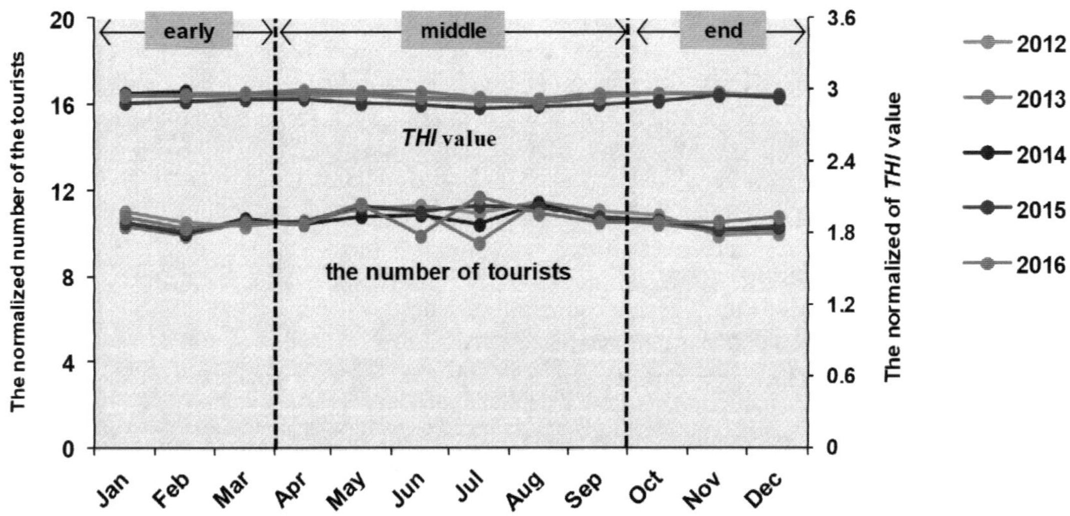

Figure 3. Proportionally comparison of the normalized THI value to the normalized number of tourists.

The contribution of the visits along 'school long holiday' (June to August) also has not sufficient to explain the dynamic. The test showed that correlation between THI and the number of visits at the peak time resulted in $r = -1$ (negative correlation and reject H_0), but with p-value $= 0.02 > \alpha = 0.01$ or has no significance.

The next assumption that there was any other microclimates unit, besides temperature and humidity (as the composer of THI) which influence to the visits, and that was rainfall [18]. The high volume of rainfall in the early-end year was assumed directly influencing the tourist interest to visit than any other. As eco-tourism destination, most tourist like to enjoy the beautiful scenery and space amenity of CBG. When heavy rainfall occurred, it was difficult to sense this comfort. This can be proven with the correlation between rainfalls to the number of visits. Based on test, the $r = -0.59$ (reject H_0), with p-value $< \alpha = 0.01$. This corresponded with the previous assumption, which increasing rainfall will be decreased of visits (negative correlation or opposite) [18]. This analysis was consistent with the fact that the high volume of rainfall which occurred in the early-end year along with the decreasing of the visits at the same periods.

This condition has to be anticipated by CBG operator in order to manage the number of tourist visits. Heavy rainfall occurred in the early-end year may not be able to manage because it is major forces, but the operator can manage space order of CBG. For example, by developing shelters both natural or man-made structure, so tourist may still to enjoy the scenery of the garden without disturbed by rainfall, or rain can be organized as a natural attraction, such as 'rain theme park' adjusted with conservation principles value. So that rainfall was not an obstacle but organized as a new attraction to pull tourists interest.

4. Conclusions

There were significantly proven the correlations between microclimates to plants survivorship, and to build space amenity of the local condition of CBG. The most microclimates unit which influencing both was rainfall. The extreme rainfall occurred in the early-end of the year influenced to reduce plants survivorship, which plants loss in this period tended to increase. The correlation was positive, that bigger rainfall directionally with bigger plants loss. Furthermore, the THI value has the negative influenced correlation to tourist visits. Nevertheless, the better microclimates unit which able to describe the dynamic of the visits over time was rainfall. Rainfall was assumed as an obstacle factor for tourist interest to visit. The correlation was negative that increasing rainfall will influence tourist visits more decreasingly. This needed the operator concern to anticipate these disadvantages by a routine maintenance and developing a better space organization.

References

[1] Gkatsopoulos P 2017 A Methodology for calculating cooling from vegetation evapotranspiration for use in urbanspace microclimate simulations*ProcediaEnviron. Sci.* 38pp 477 – 84

[2] Hampe A and Jump AS 2011 Climate relicts: past, present, future*Annu. Rev. Ecol. Evol. Syst.*42pp 313–33

[3] Maclean I M D, Hopkins J J, Bennie J, Lawson C R and Wilson R J 2015 Microclimates buffer the responses of plant communities to climate change *Glob.Ecol. Biogeogr.*24 1340–50

[4] Dobrowski S Z 2011 A climatic basis for microrefugia: the influence of terrain on climate*Glob. Change. Biol.*17pp 1022–35

[5] Hatfield J L, Boote K J, Kimball B A, Ziska L H, Izaurralde R C, Ort D, Thomson A Mand Wolfe D W 2011 Climate impacts on agriculture: implications for crop production *Agron. J.*103pp 351–70

[6] Meehl G A, Stocker T F, Collins WD, Gaye AJ, Gregory JM, Kitoh A, Knutti R, Murphy, JM, Noda A, Raper SCB, Watterson JG, Weaver AJand Zhao Z 2007 Global Climate Projections,In: *Climate Change 2007: The Physical Science Basis. Contribution of Working Group I to the Fourth Assessment Report of the Intergovernmental Panel on Climate Change*,edsS Solomon,D Qin, M Manning, Z Chen, M Marquis, K B Averyt, M Tignor and H L Miller (Cambridge and New York: Cambridge University Press)

[7] Kumudini S, Andrade FH, Boote KJ, Brown GA, Dzotsi KA, Edmeades GO, Gocken T, Goodwin M, Halter AL, Hammer GL, Hatfield JL, Jones JW, Kemanian AR, Kim S-H, Kiniry J, Lizaso JI, Nendel C, Nielsen RL, Parent B, Stöckle CO, Tardieu F, Thomison PR, Timlin DJ, Vyn TJ, Wallach D, Yang HSand Tollenaar M 2014 Predicting maize phenology: intercomparison of functions for developmental response to temperature*Agron. J.*106 pp 2087–97

[8] Smith V K, Poulos C and Kim H 2002 Treating open space as an urban amenity *Resour. Energy Econ.* 24 pp 107–29

[9] McGregor G R and Nieuwolt S 1998 *Tropical climatology: an introduction to the climates of the*

IOP Conf. Series: Earth and Environmental Science **179** (2018) 012001 doi:10.1088/1755-1315/179/1/012001

low latitudes 2nd ed (New York: John Wiley & Sons) p 352
[10] Registration Unit of Cibodas Botanical Garden 2017 *Monthly reports of dead plants of garden collection 2012 to 2016: Form D4* (Cibodas: Cibodas Botanical Garden-LIPI) [Unpublished]
[11] Underwood E C, Olson D, Hollander A Dand Quinn J F 2014 Ever-wet tropical forests as biodiversity refuges *Nat. Clim. Change*.4 pp740-41
[12] Mueller-Dombois D and Fosberg F R 2013 *Vegetation of the Tropical Pacific Islands* (New York: Springer Science and Business Media) pp 24-30
[13] Corlett R T 2014 *The Ecology of Tropical East Asia* 2nd ed (New York: Oxford University Press) pp 24-6
[14] Bruenig E F 2016 *Conservation and management of tropical rainforests: an integrated approach to sustainability* 2nd ed (London: CABI) pp 35-42
[15] Gao R, Yang X, Liu G, Huang Z and Walck J L 2015 Effects of rainfall pattern on the growth and fecundity of a dominant dune annual in a semi-arid ecosystem *Plant Soil* 389 pp 335-47
[16] Niu S, Luo Y, Li D, Cao S, Xia J, Li Jand Smith M D 2013 Plant growth and mortality under climatic extremes: An overview *J. Env. Exp. Bot.* 98 pp 13-9
[17] Spear E R, Coley P D and Kursar T A 2015 Do pathogens limit the distributions of tropical trees across a rainfall gradient? *J. Ecol.* 103 (1) pp 165–74
[18] Kilungu H, Leemans R, Munishi P K T and Amelung B 2017 Climate change threatens major tourist attractions and tourism in Serengeti National Park, Tanzania, In: *Climate Change Adaptation in Africa: Fostering Resilience and Capacity to Adapt*, eds W Leal Filho, S Belay, J Kalangu, W Menas, P Munishi, K Musiyiwa (Cham: Springer International Publishing) pp 375-92

Landscape preference among Chinese visitors to Chinese elements in a Japanese Garden

D Meng[1*], T Matsumoto[1], Y Mizuuchi[2], K Furuya[3]

[1] Graduate School and Faculty of Horticulture Chiba University, 648 Matsudo, Matsudo City, 271-8510, Japan.
[2] Assistant Professor, Graduate School of Agricultural and Life Sciences, the University of Tokyo, Japan.
[3] Prof., Dept. of Environmental Horticulture, Faculty of Horticulture, Chiba University, Japan.

*Email: viicyo@gmail.com

Abstract. Koishikawa Korakuen is an Edo-period garden in Japan that was inspired by scenes typical of Chinese gardens. This study aimed to clarify how Chinese tourists evaluate the Chinese elements at Koishikawa Korakuen and compare landscape preferences. The preferred landscapes were determined and analyzed using visitor-employed photography, GPS, and questionnaires. Fifty-seven Chinese visitors were asked to take photos in the garden and select their 10 favourites from among them. GIS was used to identify the positions of the preferred photos. The results indicate that “Full-Moon Bridge” is the most Chinese-looking landscape at Koishikawa Korakuen; the second is the lotus pond. Moreover, spatial analysis was conducted by dividing the garden into six areas. Within these six areas, seven locations showed high levels of appreciation. This study’s findings help to clarify the elements Chinese visitors prefer at Koishikawa Korakuen as well as the differences among Chinese and Japanese visitors regarding landscape preferences in Chinese-style gardens.

Keywords: chinese elements, chinese visitors, landscape preference

1. Introduction

The number of foreign tourists visiting Japan has increased in recent years. Among these tourists, the number of Chinese visitors has grown rapidly since 2013. In 2015, Chinese comprised the largest number of foreign tourists visiting Japan [1]. Gardens are popular destinations among these tourists. Experience of landscape comes from the perception of human being, so it could be different from the cultural background that they carried. Hence, it is necessary to add the comparison between different cultures into experience of landscape research. It is well known that Japanese gardens were deeply influenced by Chinese garden culture, and Koishikawa Korakuen is a representative case. By showing this aspect of the historical relationship between Japan and China, Koishikawa Korakuen can potentially trigger cross-cultural exchange. In this way, tourism not only has economic impacts but can also promote cultural communication. Therefore, it is necessary to consider foreign visitors as an important aspect of garden management.

Yorifusa began construction on Koishikawa Korakuen in 1626, and it was mostly completed by his successor Mitsukuni. Since Mitsukuni was influenced by Confucianism (between 1665 and 1682, he was advised by the Chinese Confucian scholar Chu Shun-shui [2], the garden contained many aspects of Chinese landscapes (in fact, the name Korakuen was taken from a song dynasty poem, under Chu

Shun-shui's recommendation [3]. Though some Chinese elements were destroyed as a result of natural disaster (e.g., Kara-mon and Hakke-do), many are still present in the garden today. These include Engetsu-kyo (Full-Moon Bridge) and the West Lake embankment. Moreover, Nobedan is a road paved with natural stones of various sizes, which is also a Chinese landscape technique [4].

Previous studies have clarified Chu Shun-shui's contributions to Koishikawa Korakuen [5]. Others have investigated the garden's history [7, 8, 9, 10, 11], landscape features [12, 13, 14] and use capability. Recently, Matsumoto et al. [6] used visitor photographs to investigate how Japanese people use Koishikawa Korakuen. However, there remains a need for research on foreign visitors to Koishikawa Korakuen and their evaluations of Chinese elements in the garden. Therefore, this study aimed to clarify Chinese tourists' evaluations of Chinese elements at Koishikawa Korakuen and compare landscape preferences between Chinese and Japanese visitors.

2. Methods/Experimental

2.1. Study site

Koishikawa Korakuen is located in Bunkyo, next to the Tokyo Dome and the Japan–China Friendship Center. It was designated a special historical site and scenic spot in 1952 under the Law for the Protection of Culture Properties. This double designation is quite rare, and the site has attracted many foreign tourists.

2.2. Surveys

To clarify Chinese tourists' evaluations, we used on-site surveys as well as visitor-employed photography and GPS. By synchronizing photo time codes and GPS log data, we could identify what the respondents were interested in and where the photos were taken. At the garden entrance, GPS loggers (HOLUX M-241 wireless GPS logger, v. 1.13, MTK chipset) were given to Chinese tourists (including those from Hong Kong and Taiwan). They were asked to take more than 10 photos while walking through the garden using their smartphone or digital camera. After their visit, we gathered the GPS loggers and asked them to pick 10 favorites from the pictures they took. We then conducted questionnaire surveys, which gathered the following:

1) Attributes: 1) age, 2) gender, 3) number of visits to Koishikawa Korakuen, 4) number of visits to Japan, 5) year(s) they stayed in Japan, 6) how they got information about Koishikawa Korakuen, and 7) why they visited Koishikawa Korakuen.
2) Objects of interest: their 10 preferred photos and the landscape elements they liked or disliked.
3) For each photo, 11 options were given concerning the landscape elements they focused on. The options covered the whole composition, trees, flowers, grasses, water landscapes, stones, animals, structures, buildings (outside the garden), facilities, and pavement, among others. They were asked to check "○" for items they liked and "×" for those they disliked. (Preliminary research was conducted on June 10, 2016, with four Chinese respondents. They were asked to take more than 10 photos, select their 10 favorites, and write down the objects they intended to photograph. These data were used to create the options for the questions about objects of interest.)
4) Chinese elements at Koishikawa Korakuen: there were three questions about Chinese elements in the garden: 1) a yes/no question about whether they sensed Chinese taste and flavor in the garden, 2) a question asking them to select one photo they felt reflected Chinese flavor most strongly and check the item they felt had Chinese elements from the 11 options above, and 3) an open question asking them to comment generally on Chinese taste and flavor in the garden.

2.3. Analysis

Based on whether the option "whole composition" was checked in the questionnaire, all photos were divided into two groups: point and space. "Point" refers to photographed objects focused on a single point; "space" refers to photographed content focused on expanded space. Then, based on the questionnaires, these were divided into 13 types: trees, flowers and plants, facilities, structures, buildings (outside the garden), animals, stones, pavement, water landscapes, paths, spatial trees, spatial flowers and plants, and overlook (Table 1).

Table 1. Objects of interest: types and descriptions.

Type	Interested objects	Characteristic
point tree	○	single tree is the main focus
point flower	○	interested objects are only flowers and grass
sign/facility	○	the sign stone lantern stone tablet
structures	○	structures inside the garden like bridge
buildings(outside)	○	buildings outside the garden
animals	○	interested objects are annimals
stones	○	stones are the main focus
pavement	○	the pavement of path
water landscape	■	water landscape is the main focus, pond, lake
path	■	path is main in the photo
spatial trees	■	composition of trees are valued
spatial flowers	■	composition of flowers and grasses are valued
overlook	■	distance view focused and numerous interested objects

photographed objects those focus on one point were marked with ○
photographed content those focus on space expand were marked with ■

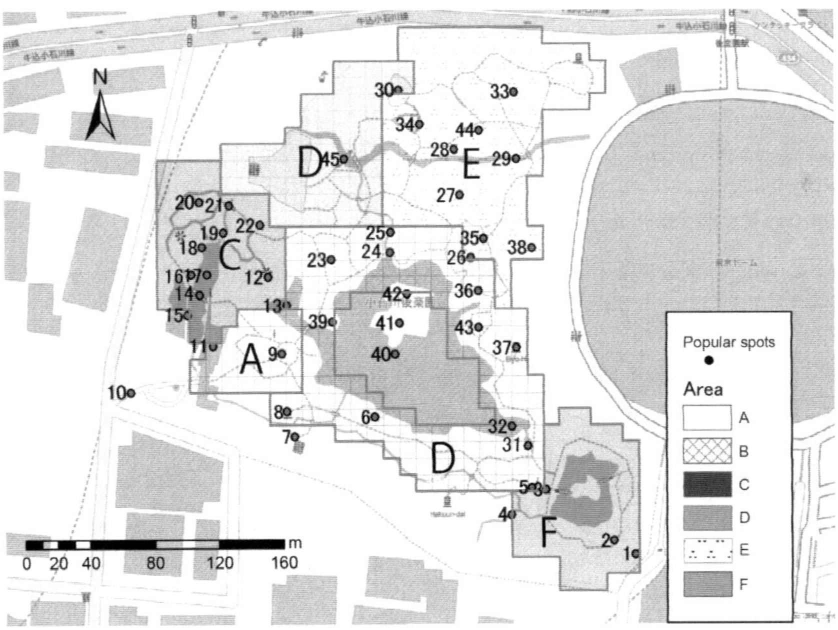

Figure 1. Distribution of each area.

The photos selected as reflecting Chinese taste and flavor were divided into the same 13 types as shown in Table 1. Based on the Chinese elements checked by the respondents, specific Chinese elements were determined (Table 4).

To identify places that interested the respondents, we conducted a two-level GIS analysis. The first was at the zone level. The whole garden was divided into six areas (Figure 1), following Matsumoto et al. (who essentially followed Yoshikawa's research); Yoshikawa noted that zoning lends spatial characteristics to Koishikawa Korakuen. ArcGIS 10.2 and js-STAR 2012 were used for the analysis. Figure 1 shows the locations of the popular spots, which are listed in Table 2.

Table 2. List of popular spots.

No.	area	popular spot	*	No.	area	popular spot	*
1	F	East gate		24	B	Sawatari	
2	F	Monument		25	B	Shiraito-no-taki	
3	F	**Former site of Chinese gate**		26	E	Pine trees	
4	F	Nezameno-taki		27	E	Irises Rice paddies	○
5	B	**Nobedan path**	○	28	E	Wisteria trellis	
6	B	Maple trees		29	E	Furosui	
7	B	Chuhosen-no-hi		30	D	**Former site of Hakke-do**	○
8	B	Former site of Saigyo-do		31	B	Naruto	
9	A	weeping cherry	○	32	B	Chikubu-jima	
10	A	West gate		33	E	Ume grove	○
11	A	Kantoku-tei		34	E	Atago slope	
12	C	**Sho-rozann**		35	E	Kuhachi-ya	
13	B	Lotus pond		36	B	Funatsuki	
14	C	Togetsukyo		37	B	Eiyo-hi	
15	C	**Seiko-no-tsutsumi(West Lake embankment)**	○	38	E	Akamon	
16	C	Byobu-iwa		39	B	Hitotsu-matsu	
17	C	Ogiwa		40	B	Tokudaiji-ishi	
18	C	Sawatari		41	B	Horai-jima	
19	C	Otowa-no-taki		42	B	Benzaiten-hokora	
20	C	Former site of Kannon-do	○	43	B	Ikei-tourou	
21	C	Tsuten-kyo	○	44	E	Yatsuhashi	
22	C	**Tokujin-do**		45	D	**Engetsu-kyo**	○
23	B	Maroya					

The items have pictures on the Koishikawa Korakuen's pamphlet are marked with ○.
The items have Chinese elements are marked in bold.

After zone-level analysis, we performed spot-level analysis. A "hotspot" is a particularly popular place for photos. The whole target area was covered by a grid comprising 10 m sections. Then, the number of photos within each square was calculated. After this, the Getis-OrdGi* statistical technique was used. Getis-OrdGi* is a hotspot analysis technique that calculates the z-score of each grid. This shows whether features have high or low spatial clustering, with each feature examined in the context of neighboring grids.

3. Results

3.1. Respondents profile

The 57 respondents were all from China (including Hong Kong and Taiwan); 23 were male and 34 were female. Thirty-five had come to Japan for a short visit, 14 were visiting Japan for the first time, and six had visited Japan more than 10 times. Twenty-two of the respondents were currently living in Japan for work or school. Forty were visiting Koishikawa Korakuen for the first time while 17 had visited the garden before.

Table 3. Whether respondents sensed Chinese taste and flavour.

answer	yes	no
count	50 (87.72%)	7 (12.28%)

3.2. Chinese elements

Table 3 shows the results for whether the respondents sensed Chinese taste and flavor at Koishikawa Korakuen; 87.72% responded affirmatively.

Table 4 shows that 50.9% of the photos selected as containing Chinese taste and flavor were categorized as structures. Among these, Engetsu-kyo (Full-Moon Bridge) was the most popular, with 40.4% of respondents saying it contained Chinese taste and flavor.

Table 4. Types of Chinese-looking photos.

Interest type	Total	Specific	count	percentage
structures	29(50.9%)	Engetsu-kyo	23	40.4%
		Uchi-niwa	3	5.3%
		Kuhachi-ya	1	1.8%
		Tokujin-do	1	1.8%
		The bridge beside Engetsukyo	1	1.8%
water landscape	11(19.3%)	Uchi-niwa	6	10.5%
		Dai-Sensui	2	3.5%
		Oigawa	1	1.8%
		Inada	1	1.8%
		Kisogawa	1	1.8%
point flower	5(8.8%)	Lotus pond	5	8.8%
spatial flower	4(7%)	Lotus pond	3	5.3%
		Inada	1	1.8%
stones	2(3.5%)	Horajima	1	1.8%
		Atago-slope	1	1.8%
spatial tree	2(3.5%)	Uchi-niwa	1	1.8%
		Shoro-zann	1	1.8%
point tree	1(1.8%)	Pine trees	1	1.8%
sign/facility	1(1.8%)	Stone lantern	1	1.8%
path	1(1.8%)	Dai-Sensui path	1	1.8%
overlook	1(1.8%)	Inada	1	1.8%

3.3 Resources of interest

All 570 photos gathered from respondents were divided into 13 types (Table 1). Hotspots. Using Getis-OrdGi* hotspot analysis, seven hotspots with different characteristics were identified (Figure 2).

Hotspots and landscape types. Hotspot 1 is around the lotus pond. Hotspot 2 is around the West Lake embankment and Oigawaarea. Hotspot 3 is near Engetsu-kyo. Hotspot 4 has popular spots such as Kohachi-ya, pine trees, and Inada. In hotspot 5, along with the revetment from Eiyo-hi to Chikubu-jima, the water landscape was the main interest. Hotspot 6 is in the Uchi-niwa. Hotspot 7 runs from the former site of Saigyo-do along the path to the maple trees, including Chuhosen-no-hi.

Photos and objects of interest in different areas. Table 5 shows the amounts of photos taken in each area and the percentage for each landscape type.

Based on the resources of interest in each area, visitors' preferred landscapes can be clarified (Table 5). In area A, the spatial flower type was the most preferred (28.57%); next was the point flower (15.71%). This result is nearly same as for hotspot 1, which is around the lotus pond in area A. We can see that the lotus pond accounted for a large proportion of interest in area A.

For area B, the water landscape was the main resource of interest, accounting for 39.47%; spatial trees accounted for 11.40%. Water landscapes also held the most interest in hotspots 5 and 7. In area

B, every type held a certain proportion of interest, indicating the while the water landscape was the main resource of interest, there were varied areas of interest.

Area C had two main resources of interest—water landscapes (32.76%) and structures (29.31%). Combined with the results for hotspot 2, which is in area C, the water landscapes are the West Lake embankment and Oigawa, and the structures are the West Lake embankment, Togetsukyo, and Tsuten-kyo.

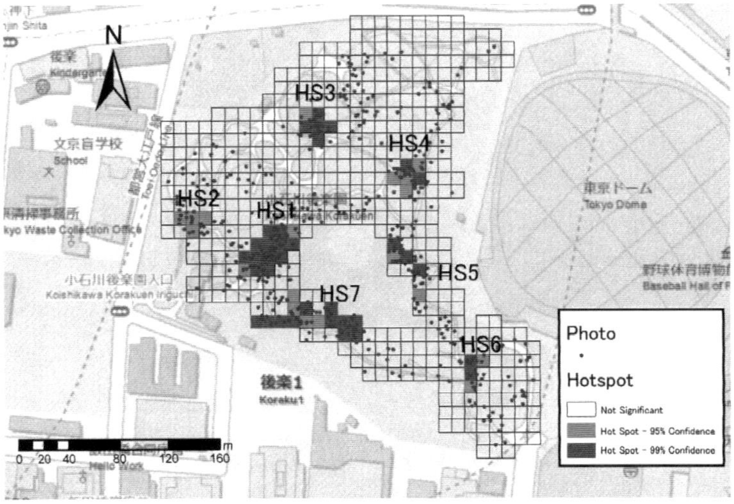

Figure 2. Distribution of hotspots.

In area D, the major resources of interest were structures (80.77%). Therefore, Engetsu-kyo was the mainly preferred landscape.

The main resources of interest in area E were structures (31.37%) and water landscapes (20.59%). All other types were below 10%.

Table 5. Resources of interest in each area.

area	total	point tree	point flower	sign/ facility	structure s	buildings (out)	animals	stones	water landscape	path	spatial tree	spatial flower	overlook	pavement
A	70	5.71%	15.71%	1.43%	2.86%	2.86%	0.00%	0.00%	14.29%	5.71%	14.29%	**28.57%**	8.57%	0.00%
B	228	7.46%	2.19%	5.70%	5.70%	2.19%	4.82%	6.58%	**39.47%**	4.82%	11.40%	5.26%	3.07%	1.32%
C	58	6.90%	1.72%	1.72%	29.31%	0.00%	0.00%	15.52%	**32.76%**	0.00%	5.17%	5.17%	1.72%	0.00%
D	52	1.92%	5.77%	3.85%	**80.77%**	0.00%	1.92%	0.00%	3.85%	0.00%	0.00%	0.00%	0.00%	1.92%
E	102	1.96%	5.88%	8.82%	**31.37%**	0.00%	3.92%	2.94%	20.59%	3.92%	5.88%	9.80%	3.92%	0.98%
F	60	6.67%	1.67%	5.00%	10.00%	0.00%	1.67%	0.00%	**48.33%**	3.33%	8.33%	15.00%	0.00%	0.00%

The interest resource types have largest number of photos in each area are marked in bold.

4. Discussion

At Koishikawa Korakuen, 87.72% of Chinese respondents reported sensing a Chinese taste and flavor in the garden. This result indicates that Koishikawa Korakuen has a high potential to promote cross-cultural exchange between Japan and China.

The most strongly sensed Chinese element was Engetsu-kyo. Twenty-two respondents selected it as the Chinese-style element from their 10 favorite photos. Eighteen respondents mentioned Engetsu-kyo in their comments, and three described Engetsu-kyo as the only object that can be recognized as a Chinese element without the need to read the signboard. Designed by Chu Shun-shui, Engetsu-kyo is a stone arch bridge that can be seen as a full moon by combining the bridge itself with its reflection in the water. Arch stone bridge is one of the Chinese han population basic traditional bridge form, it is very familiar in ChinaQ), and arch stone bridge has a long history in China, the oldest arch stone bridge that was ever recorded is Luren bridge mentioned in 'Shui Jing Zhu', which was built in AD 282D).

5. Conclusion

In considering inbound tourism in Japan, it is important to consider the evaluations of Chinese as well as other foreign visitors. This study aimed to clarify Chinese tourists' evaluations of Chinese elements at Koishikawa Korakuen and compare preferred landscapes between Chinese and Japanese visitors. The results showed that 87.72% of Chinese visitors recognized a Chinese taste and flavor at Koishikawa Korakuen. Among these Chinese elements, Engetsu-kyo was the most highly recognized.

In terms of resources of interest and preferred landscapes, there were few differences between Chinese and Japanese visitors. The analysis of hotspots showed only slight differences. This result suggests that it is not necessary to change explanations of or advertisements for Koishikawa Korakuen for Chinese visitors.

This study only compared the landscape experiences of two East Asian groups. However, Koishikawa Korakuen has many visitors from various countries, and there could be very significant differences in the way foreign visitor's experience this garden, beyond the scope of what has been studied here. It may be worth investigating, for example, whether Westerners can recognize the Chinese elements in this Japanese garden.

References

[1] JNTO Nation/Mouth The number of foreign tourists to Japan（2003～2016: http://wwwjntogojp/jpn/statistics/visitor_trends/indexhtml (2016/6/2)
[2] Yoshikawa M 1981 Koishikawa-Korakuen: Kyogaku-sha, Tokyo, 10pp
[3] LiShuhua 2002 landscape flowers taste of the Late Ming Zhu Shun and his contribution to the Tokyo Koishikawa Korakuen: Chinese Landscape 18(1): 48-51
[4] Matsumoto T, Mizuuchi Y, Furuya K 2016 A Study on Grasp of Utilization Form in Koishikawa Korakuen by Using Visitor Employed Photography and Spatial Information Technology, journal of the Japanese Institute of Landscape Architecture 79(5), 477-482
[5] Tamura T 1929 Korakuen-shi: history of Korakuen, Tokoshoin, Tokyo, 75p
[6] Yoshikawa M 1981 Koishikawa-Korakuen: Kyogaku-sha, Tokyo, 93pp
[7] Yoshinaga Y 1937 Koishikawa Korakuen (report of scenic spot survey The third album): Ministry of Education, Science and Culture
[8] Goto S, Fujii E, Shirai H 1999 A Historical Study on the Change of Water Scenery in Koishikawa Korakuen: journal of the Japanese Institute of Landscape Architecture 62(3), 272-279
[9] Ono R 2001 The Historicity of the Historic Garden Based on the Interactive Relationships between Urban Spaces and the Garden - a case study in Koishikawa Korakuen, Tokyo: journal of the Japanese Institute of Landscape Architecture64 (5) 825-830
[10] Kunii Y, Kato M 2009 Evaluation for Landscape on Garden Path by using Fractal Analysis and Positioning
 -Example by Koishikawa Korakuen Garden and Rikugi-en Garden-: Journal of agricultural science, Tokyo Nogyo Daigaku 54 (3) 182-191
[11] Li W 2005 A Research on an Act of Viewing in Koishikawa Koraku-en Garden's Early Days: journal of the Japanese Institute of Landscape Architecture 68（5），373-376
[12] Kato S, Amano K, Yokoyama K 2009 A Basic Study on the Sequential Landscape in Circuit Style Garden —Focus on the Change of Visible Space from Trail—Architecture of Infrastructure and Environment, Japan Society of Civil Engineers speeches 5107-111
[13] Fujita H, Aoki K, Aoki Y 2002 A Proposal of Carrying Capacity of Koishikawa-Korakuen : journal of the Japanese Institute of Landscape Architecture 66（1），54-56
[14] Mizuuchi Y, Son Y, Kang M, Furuya K 2015 Constructing a Survey Method for Landscape Evaluation Using Visitor Employed Photography and GPS: Landscape Research Japan

Developing online dynamic-inventory of plants database management SINDATA© of Cibodas Botanical Garden

I W Hidayat[1], N Suryana[2], and A Darmawan[3*]

[1] Researcher, Cibodas Botanical Garden-Indonesian Institute of Sciences, PO Box 19 SDL, Cipanas-Cianjur, West Java Province, Indonesia 43253

[2] Public Relations, Cibodas Botanical Garden-Indonesian Institute of Sciences, PO Box 19 SDL, Cipanas-Cianjur, West Java Province, Indonesia 43253

[3] IT and Database Technician, Cibodas Botanical Garden-Indonesian Institute of Sciences, PO Box 19 SDL, Cipanas-Cianjur, West Java Province, Indonesia 43253

[*] E-mail: agus.darmawan@lipi.go.id

Abstract. A botanical garden is an area consisted plants collection which has a high conservation value, maintained and managed as *ex-situ* principles. Other specific differences are systematically recorded of historical data of each plant and the dynamics, time over time. The study was purposed to build a valid plants database of Cibodas Botanical Garden (CBG) and the dynamics, and to manage the other related information, so these can be used widely for the public, and to efficiently of supervision for operators. The system was built by integrating *MySQL*©, *Mapserver*© and *PHP*© software to manage three basic data: inventory of plants collection and the dynamics, geospatial information, and daily microclimates conditions. This integrated system was established as *Sistem Informasi Data Tanaman* or SINDATA©. SINDATA© has managed 8,309 of outdoor (from ground cover type up to large trees) and 2,932 of indoor (i.e. orchids, cactus, herbs and *Nepenthes* spp.) plants specimen database, a set geospatial information and interactive map of CBG and daily microclimates (i.e. temperature, humidity, rainfall, solar radiation, wind velocity, and air pressure) information. This system was expected to be a representative source and scientifically valid data to disseminate the information and knowledge references of CBG for conservation-research-education purposes.

Keywords: Cibodas Botanical Garden, daily microclimates, information technology, interactive map, plants database management

1. Introduction

The difference of botanical gardens lies in embedded plants collections which having scientifically and well-documented data. Conventional plants collection data has been running over a century, while processing and documenting plant data in new digital ways just for last decades [1]. Since the introduction of digital plant data processing systems in the form of 'Botanic Garden Recorder' (BGR) applications by Botanic Garden Conservation International (BGCI) in 1996. BGR is an international standard plant data processing application used to process plant data in various botanical gardens, especially in Southeast Asia. BGR applications can only process the data that is textual and incapable to process images or maps, so that there is difficulty when will display them. BGR applications are not network-based or web-based so can only standing alone.

IOP Conf. Series: Earth and Environmental Science **179** (2018) 012003 doi:10.1088/1755-1315/179/1/012003

SINDATA© or *Sistem Informasi Data Tanaman* application is built and developed beside as a data processing media for plants which needed by Cibodas Botanical Garden (CBG), as well as a public needs for personal development and social environment. This importance as well as a part for national resilience, especially the policy of biological resources management [2] [3] [4], and for public which entitled to obtain information of plants collection in the botanical garden as a human right for the purpose of educational development and improvement of sciences. The presence of SINDATA© applications is based on a strategic botanical garden that refers to the world's conservation strategy (Global Strategy for Plant Conservation). Among which sets out ways to improve the documentation of plant data including computerized the systems for the purpose of improving collection management [3], prioritizing the provision of educational facilities to community and provide a service to the public as a source of information, especially information related to the plants. SINDATA© is a digital data processor has provided a great solution to solving problems related to inputting various types of data, such as data in the form of text or image data such as maps and photos, as well as SINDATA© applications provide solutions to the desired output by the user.

SINDATA© is an innovative application, which integrates plant data [4] [5] with microclimates condition data at CBG which updated every five minutes and spatial information data [6] [7] of CBG. SINDATA© is accessible on the internet by simply typing *http://sindata.krcibodas.lipi.go.id*. The updated microclimates data is important to describe general atmospheric condition surrounding CBG for various purposes of research and education, especially to analyze the influences to plants growth, the climate comfort and others. This information is also important as a reference to the potential tourist to visit CBG and planning their activities.

The SINDATA© application also contains plant information by using up-to-date references or literature to ensure that plant data is not outdated and valid in a scientific way and the information is always updated in accordance with the latest crop knowledge developments, using internationally-recognized references. Plants collection richness of CBG are 11,241 specimens, which consist of 8,309 of outdoor (from ground cover type up to large trees) and 2,932 of indoor (i.e. orchids, cactus, herbs and Nepenthes spp.) plants specimen. The plants collection is dynamical which mean these can be increased because of the planted addition; reduced because of plants dead (by several causes such as old ages, fallen, pests and diseases, dried, withered etc.). The dynamics are also referring to the scientifically update of the plant's name because of the development of plants taxa classification. All these dynamics are recorded in SINDATA© routinely per-month for the number of specimens and periodically per-semester for name updating if there any.

The purposes of study were build a valid plants database of CBG and these dynamics, and to manage the other related information, so these can be used widely for public and to efficiently plants collection supervision for operators. SINDATA© was also developed as an advance system of dynamic-inventory of CBG's plants database management and to construct by online to disseminate for public information and knowledge, and for research or educational purposes. The scope of the study was to develop a web-based system of SINDATA© so that widely accessible to the public. The development was also built a multi-user system that manageable by several competence operators at the same time. The system managing three basic data: inventory of plants collection and the dynamics, geospatial information, and daily microclimates conditions of CBG. The results were expected as an added value for information and knowledge-transfer to the stakeholders.

2. Methods

SINDATA© system was built by integrating *MySQL*©, *Mapserver*© and *PHP*© software to manage three basic data, which are inventory of plants collection and these dynamics, geospatial information, and daily microclimates conditions. The main plants database consists plants richness of CBG, such as number of plants specimen and the dynamics.

A plants data consists of the number of plants collection (refers to the location in the garden), the sources (self-exploration, donation, or self-propagation), the collector, the access number, scientific name (since division to species or sub-species, if any), local name (if any), the origin, type of the habitat (refers to the latitude), type of the material and type of the habitus. The access number is a series unique number which established for a plant acceptance. The code consists eleven digits, first digit is "C" code for Cibodas, four digits afterwards are code for the year, next two digit is the month in the year which plant accepted, and last four digit is the order of plant registered in that month. For example, access number C2016020016, this means the plants accepted to Cibodas in February 2016, with registered as a sixteenth order. All these information are inventoried in *Microsoft Office© Excel©* and further to be inputted to the system.

The plants dynamic consists previous month the basic number of plants specimen, addition, reduction, scientific name changes (if any) and plants photos. The inputting process was conducted daily by the operators. Up to October 2017, SINDATA© has managed 8,309 of outdoor and 2,932 of indoor plants specimen database. The outdoor plants included from ground cover type up to large trees and indoor plants consisted orchids, cactus, herbs and *Nepenthes* spp.

Furthermore, daily microclimates conditions (i.e. temperature, humidity, rainfall, solar radiation, wind velocity, and air pressure) were obtained by self microclimates tool which installed in CBG. Weather equipment which used was *Precision Weather Station Davis Instruments Vantage Pro2 Plus*™. The equipment was automatic constantly sent current data to server regarding microclimates condition of CBG. The collected data were accurately to describe weather condition surrounding equipment minimum in radius of one kilometer square. These data were uploaded by using *MySQL©* and *PHP©* software to web-based data in SINDATA© system.

Next, geospatial information which displayed previously conducted field inventory and data processing. The activities has been initiated in 2013 and finished in 2016. Geospatial information map of CBG was consisted borderlines area of CBG, elevation and contours, water body, roads, other infrastructures, and the important one is planted points of each plant specimen. These data were processed by *Mapserver©* software and serve to web-based data used *MySQL©* and *PHP©* software.

All data was serve to web-based and may widely accessed *http://sindata.krcibodas.lipi.go.id* .The data changes and dynamics were up-to-date and controlled by operators daily since 08.00 am to 16.00 pm, Monday to Friday. These information and data content of SINDATA© were expected as public knowledge and a valid scientific data for research and educational interest

3. Results and Discussion

Plants which have been rescued from their habitat, be collected and their data capable of being processed, documented and re-displayed by the botanical garden are a very urgent material for the benefit of public knowledge, research, and policies in the context of the flora conservation. SINDATA© application was managed its system and data by Registration Unit of CBG in cooperation with other units. SINDATA© users both internal and external may easily access the application of SINDATA© by typing the site, i.e. *http://sindata.krcibodas.lipi.go.id* or go first through the website of CBG at*http://krcibodas.lipi.go.id.*

3.1. Data content

SINDATA© or *Sistem Informasi Data Tanaman* to be presented on the website (figure 1a) encompassing about 8,309 of outdoor and 2,932 of indoor plants specimen database, a set geospatial information and interactive map of Cibodas Botanical Garden, and daily microclimates (i.e. temperature, humidity, rainfall, solar radiation, wind velocity, and air pressure).The main menu of were *Status IUCN*, *SIG*, *Cuaca* and *apk*(figure 1b). *Status IUCN*or IUCN status was a list of CBG plants collection that included to the IUCN Red-list. IUCN Red-list consists a plants list in the world that threatened (by over-exploited) in their natural wildlife. The *SIG* (*sistem informafi geografis*) or geographic information system was a series spatial information of CBG. *Cuaca* or weather was the atmospheric information condition surrounding CBG. The *apk* was application of SINDATA© that can be downloaded to the android©-based self phone.

The plants information were first entered into *Microsoft Office© Excel©* spreadsheets by Registration Unit of CBG. Each plant entry has all the information including species name, author citation, sub-species if any, variety if any, family, subfamily and general others description. A plants data consisted the number of plants collection (refers to the location in the garden), the sources (self exploration, donation, or self propagation), the collector, the access number, scientific name (since division to species or sub-species, if any), local name (if any), the origin, type of the habitat (refers to the latitude), type of the material and type of the habitus. The general summary of the CBG plants collection (based on family or genera or species and others classification such as genus dubious and intermediate) can be explored in the homepage system (figure 1c).The data further comprise in which the species are described.

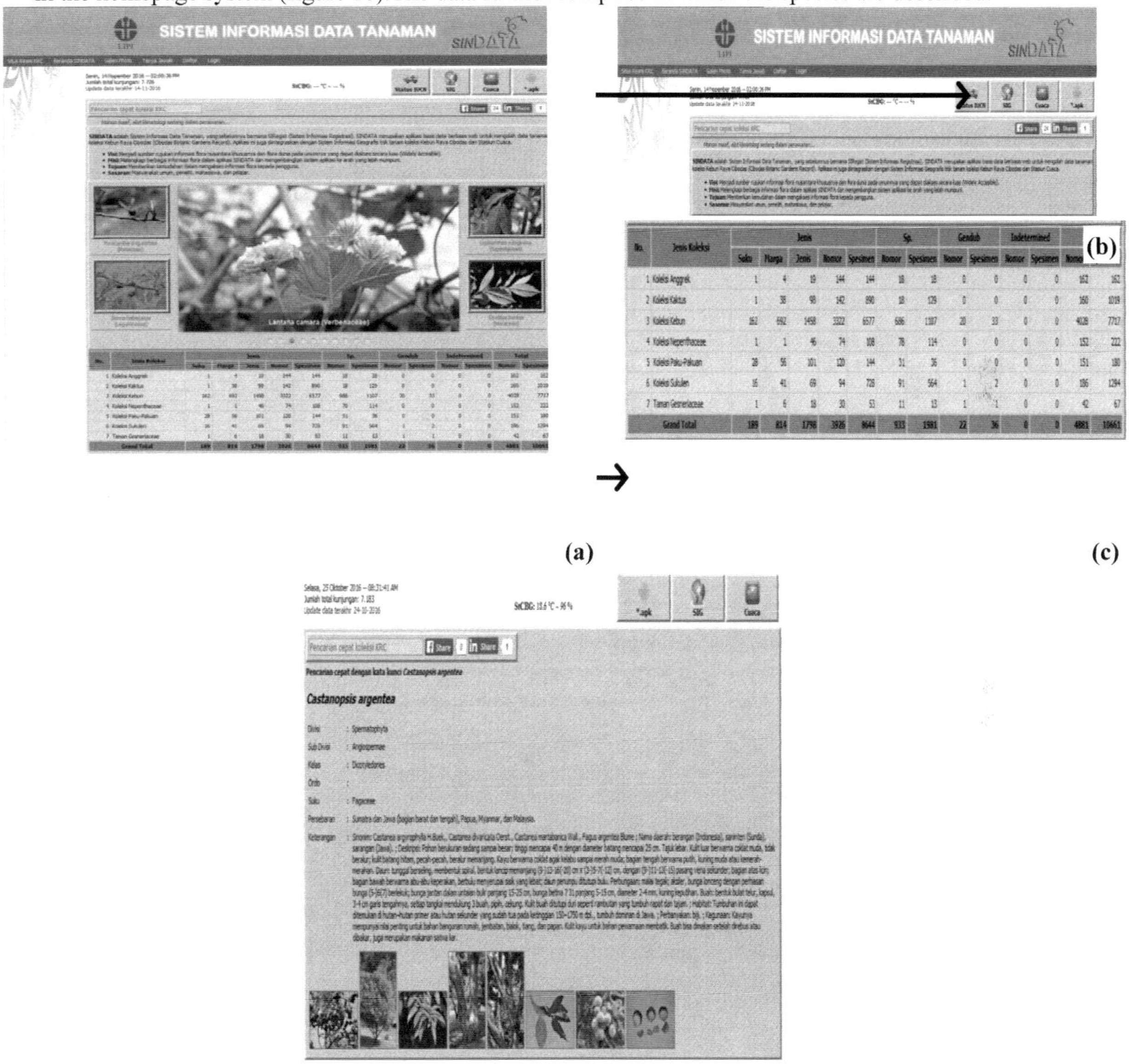

Figure 1. (a) Homepage of the SINDATA©; **(b)** Main menu of the system; **(c)** The summary of the richness of CBG plants collection; **(d)** Plant taxonomic description and images.

3.2. Additional information includes the nomenclature update according to 'The Plant List' (*http://www.theplantlist.org/*), a detailed description in plants database pages, phenology, distribution, threat status and comments on any special features of the taxon. The digitized plant specimens at a minimum resolution of 300 dpi and images of live specimens provided in the database form an information synergy on the species (figure 1d). Once entered, the data were subjected to scrutiny for accuracy and formatting consistency, and subsequently imported into *MySQL*© database tables.

3.3. 'The Weather' menu is obtained from real time weather data at CBG and updated every five minutes. The data were included temperature, humidity, air pressure, wind speed, wind direction, solar radiation, ultraviolet rays, rainfall, and evaporation. All microclimates data were subsequently imported into *MySQL*© database tables (figure 2a) and presented by *PHP*© scripting language. Furthermore, geospatial information map of CBG was consisted borderlines area of CBG, elevation and contours, water body, roads, other infrastructures, and the important one is planted points of each plant specimen. Full map of CBG in which there are points as a representation of plants collection. These data were processed by *Mapserver*© and serve to web-based data used *MySQL*© and *PHP*© database.

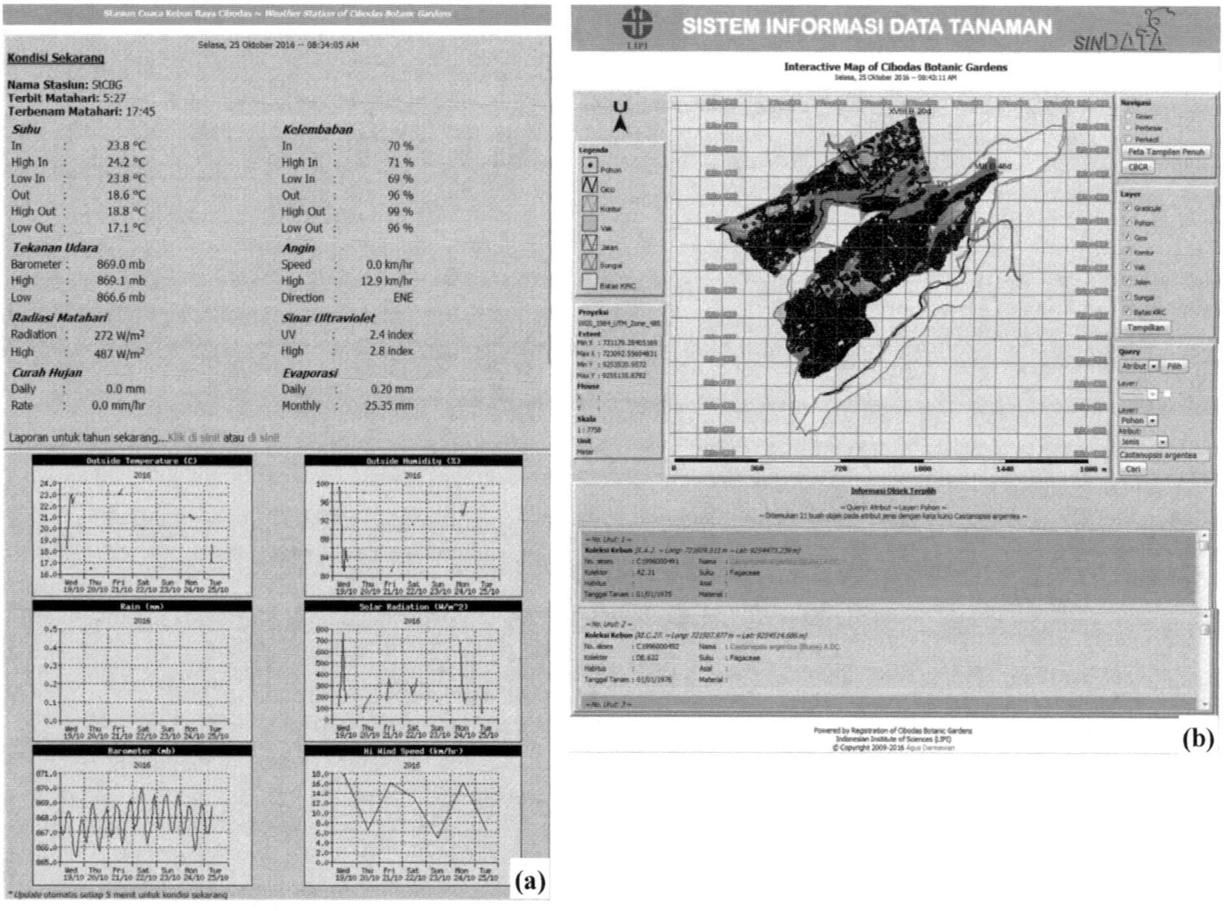

Figure 2. (a) Web-display of weather condition; **(b)** Geospatial image of CBG in SINDATA©

3.4. Website design

The website was developed using standard open-source software on the Linux operating system, including Apache, a web server (http://www.apache.org/); *MySQL*©, a database system (http://www.mysql.com/), and *PHP*©, a scripting language (http://www.php.net/). Plant taxonomic data are stored in the database, which allows rapid indexed searches to be carried out and the content to be generated dynamically. Binary (non-text) content such as high quality digital image files and documents requiring considerable storage are stored on the file system instead of inside the database, for greater efficiency.

3.5. Website user interface

An image gallery is provided that allows species to be identified by browsing images. Each image is hyperlinked to the corresponding species webpage. Alphabetically sorted species and family lists allow users to browse by species name and family respectively. The website provides basic and advanced search capabilities. Text entered in basic search is searched in all fields of the plants database, while advanced search allows text to be searched in one or more fields (e.g. genus).Furthermore, in interactive map by clicking each point on the map will appear plant data, for example in Figure 3.

Koleksi Kebun *(XI.C.27. ~ Long: 721507.977 m ~ Lat: 9254514.686 m)*

No. akses	: C1996000492	Nama	: *Castanopsisargentea* (Blume) A.DC.
Kolektor	: DE.622	Suku	: Fagaceae
Habitus	:	Asal	:
TanggalTanam	: 01/01/1976	Material	:

Figure 3. Interactive map showing point to plant collection description and its location in the garden.

3.6. Accessibility and monitoring

SINDATA© is a data processing plant application based on multi user system [5]. For internal environment of CBG, the presence of SINDATA© provided an ease in operation because each plant personal data processing can be data entry by sharing between personal and data can be validated. Data and information entries can be monitored by all data processors. Data and information users of other units within CBG such as researchers, collection units, nursery units, service and information units and other units can directly access data, and these can be used directly by their respective users.

SINDATA© can be accessed data and information anywhere and anytime. This has been proven to get a good response from users as seen from the number of external users recorded in 2017 alone reached more than 14,176 users. SINDATA© provides convenience to people who need information on plants and microclimate information of CBG. The data and information can be used as references by students, government agencies, private or public for various purposes, such as for research, scientific papers, be used in the business world and can be used as a reference to take policy at the level of government and private policy [2]. For interests of other institutions, data and information in SINDATA© can be shared with government or private institutions [5], such as Research Centre of Biology-LIPI with its InaBIF application, or with Botanic Garden Conservation International (BGCI), with the application of distance education (Digital Class System) of CBG. The presence of SINDATA© provides an ease in transferring data or data sharing with various related institutions.

The data and information completeness were continually needed includes new records, the number of dynamics data, the number of images uploaded, and the number of general information collection plants and how far the system developed. These update will be needed by internal user both CBG and external public.

4. Conclusion

Based on the display and the data that described in the system, SINDATA© has succeeded to present a user-friendly information to the public. The construction of the system was simple with a complete content to described most of all of the richness of CBG plants collection. The content of the system did not only describe the plants collection but also give the information of related data and information that important to improve the development of research and education.SINDATA© is an information system that provides access to data on plant diversity of Indonesian tropical mountain plants, especially CBG plants collection. The information content of the database is essentially built general knowledge for public and so research and educational interest. The presentation of this information implies that information specimens shown are available for examination and for further research. An integrated plants database, thus, maximizes the usefulness of the collection by providing authentic and lasting information.

Acknowledgements

Our grateful towards to the Director of CBG, Mr. Agus Suhatman, and Head of Plants Exploration and Collection of CBG, Mr. Muhammad Imam Surya, Ph.D, for supporting, advises and communications. Our grateful also towards to all Registration Unit staffs of CBG, Mr. Didi Rasidi, Mr. Habibulloh for cooperation in data compilation, and to Mr. Bahtiar, Yudi Suhendri, Dimas Ardiyanto, Mrs. Neneng Ine Kurnita, Mrs. Puji Soepriati for operating SINDATA© management. This paper was expected to be our small contribution in disseminating information and knowledge transferred of CBG's plants property.

References

[1] Allen W H 1993 The rise of the botanical database *Bioscience* 43 pp 274-9

[2] Sharrock S and Hird A 2014 Networking botanic gardens for conservation – the role of BGCI's*BGjournal*11 pp 3-6

[3] Wyatt A and Sucher R 2014 The evolution of living collections management to support plant conservation*BGjournal*11 pp7-10

[4] Poschlod P, Kleyer M, Anne-Kathrin J, Dannemann A and Tackenberg O 2003 BIOPOP - A database of plant traits and internet application for nature conservation*Folia Geobot.* 38 pp 263-71

[5] Sutherland L 2014 Integrated botanical information systems – the Australian Seed Bank Online*BGjournal*11 pp11-4

[6] Witcher E and Calonje M 2014 Using GIS to leverage plant collections data for conservation *BGjournal*11 pp 15-8

[7] Ellis E A, Nair P K R, Linehan P E, Beck H W and Blanche C A 2000 A GIS-based database management application for agroforestry planning and tree selection*Comput. Electron. Agric.* 27 pp 41-55

IOP Conf. Series: Earth and Environmental Science **179** (2018) 012004 doi:10.1088/1755-1315/179/1/012004

Spatial planning for green infrastructure in Yogyakarta City based on land surface temperature

R D Pratiwi [1*], I S Fatimah[2], and A Munandar[2]

[1] Graduate Student of Department of Landscape Architecture, Faculty of Agriculture, Bogor Agricultural University (IPB), Jl. Meranti, Kampus IPB Dramaga, Bogor 16680, Indonesia.
[2] Department of Landscape Architecture, Faculty of Agriculture, Bogor Agricultural University (IPB), Jl. Meranti, Kampus IPB Dramaga, Bogor 16680, Indonesia.

*Email: ririk.26@gmail.com

Abstract. Green infrastructure (GI) planning has developed as an important approach that aids the creation of functional, sustainable, and liveable cities. GI can play a role in climate change adaptation and mitigation through reducing air and surface temperature by providing shading and enhancing evapotranspiration. The aim of this study is to introduce a spatial analysis approach by identify priority areas thus GI can be strategically placedto improve local temperature regulation. We used a method to evaluate the local temperature regulation by green cover at land use unit level using information on land cover and LST, the mean of LST was calculated for each land use and land cover classes. LST map result shows that across the study area, LST values increased from the outskirts towards the centre of urban areas with surface ranging from 30°C to 42°C with average temperature 36.3°C. The results was shown that land use unit such as industrial, health buildings, offices, and tourism areas with built up areas land cover are become priority areas for GI planning. Awareness of this environmental issues is crucial as it can serve as a tool for cities to adapt and mitigate climate change to improve green infrastructure planning strategies.

Keywords: climate change, green infrastructure, land surface temperature, urban heat island

1. Introduction

In cities worldwide, hard surfaces can now comprise as much as 67% of land area, while 'green' areas can fall as low as 16% in some cities [1]. Urban growth, by altering cities and the surrounding countryside, presents numerous challenges for the maintenance of urban green space, and consequently also for human health and well-being [2]. Yogyakarta is one of the cities in Indonesia that experienced rapid growth and development, eventually green space has turned into built up area. Land use of Yogyakarta in 2014 is dominated by settlements that covering 1333.75 ha or 40.58% of the total area [3]. Development of Yogyakarta in the middle of DIY Province has potential to give a negative impact on the phenomenon of Urban Heat Island (UHI). UHI phenomenon is a major issue in every developing city in the world against global warming. Land Surface Temperature (LST) derived from remote sensing data is unique source of information in order to define surface UHI and it has been widely used as indicator for UHI research [4]. LST information is available from a series of satellite sensors that covering various earth surfaces. Compared with the air temperature obtained from the climatological station, the thermal image provides full spatial coverage at various temporal scales [5].

Over the past decade GI planning has developed as an important approach to green space planning that aids the creation of functional, sustainable and livable cities [6]. Green urban infrastructure can play a role in climate change adaptation through reducing air and surface temperature by providing shading and enhancing evapotranspiration, which leads to two benefits: reduced energy use and improved thermal comfort, them also affects air quality through the absorption of pollutants [7,8]. Local temperature regulation by green areas is an important urban ecosystem service, because it can reduce UHI effect so it will improve the quality of life [9, 10]. Therefore, the aim of this study is to introduce a spatial analysis approach by identify priority areas thus GI can be strategically placed to improve local temperature regulation.

2. Data and Methods

To identify priority areas for Yogyakarta GI planning, three main methodological steps were taken. First, by identifying spatial distribution LST of Yogyakarta. Secondly, by classifying land cover map. Third, estimating spatial priority areas for GI planning to improve local temperature regulation. To identifying spatial distribution LST of Yogyakarta, the method used is Split Window Algorithm. The main data used is Digital Satellite Landsat-8 OLI (Operational Land Imager) path 120-row 65was obtain from United States Geological Survey website. The satellite image was collected on 17 May 2017when the atmospheric condition was highly clear. UHI are generally stronger during the daytime, the selection of a daytime image in the summer is recommended [9]. The 2014 land use map was obtained from previous research by Ratnasari (2015) (Figure 1). DIY (Special region of Yogyakarta) province map is obtained from the Indonesian Topographic Map 2004 in the scale of 1: 25,000 by Geospatial Information Agency used as the administrative boundary and the study area.

2.1. Estimated Land Surface Temperature Distribution

The first step is determining subset area. Second, converting the Digital Number into spectral radiance. Band 10 and band 11, band 4, and band 5 converted from raw image to Top of Atmospheric spectral radiance using radiance rescaling factors [12]. The calculation formula is as follows:

$$L\lambda = M_L * Q_{cal} + A_L$$

Where: $L\lambda$ is Top of Atmospheric Spectral Radiance (Watts/ (m2*srad*µm)); $M_L$is band-specific multiplicative rescaling factor (0.0003342); $A_L$is band-specific additive rescaling factor (0.1); Q_{cal} is band 10/ band 11 image.

Third, TIRS band 10 and band 11 were taken to estimate Brightness Temperature (BT) on Kelvin by converting spectral radiance to BT for both TIR bands [12], supported by the following formula:

$$BTi = \frac{K2}{Ln\left(\left(\frac{K1}{L\lambda}\right) + 1\right)}$$

Where: K1 and K2 are band-specific thermal conversion constant from metadata; L λ is Top of Atmospheric Spectral Radiance (Watts/ (m2 * srad * µm)).

Then, estimate the value of Normalize Difference Vegetation Index (NDVI). Band 4 and band 5 are used to obtain NDVI values with the following formula [11]:

$$NDVI = \frac{NIR\ (band\ 5) - R\ (band\ 4)}{NIR\ (band\ 5) + R\ (band\ 4)}$$

Where: NIR is near-infrared band (Band 5) and R is red band (Band 4).

FVC values can be estimated using previously obtained NDVI values, as well as NDVI soil and NDVI veg values [13]. The pixel was considered as bare soil when the NDVI values are<0,2; while it would be considered as full vegetation when the NDVI values are > 0,5 [11]. FVC function is to estimate the fraction of a vegetation covered area, with the following formula:

$$FVC = \frac{NDVI - NDVIsoil}{NDVIveg - NDVIsoil}$$

Where: NDVI soil is NDVI value of soil = 0.2; NDVI veg is NDVI value of vegetation = the maximum value of NDVI (more than 0.5).

Using the average value and the difference value on Land Surface Emissivity (LSE) to estimate the LST. Estimating LSE requires the emissivity value of soil and vegetation from both TIRS band 10 and band 11 [14], with the following formula:

$$LSE = \varepsilon s * (1 - FVC) + \varepsilon v * FVC$$

Where: εs and εv are soil and vegetation emissivity values.

The combination of LSE band 10 and band 11 generates mean of LSE (ε) and difference of LSE value (Δm) [14], with the following formula:

$$\varepsilon = \frac{\varepsilon 10 + \varepsilon 11}{2}$$

Where: $\varepsilon$is mean of LSE; $\varepsilon 10$ and$\varepsilon 11$ is the emissivity of band 10 and band 11.

$$\Delta\varepsilon = \varepsilon 10 - \varepsilon 11$$

Where:Δε is difference of LSE; ε_{10} andε_{11} is the emissivity of band 10 and band 11.

The final step, estimate LST use an algorithm with the following formula [13]:

$$LST = TB_{10} + C_1(TB_{10} - TB_{11}) + C_2(TB_{10} - TB_{11})^2 + C_0 + (C_3 + C_4W)\,(1 - \varepsilon) + (C_5 + C_6W)\Delta m$$

Where: C_0 - C_6isSplit-window coefficient value; εis mean of LSE band 10 and band 11; W is Atmospheric Water Vapour Content = 0.013 [14]; Δm is difference of LSE band 10 and band 11.

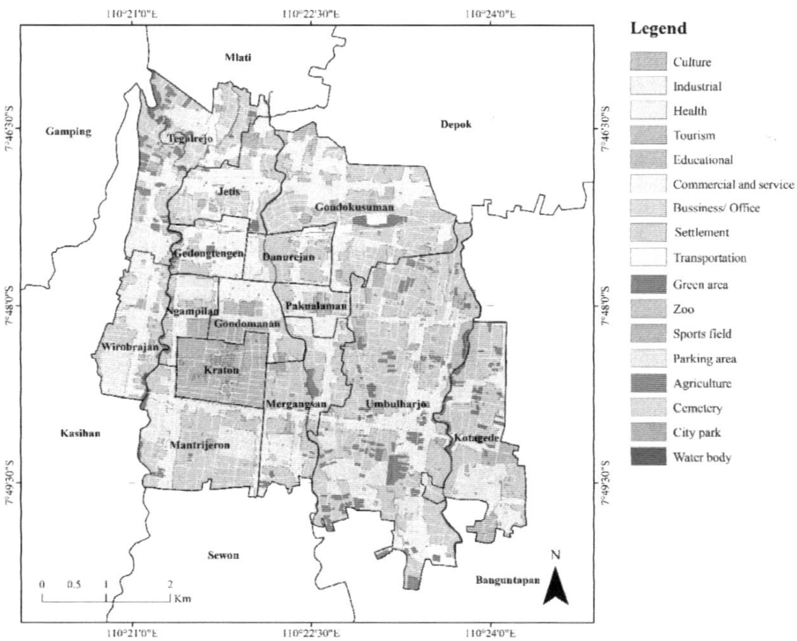

Figure 1. Land use map of Yogyakarta 2014 [3].

2.2. Land cover map classification

Procedures for image classification by supervised classification method and to classify land cover class is used sub-menu of image classification with maximum likelihood classification method in ArcMap

IOP Conf. Series: Earth and Environmental Science **179** (2018) 012004 doi:10.1088/1755-1315/179/1/012004

10.4. In land cover map analysis, the first thing to do is combining 3 bands that suitable with the spectral characteristics of each band and adjusted for research objective, in this study bands 4, 3 and 2 (natural color) was used. Then, pan sharpening process is done to sharpen the appearance of objects on the image in conducting visual analysis. The sharpening of this object is done by combining the multiband image (1, 2, 3, 4, 5, 6, 7 and 9) which has a resolution of 30m x 30m and a panchromatic band (band 8) that has a spatial resolution of 15m x 15m [15]. After that, we clipping the image that has been compiled with the administrative boundary map of Yogyakarta. The next step is to create a training area as polygons in each class that will be classified. Training area was spreaded and evenly distributed to each class that covering the entire study area. Four types of land cover identified are, (1) vegetation (including forest, street trees), (2) agriculture land, (3) bare land, and (4) built up area (commercial buildings, settlements, offices, industrial areas, etc.) In this study, the accuracy of land cover classification is classified using the reference point generated by using stratified random sampling technique. This evaluation tests visual accuracy level of the supervised classification. Accuracy is done by making contingency matrix (confusion matrix). Accuracy that can be calculated is, user's accuracy, producer's accuracy and overall accuracy.

2.3. Identify GI priority areas to improve local temperature regulation

In this study, we use method to evaluate the local temperature regulation by green cover at the land use unit level using information on land cover and surface temperature. The mean of LST was calculated for each land use and land cover classes. Based on index proposed by Schwarzet al. [10], has been developed index that establishes the influence of green cover on surface temperature for each land use unit (i):

$$index\ (i) = \frac{Mean\ temperature\ Land\ Use}{Mean\ temperature\ Green\ Cover}\ x\ percent\ Green\ Cover$$

3. Result and Discussion

3.1. Spatial distribution of land surface temperature

Based on the results of Land Surface Temperature distribution analysis, the class of temperature distribution is grouped into 12 classes (**Figure 2**). The spatial distribution of land surface temperature by land use unit exhibits the general pattern of urban heat islands, wherein temperatures tend to decrease as the radial distance from the city centre increases, with surface ranging from 30°C to 42°C. The map shows that high temperatures occur in the built up areas, while the lower temperatures occurring vegetated areas. Xiong et al. found that high temperature anomalies are closely related to built-up areas, densely populated zones, and highly dense industrial areas [16]. Land cover factors have the greatest impact on spatial distribution of UHI in Yogyakarta City. High LST values generally occur in built up areas due to the increasing population density every year in Yogyakarta, the development spread to the sub urban areas surrounding the city, some part of Sleman and Bantul Regency [13].The highest temperature occur in Gondomanan District with an average temperature of 38.4°C and the lowest temperature is in Tegalrejo District with an average temperature of 35.7°C.

3.2. Land cover map

The classification result of land cover map of Yogyakarta is shown in Figure 3. It can be seen that most of urban area is dominated by built up area, with overall presentation of vegetated area covering 25.27 ha (0.8%), bare land 450.99 ha (13.7%), agriculture land 263.82 ha (8%), and built up area 2556.86 ha (77.5%). Land use in Yogyakarta is dominated by residential areas that occupy almost half of the total area [3]. After the land cover map is classified, accuracy test obtain with overall accuracy results is 91.25%, and kappa accuracy 88.33%.

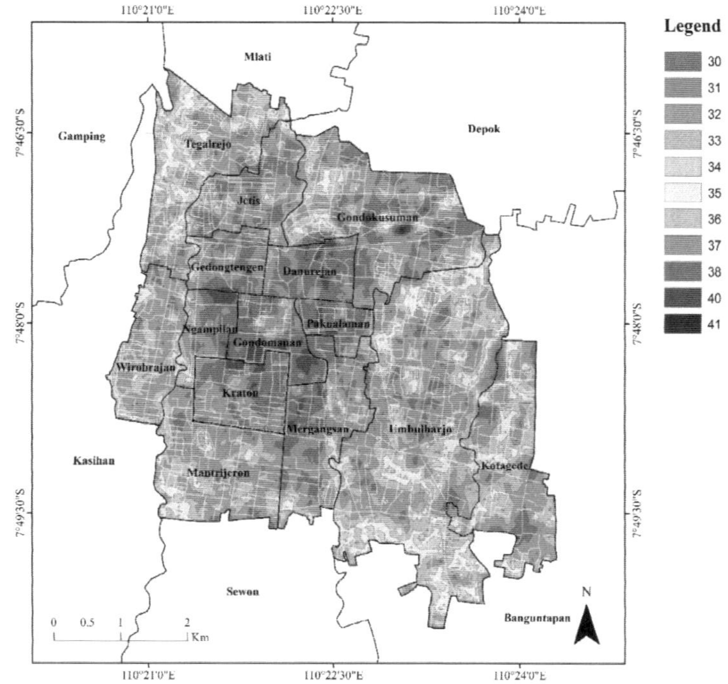

Figure 2. Land Surface Temperature map (° C) of Yogyakarta 2017.

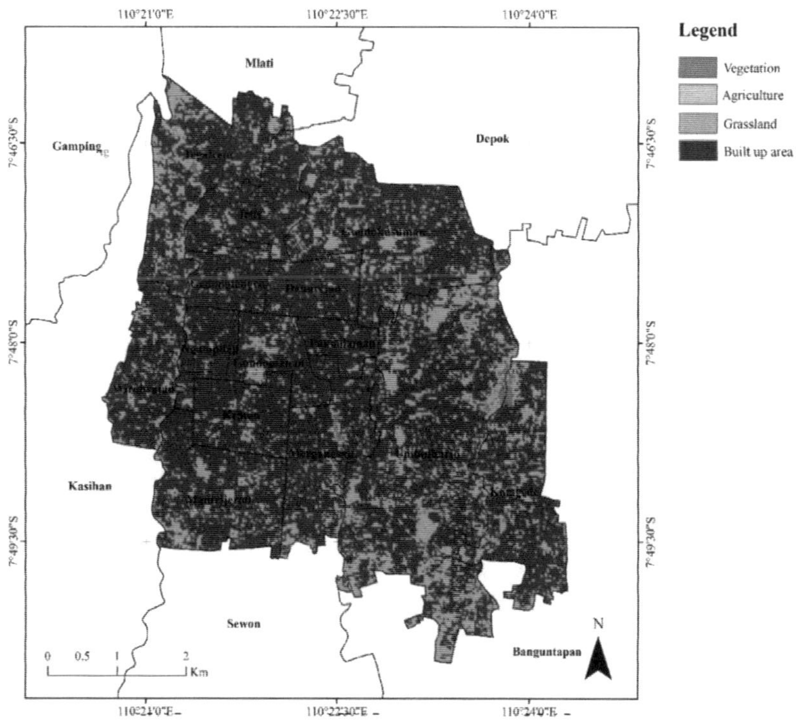

Figure 3. Land cover map of Yogyakarta 2017.

3.3. Spatial priority areas for GI planning

Table 1 summarizes surface cover information for each land use classes and means of surface temperatures for all land use and land cover classes. The results show that, the lowest average temperature in the land use class is 33.1°C in the zoo area and the highest average temperature reached 38°C occurred in health buildings area that lack of vegetation. The results confirm the expected relationship between land use/land cover characteristics and surface temperature. The land use/land cover classes related to urban fabric exhibit the highest temperature, while the lowest values occur at green covered surfaces [10]. The distribution of temperatures in urban areas is complex and depends on factors such as the pattern of urban land use, with densely built up areas being considerably warmer while green-covered surfaces are relatively cool areas [17].

Table 1. Areal percentage of each land cover class for each land use class, and list for all land use and land cover classes.

Land use classes	Land cover classes (%) Green cover classes				Built area/ impervious surfaces (%)	LST Mean (°C)
	Agricultural	Tree covered	Bare land	Total green		
Industrial	31.6	1.0	67.4	17.5	82.5	36.6
Culture	41.1	4.6	54.3	25.9	74.1	37.6
Health	23.7	1.4	74.9	17.7	82.3	38.0
Tourism	31.5	0.9	67.7	8.2	91.8	37.8
Education	39.6	1.0	59.4	29.3	70.7	36.7
Business	28.3	1.1	70.6	16.5	83.5	36.8
Office	38.5	1.0	60.5	30.2	69.8	36.7
Housing	37.2	2.4	60.4	20.9	79.1	36.2
Transportation	26.9	4.2	69.0	29.7	70.3	37.6
Agriculture	53.7	1.1	45.1	61.0	39.0	35.0
Water body	53.5	21.6	24.9	49.2	50.8	35.2
Cemetery	27.6	0.3	72.0	38.6	61.4	36.7
Green area	42.7	8.8	48.5	57.1	42.9	35.5
Zoo	39.4	45.9	14.7	83.4	16.6	33.1
Yard	18.6	0.6	80.8	50.8	49.2	37.0
Parking area	24.0	0.2	75.8	32.0	68.0	37.4
Urban park	24.1	1.4	74.5	50.9	49.1	38.0
LST Mean (°C)	36.1	35.1	36.9	36.0	36.3	

The result from Table 1 was inputted to the equation to determine the land cover index of surface temperature unit (i). The index classification is divided into 7 classes with a class interval value is 13.60 by using Sturges interval formula to determine the priority class. After classification obtained, the classes then translated into spatial forms as a priority areas map for GI planning that classified as class 1 (low priority area) that represented in green color, while class 7 (high priority area) is represented in red (Figure 4). The results shown that land use unit such as industrial, health buildings, offices, and tourism areas with built up areas land cover classes are become priority areas for GI planning. From central of the city and spreaded to the north and the south are the most urbanized areas and the most areas that need more GI.

4. Conclusion

LST spatial distribution results are ranging from 30 °C to 42°C. The highest temperature occurred in Gondomanan District with an average temperature of 38.4°C and the lowest one is in Tegalrejo District with an average temperature of 35.7°C. The lowest average temperature in the land use class is 33.1°C and the highest reach up to 38°C. The results shown that land use unit such as industrial, health buildings, offices, and tourism areas with built up areas land cover classes are become priority areas for GI planning. From central of the city then spreaded to the north and the south are the most urbanized areas and the most areas that need more GI.

Awareness of this environmental issues is crucial as it can serve as a tool for cities to adapt and mitigate climate change to improve green infrastructure planning strategies. By raising public awareness of the green infrastructure importance, will make people want to learn to improve knowledge and skills in building green spaces. Further research on community perceptions and preferences on GI will strengthen the GI planning decision support tools.

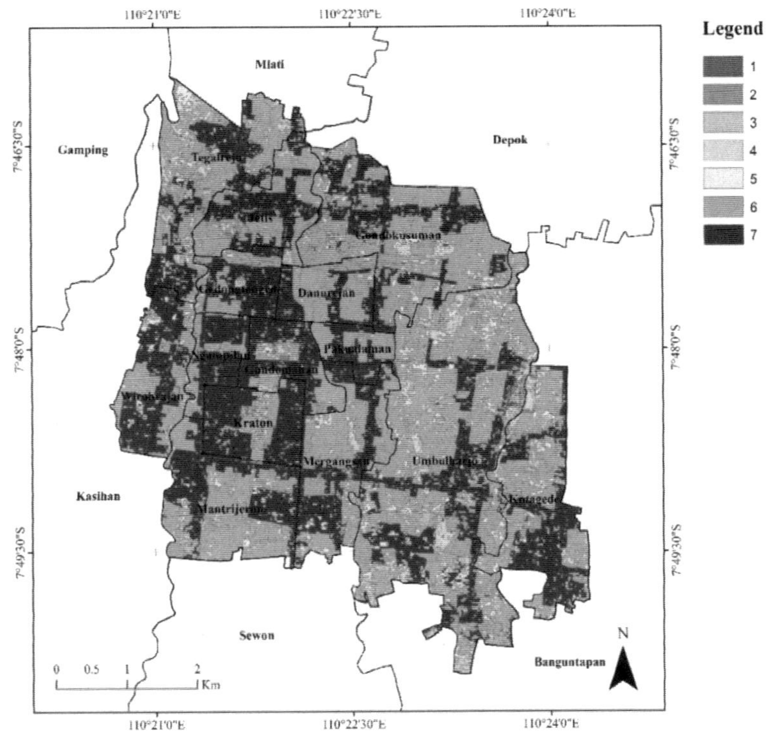

Figure 4. Priority areas of green infrastructure planning.

References

[1] Gartland L 2011 Heat Islands: Understanding and mitigating heat in urban areas (London: Earthscan)

[2] Tzoulas K, Korpela K, Venn S, Yli-Pelkonen V, Kazmierczak A, Niemela J, and James P 2007 Promoting ecosystem and human health in urban areas using Green Infrastructure: A literature review Landscape and Urban Planning 81 167–178

[3] Ratnasari A 2015 Perencanaan Kota Hijau Yogyakarta Berdasarkan Penggunaan Lahan dan Kecukupan Ruang Terbuka Hijau Institut Pertanian Bogor

[4] Tran DX, Pla F, Carmona PL, Myint SW, Caetano M, and Kieu HV 2017 Characterizing the relationship between land use land cover change and land surface temperature ISPRS Journal of Photogrammetry and Remote Sensing 124 119–132

[5] Myint SW, Wentz EA, Brazel AJ, and Quattrochi DA 2013 The impact of distinct anthropogenic and vegetation features on urban warming Landscape Ecol 28 (5) 959–978

[6] Mell IC 2010 Green Infrastructure: Concepts, Perceptions and Its Use in Spatial Planning University of Newcastle

[7] Demuzere M, Orru K, Heidrich O, Olazabal E, Geneletti D, and Orru H 2014 Mitigating and adapting to climate change: multi-functional and multi-scale assessment of green urban

infrastructure Journal of Environmental Management 146 107–115
[8] Mell I C, Henneberry J, Hehl-Lange S, Keskin B 2013 Promoting urban greening: valuing the development of green infrastructure investments in the urban core of Manchester UK Urban Forestry & Urban Greening 12 (3) 296–306
[9] Schwarz N, Bauer A,and Haase D 2011 Assessing climate impacts of planning policies–An estimation for the urban region of Leipzig (Germany) Environmental Impact Assessment Review 31 (2): 97–111
[10] Madureira H and Andresen T 2014 Planning for multifunctional urban green infrastructures: Promises and challenges Urban Design International 19 1 38–49
[11] Kurnia E 2016 Spatial modeling of land surface temperature and it's relationship with ecological factors in Bogor municipality Institut Pertanian Bogor
[12] USGS 2013 Landsat 8 Fact Sheet Amerika Serikat: US Geological Survey
[13] Guntara I 2015 Utilization of Landsat Image to estimate Land Surface Temperature through Split Window Algorithm at Bantul Regency Gadjah Mada University
[14] Latif MS 2014 Land Surface Temperature Retrieval of Landsat-8 Data Using Split Window Allgorithm-a Case Study of Ranchi District International Journal of Engineering Development and Reasearch 2 4 3840-49
[15] Wahyuni S 2015 Identifitaion of charatceristic and mapping of land cover using Landsat 8 image (OLI) at Ogan Komering Ilir Regency South Sumatera IPB
[16] Xiong Y, Huang S, Chen F, Ye H, Wang C, and Zhu C 2012 The impacts of rapid urbanization on the thermal environment: a remote sensing study of Guangzhou South China Remote Sens 4 2033–56
[17] Rizwan AM, Dennis LYC, and Liu C 2008 A review on the generation, determination and mitigation of urban heat island Journal of Environmental Sciences 20 (1) 120–8

IOP Conf. Series: Earth and Environmental Science **179** (2018) 012005 doi:10.1088/1755-1315/179/1/012005

Spatial analysis on the impact of socioeconomic vulnerability to drug abuse prevalence in Indonesia 2015

F Isnaini[1,2*], T B R Nitibaskara[1], and W Usman[1]

[1] School of Strategic and Global Studies, Strategic Review on Drugs Management, University of Indonesia, Jalan Salemba Raya No.4, Central Jakarta, 16424, Indonesia.

[2] Medical staff, Rehabilitation Center National Narcotic Board, Jl. HR Edi Sukma Desa Wates Jaya Cigombong, Kabupaten Bogor, Indonesia.

[*]Email: fitriisnaini@gmail.com

Abstract. An environment socioeconomic situation can affect drug abuse. Studies on the impact of socioeconomic vulnerability can be used in planning risk management policy on drug abuse that is more precise, effective, and efficient. This research aims to learn the impact of socioeconomic vulnerability (education level, unemployment rate, income level, gender ratio, poverty rate, and access to health service) on drug abuse prevalence and to identify the index of socio-economic vulnerability to drug abuse in each province in Indonesia, which can be used to analyze drug abuse risk. Methods used in this research are multiple linear regression statistical test and spatial analysis. The result show that F-Test sig<0.05 mean that socioeconomic factor simultaneously effect prevalence; the t-Test result shows only income level have sig <0,05 and there are 2 level of socioeconomic vulnerability index: high (9 provinces) and moderate (25 provinces), while the drug abuse risk level is in 3 level: high (5 province), moderate (9 province), and low (20 province). The conclusion is that socioeconomic factor (education, unemployment, income, gender ratio, poverty rate, and access to health service) are influence to the level of drug abuse prevalence in Indonesia, which the most significantly influence is income level.

Keywords: drug abuser prevalence, spatial analysis, socioeconomic vulnerability

1. Introduction

There is no regency or city in Indonesia that is free from drug abuse and trafficking. The drug abuse prevalence and the types of drugs that are used for the first time also vary among the provinces [1]. These variations show that environment situations affect drug abuse. One of the factors that influence the environment situation is socioeconomic factor. Socioeconomic vulnerability to drug abuse is defined as economic activities, related situations or factors of the society that may lead to drug abuse. This vulnerability is reflected in many factors, such as education level, unemployment rate, income level, gender ratio, poverty rate, and access to health service [2, 3, 4]. According to the Statistics Indonesia (BPS) [5], this factor in Indonesia vary from one province to another. Base on that statement we could hypothesis that socioeconomic factor (education, unemployment, income, gender ratio, poverty rate, and access to health service) are influence to the level of drug abuse prevalence in Indonesia.

Analyzing socioeconomic vulnerability of a certain area to drug abuse will result in knowledge that can be used by instances or institutions that are active in drug abuse prevention and eradication. The knowledge can then be used as material in planning risk management policy that is more specific, effective and efficient. Based on the definition of disaster in the Regulation of the Head of National

IOP Conf. Series: Earth and Environmental Science **179** (2018) 012005 doi:10.1088/1755-1315/179/1/012005

Disaster Mitigation Agency (BNPB) No. 02 Year 2012 [6], drug abuse is categorized as a disaster that is caused by human factor and causes casualties, property damage, and psychological impact. The risk of a disaster can be assessed and mapped using knowledge on its hazard and vulnerability [7]. Thus, the risk of drug abuse can be viewed from the relation between drug abuse prevalence, which is the hazard, and the socioeconomic vulnerability to drug abuse.

This research aims to learn the impact of socioeconomic vulnerability (education level, unemployment rate, income level, gender ratio, poverty rate, and access to health service) on drug abuse prevalence and to identify the index of socioeconomic vulnerability to drug abuse in each province in Indonesia, which can be used to analyse drug abuse risk.

2. Theoretical Review

Vulnerability are conditions determined by physical, social, economic and environmental factors or processes which increase the susceptibility of an individual, a community, assets or systems to the impacts of hazards. Hazard is defined as a process, phenomenon or human activity that may cause loss of life, injury or other health impacts, property damage, social and economic disruption or environmental degradation [7]. Socioeconomic vulnerability can be assessed using several indicators, including education level, unemployment rate, income level, gender ratio, poverty rate, and access to health services [2, 3, 4]

Vulnerability can be described using indexes and maps, where vulnerability is the result of Exposure times Sensitivity. The "assets" that are exposed include human lives (social vulnerability), economic territory, physical structure and ecological / environmental area. Sensitivity is determined indirectly through different weighting for each 'asset' or type of threat [6]. Spatial analysis is used to assess the data and components in an area. This analysis uses overlay process and other analysis techniques to seek answer or acquire useful knowledge or information [8].

3. Research Methods

This research uses literacy study method with quantitative approach that will provide details from the data of socioeconomic vulnerability that are related to drug abuser prevalence in each province in Indonesia in 2015. The dependent variable is the prevalence of drug abuser, which data is obtained from the National Narcotics Board (BNN). The independent variable is socioeconomic vulnerability that is related to drug abuse, using several indicators, which are education level, unemployment rate, income level, gender ratio, poverty rate and access to health services. Data for the independent variable are obtained from the Statistics Indonesia (BPS).

Education level indicator in this research is using the number of graduates of Primary School (SD), Junior High School (SMP), and Senior High School (SMA) in 2015. Unemployment rate indicator draws data from the percentage of the number of unemployment against the number of work force or known as the Open Unemployment Rate in 2015. Income level indicator is using data of monthly expenditure per capita of population in each province in 2015. Gender ratio indicator is derived from gender ratio data per province in 2015. Poverty rate indicator draws data from poverty percentage per province, to learn the percentage of poor people from the population of an area. Access to health service indicator is using data of ratio of local health center (*Puskesmas*) per 30,000 people.

The analysis methods used are the statistical analysis and spatial analysis. The first part is the linear regression statistical analysis that is used to determine the impact of the socioeconomic vulnerability on drug abuse prevalence. This analysis is made using statistic software SPSS version 24. Using the following formula:

$$Y=bo+b1X1+b2X2+b3X3+b4X4+b5X5+b6$$

Remarks:
Y: the prevalence of drug abuser

bo: constant
X1: education level; b1: constant of X1
X2: unemployment rate; b2: constant of X2
X3: income level; b3: constant of X3
X4: gender ratio; b4: constant of X4
X5: poverty rate; b5: constant of X5
X6: access to health services; b6: constant of X6

The second part is spatial analysis using Geographic Information System (GIS) to compose socioeconomic vulnerability index using ArcGIS 10.3 software, which helps in classifying, scoring and weighting, as well as conducting overlay analysis to create a map.

Score for each criteria is obtained from dividing the data value of each indicator into low (0.333), moderate (0.666) and high (1), according to the normal curve. The weighting are determined from statistical test results, by taking into considerations the coefficient value of each indicator. The coefficient value is later converted into percentage, with all socioeconomic vulnerability indicators totaling to 100%. Afterwards, the socioeconomic vulnerability index of each province is calculated using the following formula:

...= (X1 Weight * X1 Score) + (X2 Weight * X2 Score) + ... + (X6 Weight * X6 Score)

The result of the socioeconomic vulnerability index is later categorized into three level: low (0.333), moderate (0.666), and high (1). Than the result is mapped.

The third step is classifying drug abuser prevalence data in each province into low (0.333), moderate (0.666), or high (1) level. The result is then turned into provincial-based drug abuse prevalence map. The fourth step is determining the drug abuse risk level by reviewing the relation between drug abuser prevalence level and index of socioeconomic vulnerability to drug abuse, based on the matrix seen in Figure 1.

RISK LEVEL		SOCIOECONOMIC VULNERABILITY INDEX		
		LOW	MODERATE	HIGH
DRUG ABUSER PREVALENCE	LOW			
	MODERATE			
	HIGH			

Figure 1. Risk level matrix.

4. Results

The linearity test result shows a linear pattern for the independent variables. Normality test results using the Kolmogorov-Smirnov test, Histogram of the Residual, and Normal Probability Plot show normal distribution regression model. Autocorrelation test result using Durbin-Watson test shows that this research contains no autocorrelation. The R2 coefficient of determination result reveals that 63.9% variants of the independents variables can explain the dependent variable, which shows strong autocorrelation. The result of heteroscedasticity test using Scatterplot test shows that the curve does not show regulated pattern, which means the regression function of this research does not appear because of different variants. Based on the results of those tests, which are classic assumption test, the criteria to run multiple linear regression test have been met. The result of the multiple linear regression test can be seen in table 1.

IOP Conf. Series: Earth and Environmental Science **179** (2018) 012005 doi:10.1088/1755-1315/179/1/012005

Table 1. Multiple regression test.

Model		Unstandardized Coefficients B	Unstandardized Coefficients Std. Error	Standardized Coefficients Beta	t	Sig.	Colinearity Statistics Tolerance	Colinearity Statistics VIF
1	(Constant)	.256	.140		1.829	.079		
	Education	.065	.099	.094	.657	.516	.650	1.539
	Unemployment	.108	.070	.199	1.539	.135	.796	1.256
	Income	.638	.151	.597	4.218	.000	.667	1.500
	Gender	-.221	.086	-.315	-2.551	.017	.877	1.140
	Poverty	-.021	.104	-.033	-.206	.839	.509	1.966
	Access to Health Service	-.119	.115	-.174	-1.030	.312	.467	2.143

Coefficients[a]

a. Dependent Variable: Prevalence

The F-test result shows that independent variables simultaneously impact dependent variable. The t-test result shows that out of six independent variables, only Income Level gives significant impact on dependent variable. After completing the statistical tests, spatial analysis is conducted using scoring and weighting, classifying and creating the map. The regulation for scoring and weighting are shown in Table 2.

Table 2. Regulation for scoring and weighting.

No	Indicator	Class (score) Low (0,333)	Class (score) Moderate (0,666)	Class (score) High (1)	Weight
1	Education Level	25,634 – 254,552	254,553 – 627,942	627,943 – 1,554,791	5.546
2	Unemployment Rate	1.990291 – 5.147453	5.147454 – 7.504688	7.504689 – 9.932363	9.215
3	Income Level	533,891 – 840,695.9	840,696 – 1,193,642	1,193,643 – 1,773,431	54.437
4	Gender Ratio	94.2 – 101	101.1 – 105.5	105.6 – 113.2	18.857
5	Poverty Rate	0.003622 – 0.008466	0.008467 – 0.014125	0.014126 – 0.028520	1.792
6	Access to Health Service	0.58 – 1.35	1.36 – 2.46	2.47 – 5.20	10.154

Socioeconomic vulnerability index can be calculated by adding the multiplication values between weight and score of each indicator. The result shows that there are nine (9) provinces (26.47%) with high socioeconomic vulnerability index and 25 provinces (73.52%) with moderate socioeconomic vulnerability index. These data are mapped into Figure 2.

Most Indonesian territories are not high vulnerable areas, but without proper supervision, those areas might turn into high socioeconomic vulnerability index areas. Provinces with high level are mostly located in western Indonesia. Only one of such provinces is located in eastern Indonesia, while the other two are in middle Indonesia. Analysis on provinces with high socioeconomic vulnerability index shows that the dominant indicators are unemployment rate and access to health service. Comparing the socioeconomic vulnerability index of those provinces with their Human Development Index (HDI) reveals that only four provinces have high HDI. Thus, we cannot assume that high HDI also means high index of socioeconomic vulnerability to Drug Abuse, because there are several indicators that differ in the measurement.

Upon examining the location of the provinces with high socioeconomic vulnerability index, it is revealed that five out of nine provinces with high socioeconomic vulnerability index are located in Indonesia's border areas and one of them is bordering the high seas. This shows that provinces located in border areas and close to high seas are more vulnerable to drug abuse.

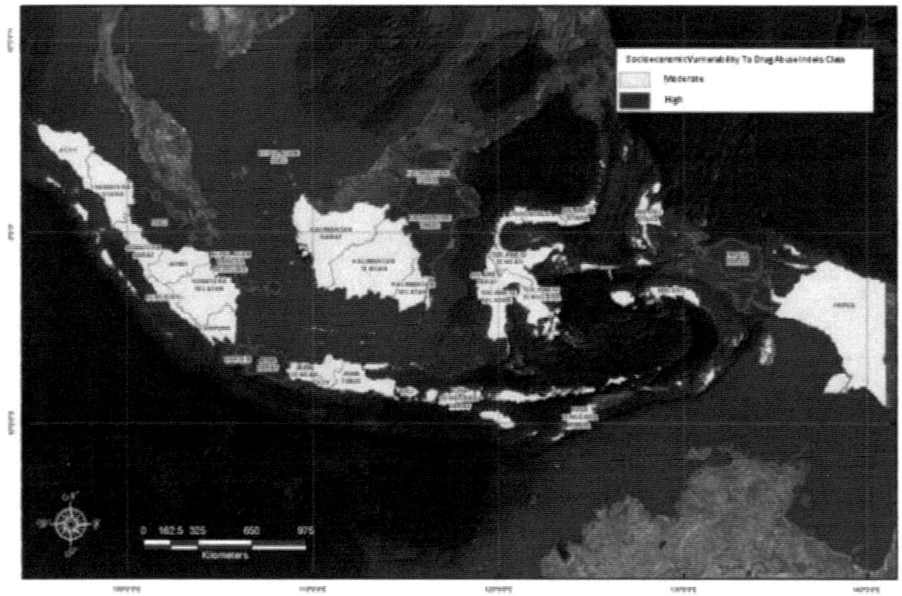

Figure 2. Distribution of socioeconomic vulnerability to drug abuse index.

Unemployment may lead to many issues, including drug abuse. Busaglia and Dijk [9] mentions that unemployed are the manpower resource for organized crimes activities This requires attention because based on the Indonesian Statistics (BPS) projection, Indonesia has now reached the demographic bonus, with the percentage of working age population predicted to be increasing until 2035. In 2010, the number of working age population was 66.5%. The number is predicted to increase to 67.7% in 2020 and 68.1% in 2030. This momentum has to be supported with employment availability; otherwise, the high number of unemployment might be exposed to illegal businesses.

Other than unemployment rate, access to health service also contributes to the high socioeconomic vulnerability index. Easy access to health services is related to the drug abuse vulnerability [4]. Mendoza and Booth [3] states that characteristics of an area, such as the availability of therapy facilities or alcoholic beverages stores, are related to the success of therapy and behavior of substance abuse. Upon examining the provinces with high socioeconomic vulnerability index, 6 out of those 9 provinces have high access to health service indicator, which means population in those provinces is experiencing difficulties in accessing health services, including early health care to drug abuse. Data on drug abuser prevalence shows there are more drug abusers in western Indonesia then other parts of Indonesia. Limited number of drug rehabilitation centers in western Indonesia is hampering the drug abusers from receiving health services. This will eventually increase the vulnerability in the areas. Limited access to health services should be overcome so that drug abuse prevention and drug abuser rehabilitation programs can reach remote areas and benefit all Indonesian citizens.

4.1. Drug abuse risk level

Analysis on drug abuse risk level and its risk mapping can help in giving strong arguments for the government in planning drug abuse mitigation plan in Indonesia. The enormous area of Indonesian territory and the limited human and financial resources in drug abuse mitigation call for a drug abuse risk level analysis in Indonesian territories in order to plan effective and efficient programs. Failure in managing drug abuse risk properly will end in real disturbances on and threats against the national defense and security.

This risk level is determined by relating the drug abuse prevalence level with the index of socioeconomic vulnerability to drug abuse, with the requirements mentioned on figure 1. The result shows that there are five (5) provinces (14.4%) with high risk level, nine (9) provinces (26.47%) with moderate risk level, and 20 provinces (58.8%) with low risk level. These data are mapped into Figure 3.

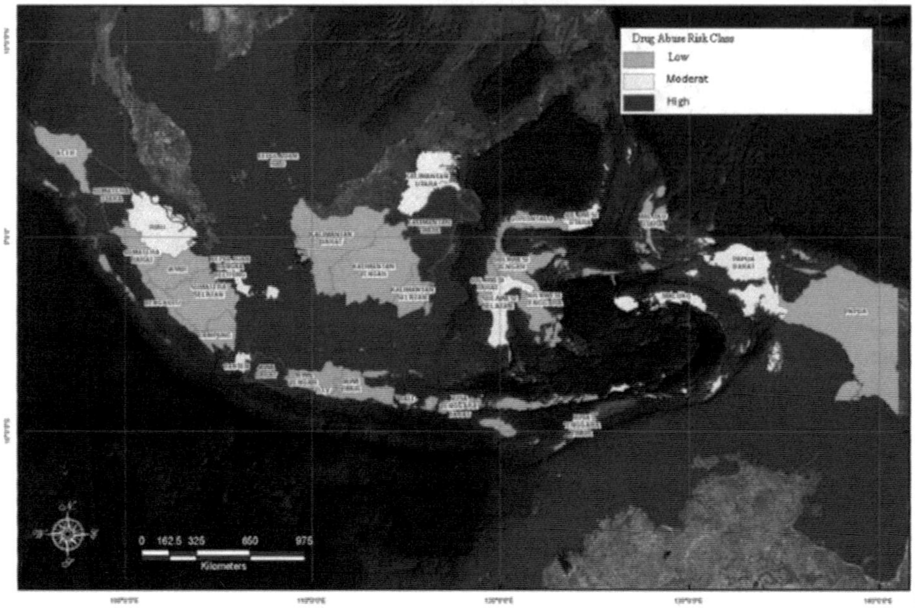

Figure 3. Distribution of drug abuse risk level.

The results of the drug abuse risk analysis based on socioeconomic and drug abuser prevalence show that Indonesian territories have various risk level. This research shows that 58.8% of Indonesian provinces are in low risk level, but without proper supervision and management, the risk level may increase. This prediction is based on two factors, which are the haste of drug trafficking, which will impact the drug abuser prevalence, and the Indonesian population socioeconomic conditions, which are very dynamic. Drug abuser risk level will affect Indonesian national resilience because the drug abuser prevalence will impact Indonesian human resources, which is the natural aspect of national defense, while the socioeconomic condition is part of the social aspect of the national defense.

Taking economy point of view, drug trafficking can be considered as a company business, where the business owners are looking for the most profitable location for the production, distribution and market or consumers [10]. Indonesia is in the brink of potential social and economic growth, considering the dynamics of Indonesian demography that is predicted to reach 305 million people in 2035, with 68.1% workforce in 2030. This, however, will also cause several vulnerabilities, including drug abuse.

Drug dealers have already eyed the opportunities in Indonesia's demography, which can be seen in the high number of students being targeted in drug trafficking. Students are the potential long-term consumers. They will one day have their own income, which means increased purchasing power on drugs. This research reveals that income level has impacts on drug abuser prevalence partially. This mean that income level could led to the higher drug abuse prevalence, but we could not judge that the higher income level the higher the drug abuse risk level. It is because the drug abuse risk level is determined by reviewing the relation between index of socioeconomic vulnerability to drug abuse, which the income level is a part of it and drug abuser prevalence level.

Employment availability should also be anticipated; otherwise, the number of unemployment will soar. This will eventually increase the drug abuse vulnerability. The unemployed will easily be swayed into illegal business, including drug abuse, which will increase the drug abuser prevalence. This will result in the increase of drug abuse risk and will eventually be affecting Indonesian defense and security.

5. Conclusion

The socioeconomic factor (education, unemployment rate, income, gender ratio, poverty rate, and access to health service) influence the level of drug abuse prevalence in Indonesia, which the most significantly influence is income level. Index of socioeconomic vulnerability to drug abuse consists of two level, high (9 Province) and moderate (25 provinces). Provinces with high socioeconomic vulnerability index are Riau Islands, DKI Jakarta, West Java, Bangka Belitung Islands, East Kalimantan, Banten, West Papua, Riau, and North Kalimantan.

Drug abuse risk which can be viewed from the relation between drug abuse prevalence and the socioeconomic vulnerability to drug abuse, in Indonesia is classified into 3 level, which are high (5 provinces) moderate (9 provinces) and low (20 provinces). High levels are in DKI Jakarta, East Kalimantan, North Sumatera, West Java, and Riau Islands.

6. Suggestions

More researches are needed to study the relation between drug abuse and Indonesian territorial condition, including the addition of more socioeconomic vulnerability indicators, in order to obtain more comprehensive knowledge on the cause of drug abuse. Researches with smaller analysis unit, such as regency, district, and village, are also necessary to unveil description or information that are more suitable to the condition of socioeconomic vulnerability to drug abuse in each area.

References

[1] BNN 2015 Laporan Kinerja Badan Narkotika Nasional Tahun 2014 Jakarta: BNN
[2] Elizabeth T 2008 Socioeconomic Factors and Illicit Drug Demand Jurnal The Park Place Economist Volume XVI
[3] Mendoza N and Jamie B 2013 Using GIS to describe risk and neighborhood-level factors associated with substance abuse treatment outcomes Journal of Psychology September 2013
[4] United Nations Office On Drugs and Crime 2016 World Drug Report 2016 Vienna: United Nations Publication
[5] Badan Pusat Statistik (BPS) 2016 https://wwwbpsgoid/linkTableDinamis/view/id/1119 Access on 12 Desember 2016 at 06.00 wib
[6] Badan Nasional Penanggulangan Bencana (BNPB) 2012 Peraturan Kepala Badan Nasional Nomor 02 Tahun 2012 tentang Pedoman Umum Pengkajian Risiko Bencana BNPB Jakarta
[7] United Nations Office For Disaster Risk Reduction (UNISDR) 2009 Terminologyhttps://wwwunisdrorg/we/inform/terminology access on 29 April 2017
[8] ESRI 2016 GIS Dictionary
[9] http://supportesricom/other-resources/gisdictionary/term/spatial%20analysis Acsesed pada 10 November 2016 Pukul 02.00 WIB
[10] Edgordo B and Dijk J 2003 Controlling Organized Crime and Corruption in The Public Sector Forum on Crime and Society Vol 3 No 1 and 2 Desember 2003
[11] Phil W dan Roy G 2002 Anticipating Organized and Transnational Crime Crime, Law and Social Change Journal Ed 37 Netherlands: Kluwer Academic Publishers

Disaster risk and travel decision of Middle Eastern tourists to natural destination in Indonesia

F Rahmafitria[1*] and Misran[2]

[1] Resort and Leisure Management Program, Indonesia University of Education, Bandung, Indonesia
[2] School of Tourism, Bandung, Indonesia

[*]Email: rahmafitria@upi.edu

Abstract. Disaster risk perception of tourists become one of prominent things to be concerned as a parameter to drive tourist willingness to visit, especially for a destination with high proneness natural hazard. Since the Middle Eastern tourist become one of the potential international tourists in many country, their perceived risk is also very important. This research aims to analyse Middle Eastern tourist's perception on safety and attractiveness of natural destination in Indonesia. According to a set of questionnaires on risk perception, most of the respondents said that Indonesia has low criterion as a prone disaster destination. However even though they do not recognize the mitigation system and safety procedure in case of the disaster occur, they still tend to visit again. The rare frequency on disaster evidence and the good reputation of tourism sector in Indonesia may cause their high credibility and the low perceived risk on disaster evidence. The theory of familiarity in influencing perceived risk also has an implication of their low perceived risk. This phenomenon means that the management should developed a well-planned mitigation system to eliminate risk and lowering their expectation to avoid disappointed in visiting the destination.

Keywords: disaster preparedness, middle eastern tourist, natural destination, perceived risk

1. Introduction

One of the challenges in managing nature based destination is the preparedness of mitigation system, as the most unique the nature is, the most hazard and high prone it is [1,5,15].. The level of risk on nature-based destination, will be influenced by hazard level and its vulnerability, both on social and management context [20]. Thus, reducing social vulnerability, particularly for the nature-based destination, could be an alternative to reduce risk, which will much relate with tourist's knowledge and their understanding on destination proneness. However, tourist's perception on destination risk and decision to visit will be very important in risk analysis because many individuals do not consider risk factors when selecting area to visit [17].

Tourist's perceived risk and its influence to visit decision, is very important in developing mitigation system in natural destination. Learning about their perception will be very beneficial in determine planning and marketing strategic of the destination. Since the Middle Eastern tourists become one of the targets of international tourists in Indonesia, then their preference is very important to define.

Destination preparedness to overcome and dealing with natural disaster will be the key of tourists' safety and security feelings, developing their loyalty to revisit and promote to others.

Things to be the once importance of this research is the growth of Middle Eastern tourists in Indonesia. According to the data from the ministry of Tourism in Indonesia, the number of Middle Eastern tourists in 2016 was increasing 66.14% (198,434 tourists) from its number on 2015. The data also indicated that 60% of them are likely to visit Indonesia for its natural and cultural attraction, and doing shopping. About 35% of them are likely to visit nature-based destination in West Java, East Java, Central Java, Jakarta and South Sulawesi. Thus, this condition shows us the importance of managing nature-based destination properly, to attract more outbound tourists. On the other side, it is also necessary to construct the safety and security system on natural destination by developing mitigation system for tourists.

Since the number of Middle Eastern tourists to visit Indonesia is getting increase, the preparedness of the destination become an important action. Destination preparedness is not only about facilities and infrastructures for tourists' activities, but also about their safety and security aspect, including the mitigation system regarding the high proneness of natural destination in Indonesia. This research shows the important of understanding tourists' perception on mitigation system which covering the stages of before the disaster happen, and analyze their willingness to revisit the destination [3]. Research in destination mitigation preparation mostly focused on infrastructure and mitigation model, and still small number of tourists perceive risk.

2. Literature Review

A decision to buy a product related to travel and tourism does not take place instantly. Once a potential tourist is interested to buy a product, s/he needs to have awareness about advantages and disadvantages of the product first. This is a very important process before s/he then decides to try it and buy it [2]. However, as the travel and tourism industry includes in the service sector the trial process only happens by buying the product [6].

This process obviously occurs chronologically without any disruption. And in the travel and tourism industry, the risk may disconnect this process as something may happen behind our design, whereas natural disaster categorized [8]. In fact, scholars then define that the real issue lies in the tourist perception about the disaster, not the disaster itself, and therefore decision to travel may change or not change according to this perception [7].

Therefore, as a high-risk industry from the natural disaster it needs to consider the perception of the tourist [7]. It means that the personal tourist decision is sensitive with reconsideration [2]. However, the decision review depends on the personal that has the capability to change the decision. Individual perception is significant and may become a substitution of the decision [2]. While the tourist who follows the group of the family, the group leader has the most powerful decision, including a respectable woman , such as a wife or a mother.

In addition, this perception is also various based on the cultural background, where most of the visitor prefers to change their plan and decision after the information of natural disaster event [16]. In most countries, probably majority of natural disasters are not become a problem, but the people perception to the natural disaster [7].

Regarding on the disaster risk that generally understood widely,t hen risk controlling is become an important thing. Luckily, there are some effective ways for managers to minimize the risk. Firstly, seeing on how familiar the tourist with the offered product, as tourist with more familiarity to the destination may reduce the risk . In addition, the risk may decrease by lowering the expectation to the product. Rising expectation tends has probability to the disappointed tourist. On the contrary, lower expectation produced the tourist are ready to accept the risk. Collecting more information before buying a product may also eliminate the risk, therefore the tourist who has more knowledge to the product will easily accept the predictable risk [2].

Furthermore, the tourist decision to visit is influenced by many factors. As well as branding positive image may reduce this sensitivity [14]. There are some important points according to competition and sustainability between other similar destinations. These important steps to create competitive image are: a) putting a strong image inside the potential tourist memory, b) differentiating the image to create a unique destination other than another destination, c) and performing a positioning to the destination based on the previous image [13].

3. Methodology

The research was focus on tourist's perception regarding their preference on perceived risk with
1) tourist's knowledge on natural disaster as independent variable, and 2) their visit intention as dependent variable. A set of Arabic questionnaire were spread through a survey on the destination, and assisted by local guides who trusted already by the tourists. Some literature had mentioned that most Middle Eastern people seem to be more introverting, suspicious hard to please (Visit Britain, 2010). Thus, the role of senior local guides was very important to re-enforce the process of the survey. The questionnaire was using a Likert scale ranging from 1 (very low) to 6 (very high), which also defining perceived risks as very low to lack of knowledge on disaster and their willingness to visit, and vice versa. Thus, the analysis was using ANOVA, to find the influence between the variables.

The research conducted on August 2017, in some nature destinations in West Java, Indonesia, which become popular for most of Middle Eastern tourists. Puncak area in Cipanas and Sariater Resort in Subang, West Java are the sample of region for this research. There were71 respondents participated in this research, due to the difficulties to propose Middle Eastern tourists to full fill the questionnaire. The respondents presented by open questionnaires.

4. Result and Discussion

Since the travel pattern of Middle Eastern tourist become frequently increase, Indonesia's government has set a specific target, which in 2019, there will be 5 million Moslem tourist visit Indonesia (a quarter of total target of international tourist). Therefore, four aspects of destination preparedness is setting up, namely product, human resources, institutions, and promotion. Regarding with destination readiness (product), tourist preference become one of the indicators to identify what kind of tangible product and services that attract them.

From the survey of Middle Eastern tourists, most of the respondents come from Saudi Arabia (93.75%), and all of them are Moslem. Their age is between 25 – 45 years old (65.22%), having bachelor degree for education (57.14%) and are categorized as high spending power tourists with 71.83% are having 1,000 – 3,500 US $ for income. For their travel pattern, it is found that they likely come with a small group number, about 3-5 person (78.87%), and 63.38% are staying for quite long time (1-2 weeks). The peak of their visit is on summer holiday, on July and August. Tourist travel motivation indicated that most of them are likely to enjoy the freshness and aesthetic of natural tropical landscape (66.20%) and mountainous landscape (46.48%). Recreation and spending holiday are their internal motivation to visit Indonesia (75.57%). The demographics and travel pattern of the respondents is mention in Table 1.

Table 1. Demographics and travel pattern

Nationality	Saudi Arabia		Riyadh		Bahrain	
	93.75 %		4.69%		1.56 %	
Religion	Islam			Non Islam		
	100 %			0.00		
Address	Mecca	Jeddah	Dammam	Riyadh	Taif	Muscat
	21.43 %	28.57 %	10.71 %	21.43%	7.14 %	10.71 %
Age	<17 years old	17 – 25 years old	26 – 35 years old	36 – 45 years old	46 – 55 years old	> 56 years old
	2.90 %	21.74 %	39.13%	26.09 %	7.25 %	2.90 %

Education	High school	Bachelor	Master	Doctor	Others	
	14.29 %	57.14%	17.14%	8.57%	2.86%	
Income	<500 USD	500 – 1,000 USD	1,001 – 2,000 USD	2,001–3,500 USD	3,501 – 5,000 USD	> 5,000 USD
	12.68 %	8.45%	25.35%	22.54%	23.94 %	7.04 %
Length of stay	<1 week		1-2 weeks		> 2 weeks	
	7.04%		63.38%		29.58%	
Number of person	<3 person		3-5 person		> 5 person	
	35.21 %		43.66%		19.72 %	

Regarding the risk perception (Table 2), most of the Middle Eastern tourists are not having enough information on proneness level of the destination. The attractiveness of tropical landscape of Indonesia has attracted the tourist to visit the destination. They thought that Indonesia is not a prone destination (Mean = 2.56). In this case, we have to highlight that proneness is different with disaster happened. Most of the tourists do not have enough information regarding disaster proneness of Indonesia. The rare frequent of disaster in the destination has influence their preference that Indonesia has no warning for disaster. While the promotion of Indonesia's tourism never socialized about this kind of hazard, as the content of Indonesia's promotion is about the naturalness, nature and culture diversity and its authenticity. Therefore, they likely to feel secure while visiting nature destination in Indonesia (Mean = 4.09). In one side, this kind of perception could be a strong potential as market target. The good perception on the destination will positively influenced their decision to visit. In another side, if their expectation do not same with the face reality, it could be disappointing [11].

However, despite of having few information regarding destination proneness, they always well preparing their traveling plan (Mean = 4.42). Some researchers found that careful planning is indicating the feeling of secure and a well prepare process. The well prepare process before travelling, will decrease risk and escalate self-secure [8]. Unfortunately, most of the respondents do not understand some clue of disaster portent (Mean = 2.53), and do not have proper knowledge related with disaster and rescue technique (Mean = 2.63). This phenomenon should overcome by the readiness of mitigation system on the destination. The specific information and knowledge about local hazard, avoided area, forbidden behaviour will educate tourists to be more aware on their safety. Thus, signage about what do they should do in case the disaster is happened, also play an important role in bridging tourists understanding and knowledge on the mitigation system [4,9,12]

Table 2. Risk perception of Middle Eastern tourists

Risk Perception	**N**	**Mean**	**Std. Deviation**
Indonesia is a prone destination (land slide, earth quake, eruption)	71	2.56	1.39
Tourists should have knowledge on destination proneness	71	3.39	1.58
Are you familiar with disaster portent?	71	2.53	1.55
Do you well understand about self-rescue technic in case of disaster?	71	2.63	1.48
Do you prepare well enough all the safety equipment while you were travelling?	71	4.42	1.87
Do you feel secure while visiting nature-based destination in Indonesia?	71	4.09	2.04

Another variable to be analysed in this research is decision to visit, as an important indicator in determining the influence of tourists' risk perception. It is found that among Middle Eastern tourist, Indonesia has a good reputation as tourism destination country (Mean = 5.08). The characteristic of Middle Eastern tourist behavior believes more on worth of mouth rather on other media of promotion. The good experience of previous tourists has play an important role in putting up other Middle Eastern to visit Indonesia and also become a strong stimulator, as enhancement of kinship and social interaction are very important to them (Michael, et al. 2011; UNWTO 2012). Therefore, their willingness to re-visit some nature destinations in Indonesia is strong (Mean = 4.77). Other criterion of decision to visit is the willingness to persuade other to visit the destination. In this study, the respondents are likely to inform the beauty of Indonesia (Mean = 4.94), according to what they had experienced before.

Table 3. Decision to visit Indonesia of Middle Eastern tourists

Decision to visit	N	Mean	Std. Deviation
Are you willing to re-visit some nature destination in Indonesia?	71	4.77	1.77
Do you think that Indonesia have a good reputation as tourism destination country?	71	5.08	1.59
Are you willing to inform others (friends and relatives) about Indonesia as tourism destination country?	71	4.94	1.70

The influence of tourists risk perception on the destination to decision to visit analysed in this research. From the constant of the model it is indicated that the better their understanding on risk perception, the more it is influenced their decision to visit. However, their perception on Indonesia as an attractive tourist destination is quite high, since the model indicates that without any knowledge on destination proneness and preparedness, they attend to visit Indonesia for 2.011 point. This model has described 14.7% (R square) of the total factors that influence visit intention.

Table 4. The regression of tourists' risk perception on their decision to visit

Coefficients[a]

Model		Unstandardized Coefficients		Standardized Coefficients		
		B	Std. Error	Beta	T	Sig.
1	(Constant)	2.011	.403		4.995	.000
	Decision to visit	.269	.078	,384	3.453	.001

a. Dependent Variable: Tourists' risk perception

Table 5. Model summary of Middle Eastern tourist decision to visit

Model Summary[b]

Model	R	R Square	Adjusted R Square	Std. Error of the Estimate
1	.384[a]	.147	.135	1.00521

a. Predictors: (Constant), Decision to visit

b. Dependent Variable: Knowledge on disaster prone

5. Conclusion

The fast growth of tourism sector in Indonesia has driving the government to set a high target of international tourists' visit, including for the Middle Eastern tourists whose being targeting for 5 million tourists in 2019. According to the result, the destination preparation becomes one thing that government need to pay attention for, because the tourist decision to visit is influenced by their perception on the destination. Tourist perception is built by their experience and knowledge, therefore it is very crucial to share the proper and open information about the destination in the promotion program, and in parallel with that, the destination readiness on infrastructure and mitigation system are also developed.

Especially for nature-based destination, which has its specific sensitivity and proneness, the mitigation system is not a management option, but should be an obligation.

Even though the Middle Eastern tourist perception on risk is low, but they still have a high willingness to visit Indonesia. Their trust on Indonesia has leading them to put their decision in enjoying tropical landscape in Indonesia. The proper on site information through interpretation media will be very effective in enriching their knowledge regarding the destination's proneness and strengths. Information about safety procedures and the way to evacuate in case any disasters happened, are also very important, as they still has no insight about the mitigation procedure in a tropical landscape. The more familiar we are with the destination, the more the risk will be easy to reduce.

References

[1] Highfield, W. E., Peacock, W. G., & Van Zandt, S. (2014). *Mitigation planning: Why hazard exposure, structural vulnerability, and social vulnerability matter.* Journal of Planning Education and Research, 34 (3), 287–300

[2] Holloway, J. C. (2004).*Marketing for Tourism*. Fourth, *Marketing for Tourism*. Fourth. Prentice Hall. doi: 10.1007/978-1-349-21946-9.

[3] Hudson, S. (2008).*Tourism and Hospitality Marketing: A Global Perspective*. First edition. Sage.

[4] Hughes, M., & Saunders, A. M. (2005). *Interpretation, activity participation, and environmental attitudes of visitors to Penguin Island,* Western Australia. Society and Natural Resources, 18,611- 624.

[5] Kim, J et al. (2016). J. *Amenity or hazard? The effects of landslide hazard on property value in Woomyeon Nature Park area, Korea* Landscape and Urban Planning 157 (2017) 523–53

[6] Korczynski, M. (2002). 'Service work : the customer- oriented bureaucracy', in *Human Resource Management in Service Work*, pp. 58–83.

[7] Kozak, M., Crotts, J. C. and Law, R. (2007) 'The impact of the perception of risk on international travellers', *International Journal of Tourism Research*, 9(4), pp. 233–242. doi: 10.1002/jtr.607.

[8] Law, R. (2006) '*The perceived impact of risks on travel decisions'*, International Journal of Tourism Research, 8(4), pp. 289–300. doi: 10.1002/jtr.576.

[9] Madin, E., & Fenton, D. M. (2004*). Environmental interpretation in the Great Barrier Reef Marine Park: an assessment of programme effectiveness*. Journal of Sustainable Tourism, 12,121-137.

[10] Michael, I., Armstrong, A., Badran, B., & King, B. (2011). *Dubai outbond tourism : An exploratory study of Emiratis and exspatriates*. Journal of Vacation Marketing, 17(1), 83-91.

[11] Michalko, G., Irimias, A., & Timothy, D.J. (2015). *Dissapointment in tourism : Perspective on tourism destination management*. Tourism Management Perspective. 16 (2015), 85-91.

[12] Powell, R. B., & Ham, S. H. (2008). *Can ecotourism interpretation really lead to proconservation knowledge, attitudes and behaviour? Evidence from the Galapagos Islands*. Journal of Sustainable Tourism, 16, 467-489.

[13] Ritchie, J. R Brent & Crouch, G. L. (2003) *The Competitive Destination: A Sustainable Tourism Perspective*. CABI Publishing.

[14] Schwaighofer, V. (2014) *Tourist Destination Images and Local Culture*. Springer Gabler.

[15] Smith, K. (2013). Environmental hazards: assessing risk and reducing disaster. New York, NY: Routledge.

[16] Tasci, A. D. A. and Boylu, Y. (2010) 'Cultural comparison of tourists' safety perception in relation to trip satisfaction', International Journal of Tourism Research, 12(2), pp. 179–192. doi: 10.1002/jtr.745.

[17] Trumbo, C., Lueck, M., Marlatt, H., Peek, L., (2011). The effect of proximity to Hurricanes Katrina and Rita on subsequent Hurricane outlook and optimistic bias. Risk Analysis 31 (12), 1907-1918 A reference

[18] UNWTO. (2012). The Middle East outbond travel market with special insight into the image of Europe as a destination. Report by the UNWTO: Spain.
[19] Visit Britain (2010). UEA market and trade profile. Report by visit Britain, UK.
[20] Wisner, B., Blaikie, P., Cannon, T., Davis, I., (2004). At risk.Natural hazards, people's vulnerability and disasters. Second edition. Taylor and Francis Group. London and New York. 470 p.

City expansion and urban heat island in Bogor

A Nurwanda*

Graduate School of Horticulture, Chiba University, Matsudo-shi, Chiba, Japan

*Email: atiknurwanda@chiba-u.jp

Abstract. Bogor is a satellite city of Jakarta Metropolitan that experience rapid population growth and urban development. Urban landscape change for urban development affects the city expansion, development of urban heat island (UHI), and the increasing of land surface temperature (LST). The objectives of this study are to analyze the urban landscape changein1990 and 2017, to analyze the LST change caused by city expansion, and to analyze the development of the UHI phenomenon. We used Local Climate Zone (LCZ) which is a classification system of urban surfaces. In LCZ classification, we used three categories in built-up area, three categories in vegetated area, bare soil, and water. Landsat 5 TM in 1990and Landsat 8 OLI/TIRS in 2017 were used in this study. In 1990, the urbanized area was 9.38% and the vegetated area was 83.98%. The urbanized area continuously increased and become 27.24% in 2017 and the vegetated area decreased to 58.85% in 2017. By the comparison of LST in 1990 and2017, the areas that show high LST is increased. As the result of the analysis, the changes from non-urban to urban surface highly affected the UHI in Bogor City through the rise of LST.

Keywords: city expansion, land surface temperature, local climate zone, urban heat island

1. Introduction

Urban Heat Island (UHI) is a condition or phenomenon where the atmospheric in the city is warmer than suburban area. Surface and atmospheric modifications due to urbanization generally lead to a modified thermal climate that is warmer than the surrounding non-urbanized areas [1]. To examine the UHI, UHI phenomenon can be observed by using remote sensing data [1, 2, 3]. Voogt and Oke [1] clearly reviewed and described the UHI phenomenon by utilizing remote sensing data.

UHI mainly caused by urban landscape change for urban development which resulted higher land surface temperature (LST) [4]. There are many factors which affect the surface change in urban areas: physical characteristics of the surface (albedo, thermal, and heat capacity), replacement vegetation by asphalt and concrete, urban land cover type [5]; vegetation abundance [6]; and landscape configuration [7]. Others researchers who concern in UHI studies also showed the relationship between LST and urban landscape changes which affect the UHI [6, 8, 9].

Interaction of the urban surfaces with the atmosphere is governed by surface heat flux [5, 10].The surface energy balance is determined by downward and upward radiation flux, latent, and sensible heat fluxes. The downward flux is determined by the fraction of total radiation reflected and absorbed by the surface. The upward flux follows the Stefan–Boltzmann expression and is determined by the surface's temperature and emissivity. The energy balance of the surface also determined the LST [10].

Based on the characteristic of the urban surface and its climate interaction, the urban surface is classified with new classification system: "Local Climate Zone" (LCZ) classification by Stewart and Oke [11]. They classified the urban surface based on the surface structure and surface cover characteristic related to the climate interaction which was explained previously. Surface structure affects the local climate through modification of airflow, shortwave and long wave radiation balances, and atmospheric heat transport. Surface cover modifies albedo, moisture availability, and heating/cooling potential of the ground.

Regarding the urban development process, population growth has an important role as an internal driving factor of city expansion. Population growth may lead the changes of social characteristics and the intensity of its activities led to rapid changes in urban utilization. Certainly, the rapid changes for urban development affected the higher LST, which then cause the development of UHI. UHI occur in almost all urban areas – large or small size city [11].

As a primary of residential purposes, Bogor City has a strategic location and comfortable climate. In 1745, the Dutch East Indies Governor (Baron Van Im hoff) built the Bogor Palace along with the construction of Daendels Highway which connecting Bogor with Batavia (now is Jakarta metropolitan). That period is the beginning of the development of Bogor City[12,13].As a satellite city of Jakarta, Bogor City experienced population growth and urban development rapidly in 1990, the populations 272 thousand and in 2015 the population achieved more than one million[14]. This issue brings Bogor city into new environmental challenges; one of them is city expansion which may affect the UHI.

This study focused on utilizing remote sensing to investigate the city expansion and the UHI that reflected by the LST change in different period. The objectives of this study are (1) to analyzed the urban surface change in 1990 and 2017;(2) to analyze the LST change caused by the city expansion; and (3) to analyze the development of the UHI phenomenon. To obtain the LCZ and the LST maps, Landsat images were used in this study.

2. Method

2.1. Study area and materials

The analyzed area is shown in Figure 1. Landsat 5 TM (July 9th1990) and Landsat 8 OLI/TIRS (July 19th2017) were obtained from United State Geographical Survey web page (https://earthexplorer.usgs.gov/), path 122 and row 065. Band 1-5&7 (Landsat 5 TM) and band 1-7 (Landsat 8 OLI) were used for LCZ classification. In addition, band 6 (Landsat 5 TM) and band 10 (Landsat 8 TIRS) were used to analyze the LST. We used projection WGS-84 datum UTM zone 48 South.

2.2. Local Climate Zone (LCZ) classification system

We used LCZ classification system as urban surface categories. LCZ classification was developed and described by Stewart and Oke [11]. LCZ comprises a new and systematic classification of field sites for UHI studies. The classification divides urban and rural landscape, each defined by structural and land cover properties. Surface structure affects the local climate through modification of airflow, atmospheric heat transport, shortwave and long wave radiation balances. Surface cover modifies the albedo, moisture availability, and heating/cooling potential of the ground.

We conducted the classification process in Quantum GIS with Semi-Automatic Classification plug-in. Considering the purposes of this study, we divided LCZ classification into nine categories: (1) Compact low-rise building, (2) Large low-rise building, (3) Roads and paved, (4) Dense trees, (5) Scattered trees and bushes, (6) Low plants, (7) Bare soil, (8) Water, and (9) Cloud. To determine the categories, maximum likelihood classification (MLC) algorithm was used. MLC is a method for determining a known class of distribution as the maximum for a given statistic. Algorithm builds the probability density functions for each category. At classification process, unclassified pixels are assigned membership based on the relative likelihood of that pixel within each category's probability density function.

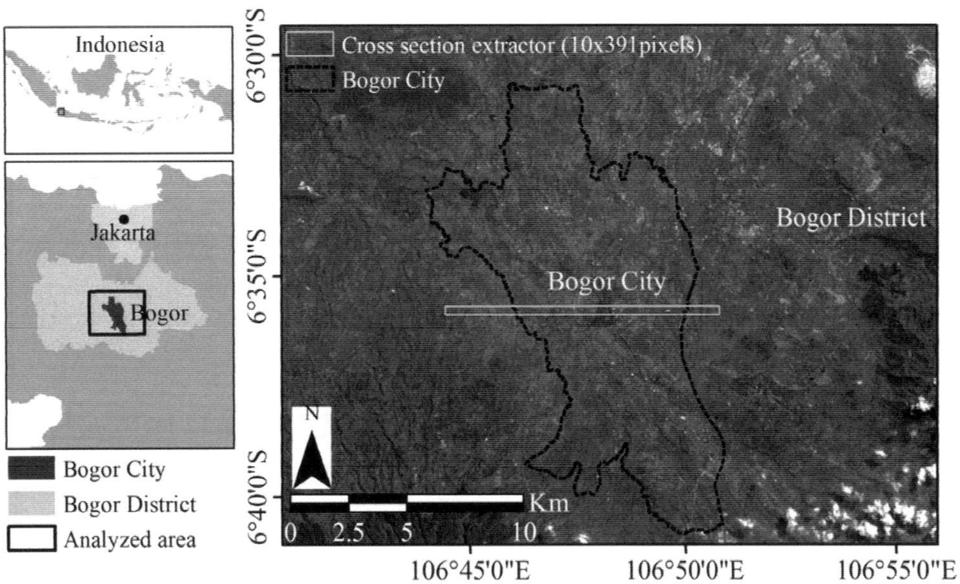

Figure 1. Analyzed area. The map is a false color visualization of Landsat 8 OLIon July 19th 2017. Cross section extractor (10x391 pixels) across the central city is used for investigating the UHI development due to city expansion.

2.3. Data post-processing and accuracy assessment of LCZ classification

Compact low-rise building, large low-rise building, roads and paved are merged as urbanized category. Dense trees, small trees and bushes, and low plants are merged as vegetated area. To identify the errors between categories, we conducted accuracy measurement of LCZ classification. Accuracy assessment of LCZ classification used error matrix which contain actual and predicted classification. Classification of an individual class is user's accuracy and producer's accuracy. User's accuracy is the accuracy obtained by dividing the number of correct pixels of each category by the total pixels that classified in that category. Meanwhile, producer's accuracy is the accuracy obtained by dividing the number of correct pixels of each category by the total pixels of training data in that category. Then, overall accuracy can be obtained. The formula of producer's accuracy, user's accuracy, overall accuracy, and kappa coefficient are described by Pal et al. [15].

2.4. LST measurement from the thermal band and UHI development

LST distribution from Landsat image was obtained by the following the steps below. The formula was obtained from USGS web page (https://landsat.usgs.gov/) and was clearly described in Pal et al. [15]. UHI intensity is calculated as a difference between the spatial temperature obtained from the LST maps. In order to evaluate whether the city expansion influence the UHI, the cross section area was extracted from LCZ and LST change between 1990 and 2017 with 10 x 391 pixels (see Figure 1). Then, the correlations of the LST change and the urban change in 1990 and 2017 was made.

3. Result and Discussion

3.1. Accuracy evaluation of LCZ classification

Table 1 shows the overall accuracy and kappa coefficient of LCZ classification. The overall accuracy of the classified images (1990and 2017) respectively are 95.02% and 98.94% with kappa coefficient 0.83and 0.97.Overall accuracy for each period is above 90%, it indicates that the accuracy is high. Overall accuracy in 2017 has the higher value, it is because of the availability of more detailed and higher resolution reference maps. On the other hand, kappa coefficient is above 0.8 for all period classified images. When kappa coefficient is above 0.8, this value is classified as almost perfect [16] or very good agreement of the accuracy [17].

Table 1. Accuracy assessment of LCZ classification

Year	Overall accuracy (%)	*Kappa coefficient
1990	95.02	0.83
2017	98.94	0.97

*0-0.20 (slight), 0.21-0.40 (fair), 0.41-0.60 (moderate), 0.61-0.80 (substantial), 0.81-1.00 (almost perfect) [16].

*<0.4 (poor or very poor agreement), 0.4-0.55 (fair agreement), 0.55-0.70 (good agreement), 0.70-0.85 (very good agreement), and >0.85 (excellent agreement) [17].

3.2. LCZ analysis

LCZ classification maps over the entire period is shown in Figure 2 and LCZ statistical analysis in a different year is shown in Table 2. In 1990, vegetated area was predominant land use type (83.98%), but it continuously decreased to 58.85% in 2017. The annual rate of urban expansion is 0.7% (438.69 hectares per year) from 1990 to 2017. Dense trees, farm land, and bare soil are the main resource for city expansion. The city expansion was intense in the center of Bogor City. In suburban area, the city expansion was sparsely and broadly distributed.

Figure 2. LCZ maps of Bogor

Table 2. LCZ statistical analysis in different years

LCZ	1990		2017	
	hectares	%	hectares	%
Urbanized area	5846.85	9.38	16979.4	27.24
Vegetated area	52337.52	83.98	36677.79	58.85
Bare soil	1881.45	3.02	3284.91	5.27
Water	2068.38	3.32	3802.23	6.10
Cloud	188.28	0.30	1578.15	2.53

Total area = 62322.48 hectares; A. Urbanized area is a merge of built-up areas (compact low-rise building, large low-rise building, roads and paved); B. Vegetated area is a merge of green areas (dense trees, small trees and bushes, and low plants).

3.3. LST maps

The spatial pattern of the LST in two periods is shown in Figure 3.The UHI (areas with relatively high LST) scattered in Bogor City. In 1990, mean LST of Bogor City was 23.28 °C and Bogor sub urban (the area without Bogor City in the analyzed area of Figure 1) was 22.51 °C. In 2017, the area that show high LST is increased. Mean temperature of Bogor City was 25.44 °C and Bogor sub urban was 23.75 °C. It indicated that the urbanized area became warmer than other areas and the LST is an impact of urbanization.

Comparing the LST maps with the LCZ maps (Figure 2), the LST change followed the pattern of the LCZ change. Urban heat island mainly concentrated in the central city. From 1990 to 2017, average LST has increased and the temperature difference between Bogor City and Bogor District (suburban) has also increased from 0.77 °C to 1.69 °C. Based on the LST maps, UHI phenomenon in 1990 already occurred at small area inside Bogor City. Thus, UHI phenomenon in 2017 almost covered Bogor City area and spread outside to Bogor sub urban area.

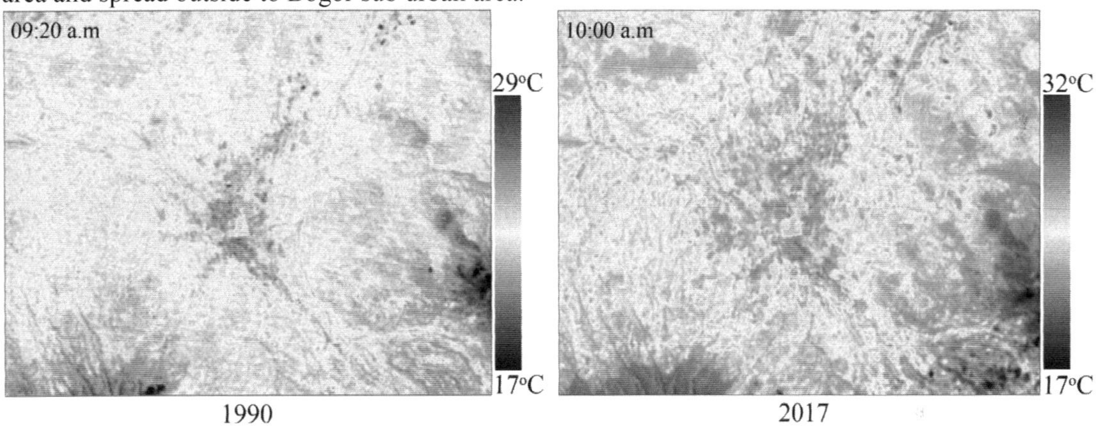

Figure 3. LST maps of 1990 and 2017.

3.4. UHI expansion

The results of the urban change and the LST change are shown in Figure 4. Figure 4A shows the urban change between 1990 and 2017.The horizontal axis is a number of pixels and vertical axis is the degree of non-urban to urban change. The value of 10 vertical pixels is averaged to give the degree. Zero means non-urban in 10 pixels and one means 100% urban in 10 pixels. In 1990, the city area was located near Bogor Botanical Garden, especially in part X and part Yin Figure 4A. Then, there has been the change of the urban surface with the degree of change. The gap between the blue line (1990) and the red line (2017) on part of Z1&Z2are the example that shows the biggest change of the surface from non-urban to urban. It indicated that the degree of changes from non-urban to urban in the old city center near Bogor Botanical Garden has saturated and it went out to the outer part of the city.

Figure 4 B describes the LST change between 1990 and 2017 (LST of 2017 – LST of 1990). When the gap between urban change degrees was bigger, the LST change increased (see Figure 4B on part of Z1&Z2, and compare with Figure 4A). Z part indicated that in those areas experienced the highest LST change.

The comparison of the LST change with the urban change (from non-urban to urban) from 1990 to 2017 is shown in Figure 4C. By this comparison, the LST increased following the land conversion from non-urban to urban. This result indicated that the surface UHI development has occurred with the city expansion and the increase of the LST.

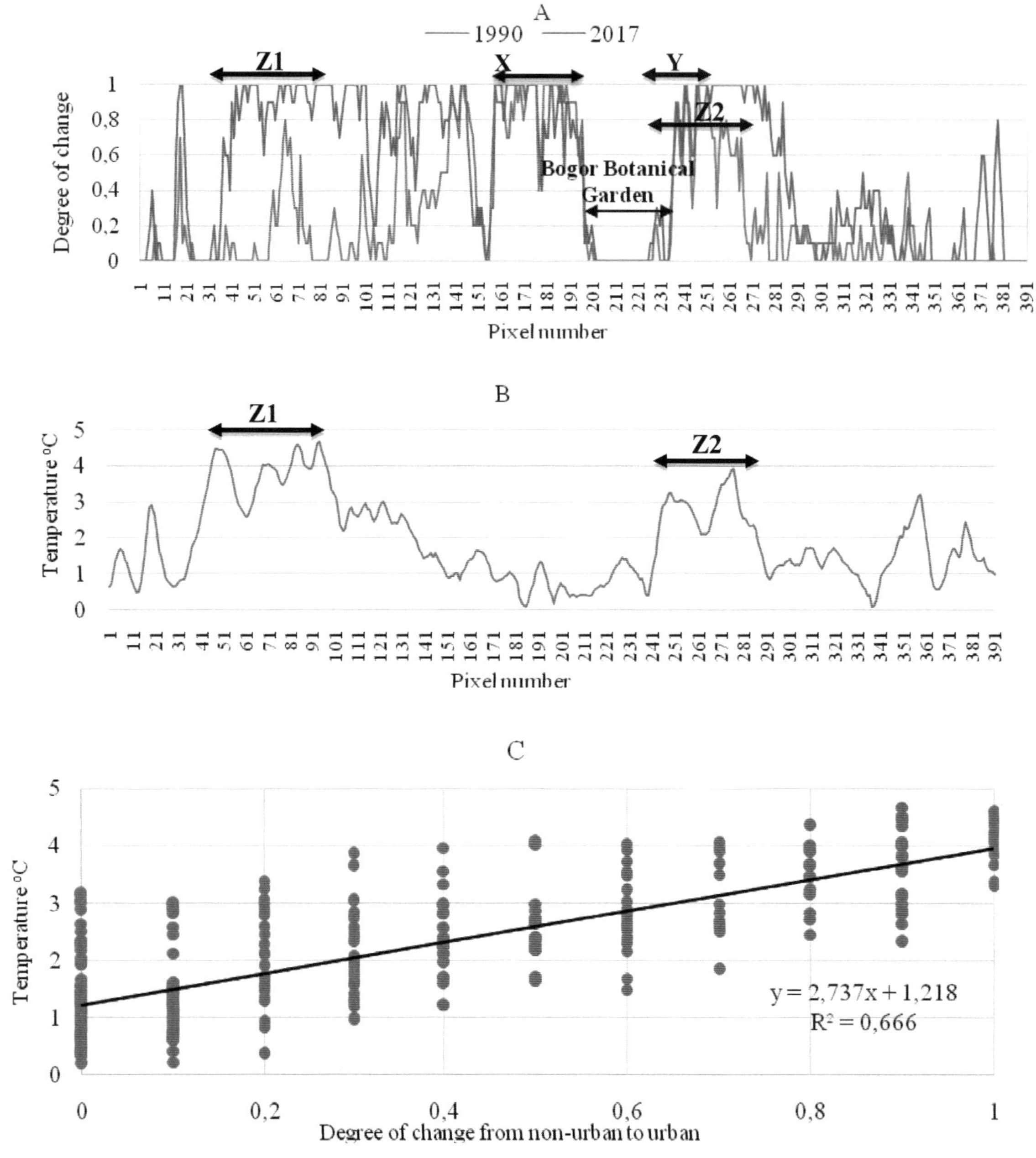

Figure 4. Urban change and LST change. (A) Urban status in a degree, (B) LST change between 1990 and 2017, (C) Scatter plot of LST change and urban change (from non-urban to urban)

4. Conclusion

Based on the analysis of LCZ and LST, it is found that Bogor has experienced rapid city expansion in the last three decades. In 1990, urbanized area was a little part inside of Bogor City. From 1990 to 2017, the urbanized area continuously increased and the vegetated area decreased. By the comparison of LST between 1990 and 2017, the area that shows high LST increased and it is highly correlated with the city expansion. The changes from non-urban to urban surface highly affected the surface UHI in Bogor City through the rise of LST.

References

[1] Voogt, J. A., & Oke, T. R. (2003). Thermal remote sensing of urban climates. *Remote Sensing of Environment*, *86*(3), 370–384. https://doi.org/10.1016/S0034-4257(03)00079-8

[2] Taylor, P., Roth, M., Oke, T. R., & Emery, W. J. (1989). International Journal of Remote Sensing Satellite-derived urban heat islands from three coastal cities and the utilization of such data in urban climatology, (November 2014), 37–41. https://doi.org/10.1080/01431168908904002

[3] Streutker, D. R. (2002). A remote sensing study of the urban heat island of Houston, Texas. *International Journal of Remote Sensing*, *23*(March 2015), 2595–2608. https://doi.org/10.1080/01431160110115023

[4] Estoque, R. C., Murayama, Y., & Myint, S. W. (2017). Effects of landscape composition and pattern on land surface temperature: An urban heat island study in the megacities of Southeast Asia. *Science of the Total Environment*, *577*, 349–359. https://doi.org/10.1016/j.scitotenv.2016.10.195

[5] Dousset, B., & Gourmelon, F. (2003). Satellite multi-sensor data analysis of urban surface temperatures and landcover. *ISPRS Journal of Photogrammetry and Remote Sensing*, *58*(1–2), 43–54. https://doi.org/10.1016/S0924-2716(03)00016-9

[6] Weng, Q., Lu, D., &Schubring, J. (2004). Estimation of land surface temperature-vegetation abundance relationship for urban heat island studies. *Remote Sensing of Environment*, *89*(4), 467–483. https://doi.org/10.1016/j.rse.2003.11.005

[7] Patrick, J., Christopher, C., & Chow, W. T. L. (2013). Landscape configuration and urban heat island effects : assessing the relationship between landscape characteristics and land surface temperature in Phoenix , Arizona, 271–283. https://doi.org/10.1007/s10980-012-9833-1

[8] Chen, X. L., Zhao, H. M., Li, P. X., & Yin, Z. Y. (2006). Remote sensing image-based analysis of the relationship between urban heat island and land use/cover changes. *Remote Sensing of Environment*, *104*(2), 133–146. https://doi.org/10.1016/j.rse.2005.11.016

[9] Rhee, J., Park, S., & Lu, Z. (2014). Relationship between land cover patterns and surface temperature in urban areas. *GIScience & Remote Sensing*, *51*(5), 521. http://doi.org/10.1080/15481603.2014.964455

[10] Benali, A., Carvalho, A. C., Nunes, J. P., Carvalhais, N., & Santos, A. (2012). Remote Sensing of Environment Estimating air surface temperature in Portugal using MODIS LST data. *Remote Sensing of Environment*, *124*, 108–121. https://doi.org/10.1016/j.rse.2012.04.024

[11] Stewart, I. D., &Oke, T. R. (2012). Local climate zones for urban temperature studies. *Bulletin of the American Meteorological Society*, *93*(12), 1879–1900. https://doi.org/10.1175/BAMS-D-11-00019.1

[12] Indra prahasta, G.S. (2004). The role of sustainable community on developing sustainable city (a case study on Kampoeng Bogor in Bogor City). Bogor: Center for Regional System Analysis, Planning, and Development

[13] Irianti, EF. (2008). The alteration of land use, land cover, and green open space in Bogor City 1905-2005. [Thesis]. Bogor (ID): Bogor Agricultural University

[14] BPS-Statistics of Bogor City. (2017). Bogor City in Figures 2017, 244. https://doi.org/32.710.16.01. Available online: http://bappeda.kotabogor.go.id/images/slidertabs/90f869bc2bc1549f849cef02c06f84cd.pdf (accessed on August 19th 2017)

[15] Pal, S., &Ziaul, S. (2017). The Egyptian Journal of Remote Sensing and Space Sciences Detection of land use and land cover change and land surface temperature in English Bazar urban centre q. *The Egyptian Journal of Remote Sensing and Space Sciences*, *20*(1), 125–145. https://doi.org/10.1016/j.ejrs.2016.11.003

[16] Landis, J. R., & Koch, G. G. (1977). The measurement of observer agreement for categorical data. Biometrics, 159-174.

[17] Monserud, R. A., & Leemans, R. (1992). Comparing global vegetation maps with the Kappa statistic, *62*, 275–293.

IOP Conf. Series: Earth and Environmental Science **179** (2018) 012008 doi:10.1088/1755-1315/179/1/012008

Slum upgrading based on flood mitigation for resilience of Manado City

W J Mononimbar

Urban and Regional Planning Study Program, Sam Ratulangi University

*E-mail: wmononimbar@yahoo.com

Abstract. The resilience of a city is created by the resilience of all parts of the city that are able to adapt on occurred changes, whether climate change or natural disaster, unexpected development changes and social and economic turmoil of urban society. One area of the city that has a high level of vulnerability to these changes is the settlement that is located around the riverbank area. It is generally identified as slums that have irregular settlement patterns, unsuitable housing conditions and not equipped with infrastructure and facilities of settlements adequate, which cause the settlements can't survive when the floods occurred. Manado City as the capital of North Sulawesi Province, according to Decree of Mayor of Manado Number 163 of 2015 on Determination of Slum Area, there are 13 slum areas around the river border with high flood vulnerability level. Recorded in January 2014 there has been a massive flood in Manado that caused damage and huge losses and even casualties in these areas. This is due to irregular settlements and lack of infrastructure. This study was conducted in one of the slum-heavy at Sindulang Satu District covering aspects of land use analysis, river and coastal boundary conditions, building conditions and disaster mitigation infrastructure systems (river embankments, drainage channels, circulation systems and evacuation routes, green open spaces, electricity, etc). The results show that the boundary area of the river needs to be restructured, the river embankment and drainage channel are repaired, the high density of the building needs to be reduced to provide sufficient open spaces, the need to build disaster evacuation routes, improved building conditions and provide an alternative source of electricity and clean water so that if there is a disaster then the area can survive.

Keywords: flood mitigation, resilience, slum upgrading

1. Introduction

1.1. Background

Urban resilience is the capacity of individuals, communities, institutions, enterprises and systems within a city to survive, adapt and grow with various shocks and pressures experienced both physically and socially. It means that the resilience of the city encompasses all aspects of the physical area to the social and economic aspects. City resilience is also created from the resilience of every part of the region that is capable to survive. The resilience is decreasing survive when the physical and social pressures are slum areas. According to the Law of the Republic of Indonesia Number 1 of 2011 about Housing and Settlements, slums are unfit settlements due to building irregularity, high building density and quality of buildings and infrastructure that are not eligible. This condition causes the residential area to be vulnerable to disaster and to other physical pressures. When a large scale disaster occurs, it can cause damage. Though a resilience city is when all parts of the city can survive and adapt to all kinds of physical and social pressures.

Manado City as the capital of North Sulawesi Province has a number of slum areas located in the river border area. Based on Decree of Mayor of Manado Number 163 of 2015 about Stipulation of Slum Area in Manado City, there are 13 slum areas located in Tondano watershed area, with high level of disaster risk, this is due to high population and building density, house condition is mostly non-permanent, irregular location, almost no distance between buildings, unclear circulation system, lack of drainage and river embankment and lack of space open green.

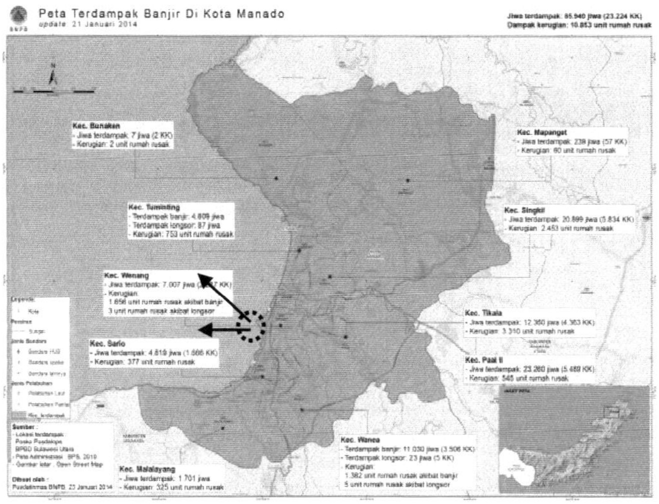

Figure 1. Map of Flood Affected Area in Manado City
Source: Regional Disaster Management Agency of North Sulawesi, 2014

1.2. Objectives

Objective of study is to analyze the slum upgrading based on flood mitigation in Sindulang Satu District located between the mouth of Tondano watershed and the coastal area. It is expected the risk of flood disaster can be minimized and can create a resilient area.

1.2.1. Understanding of City Resilience, Slum Upgrading and Flood Mitigation System

City resilience describes the capacity of cities to function, so that the people living and working in cities - especially the poor and vulnerable - survive and thrive no matter what stresses or shocks they encounter [10]. The resilience is created from the resilience of all parts of the city that is able to change the happenings, whether the climate change or natural disasters, unexpected development changes and social and economic turmoil of urban society. One area of the city that has a high level of vulnerability to these changes is the slum.

Defines slum is informal settlements within cities that have inadequate housing and squalid, miserable living conditions [9]. They are often overcrowded and lack basic municipal services such as water, sanitation, waste collection, storm drainage, street lighting, paved sidewalks, roads for emergency access and durable housing that protects against extreme climate conditions. Many slums are built on the river border area and has a very high flood vulnerability, as it is usually not equipped with durable housing and adequate infrastructure, so that when the floods come the settlements can not survive [5].

This condition has a negative impact on the whole city because it deals with city issues by containing environmental degradation. Therefore it is necessary to slums upgrading. According to Cities Alliance, slum upgrading is a process through which informal areas are gradually improved, formalized and incorporated into the city itself, through extending land, services and citizenship to slum dwellers. It involves providing slum dwellers with economic, social, institutional and community services available to other citizens. These services include legal (land tenure), physical (infrastructure), social or economic. In correlation with flood mitigation, the above aspects can be categorized into two aspects of flood mitigation system, including structural flood mitigation, ie physical structures that are built or modified to reduce flood impact such as embankment infrastructure, drainage system, etc., and flood mitigation non-structural such as land use planning controls, etc. These aspects are included in the framework of resilience city especially on infrastructure and environment aspect [10].

2. Method

2.1. Study Field

The field of study is in Sindulang Satu District located in Manado City. Sindulang Satu District located right at the mouth of Tondano watershed (between Tondano watershed and coastal), where floods flooded to a height of 2 meters so that many houses and businesses which was damaged and was washed away by the flood.

2.2. Aspects of Analysis

Aspects of analysis are focused on infrastructure and environmental aspects related to flood disaster mitigation infrastructure, with the following description:

Table 1. Aspect of Analysis

Main Aspects	Sub Aspects
Condition of river and coastal border area	
Land use	
Buildings conditions	Density and distance of buildings Structure and construction
Flood infrastructure system	Embankments Drainage channels Roads and evacuation system Signage system Open space Rescue/escape hill Electricity

2.3. Method of Collecting Data

The primary data collection method is observation to see the condition of settlements such as buildings condition, availability of flood infrastructure and river border area and survey through structured interview to respondent (dwellers) with purposive random sampling technique to see their preference for slum upgrading. Secondary data collection is through the collection of government information and policies such as regulations of river borders, etc. The research instruments are questionnaires, maps, photographs, drawings and documents on the condition of the study sites.

2.4. Data Analysis Method

The method of analysis is qualitative to describe the existing settlement conditions and then can be recommended to improve the settlement better than the previous condition. For spatial analysis and map plan using ArcGis 10.3 software.

3. Results and Discussion

To analyze the slum upgrading it is necessary to see the condition of the whole settlement including the condition of river border area, land use, buildings condition and flood infrastructure system.

Regional Overview

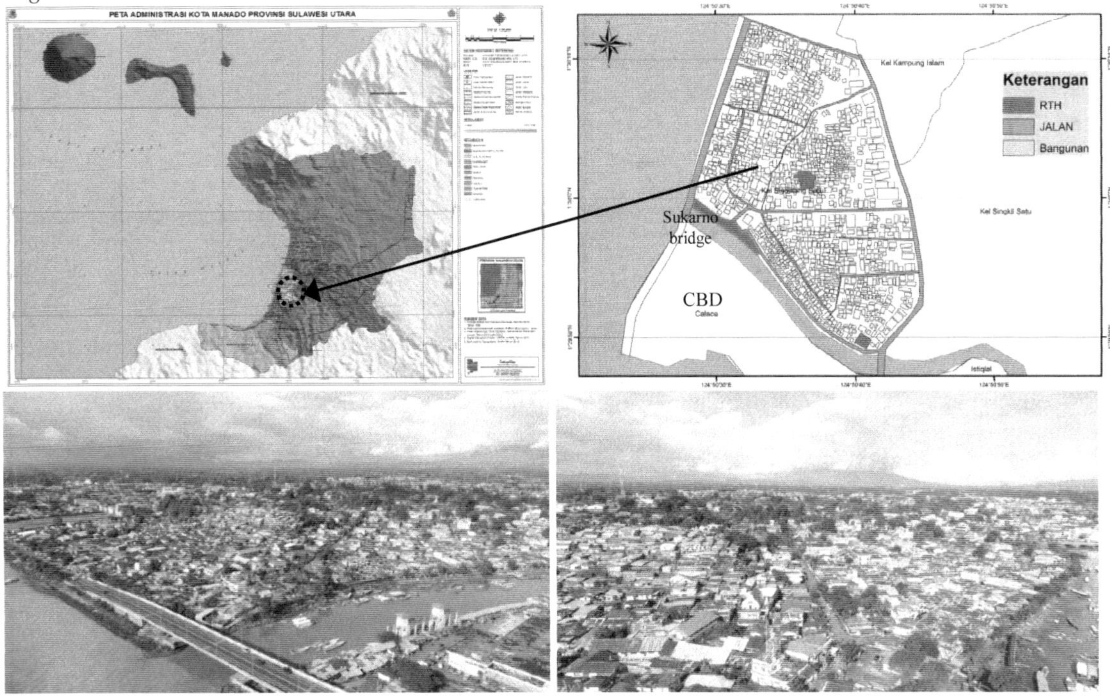

Figure 2. Map of Manado City and Sindulang Satu District (top) and situation of the site (bottom)

Sindulang Satu District is one of the most strategic parts of Manado City because it is located at the mouth of Tondano watershed, called Kuala Jengki, which is the largest river in Manado City. Thus the area lies between Tondano and coastal areas. In addition, this area is also located between densely populated settlements in Tuminting District with a central business district of Manado City in Wenang District. There are several important urban infrastructure namely Sukarno Bridge (icon of Manado City), Megawati Bridge and Boulevard II street (coastal areas). Sindulang Satu District has 5 Neighborhood with an area of 24.9 ha, the population of 8,306 people and a high population density of 334 people/hectare.

3.1. River and Coastal Border Condition

According to data from the Regional Disaster Management Agency (BPBD) of Manado City in 2017, Sindulang Satu District is one of flood-prone areas with a risk of flood height of 1-4 meters. Based on the analysis, this is cause by the topography of the sloping area (0-8% sloping), with elevation elevation <0,5-4 m asl. Measured from the height of the river that is about 15 cm-0.5 meters. This means that almost all settlements are at low levels of sea and river levels, especially in the Neighborhood II and Neighborhood IV (riverbank area) and Neighborhood I & Neighborhood III (coastal area / Boulevard II street). This is due to the last few years there is an increase in river and sea surface altitude consequently if the rain water intensity is high enough and overflow of the dike the floods directly occur in the area, especially because street in the coastal areas and river borders are built higher than settlements. The river has a width of about 40 meters and the depth of the river ranges from 1-2 meters. River condition is not very good because there has been silting by mud sediment and pile of garbage and constriction of river body. As for the coast is good enough because the street along the coast (Boulevard II street) has a width of 12 meters and has been built dikes with a height of 2 meters.

Most of the side of the river has been built dikes with a height of 1.7 meters, with rebound concrete material. But at some point the condition is damaged so it needs to be fixed. River border distance does not meet the standard / very minimal i.e. 0-2 meters and has been built inspection line with a width of 2 meters but almost along the river border and the line inspection has been placed boats, where raising livestock, porch, motorcycles, a place to store goods, dry clothes, place to sell and others. This led to the border area as a river basin and the protective settlement area almost ceased to exist. Whereas based on Spatial Planning Regulation of Manado City of 2014-2034, Kuala Jengki which is a Tondano watershed including large river with embankment in urban areas that must have at least 5 meters border in the right-left body of the river.

Figure 3. The condition of river border area, damaged and misuse of embankment

3.2. Land use

Based on the analysis of the existing condition, from the area of 24.9 hectares approximately 87.8% (21.86 hectares) has been utilized as an area built with occupancy and mix use (residential and business ie shop/stall, restaurant, shophouse, etc), about 10% is street designation and the rest about 2.2% (0.54 hectares) is the designation of public open space. This means the designation of open space is very small. Spatial Planning Regulation of Manado City of 2014-2034 requires green open space is 30% of the area, therefore it is necessary to reduce the density of buildings by 27.8% by way of relocation of buildings in high density areas and uninhabitable on the river banks (Neighborhood IV and part of Neighborhood II) and hilly areas with slope ≥ 25% (part of Neighborhood V).

From the survey results, 100% land ownership status is illegal (no land certificate) in both areas. Based on data from Head of Lurah and Head of Neighborhood II and IV, in 2016 there has been a housing relocation program in river border to Pandu District (10 houses in Neighborhood II and 40 houses in Neighborhood IV). From the interviews of 25 respondents in Neighborhood II and IV, all are willing to relocate. Likewise with respondents in Neighborhood V states ready to be relocated. Building density arrangement can be done with land sharing system and flat development to be available enough land for public open spaces (Mononimbar, 2014). This has been in accordance with respondents' preferences, ie settlers who have not been included in the 2016 relocation program can be created flats and the building density reduction can be used for the construction of public open spaces (parks, children's playgrounds, sports fields, etc.).

Table 2. Land use of existing and planning

Existing	Large (hectare)	Planning	Large (hectare)
Buildings	**21,86**	Buildings	**15,69**
Roads	**2,5**	Roads	**2,5**
Open Spaces:	**0,54**	Open Spaces:	**6,71**
1. Open space in Neighborhood II (behind Polairud office)	0,33	1. Open space in Neighborhood II (behind Polairud office)	0,33
2. Open space in Neighborhood V (hilly area/Sindulang Park)	0,14	3. Open space in Neighborhood V (hilly area/Sindulang Park) with slope $\geq$ 30%	1,13
4. Sport field in Neighborhood IV	0,07	2. Sport field in Neighborhood IV	0,07
		3. River border area (5 meters)	0,43
		4. Housing in the river border area (40 house in Neighborhood IV, 10 house in NeighborhoodII)	0,77
		5. Border of Boulevard II street & Hasanudin street (3 meters)	1,18
		6. Border of coastal areas (30 meters)	2,8
Percentage of open space	**2,2%**	**Percentage of open space**	**27%**

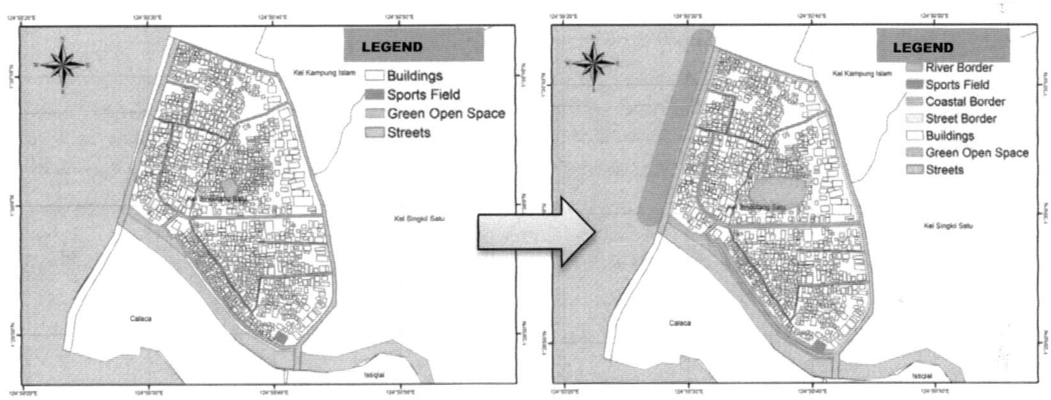

Figure 4. Map of land use existing (left) and land use planning (right)

3.3. Buildings condition

- *Density and distance between buildings*

Vulnerability to flood hazard is also influenced by the very high building density of 334 buildings/hectares with a distance between buildings about 0-1 meters. Building density causes loss of water catchment areas and green open spaces in the area and the absence of flood evacuation routes. In addition the position and orientation of the building is not clear because it develops sporadically. It is necessary to relocate for non-habitable buildings in the border area of the river in order to construct green open spaces and construct residential areas outside the river boundary with riverfront or waterfront settlement concept.

- *Structure and Construction*

Overall, the structure and construction of buildings in Sindulang Satu District is in ratio of 20:80 (permanent: non permanent), where non-permanent buildings are mostly found in Neighborhood II and IV (riparian area) and Neighborhood V (hilly area). This leads to non/semi permanent buildings around the river border at high risk of being damaged or washed away by floods. In Neighborhood I and III (coastal area / Boulevard II street) the structure and construction of about 70% buildings is permanent.

IOP Conf. Series: Earth and Environmental Science **179** (2018) 012008 doi:10.1088/1755-1315/179/1/012008

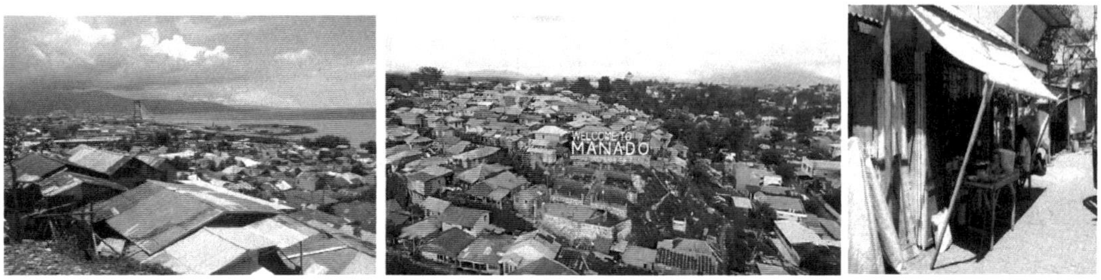

Figure 5. Land utilization with high building density and non-durable building conditions

3.4. Flood infrastructure system

- *The embankment*

It has been built with concrete rebound material, with a height of about 1.7 meters along the river side. The condition of most of the embankment has been damaged, and the function is diverted so that the benefits of the embankment as a safety of the river border become less.

- Drainage channels

At some point locations there are drainage channels with varying channel dimensions: 30-40 cm high, 20-30 cm wide, 20-30 cm wide, concrete material and partial condition not working because it is covered with soil sediment, grass, garbage or converted function into parking lot, terrace, and others.

- Roads and circulation or evacuation system

The main (primary) street on study sites are 12 meters wide (Boulevard II street) and 5 meters (Hasanudin street). Secondary local street have widths varying between 3-5 while tertiary local streets are 0.5-2 meters so that they can only be accessed by motor on certain street segments. The circulation or access system is not clear because there are many forbidden street. This makes it difficult for residents to evacuate in the event of a flood. The material on primary and secondary local streets is asphalt, whereas tertiary streets are paving and some are still compacted soils.

- Signage system

Signal systems or disaster mitigation signs are not located at the study site. In accordance with the standards for disaster-prone residential areas, signage system should be built for environmental orientation and disaster evacuation to minimize loss of life and material. Disaster mitigation signs include evacuation route information boards, hazard alerts, river level monitoring posts and others.

- Open space

As described above, the existing open space is 2.2% (0.54 hectares) consisting of open space in the hilly area (Sindulang Park) and open space behind the Polairud office. This means that the availability of open space is very less, so it needs to add open spaces in residential areas, in the form:

1. Green open space in the form of park or urban farming. Urban farming can be one source of economic improvement of citizens. The suitable location for the park or urban farming is in the open space behind Polairud's office (Neighborhood II) and on the hill of Sindulang Park (Neighborhood V).
2. Water catchment area. It is necessary to build a water catchment pond and plant water conservation plants and soil conservation such as bamboo or banyan or pine or other types of tree stands as well as grasses such as grass roots that can absorb water. Both plants can also function as water retention and can neutralize river water from contamination / water pollution. Proposed location in an open space in a riparian area (behind Polairud's office).
3. Green line along the border of streets and rivers.
4. Sports field or children's playground. The proposed location for the sports field is on the existing sports field (Neighborhood IV) while the child's playground can be built on the hill of Sindulang Park (Neighborhood V) or behind Polairud's office.

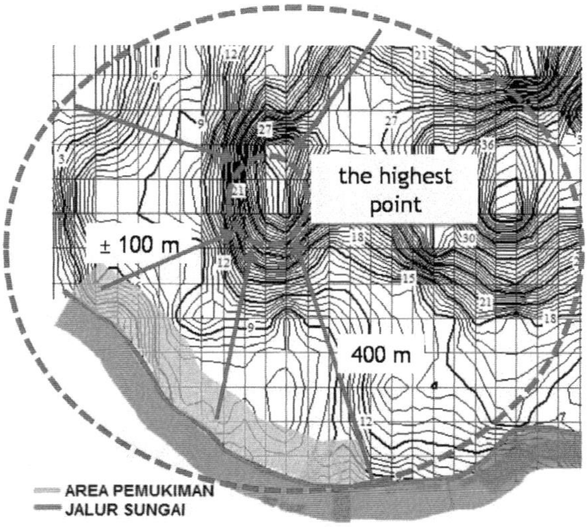

Figure 6. Natural hill in centre of district can be used as a rescue hill

- Rescue/escape hill

One of the required disaster mitigation infrastructure is escape or rescue hill, as a place for disaster evacuation. There is a natural hill (Sindulang Park) located in the center of district with a height of 10-30 meters and can be reached from many ways. This hill can be used as a rescue/escape hill.

- Electricity

100% occupancy and streets using electricity source from State Electricity Company. However, to anticipate the electricity crisis when disaster occurs, it is necessary to build solar cell system in every neighborhood.

4. Recommendation

Based on the above analysis it is proposed the concept of slum upgrading to overcome or minimize the risk of flood in the study location, as follows:

- Setting the boundary area of the river by improving the embankment and normalizing the river.
- Restoration of land use by increasing the percentage of open space up to about 30% through controlling or restructuring of river boundary area and reduction of building density through relocation of houses in riparian areas and hills with slope $\geq$ 25%.
- Build flats for people who have not been included in the relocation program to Pandu District in 2016.
- Application of riverfront or waterfront settlement concept. This is so that the river can be kept clean. However, this should be monitored (the river border utilization control system must be assertive) so as not to occur over the river border function.
- Improving the quality of infrastructure systems including improving drainage channels, repairing damaged dikes, adding to the construction of tertiary local streets that can serve as evacuation routes, building environmental signage systems and evacuation signs, adding green and non-open spaces green, arranging hilly area in Sindulang Park as public open space or playground or escape hill and building solar cell system in each Neighborhood.

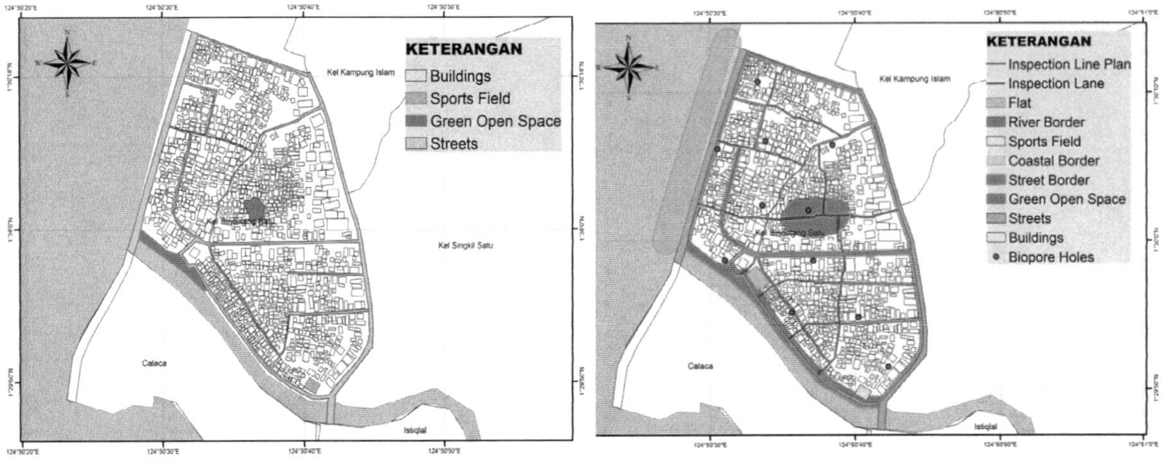

Figure 7. Map of spatial existing (left) and planning existing (right)

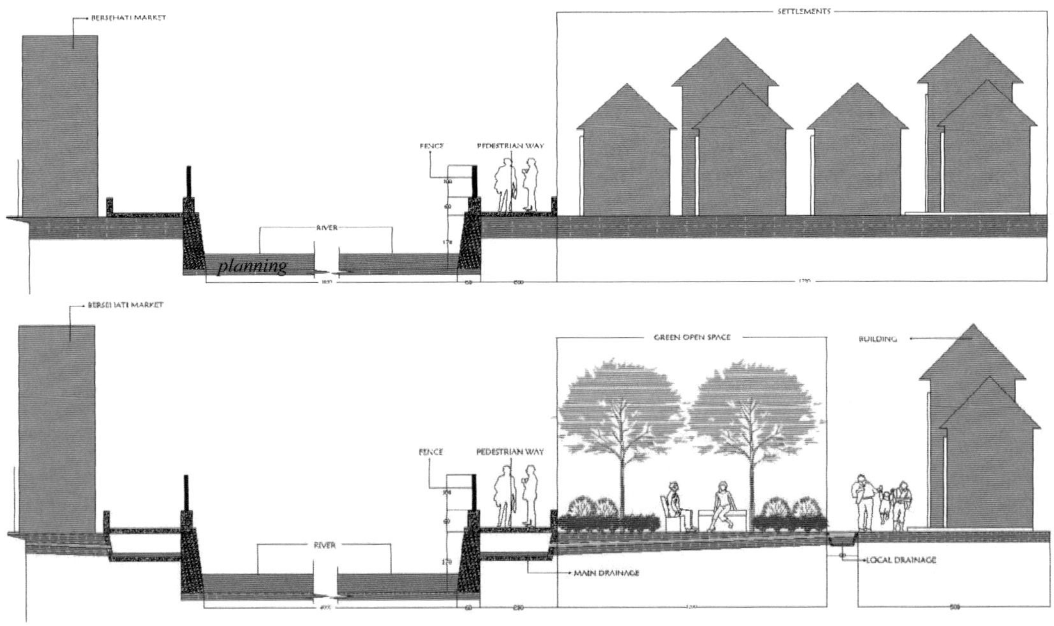

Figure 8. Section of existing (top) and restructuring plans of Tondano River border area (bottom)

5. Conclusion

The results of the study in Sindulang Satu District are irregular settlement conditions with very high building density, non-durable houses, unclear circulation system, improper river bank embankment, inadequate drainage network and lack of green open space. Therefore, improvement efforts need to be taken to restructure the river border area, arrangement the overall settlement utilization, improvement the condition of the buildings and the quality of infrastructure based on flood mitigation system such as improvement of river embankment, drainage channel, construct evacuation path and green open spaces

or rescue hill, etc. It is expected that the results of this study can be one of the solutions to realize the resilience of the region from various vulnerable physical pressures. Thus can realize the resilience of Manado City as a whole.
This result needs to be followed up with studies on other aspects / areas, such as social, culture and economics studies. It aims to make slum upgrading efforts to be integrated and comprehensive.

References

[1] Bakornas PBP 2002 Direction of Urban Disaster Mitigation Policy in Indonesia
[2] Brown A and Junmookda K 2014 *How Urban Resilience Can Make Cities and Nations Safer from Disasters* The Rockefeller Foundation New York
[3] Directorate of River and Coastal Directorate General of Water Resources Ministry of Public Works in collaboration with JICA 2012 Action Guidance and Flash Flood Disaster Mitigation System Semarang
[4] Hakim, Rustam 2004 Landscape Architecture, Human, Nature and Environment. Jakarta: FALTL Trisakti University
[5] Mononimbar, W.J. 2014 *Handling Flood-prone Settlements in River Border Area.* Scientific Journal of Media Engineering Vol. 4 No.1, 26-31
[6] Regulation of the Minister of Public Works of The Indonesian Republic No. 63 / PRT / 1993 on River Border Lines, River Use Areas
[7] Government Regulation of the Republic of Indonesia No. 38 of 2011 on River
[8] Regional Regulation of Manado City No. 1 of 2014 on RTRW of Manado City 2014-2034
[9] The Cities Alliance 2016 *About Slum Upgrading* UNOPS Brussels, Belgium
[10] The Rockefeller Foundation 2014 100 Resilient Cities New York
[11] The Laws of The Republic of Indonesia Number 1 of 2011 on Housing and Settlement Areas

Re-design of Tegalega Park in Bandung for flood evacuation park

C Handayani[1*] and F N H Utami[2]

[1] Land and Residential Office of Bogor Regency, Cibinong ,Indonesia
[2] Department of Landscape Architecture, Faculty of Agriculture, Bogor Agricultural University,Indonesia

*Email : citradly@gmail.com,

Abstract. The Southern part of Bandung city Indonesia is often flooded in rainy season. The frequent flooding in this area is the main reason for Bandung municipality to initiate mitigation plan. The selected place for mitigation due to evacuation space provision was Tegalega Park. The mitigation plan was contained in Bandung City Planning 2011-2031. The Tegalega Park was considered as an evacuation park, which has been considered as one of several mitigation efforts in providing evacuation space. Evacuation parkhas functions as a city park on a typical day and a muster point in the event of a disaster. This study aims to design an evacuation park that meets both functions. Methods applied were spatial, descriptive and quantitative analysis. Spatial analysis was used to determine the appropriate area for the evacuation park (not located in flood prone area, far from hazardous area, and near with hospitals). Descriptive analysis was used to analyze social and cultural aspects and history of flood disaster in Bandung. Furthermore, quantitative analysis was used to calculate evacuation path distance (Mohammad Toha Street 6.35km and Cibaduyut Raya Street 6.74 km), optimum thermal temperature (without modification), and carrying capacity. The results of the analysis show that Tegalega Park can be used as evacuation site. The design of Tegalega Park took Bandung Lautan Api stilation as a design concept which were described in site plan and 3D perspective.

Keywords: Bandung City, Tegalega Park, design, fload, evacuation park

1. Introduction

West Java Province is aprovince in Indonesia where flood often occur during rainy season. Based on the West Java Province statistical data on 2002-2010, Bandung city is the most flood event occur with 108 incidents [1]. Meanwhile, the landform ofSouthern area of Bandung city is mostly concave landform where sensitive to flood disaste [2]. As a mitigation effort to flood disaster in South Bandung, the municipal government planned Tegalega Park as evacuation area for flood disaster [3], stated in the Bandung CityPlanning2011-2031. The more in-depth research is needed to identify potential aspect of Tegalega Park as a flood disaster evacuation area. In addition, the design of Tegalega Parkis need to be improved in order to integrate functions of Tegalega Park as a city park an evacuation area. Evacuation park is a city park concept that has been widely developed in countries to deal with natural disasters. Japan is a country that already establish evacuation parks to face earthquake and tsunami.

The purpose of this research is identifying flood area in Southern partof Bandung city analyzing potential aspects of Tegalega Park as flood disaster Evacuation Park, and designing Tegalega Park as evacuation Park to face flood disaster. The benefits of this research are become input and consideration for Bandung City Government in developing Bandung Spatial Plan 2011-2031 in relation with mitigation and evacuation of disasters.

2. Methods

2.1. Study site

The research was conducted in Tegalega Park,Bandung city, Indonesia (Figure 1). the total area is160.630m^2.

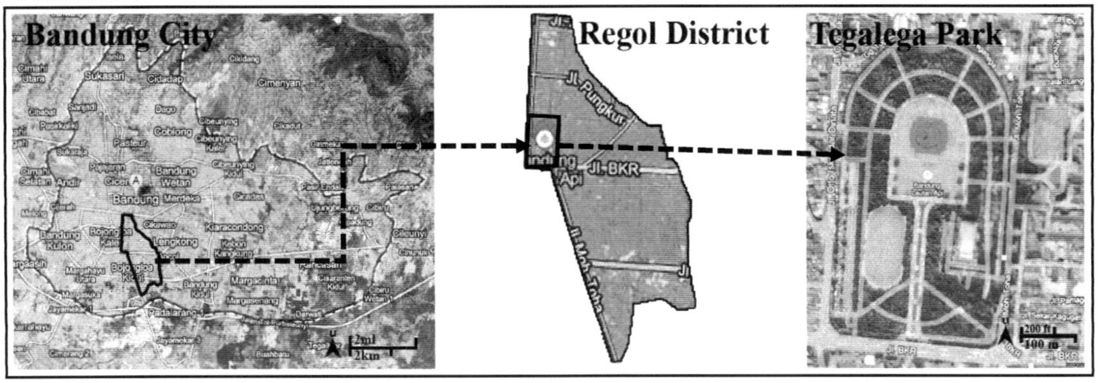

Figure 1. Orientation map of study area

2.2. Research methods

2.2.1. Spatial analysis.

The spatial analysis was imlemented to determine suitability of Tegalega Park as an evacuation site. The physical and social aspects were analyzed to determinel and suitability (Tabel 1).

Tabel 1. Suitable Aspects for Evacuation Site

Aspects	Explanation	
	Suitable	Unsuitable
Flood zone [4][5]	Not located in flood prone area	In flood prone area
Evacuation route [5][6]	The distance of evacuation route is≤ 8 km	The distance of evacuation route is > 8 km away
Safety location [4]	≥ 1.6 km from industrial area	< 1.6 km from industrial area
Health facilities [7]	≤ 8 km from hospital or clinic	> 8 km from hospital of clinic

2.2.2. Descriptive analysis.

The descriptive analysis was implemented to identify social and cultural aspects as well as history of flood disasters in Bandung. The social and cultural aspects consist of population who were potentially affected by flood, including their behavior and hopesas well as legal aspect.

2.2.3. Quantitative analysis.

The quantitative analysis wasused to determine evacuation route, optimum thermal temperature and carrying capacity. Carrying capacity is an important aspect to be considered in identifying space and facilities should be available on the site. The carrying capacity (K) wascalculated by dividing area (l) to needed standard (p). The formula was calculated as below[8]:

$$K = l / p$$

Evacuation routemust betheshortest one. Therefore to determine which route that optimumoptimumfor evacuation route, the maximum reachable distance of refugee to the evacuation site should be considered.The maximum distance (D max) was measured by multiplying minimum warning time (T min) to travel speed (a). This research assumedthat evacuees travel without vehicel to get maximum time. The formula was calculated as follow:

$$D\ max = a \times T\ min$$

Evacuation space must have good quality of atmosphere and needs comfortable for evacuees.Optimum temperature wasan indicator that affect human's comfort. The optimum temperature of evacuation space was calculated formulas follow [11]:

$$\text{Optimum Temperature} = 17.8 + (0.31 \times T_{a(out)})$$

where $T_{a(out)}$ isapproximate temperature each month. The result of optimum temperature then will be subtracted by average temperature of Tegalega Park. If the deviation number is occur in2.5^{0}C or -2.5^{0}C, then microclimate modificationisneeded. The microclimate modification can be done by using vegetation shade and a good ventilation of building.

3. Result

3.1. South Bandung

South Bandung was ranked as region with the highest flood disaster in West Java. The flood is not only caused by the overflow of Citarum River but also because of the concave landform of Southern part of Bandung city. There were three sub-districts in Southern part of Bandung city which were affected most by flood, include Baleendah, Dayeuhkolot, and Bojongsoang sub-districts.

Based on Citarum watershed study in 2011[12], 94% of Dayeuhkolot area were highrisk to befloodedannuallyand also the most number of population. Based on recapitulation of flood events in 2013 until Maret 2014, Dayeuhkolot had 50-200cm water levelof flood with 21,635 household suffered, and 884 households evacuated.It is possible to evacuate refugee from Dayeuhkolot sub-district to Tegalega Park because of closest distances than the other sub-district.

3.2. Tegalega Park

The flood analysis was done by overlaying flood threat of Citarum watershed map from National Board for Disaster Management of Indonesia (figure 2). Based on the spatial analysis, the Tegalega Park was not located in the flood prone area. Moreover, Tegalega Park waslocated far from industrial area that produce hazardous materials (figure 2). Furthermore, the park wasalso located close to hospitals; Immanuel Hospital within 1.25km far)and Bhayangkara Sartika Asih Hospital within 2.37km)(figure 2).Futhermore, the south side of Tegalega Park was bordered by BKR Street asevacution route stated Bandung City Planning.

There are three main roads that can be used to evacuate from Dayeuh kolot to Tegalega Park, they are Mohammad Toha Street (6.35km length), Cibaduyut Raya Street (6.74kmlength), and Kopo-Soreang Street (9.28kmlength).Based on evacuation route calculation (the travel speed of evacuees is estimated at 4 km/h[9] and the minimum warning time for flood risk in indonesia is approximately1-2 hour [10]), the maximum distance for the evacuees to travelis 4-8km. Thereforethe selected road asmain evacuation route is Mohammad Toha Street and the best alternative of evacuation route is Cibaduyut Raya Street (figure 2).

Optimal temperature of Tegalega Park based on calculation was 25.1^o C and the average of temperature in Tegalega Park was 23.5^o C.The deviation of optimal temperature and average of temperature was 1.6^{o}C. The result indicate that Tegalega Park does not need microclimate modification. Tegalega Park also has a good drainage system. The slope ofTegalega Park is approximatelly 0-15%, so the runoff possible toflow following the slope with direction to the south of Tegalega Park and ends at Ciateul River.

As a city park, Tegalega Park has many facilities and elements. The area of Bandung Lautan Api Monument, The Asia-Africa Conference Forest, and Tirta Lega Swimming Pool should be preserve butother areas can be planned for new function. Area ofevacuation will be located close to road system and parking area. The area should be far from the preservation area to avoid vandalism.

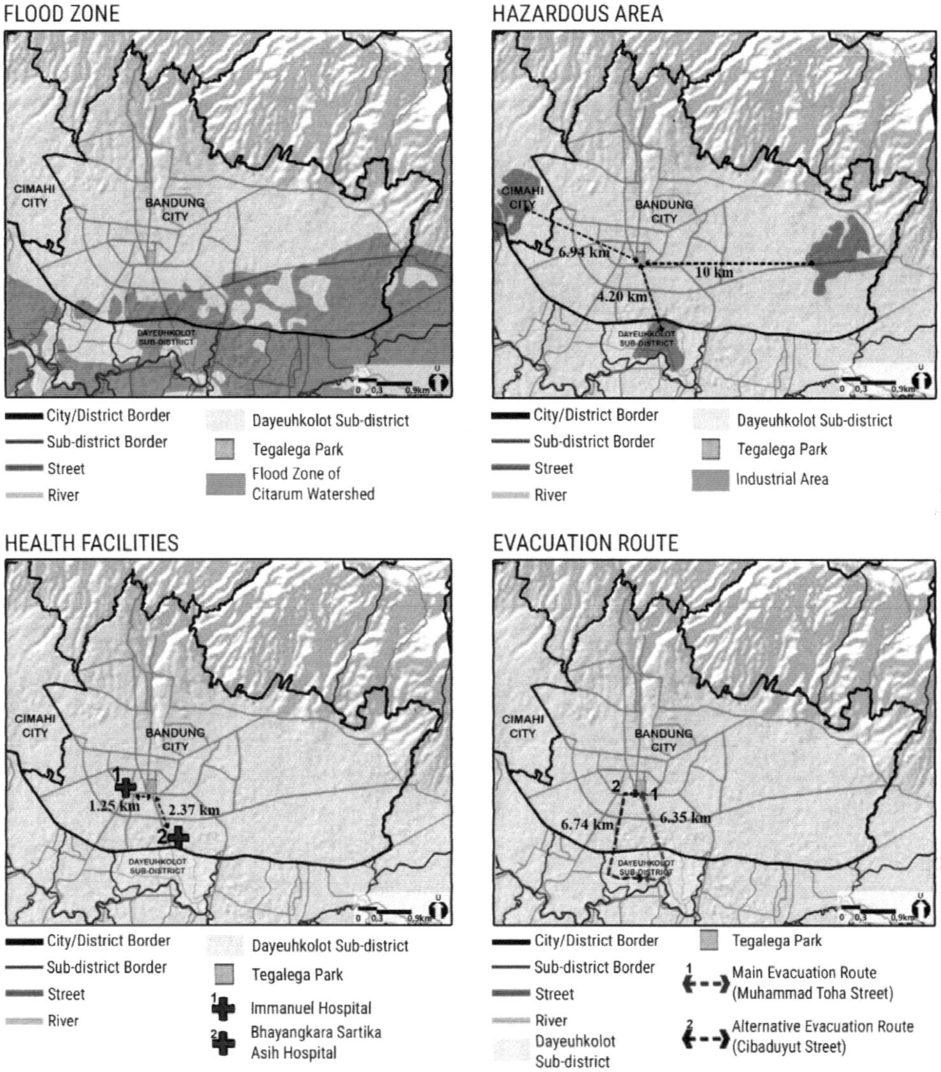

Figure 2. Spatial analysis of Flood Zone, Hazardous Area, Health Facilities, and Evacuation Route

3.3. Design of Evacuation Park

The re-design of Tegalega Park isexpected to meet four values; resilient mitigation, historical, ecological, and social. As mentioned above, the Tegalega Park will be designed as an evacuation site during flood disaster according to Bandung City Planning 2011-2031. It aimedto help refugees in flooding period.The evacuation site will be considered as an evacuation park.

Tegalega Park has Bandung Lautan Api Monument. The monument has historical value that needs to be preserved. The monument was established to commemorate the struggle of the Indonesian independence and patriotic values. Furthermore Tegalega Park had The Asia Africa Conference Forest as ecological value. That forest also needs to be preserved as public green open space for Bandung. On the other hand, Tegalega Park as a city park also has social functions such as recreational and sport avtivities.

3.3.1. Design Concept.
The design concept is taken from the shapes ofBandung Lautan Api Stilation. Bandung Lautan Api Stilation is triangular prism shaped the smallmonument, which spread over ten points in Bandung. Each stilation shows trace of Bandung Lautan Api event. Design concept diagrams are presented in figure 3.

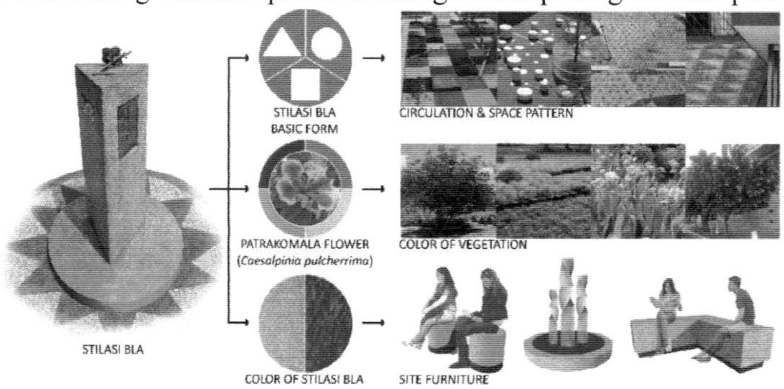

Figure 3. Diagram of Design Concept

3.3.2. Concept Development.
Concept development consists of spatial concept (figure 4), concept of circulation (figure 5), concept of vegetation, and concept of facilities and utilities.

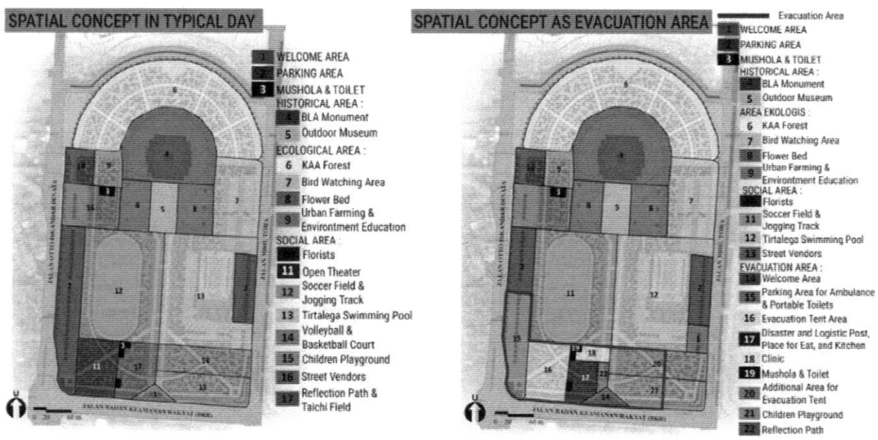

Figure 4. Spatial Concept of Tegalega Park

IOP Conf. Series: Earth and Environmental Science **179** (2018) 012009 doi:10.1088/1755-1315/179/1/012009

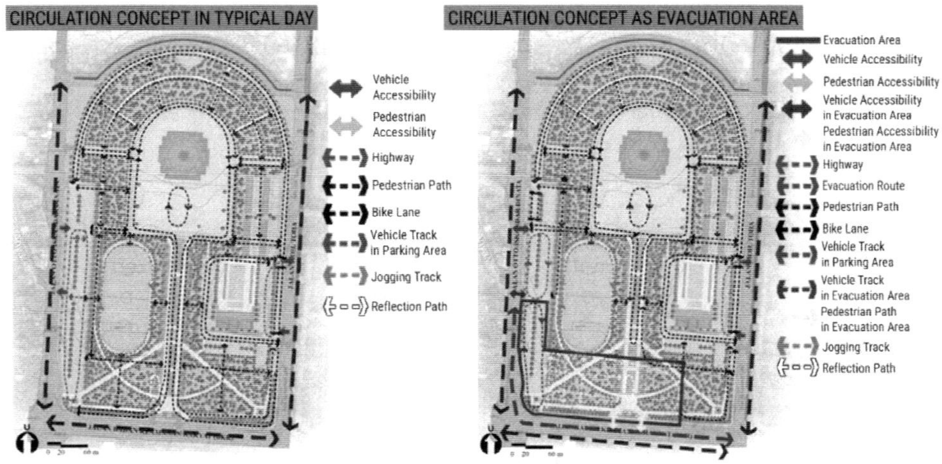

Figure 5. Concept of Circulation in Tegalega Park

3.3.3. Siteplan and Perspective.

The detail of siteplan and perspective will be focused in evacuation site. The evacuation site is also can be utilized as recreational area when eve rthe park is not used as evacuation area. The total area of evacuation site was25,411.53m^2. The area waschosen as an evacuation site because it is more accessible and not in an area with historical value. There were some facilities will be provided in evacuation are, such as tent, disaster post, health center, public kitchen and dining area, children playground and reflection path, parking area, and generator house. The siteplan werepresented in figure 6 (b) and figure 7.

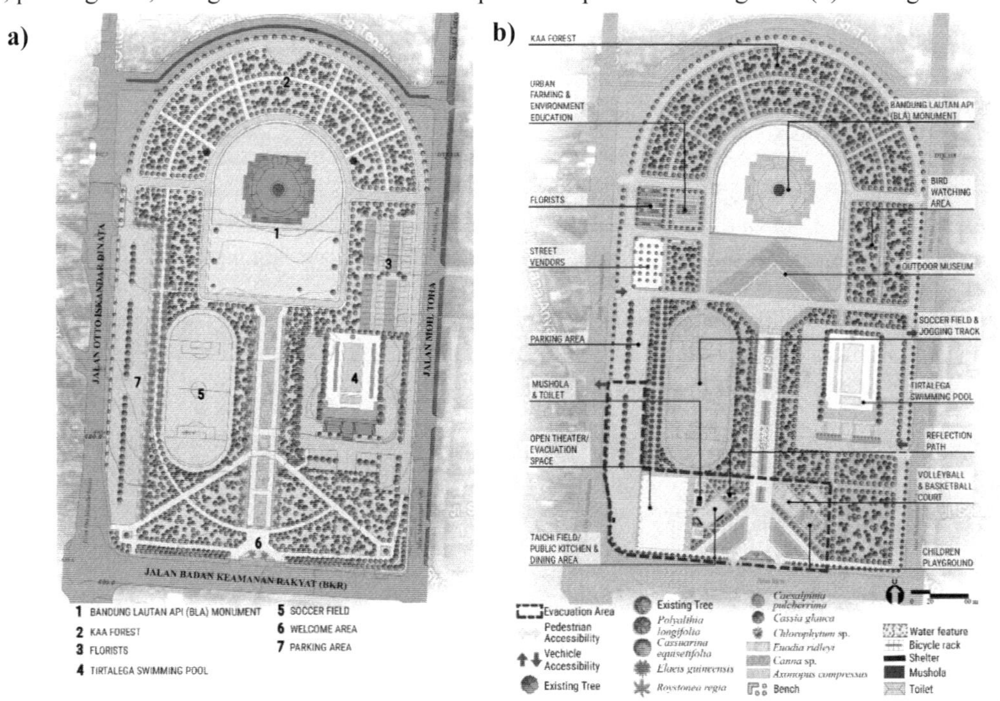

Figure 6. a) Basemap (before re-design); b) Siteplan of Tegalega Park

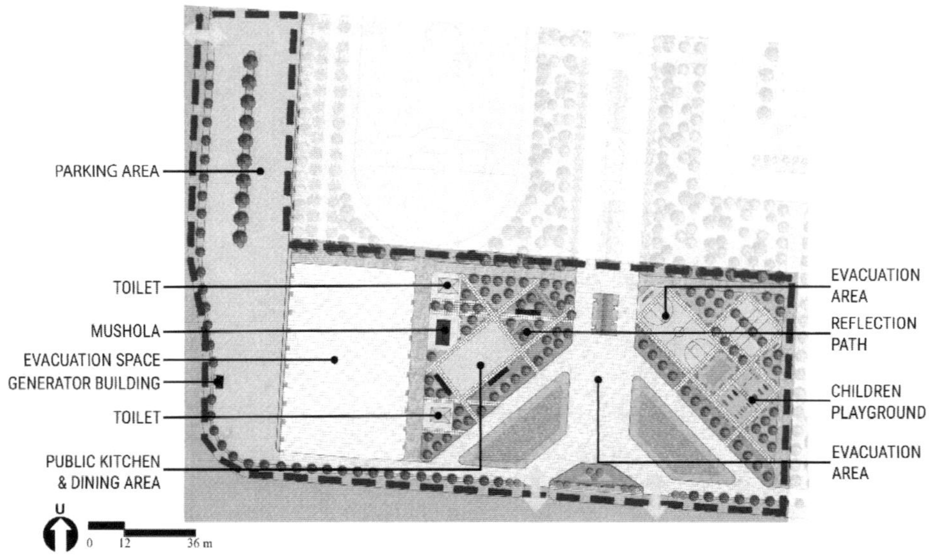

Figure 7. Siteplan of Evacuation Area

The open theater space will be functioned as tent area in flood period. Each tent'swidth is5x12m^{2}withcapacity of 30 refugees. The area was also plannedas disaster post and health center to accommodate the refugees. Velbed or folding bed is prepared in the health center for sick refugees. The area for evacuation tent use concrete pavement material with*Euodia ridleyi* as a border vegetation.This area has a total area of 3.415,58m^2.

The evacuation area was also planned to accommodate utility area for public kitchen and dining area. This area will have a total area of 420m^2 and used as an common field in a non-evacuationday. The base of eachgazebo will be used as a logistic storehouse.Children playground and reflection path were planned as attractive recreational facilities for children and adults. Those facilities wereprovided for entertaining the refugees and eleminate post-disaster trauma experienced by refugees.

Mushola (praying room) and toiletswere located beside the evacuation tent area. The Mushola was55m^2 and used by refugees for praying. There were two toilets bprovidedfor women and men, each consists of eight rooms. Based on UNHCR Emergency Standards and Indicators [13], one toilet is estimated to be used by 20 people in turn.

Parking area in evacuation site is used to park the ambulance, water wagon, and garbage container truck. This area is also used to put the portable toilets. There is emergency generator building near the parking area. This is used as additional electricity if is needed.The perspectives of every facilities are presented in figure 8.

Figure 8. Design Perspectives of the Evacuation Facilities

3.3.4. Carrying Capacity.
The carrying capacity calculation was used to estimate how much people or units can be accommodated in the evacuation area. The standard requirements of evacuation facilities refers to the standard preparation of mitigation space from Garden and Funeral Office of Jakarta [14] and United Nations High Commissioner of Refugees (UNHCR) [13] with modification. The result of carrying capacity calculation, facilities, and standard needed are presented in tabel 2.

Tabel 2. Carrying Capacity of each Facilities in Evacuation Area

No.	Facilities	Area(m^2)	Quantity	Total Area (m^2)	Standard(m^2/person) / (m^2/unit)	Carrying Capacity
1	Evacuation Tent Area	3415.58	1	3415.58	84	40 tents
	Refugee Tent	60	38	2280	2	1140 persons
	Disaster Post	60	1	60	2	30 persons
	Health Center	60	1	60	2.5	24 persons
2	Basketball Court*	420	1	420	84	5 tents
	Refungee Tent*	60	5	300	2	150 persons
3	Front Plaza of Wellcome Area*	1008	1		84	12 tents
	Refugee Tent*	60	10	600	2	300 persons
	Disaster Post*	60	1	60	2	30 persons
	Health Center*	60	1	60	2.5	24 persons
4	Dining Area	420	1	420	1	336 persons
5	Public Kitchen	20	1	20	2	8 persons
6	Children Playground	248.1	1	248.1	10	24 persons
7	Reflection Path	20.25	1	20.25	10	2 persons
8	Mushola	55	1	55	1	55 persons
9	Toilet	44	2	88	4	16 toilets
10	Parking Area	5142.34	1	5142.34		
	Army Truck	25.5	10	255	25.5	10 units
	Ambulance	10	3	30	10	3 units
	Water Wagon	18	3	54	21	3 units
	Portable Toilet	4	15	60	4	15 units

*Additional spaces for evacuation tent in case the refugee increase

4. Conclusion
Tegalega Park is suitable area for evacuation because it is not located in flood prone zone, far from hazardous area, and near with hospitals. Evacuation route from Dayeuhkolot sub-district to Tegalega Park can be reached by Mohammad Toha Street as a main route with 6,35km distance and Cibaduyut Raya Street as an alternative route with 6,74km distance.Tegalega Park as evacuation areacan accommodate 1.590 people of refugees.

The re-design of Tegalega Park was aimed to establish a city park which not only functions as recreational, historical education, and public green open space but also for resilient mitigation, that is evacuation space for flood disaster refugee. There were facilities designed to fulfill the functions of Tegalega Park as an evacuation site, such as evacuation tent area, disaster post, health center, public kitchen and dining area, children playground and reflection path, parking area, and generator house.Last but not least, the design concept of Tegalega Park is taken from shapes of Bandung Lautan Api Stilation which aims to strengthen historical value of Bandung Lautan Api history.

References

[1] Suprapto2011,*Penanggulangan Bencana. Statistik Pemodelan Bencana Banjir Indonesia(Kejadian 2001-2010)*. 2(2) p. 34-37.

[2] Pemerintah Kota Bandung 2005 *Profil Kabupaten/Kota*(Bandung: Pemerintah Kota Bandung)

[3] Pemerintah Kota Bandung 2011*Rencana Tata Ruang Wilayah Kota Bandung Tahun 2011-2031* (Bandung: Pemerintah Kota Bandung)

[4] American Red Cross2002*Standards for Hurricane Evacuation Shelter Selection*. http://www.floridadisaster.org/Response/engineers/documents/2007SRR/2007SRR-AppxC.pdf.

[5] Federal Emergency Management Agency 2006 *Design and Construction Guidance for CommunityShelters*.http://www.fema.gov/media-library-data/20130726-1508-20490-8283/fema_p_361.pdf.

[6] Cova TJ and Church RL 1997 *International Journal of Geographical Information Science. Modelling community evacuation vulnerability using GIS***11** 763

[7] Kar B and Hodgson ME, 2008, *Transactions in GIS. A GIS-Based Model to Determine Site Suitability of Emergency Evacuation Shelters***12** 227

[8] Nurisjah S, Pramukanto Q and Wibowo S 2003 *Daya Dukung dalam Perencanaan Tapak*. (Bogor: Program Studi Arsitektur Pertamanan, Jurusan Budidaya Pertanian, Fakultas Pertanian. Institut Pertanian Bogor)

[9] Blom E2013, *Construction technologies for flood-proofing buildings and infrastructures: Conceptsand Technologies forSmartShelters*.http://www.floodprobe.eu/partner/assets/documents/D4.3bConceptsandTechnologiesSmartSheltersfinal.pdf.

[10] Kementerian Pekerjaan Umum 2012, *Pedoman Penyusunan Sistem Peringatan Dini dan Evakuasi untuk Banjir Bandang*(Jakarta: Kementerian Pekerjaan Umum)

[11] Pontangaroa R and Hynds M. 2008. *Thermal Comfort Tools for Emergency Shelter in Major Disaster* http://www.irbnet.de/daten/iconda/CIB11441.pdf.

[12] Badan Penanggulangan Bencana Daerah Kabupaten Bandung, 2014, *Rencana Kontijensi Banjir Kabuapten Bandung*(Kabupaten Bandung: Badan Penanggulangan Bencana Daerah Kabupaten Bandung)

[13] UNHCR e-Centre 2000 *A Handy Guide to UNHCR Emergency Standards and Indicators*. http://www.the-ecentre.net/resources/e_library/doc/handyGuide.pdf.

[14] Dinas Pertamanan dan Pemakaman DKI Jakarta, 2013, *Penyusunan Standar RTHPertamanan dan Pemakaman sebagai Ruang Mitigasi Bencana*(Jakarta: Dinas Pertamanan dan Pemakaman DKI Jakarta)

An evaluation of visual aesthetic quality of pedestrian pathways based on ecological network corridor within campus landscape

A V Rahmandari[1*], A Gunawan[2], and W Q Mugnisjah[2]

[1] Graduate Students of Landscape Architecture, Department of Landscape Architecture, Faculty of Agriculture, Bogor Agricultural University (IPB), Jl. Meranti, Dramaga, Bogor 16680, West Java, Indonesia.
[2] Department of Landscape Architecture, Faculty of Agriculture, Bogor Agricultural University (IPB), Jl. Meranti, Dramaga, Bogor 16680, West Java, Indonesia.

[*]Email: vidya.rahmandari@gmail.com

Abstract. Campus landscape reflects the identity and character of a college. The need of public accessibility in the form of pedestrian pathway in campus should be provided particularly to realize a campus life that is environmentally friendly and sustainable. In addition, the presence of a green way on the edge of the pedestrian track in the campus landscape can provide benefits as ecological network corridors that facilitate wildlife movement, linking wildlife habitats, and improve the aesthetic quality of pedestrian pathways. This study aimed to evaluate the visual aesthetic qualities of pedestrian pathways based on ecological network corridors within campus landscape. This study was conducted at campus of Bogor Agricultural University (IPB) Dramaga. The used method in this study was descriptive quantitative with analyses of Scenic Beauty Estimation (SBE), Shannon Wiener Index (H), and correlation analysis with Rank Spearman. The results showed that the aesthetic quality of pedestrian pathways based on ecological network corridors at IPB Dramaga campus overall showed good results which is characterized by the neatness of pedestrian characteristics and dominant of trees presence. This was indicated by a very high SBE value with a moderate H value. However, there were inadequate pedestrian pathways that have medium SBE value with low H value which is characterized by misused of pedestrian pathways, surrounding building appearance that were not neat, and minimum of trees presence. The correlation test showed positive correlation between SBE aspect and trees diversity aspect (0.335) though there is no significant correlation between the two aspects.

Keywords: campus landscape, corridor, ecological network, landscape aesthetic, pedestrian, scenic beauty estimation

1. Introduction

Campus landscape is a landscape that reflects the identity and character of a collage rather than just a left-overspace between buildings. Campus landscape can be as same as a small town because in it there are various activities such as work, study, and business with a high number of civitas academicians. A good campus environment is essential for educational missions as well as users' well-being. Therefore, the needs of accessibility for the community within the campus needs to be provided primarily to create a sustainable and environmentally friendly campus life [1]. One of the accessibility needs is pedestrian pathways to support the effectiveness of campus mobilization between buildings. The existence of pedestrian pathways is also useful to support the Green Campus concept, which is currently widely applied in several campuses in Indonesia to realize the environmentally friendly and sustainable campus. One campus that applies the Green Campus concept is Bogor Agricultural University (IPB) Dramaga which has an area of approximately 250 ha with wide green open space and high biodiversity that supports the life of various types of flora and fauna. The Green Campus concept that applied in Campus IPB has a high commitment on improvement of high energy efficiency, resource conservation, and environmental quality improvement to create a healthy life and a conducive learning environment on a sustainable basis [2]. In Green Campus concept there is a Green Transportation system that applied in IPB Dramaga campus that is developed to reduce environmental impact due to the use of fossil fuel. This system encourages people to return to the culture of walking and cycling and restrict the use of motor vehicles. Walking has an advantage that is low on speed, so pedestrians can observe the environment and objects in detail and can easily aware of the surrounding environment [3]. The convenience and safety pedestrian pathways can also be supported by the existence of green open space in the form of greenway along it.

The development of a greenway along the pedestrian track in a campus environment can enhance the user's comfort and safety, in addition in creating an environmental balance [4]. The vegetation on the greenway can provide aesthetic value, provide functions of visual control, physical barrier and climate control (temperature, sun, wind and humidity), and also provide ecological functions as animal habitat [3, 5, 6]. The existence of a green way along the pedestrian pathways can also be part of an ecological network corridor. Ecological networks are a concept derived from Europe that is useful for maintaining ecosystem functions, facilitating conservation of species and habitats, and preserving species and habitats in order to survive in human-dominated landscapes [7, 8]. Ecological corridors is one of ecological networks component that serve to provide connectivity in the form of linking habitats, increase the number of habitats, provide essential elements for animals such as food, shelter, and protection from predators [9, 8]. The existence of these ecological network corridors is important especially for wildlife movements which habitat is fragmented in the campus landscape as green open spaces within the campus landscape continue to decline due to infrastructure development such as buildings, roads, and parking areas to support and improve the learning process.

The existence of pedestrian pathways and ecological network corridors along the pedestrian track within the campus landscape can link the human and animal movement, and also can provide recreational opportunities for the community. IPB Dramaga campus is one of the campus that can support the concept of pedestrian pathways based on ecological network corridor. This can be seen from the high biodiversity within the campus that needs to be preserved. This study aims to evaluate the visual aesthetic qualities of pedestrian pathways based on ecological network corridor within campus landscape at IPB Dramaga Bogor Campus. The basis of this research is the lack of concern about the visual aesthetic element in the development of campus landscape on pedestrian path ways by paying attention on ecological network corridor. The results of this study are expected to be useful as an input for planners and landscape designers in order to give more attention on aesthetic qualities in designing campus landscapes, such as improving the quality of life, maintaining biodiversity, facilitating the interaction between nature and humans, and support sustainable development.

2. Methods

This research was conducted at Bogor Agricultural University (IPB) Dramaga, Bogor, precisely on the Main Academic Road, IPB Dramaga Campus, with the total length is 6.3 km. The research was conducted in September–December 2016. The study sites were divided into 14 pedestrian pathways segments (Figure 1) and the condition of every pathways segments can be seen in Figure 2. The pathways segment is divided based on the name of the road, and each segment is subdivided according to the conditions and characteristics of pedestrian and trees [5] (Table 1). The method used in this research is descriptive quantitative.

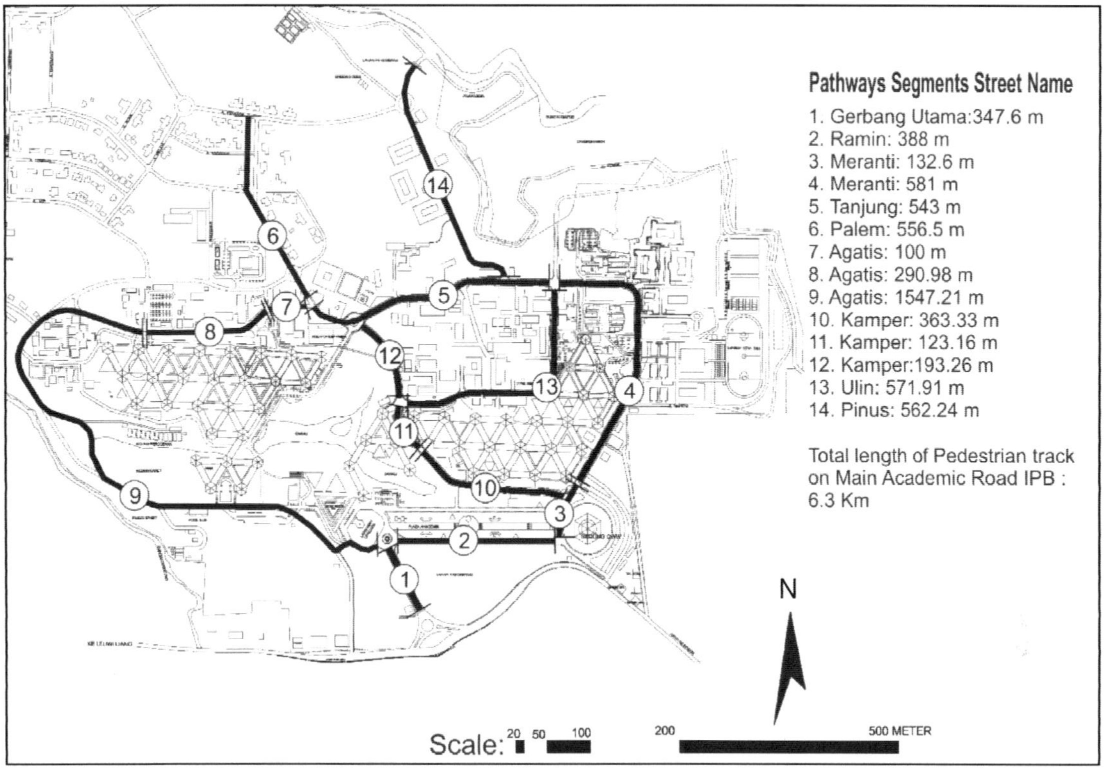

Figure 1. Map of the study area, showing the pedestrian pathways segments in the Main Academic Road in IPB Dramaga Campus Bogor.

2.1. Visual aesthetic analysis

Visual aesthetic qualities were analyzed using Scenic Beauty Estimation (SBE) which consist of three main stage [10]. The first stage of SBE is site visit and observe the study area to determinate the vantage points for landscape photograph. In every 14 pathways segments were defined several vantage points that represents the various landscape characters presented in Table 1. The total number of vantage points is 96 photos which represents the dominant landscape characters that is spread over the 14 pathways segments. The photos are selected based on the quality of the picture, the suitability on the purpose of the study, and the representation of landscape elements [5]. The landscape setting in the SBE analysis perceives the pedestrian conditions alongside the greenway as ecological network corridors, so the analysis is limited by looking at the character of the pedestrian pathways and the trees character alongside the pedestrian.

The second stage is the visual aesthetic quality assessment by 34 respondents. The respondents are students from Landscape Architecture, Bogor Agricultural University, because they have environmental knowledge [10]. Landscapes photos are prepared in the form of slides in Microsoft Power Point, which then presented in front of the respondent for 8 seconds per slide. Respondents were asked to rate each slide on a scale of 1 to 10, where point 1 shows the lowest landscape quality, while the rating scale of point 10 indicates the highest landscape quality [10].

The third stage is the visual aesthetic quality analysis that is assessed using normal Z distribution average. The formulation used in the SBE data processing is as follows.

$$SBE_X = [Z_{LX} - Z_{Ls}]X\ 100$$

Where SBE shows the aesthetic quality value toward x-landscape (x = 1,2,3, ..., 96), Z_{LX} shows the average Z value of landscape x, and Z_{LS} shows the average Z value of the landscape used as standard (Z

= 0) [10]. Then the results of SBE are classified into four category where the SBE value <-20 is low, the SBE value between -20 to 20 is moderate, the SBE value above 20 is high [10]. However, the value of SBE> 60 is considered as a value of sanction so it categorized as very high.

Figure 2. The condition of every pathway segment.

Table 1. Pedestrian pathways characteristic in every segment.

Pathways Segment	Pedestrian Characteristic											Trees Characteristic					
	Pedestrian Materials					Drainage		Greenway between Pedestrian and Road		Pedestrian and Bicycle Lane		Main Greenway		Trees Planting		Trees Canopy	
	Concrete	Asphalt	Paving block	Split Stone	Coral Stone	Open	Closed	Available	Not Available	Joined	Separate	Available	Not Available	Adjacent	Apart	Dense	loose
1			●			●		●			●	●		●		●	
2			●			●			●		●	●		●		●	
3			●				●		●		●	●		●		●	
4	●					●		●		●		●			●		●
5				●			●	●			●	●		●		●	
6		●				●			●	●		●		●		●	
7		●					●		●		●		●		●		●
8		●				●			●	●		●		●		●	
9		●				●			●	●		●		●		●	
10	●						●	●		●		●		●		●	
11	●					●		●		●			●		●		●
12			●				●		●		●	●		●		●	
13		●				●			●	●		●		●		●	
14					●		●		●		●	●		●		●	

2.2. Trees diversity analysis

The analysis of trees diversity was done by using Shannon Wiener's Diversity Index [11] to determine the ecological network corridor conditions. The initial phase of the trees diversity analysis is done by collecting trees data in the main greenways alongside the pedestrian pathways. The data collection is done by using Line Transect method with plot sample size is 20 x 20 m and skip with a distance of 10 m in path and then make another plot sample so that along the observation line has the same spacing distance [12]. The data collection is done by recording all existing tree data including the species name and number of individuals contained in the plot. After that, we analyze the diversity of trees using Shannon Wiener's diversity index. The higher value of the diversity in a corridor shows a diverse trees composition that can provide far greater benefits for more animal species who use the ecological network corridors [9, 13].

2.3. Correlation analysis

Correlation analysis is done to analyze the relationship between the visual aesthetic qualities (SBE) variable and ecological network corridor variables (trees diversity). Correlation analysis was done by using Rank Spearman test [14].

3. Result and Discussion

3.1. Visual aesthetic quality

The result showed different levels of aesthetic quality of the pedestrian pathways in the campus landscape. The results of SBE value on every segment can be seen in Figure 3 with values ranging from

14.4 to 75.0 which indicates that the aesthetic quality of the pedestrian pathways in IPB Dramaga Campus landscape has medium to very high quality.

Landscape with a very high aesthetic quality is found in segment 1 and 2 with SBE value of 67.0 and 75.0 respectively. The very high quality of the landscape character can generally be seen in Table 1. The very high aesthetic quality is characterized by the neatness of pedestrian characteristics and the dominant of tree presence. Adjacent tree stands with dense tree canopies can improve the aesthetic quality of the landscape and provide character and identity to the campus environment. This is supported by the statement of Simonds and Starke [4] that the most interesting part of the tree is the tree canopy because it can provide an identity and character to the environment. Gunawan and Yoshida [15] and Gunawan [5] also stated that landscapes with dominant tree stands are the most preferred and highly esthetic landscapes.

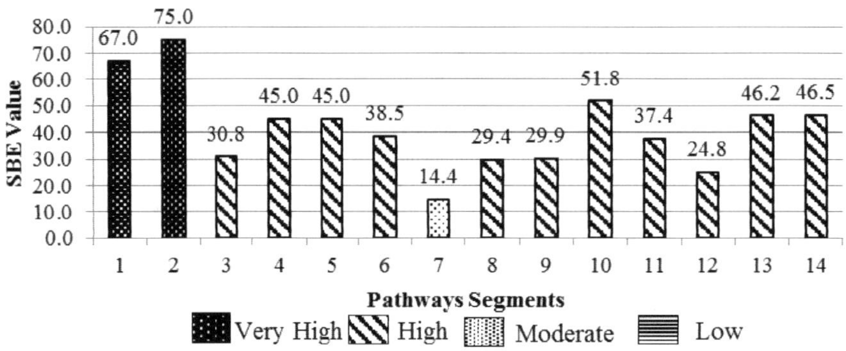

Figure 3. SBE value in every path way segment.

The high aesthetic landscape is found in 11 pathways segments shown in Figure 3 with SBE values ranging from 24.8 to 51.8. Seen as a whole in Table 1, the high aesthetic pathways segments have various differences in pedestrian characteristics and trees characteristics. In this case, the characteristics of the pedestrian and the characteristics of the trees are assessed aesthetically and the condition of each of these segments does not contain any disturbing human activity. This is in line with Yang *et al.* [16] statement that landscape values are related to landscape attributes and the intensity of human activity.

In contrast, landscapes that have the lowest aesthetic quality (moderate category) are in segment 7 with SBE value 14.4. Based on Table 1 this segment has characteristics such as unavailability of the main greenway, distant of trees planting distance, and the loose tree canopy. In addition, in this segment there are buildings on the side of the pedestrian pathways which causes arid, and the condition of this pedestrian pathways is used as a parking area of the motorcycle that can interfere the walking activities of the pedestrians and lowered the value of the aesthetic quality. According to Gunawan [5], the variety of size, shape, and tidiness of building architecture can make the aesthetic quality of the landscape low. No or least vegetation in the form of tree stands, shrubs, or soils is also the least favorable type of landscape, as buildings that dominate the landscape can give the impression of arid and intense heat.

3.2. Ecological network corridor

Ecological network corridors can be functionally defined to show connectivity between areas and physical structures to indicate connectedness [8]. The condition of the ecological network corridors in each pathways segment can be determined from the results of the Shannon Wiener (H) diversity index on the greenway alongside the pedestrian pathways. Based on Figure 4 it can be seen that the value of H diversity ranges between 0 and 2.86 which indicate from low to moderate trees diversity. The highest value of H is in segment 5 with a value of 2.86 which show moderate trees diversity. While the lowest value of H is in segment 11 with a value of 0 which show low trees diversity. Low H values can lead to disconnection of animal connectivity and affect wildlife movement due to disconnected greenway [9].

The condition of the green way along the pedestrian pathways in the Main Academic Road, IPB Dramaga Campus, in most other segments have been connected continuously so it can facilitate the movement of animals. The existence of linear greenway into a network corridor located along the edge of a pedestrian pathways within the campus landscape can enhance the comfort and safety for the community and can serve as biodiversity conservation [9, 17].

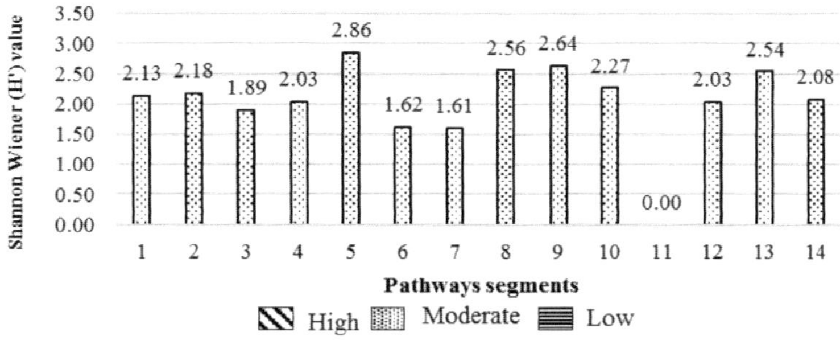

Figure 4. Shannon Wiener's diversity index value of tree on each pathway segment.

3.3. Correlation between visual aesthetic and ecological network corridor

The evaluation of visual aesthetic qualities of the pedestrian pathways based on ecological network corridors within the campus landscape can be seen by correlate the aesthetic value (SBE) variable with the ecological network corridor variables (trees diversity). Based on Rank Spearman test result between the SBE values and the trees diversity value (correlation significant at the 0.05 level) has Sig result. (2-tailed) of 0.279, this indicates that there is no significant correlation between the two aspects. However, correlation coefficient result is 0.335, this shows a positive number which indicates a positive correlation between aesthetic aspects and trees diversity aspects. It can be interpreted that the increasing ofvisual aesthetic value of pedestrian pathways can be influenced by the increasing of trees diversity value. In contrast, the lower aesthetic quality value of the pedestrian pathways, the lower value of trees diversity. This result is supported by the statement of Klein *et al.* [15] who found a statistically significant increase trend in visual aesthetic preferences for landscape structures that have larger amounts of vegetation diversity as well as indicated an increase in ecological function. Although from the results of this study the influence between the two aspects is very small.

3.4. Design concept implementation for pedestrian pathways based on ecological network corridor within campus landscape

In general, to improve the visual aesthetic qualities of pedestrian pathways based on ecological networks corridor within the campus landscape can be done by taking into account the conditions of safety, convenient, easy and humane pedestrian pathways. Based on Regulation of Minister of Public WorksNumber03/PRT/M/2014 [19], the ideal standard for pedestrian pathways planning for educational areas, the minimum width of the pedestrian pathway is 115 cm to 200 cm which is the minimum service standard B. The dimensions width of the pedestrian pathways which is combined with a bicycle lane without buildings on either side is 300 cm, with 150 cm for the pedestrian pathways and 150 cm for the bicycle lane. The dimensions width of the pedestrian pathways where one of the side is adjacent to the building is 325 cm, where the path that is adjacent to the building for the pedestrian path way nor the bicycle lane is 175 cm and the other path is 150 cm. The dimensions width of the pedestrian pathways where both sides are adjacent to the building is 350 cm, where the pedestrian pathways and the bike lane is 175 cm wide. The standards to improve the quality of ecological network corridors can be done by (1) increasing vegetation diversity, (2) making variation of vegetation structures such as ground cover, shrubs, and tree layers to provide greater benefits for more species, (3) providing a continuous corridor form, although some other species have no effect on the cut-off corridors, and (4) planting random vegetation to make it more attractive for animals and humans [20, 13, 9, 21]. The landscape visual aesthetic quality value will increase in line with the increase of diversity, the level of naturalness, and the presence of patches or leafy corridors [18]. In addition, it is also necessary to consider the tidiness of building architecture, as well as managing human activities that can interfere visually.

Specifically, for the IPB Dramaga Campus, almost every pathway segment already has a good visual aesthetic quality of pedestrian pathways based on ecological network corridors. Viewed from ecological network corridor side, the tree planting alongside the pedestrian pathways has various combinations of randomly arranged trees. The width of the existing pedestrian pathways has fulfilled the standard criteria of having a width of at least 115 cm to 200 cm. Where the current pedestrian pathways in IPB Dramaga Campus ranges from 120 cm to 230 cm. For pedestrian pathways that combined with bicycle lane in segments 4, 10, and 11 do not have minimum standards of 300 cm, where the existing width is only 210 cm wide in segments 4 and 200 cm wide in segments 10 and 11. In segments 6, 8, 9, and 13 have no pedestrian pathways, but these segments are visually assessed aesthetically and have a continuous greenways as ecological network corridor. The feedback for this pathway, it is better to make pedestrian pathways without interrupting existing trees with a width of at least 115 cm so that pedestrians are safe from motor vehicles and can facilitate movement between areas. The low quality of aesthetics in segment 7 is caused by poor management, causing pedestrian pathways is used as motorcycle parking, whereas in terms of trees diversity it is able to support as ecological network corridor. For that, it needs good parking management in segment 7 to improve the aesthetics quality. Segment 11 has moderate SBE quality, but low H diversity value, this has resulted in inability to provide continuous ecological network connectivity. Overall IPB Dramaga Campus landscape needs to add different types and species of trees in segments with low H diversity so that the value of trees diversity becomes higher and can connect the disconnected ecological network corridor and also add aesthetic value on pedestrian pathways. In addition, vegetation planting structures also needs to be combined from ground cover vegetation, shrubs, to trees. The existing trees in each segment of the pathways needs to be maintained and managed well. The selected trees diversity needs to be adjusted to the needs of animals that will use ecological network corridors. Considering aesthetic qualities can help anticipating in landscapes change and environmental impacts, while paying attention in ecological quality can also be affected by the aesthetic value that felt in the landscape [22]. If the pedestrian pathways facilities within the campus have a good condition, comfortable, and aesthetic it is expected to increase the desire of the community to walk. Thus, it can also support the implementation of Green Transportation system in IPB Dramaga Campus and encourage people to walk. According to Ignatieva *et al.* [17] the design features of ecological networks are by combining ecological features (species and biodiversity movement), provide recreational functions (bicycle and pedestrian pathways), provide aesthetic (beauty) elements, and become a cultural identity that can create a sustainable environment.

4. Conclusion

The good visual aesthetic qualities of the pedestrian pathways based on ecological network corridors within the campus landscape is the one that has very high SBE values with moderate to high trees diversity values. Inadequate pedestrian pathways based on ecological network corridors within the campus landscape are those with moderate SBE values and low trees diversity values. Implementation of the concept of pedestrian pathways should generally accommodate the simultaneous movement of people and animals, provide benefits as biodiversity conservation, provide community recreation functions, provide aesthetic elements, and become the identity and character of a campus landscape.

References

[1] Shekari ZA, Moeinaddini M and Shah MZ 2014 A pedestrian level of service method for evaluating and promoting walking facilities on campus streets Land Use Policy 38pp175-193
[2] Tim Implementasi Green Campus IPB 2016 [internet] [accesed 2016 November 26] Retreived from http://greencampusipbacid
[3] Iswanto D 2006 Pengaruh Elemen-Elemen Pelengkap Jalur Pedestrian Terhadap Kenyamanan

Pejalan Kaki (studi kasus: penggal jalan pandanaran, dimulai dari jalan randusari hingga kawasan Tugu Muda) J Perencanaan Kota dan Permukiman 5(1)pp 21-29

[4] Simonds JO and Starke BW 2013 Landscape Architecture, A Manual of Environmental Planning and Design (New York: McGraw-Hill Book Company)

[5] Gunawan A 2005 Evaluasi kualitas estetika lanskap Kota Bogor Jurnal Lanskap Indonesia 1(1) pp 21-24

[6] Nikita O and Gunawan A 2016 Pengaruh Komposisi Elemen Taman dan Kriteria Hemat Energi terhadap Kualitas Estetik Visual in: Gunawan A (Ed) Estetika Ekologis: teori dan konsep untuk desain lanskap danl ingkungan (Bogor: IPB Press)

[7] Bennett G and Mulongoy KJ 2006 Review of experience with ecological networks, corridors and buffer zones (Canada: Secretariat of the Convention on Biological Diversity)

[8] Jongman R and Pungetti G 2004 Ecological Networks and Greenways: Concept, Design Implementation (Cambridge: University Pr)

[9] Hilty JA, Jr Lidicker WZ and Merenlender AM 2006 Corridor Ecology: The Science and Practice of Linking Landscape for Biodiversity Conservation (Washington: Island Pr)

[10] Daniel TC and Boster RS 1976 Measuring Landscape Aestethics: The Scenic Beauty Estimation Method (Colorado: USDA Forest Service)

[11] Ludwig JA and Reynolds 1988 Statistical Ecology: A Primer on Methods and Computing (New York: John Willy & Sons)

[12] Kusmana C 1997Metode Survei Vegetasi (Institut Pertanian Bogor:Bogor)

[13] Fleury AM and Brown RD 1997 A Framework for the design of wildlife conservation corridors with specific application to southwestern Ontario Landscape and Urban Planning 37pp 163-186

[14] Santoso S 2015 Menguasai SPSS22 from Basic to Expert Skills (Jakarta: PT Elex Media Komputindo)

[15] Gunawan A and Yoshida H 1994 Visual judgment on landscape and land uses of Bogor municipality Bul Kyoto Univ Forest 56pp 119-131

[16] Yang D, Luo T, Lin T, Qiu Q and Luo Y 2014 Combining Aesthetic with Ecological Values for Landscpe Sustainability J Plos One 9 (7)pp 1-8

[17] Ignatieva M, Stewart GH and Meurk C 2010 Planning and design of ecological network in urban area Landscape and EcolEng 7pp 17-25

[18] Klein LR, Hendrix WG, Lohr VI, Kaytes JB, Sayler RD, Swanson ME, Elliot WJ and Reganold JP 2015 Linking ecology and aesthetics in sustainable agricultural landscape: Lessons from the Palouse region of Washington, USA Landscape and Urban Planning 134pp 195-209

[19] Peraturan Menteri Pekerjaan Umum Nomor 03/PRT/M/2014 tentang Pedoman Perencanaan, Penyediaan, dan Pemanfaatan Prasaranadan SaranaJaringan Pejalan Kaki Di Kawasan Perkotaan

[20] Bloemmen M and Van der Sluis T 2004 European Corridors – Example Studies for The Pan-European Ecological Network (Netherlands: Wageningen University and Research Centre, Biological Conservation)

[21] Jongman RH 2007 Ecological Network, from Concept to Implementation - Landscape Ecological Application in Man-influenced Area (Netherland: Springer)

[22] Gobster PH, Naussauer JI, Terry CD and Fry G 2007 The shared landscape: what does aesthetic have to do with ecology Landscape Ecol 22pp 959-972

IOP Conf. Series: Earth and Environmental Science **179** (2018) 012011 doi:10.1088/1755-1315/179/1/012011

Permaculture model for fulfilling nutritious food needs students of IPB dormitory

R Nabilah[1*], W Q Mugnisjah[2], and A Gunawan[2]

[1] Graduate Student of Landscape Architecture Department, Faculty of Agriculture, Bogor Agricultural University, Jl. Seruni Dramaga Campus, Bogor 16680, Indonesia.
[2] Landscape Architecture Department, Faculty of Agriculture, Bogor Agricultural University, Jl. Seruni Dramaga Campus, Bogor 16680, Indonesia.

*Email: rzknabilah@gmail.com

Abstract. Food quality is determined by food system, from the food production process to food consumption. The production process is related to the agricultural landscape planning. Bogor Agricultural University (IPB) as the centre of excellence of agricultural science and technology, can be a potential sample for implemantation of agricultural planning for production. Thus, the concept of permaculture, which is cover agricultural productivity based on socio-economic, and ecological has a big chance to be applied. The objective of this research is to compose permaculture model for fulfilling the nutritious food needs of IPB's dormitory students. The potential research surveyed in 8 campus experimental gardens and 4 business development sites. All of the site samples can be used as the potential of commodity business. The result showed the nutritious food needs row materials for IPB's dormitory students as much 2 299.14 tons for plant resources production per year and 406.1 tons for animal resources production per year which considered of 18 commodities. The results showed that there is a potential about 33.53% plant resources and 13.25% animal resources to supply nutritious food needs for students of IPB dormitory. This approach could be used by IPB to created food self-sufficiency model by taking advantage of the land owned.

Keywords: agriculture landscape, IPB's dormitory students, nutritious food, permaculture

1. Introduction

Bogor Agricultural University (IPB) is a college that provides a dormitory for students at first year. However, based on average calorie needs showed 12.5% underweight and 23% very thin, that expected unbalance between student activity and food consumption [1]. According to the World Health Organization the common risk factors for illness are unhealthy food consumptions. Thus, should be consideration of nutrition fulfillment for IPB's dormitory students [2]. IPB have a purpose Tridharma of these higher learning institutes, such as education, research, and community extension service. Then, IPB having supporting facilities for all these activities, that includes experimental gardens, ponds, marine laboratories, cattle grazing fields, and forest education [3]. Therefore, IPB as a centre of excellence based on agricultural science and technology has a big chance for applied agricultural model. In addition, IPB has cooperation with some villages surrounding IPB campus as a part of collaboration community services. Agricultural model releated with food sufficiency, and if IPB can making the model, is represented food self sufficiency in the scope of institution. Therefore, the model of permaculture can be carried out in several experimental garden and some villages surrounding IPB campus. Permaculture is a combination of words permanent and agriculture, which means agriculture technique from involves dynamics of the existing practice about traditional and scientific knowledge. Permaculture concept are integrating between plants, animal, human, and landscapes. Permaculture planning based on science observation and patterns from nature, then an integrated science and arts [4]. The main aim of this study was to explore permaculture model on the campus sites to fulfiling nutritious food needs student of IPB dormitory. The benefits of that aim, to know about suply and demand at the institution scope. Then, product from self sufficiency from agriculture business can make the good quality product [5].

2. Method

2.1. Location and time of study

The location of research is located in IPB Experimental Garden and some villages surrounding IPB campus in Bogor District and Bogor City, West Java Province. The land is located in 14 Experimental Gardens and 10 villages surrounding IPB campus. The experimental gardens are Babakan, Sawah Baru, Cikabayan, Cikarawang A and B, Leuwikopo, Pasirkuda, Sindang Barang, Experimental Ponds BDP FPIK, Sabisa Farm, ADS Cikarawang, Agribusiness Development Station (ADS) Cikarawang, FAPET Field Laboratory, and Sabisa Farm. 14 villages surrounding IPB campus is located in Dramaga and Ciampea Sub-districts. The villages in sub-district of Dramaga are Cikarawang, Babakan, Dramaga, Ciherang, Sinarsari, Neglasari, Petir, Purwasari, Sukawening, and Sukadamai villages. The villages located in Sub-district of Ciampea are Cihideung Ilir, Cihideung Udik, Cibanteng and Benteng villages. Furthermore, data collection of IPB's students dormitory and interviewees was conducted at Dramaga IPB Campus (Figure 1). The study was conducted from January to August of 2017.

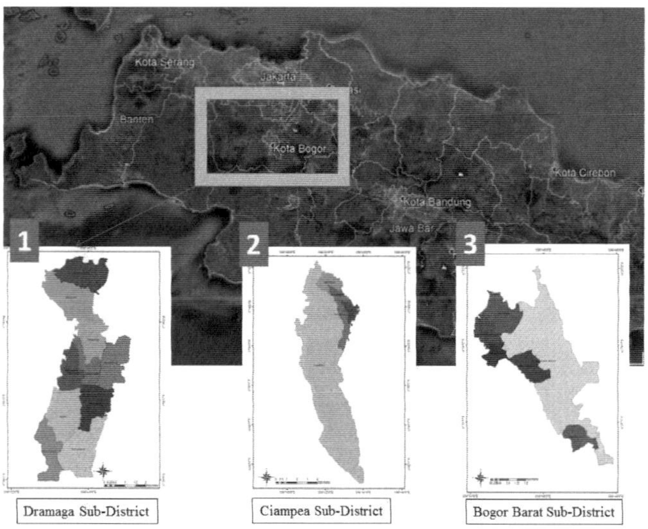

Figure 1. Research location.

2.2. Stages of data collection

The research method was clearly by using mixed method, that is the method whose compare quantitative and qualitative data analysis [6]. Planning of permaculture model based on the planning stages from Gold 1980 [7]. The planning stages include the pre-research, inventory, analysis, synthesis, and planning. The research conducted by interviews, questionnaire, and involving several research objects. The first object is nutritional status of students. The student respondents was 25 male and 25 female from IPB's students dormitory. The second object is the person who responsible for the management agriculture landscape from several fields. That person were 14 field managers from the research location and used stratified random sampling [8]. The third research objects were 20 practitioners of agriculture in 3 Sub-districts, which is done by random sampling method.

2.3. Food needs analysis and nutrition status of IPB's dormitory students
The nutritional status analysis was done by direct research, by distributing questionnaires to 25 male and 25 female students. Nutritional status data obtained from measurement of body weight and height. Analysis of nutritional status is based on data form of identity of respondents covering gender, height, age, and type of physical activity tabulated, then analyzed descriptively. The energy needs of IPB boarding students can be estimated by grouping physical activity by weight or lightness of activity. The nutritional needs rate, food requirement, commodity requirement, and the area of land needed for planning are obtained from the 2006 research data with modification [1].

2.4. Analysis of agricultural potency in experimental garden and village surrounding IPB Campus
The analysis of the location of the research to find out boundary treads and geography, climate, soil type, topography, rainfall, vegetation, animals, facilities and accessibility. Each shows the general situation of the research site.

2.5. Analysis of the land area needs for permaculture model planning
Analysis of land area can be interpreted from the calculation of commodity needs in one year, cropping pattern, and land condition.

$$LLdR = \frac{KKt \times 365}{PPKt \times D} \qquad (1)$$

Remarks:
LLdR = land area of design (ha)
KKt = total requirement commodities per day IPB's dormitory students (g)
PPKt = potential production of related commodities (kg / ha)
D = number of cycles per year

2.6. Analysis of cropping pattern
Analysis of plant and animal requirements was analyzed using a simple analysis, the calculation of each livestock and fishery business with the number of commodities produced.

$$RL = \frac{LPt}{Ln} \times 100\% \qquad (2)$$

Remarks:
RL = Ratio percentage of potential land to land area requirement (%)
LPt = Potential agricultural land for plants resources (ha)
Ln = Overall land requirement (ha)

2.7. Synthesis of food sufficiency for students of IPB dormitory from permaculture model
Food sufficiency is synthesized through the data previously obtained. Then, commodity needed adjusted with current condition at IPB Experimental Garden. That aims to find out how the contribution of IPB Experimental Garden in supporting the fulfiling of nutritious food needs. The analysis of technical assumptions in farming efforts in some research sites that have been determined. The main concept of permaculture is energy can not be created and destroyed, but only changes from one to another. Energy flows are used for waste management. The concept of permaculture model included the main crop, plants, and animals at surrounding of the lands.

2.8. Analysis model of permaculture

The entire data were analysed by mixed method. This method combines several methods in a phenomenon [9]. Use of this method to search for the same object from different angle of data retrieval [10]. Mixed methods produce combined knowledge, which arises through the integration of various types of information [11].

3. Results and Discussions

3.1. Nutritional food needs of IPB's dormitory students

The result from observation on 25 male and 25 female students showed that the percentage of students with normal category is 64%, while the lowest percentage of 2% goes into the category of students with nutritional status is very fat. Students with nutritional thin status are 24% and the status was very thin with 6% (Table 1).

Requirement energy in a day can be estimated from Basal Metabolic Rate (BMR) which is analyzed from body weight and height variable. The energy needs component is different for each person based on age, sex, body size, and health level. The results showed that the BMR score of the male students was 1 352.95 kcal / day and the female students was 1 273.75 kcal / day with an average calorie of 1 313.35 kcal / day. From the calculation of caloric requirement per day, the results obtained for male 2 410 kcal and women of 2 311 kcal with the overall average caloric requirement is 2 360 kcal (Table 2).

The results showed that 86% of students agree if the food supply is managed by IPB. Furthermore, 94% of the students approved the catering business managed by IPB (Table 2). Nutritional adequacy, energy needs (calories, protein, fat, and carbohydrate), food fulfillment needs, and the area of land needed for planning were obtained from 2006 data with modifications to the needs of 3 500 students (Table 5).

The results showed that 86% of students agree, if the food supply is managed by IPB. Furthermore, 94% of the students approved the catering business managed by IPB (Table 3). According to health is closely related to the nutritional adequacy obtained from quality food consumption.

Table 1. Body mass index and nutritional status of dorm students.

Nutritionals status	Body Mass Index	Male (individual)	Female (individual)	Percentage (%)
Very thin	<17.0	3	-	6
Thin	17.0-18.4	5	7	24
Normal	18.5-25.0	15	17	64
Fat	25.1-27.0	1	1	4
Very Fat	>27.0	1	-	2

Table 2. Calories of IPB's dormitory students.

Gender	BMR (kcal/day)	Calories needs (kcal)
Male	1 352.95	2 410
Female	1 273.75	2 311
Average	1 313.35	2 360

Tabel 3. Preferences of IPBs dormitory students based on food needs.

No.	Questions	Precentage	
		Agree	Disagree
1.	Food supply is managed by IPB	86%	14%
2.	Food quality affects the health	100%	0%
3.	Food supply from food catering for daily food needs	94%	6%
4	Prices of food more expensive with better quality	54%	46%

3.2. Agricultural potential in IPB experimental garden and village surrounding IPB

Potential land is obtained from synthesis data, searching from literature and bio-physical land. Thus, there are 12 locations that are divided into two groups of food supplements. There are consisting of 8 IPB's experimental sites as permaculture model and 4 locations of IPB business development. The selection of the location is also determined from the distance proximity of the IPB campus, especially from dormitory students (Table 4). Agricultural commodities needed from the conversion of the student menu are divided into 19 commodities [1]. These commodities are rice, soybeans, corn, papaya, chicken meat, kale, red peppers, spinach, beans, tomatoes, chicken eggs, fish, leek, and celery (Table 5).

IPB Experimental Garden selected to be planned as experimental garden is Babakan experiment Garden, Sawah Baru, Cikarawang A + B, Cikabayan A + B, Leuwikopo, and Sindang Barang and Pasir Kuda. The area that can be utilized for the design of the experimental garden is only 16.75 ha of permaculture model planning and 15 hectare utilization of cooperation with the location of IPB business development. Total land owned by IPB for permaculture landscape planning as a commodity to meet dormitory student food requirement is 16.75 hectare.

Tabel 4. Area of potential permaculture planning.

No.	Experimental Field	Distance from dormitory's IPB for male (km)	Distance from dormitory's IPB for female (km)	Potential area for Planning (ha)
1.	Babakan	1.37	0.88	2.49
2.	Sawah Baru	1.35	0.85	2.07
3.	Experimental Garden of BDP FPIK	0.82	1.05	1.5
1.	Cikabayan	1.42	1.61	2.44
2.	Cikarawang A+B	0.38	0.63	3.68
3.	Agribussiness Development Station (ADS)	0.49	0.74	4.00
4.	Leuwikopo	1.11	1.57	3.19
5.	Sindang Barang-Pasir Kuda	8.76	8.25	1.02
6.	Pasirkuda	8.74	8.20	0.88
7.	Sabisa Farm (Sindang Barang)	5.85	5.45	0.05
8.	Stasiun Peternakan Fapet	1.20	1.60	10.00

The location of the experimental garden is located in Sindang Barang, Pasir Kuda, and Sabisa Farm were located in West Bogor District. Sindang Barang-Pasir Kuda is about 8.76 km to reach dormitory's IPB for male and 8.25 km to reach the dormitory's IPB for female. Then, Sabisa Farm is 5.85 km to reach dormitory's IPB for male and 5.45 km to reach dormitory's IPB for female (Table 4). By using a vehicle, the distance can be reached with 45 minutes-60 minutes journey in the busy road conditions. The location of experimenal garden most closest is Cikarawang A + B which is just 0.38 km to reach the dormitory's IPB for male and 0.63 km to reach the dormitory's IPB for female. By using a vehicle, the distance can be reached with 15-20 minutes journey in the crowded condition streets (Table 4).

3.3. Food sufficiency students of IPB dormitory

Determination of land for planning is obtained from the meal menu data which is converted into several agricultural commodities. Determination size of the land in the design will vary for each commodity. This is the different-potential production of each commodity [1]. Based on the character of IPB Experimental Garden including lowland and some highland commodities are eliminated. Thus, agricultural commodities that can not be incorporated into the planning of the permaculture landscape are carrot, potato, and cabbage commodities (Table 5).

The food needs from vegetable resources is divided into 14 agricultural commodities i.e. rice, soybeans, corn, papaya, kale, chilli, spinach, beans, tomatoes, leek, celery, potatoes, carrots and cabbage. The total food needs from vegetable resources per year for students of IPB dormitory is 775.08 tons of rice, 189 tons of soy beans, 162 tons of corn, 126 tons of papaya, 126 tons of kale, 1.26 tons of chilli, 126 tons of spinach, 94.5 tons of beans, 126 tons tomatoes, and celery 18.9 tons. Animal food resources for students of IPB dormitory are 126 liters of cattle, chicken eggs 78.75 tons, carp 350 tons, and 210 tons of chicken meat. The location to supply the food needs of students of IPB dormitory is divided into 3 points supply. The first point is obtained from the planned permaculture model planning at 8 IPB Experimental Gardens.

Table 5. Determination of commodity type and area of land in design.

No.	Type of Food	Needs for serving	Unit	Conversion	Food needs per day (3 500 people)[5] (kg)	The Area of Land needs (m^2/cycle)[6]
1	Rice[3]	400	Gram	Paddy	2 153	583 000
2	Tempe[4]	25	Gram	Soy	175	320 800
3	Tofu[4]	25	Gram	Soy	350	
4	Corn	150	Gram	Corn	450	96 250
5	Papaya	100	Gram	Papaya	350	64 200
6	Chicken meat	60	Gram	Chicken meat	210	560
7	Kale	100	Gram	Kale	350	14 000
8	Chili	1	Gram	Chili	3.5	420
9	Spinach	100	Gram	Spinach	350	21 000
10	Beans	75	Gram	Beans	262.5	47 800
11	Tomatoes	100	Gram	Tomatoes	350	39 700
12	Chicken eggs	1	Buah	Chicken eggs	3 500[1])	280
13	Fish	100	Gram	Fish	350	10
14	Potatoes	150	Gram	Potatoes	525	-
15	Carrots	150	Gram	Carrots	525	-
16	Cabbage	100	Gram	Cabbage	350	-
17	Leek	40	Gram	Leek	140	5.6
18	Celery	15	Gram	Celery	52.5	10.5
19	Cow milk	100	Ml	Cow milk	350[2])	350
				Total	6 386.5	1 188 0361

Source: (Priandono 2006) [1] with modification
Information : [1]) items
[2]) liter
[3]) rendement of milled rice = 65%
[4]) 1 kg tempe = 2 kg soy
[5]) Total food needs for 3 500 students / day
[6]) this land area became a reference for permaculture planning

The second supply point is located at the IPB business development location consisting of 4 locations, namely ADS Cikarawang, Sabisa Farm, FAPET IPB Field Laboratory, and BDP FPIK Experimental Pond. The food supply in the third point, obtained from village surrounding IPB campus, which is Cikarawang Village (Table 4). The amount of supply that can be fulfilled from the permaculture model planning is 41 tons of rice, 31 tons of soybeans, 171 tons of corn, 116 tons of papaya, 99 tons of kale, 17 tons of chili, 41 tons of spinach, 82 tons of beans, 76 tons of tomatoes, 70 tons of onion, and 27 tons of celery. The total supply from permaculture model is 771 tons which is divided into 14 vegetable food commodities. This means that there are only 33.53% that can be supplied from model of permaculture. The result is still very less, considering that the land used is only 13% of the total assets owned by IPB experimental garden (Figure 2).

The amount of supply obtained from the IPB business development site, which includes ADS Cikarawang, Sabisa Farm, Fapet Experimental Field, and BDP FPIK Production can supply 14.78 tonnes of foodstuff consisting of vegetable commodities of kale spinach, spinach, tomato and papaya. That is, there are only about 0.643% of all the food needs of students IPB dormitory.

The amount of supply obtained from the IPB business development location for animal food resources commodities are 360 liters of cow milk, 0.526 tons of chicken eggs, 30 tons of carp, and 18 tons of chicken meat. That is, there is 13.2% of supply from IPB business development location to supply animal food to IPB dormitory. The amount of supply for the needs of animal food from permaculture planning is 4.23 tons of carp, and 3 tons of chicken meat. That is, there are only 2% of the animal food supply of the permaculture landscape model (Figure 3).

Cikarawang village has a lot of mixed gardens land. Lack of land for the supply of some commodities from dry land, can be inadequate from some mixed gardens in the village Cikarawang. Overall, 61,195 ha of dryland is required for some commodities. There are already 16.73 ha of IPB land that can be used. That is, there is a shortfall of about 44.46 ha. The shortage of dry land can be selected from mixed gardens in Cikarawang Village (Figures 2 and 3).

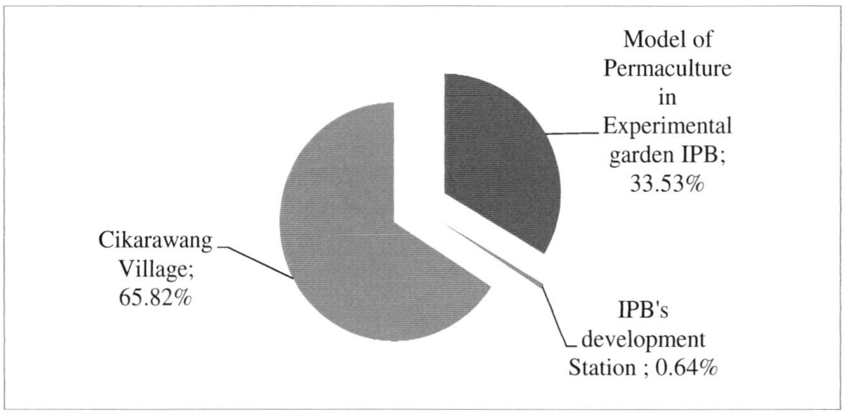

Figure 2. Percentage of fulfillment of vegetable food requirement for IPB dormitory students.

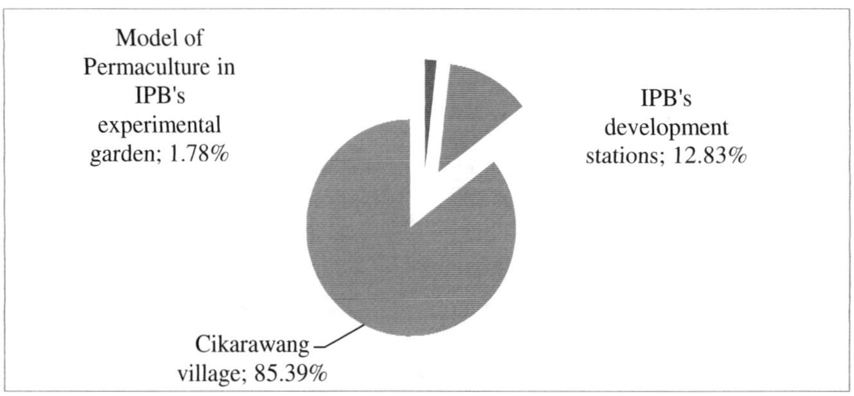

Figure 3. Percentage of fulfillment of animal food requirement for IPB's dormitory students.

3.4. Permaculture model

The main characteristic in this concept is to conduct highly efficient and intensive cultivation on a farm. For example, on the site it is planned to have at least 4 blocks in each production garden. This is related to efficient land use and the rise of plant diversity (figure 4). Permaculture concept of mutual integration between plants, animals, humans, and landscape into the product symbiotic system of one element can meet other needs. Thus, it can be used to design ecosystems for the needs of the human population without neglecting the natural environment [12]. The most influential factors in permaculture landscape planning are micro and macro climates, land formation, water, vegetation and animals, buildings and infrastructure, zoning, land use, soil following fertility and management, and aesthetics. Permaculture systems can be applied using minimal materials, energy, and labour. The way of recycling is the source can go back into the system as well as minimize pollution [13].

Permaculture model characteristics are the diversity and integration of several agricultural enterprises including agriculture, livestock and fisheries. Thus, can create synergies between one another. The synergies can be seen in the utilization of waste generated from each of these businesses can be complementary.

The main characteristic that emerges in this concept is to conduct highly efficient and intensive cultivation on a farm. For example, on the Cikarawang A Experiment Garden site, Pasirkuda Experimental Garden, and New Rice Field Experiment are planned to have at least 2 land shares in each experimental garden. Furthermore, the Cikabayan Experimental Garden, and Cikarawang B are planned to have 4 land divisions in each experimental garden. This is related to efficient land use at a time. The management approach should meet the global demands for food production and pay attention to ecosystems in the context of resources on land, water, climate change, and extensive ecosystem degradation [14].

8 experimental garden are choosen for permaculture model. There are Cikabayan, Leuwikopo, Babakan, Sawah Baru, Cikarawang A, Cikarawang B, Pasirkuda, and Sindang Barang. The division of the location consisting of four part those are padddy field, horticulture, palawija, cettle pen, and fish pond. The concept permaculture model ar evident from combination between agriculture and livestock. Then, planting patterns which determined is a plant diverse. In addition, the edge plants that had been selected based on native plants from each location. The next, the plants selected can serve as barier pest. The most important, election agriculture crops using overlapping cultivation system.

In this concept, several IPB Experimental Gardens will be developed into an agricultural production area with the selection of several agricultural commodities to meet food needs. This concept is used to maximize the potential of land with waste that is minimally disposed to the surrounding environment. Disadvantages of the entire land used from the garden IPB will be supplied from the Village Cikarawang as Lingkar Village Campus IPB.

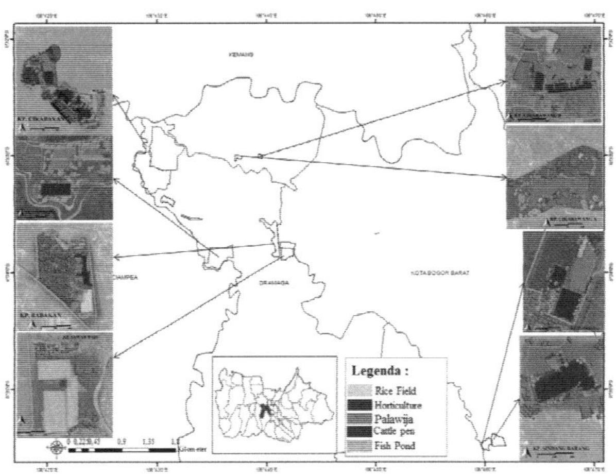

Figure 4. Model of permaculture.

4. Conclusion

The average calorie requirement of IPB male and female dormitory students based on basal metabolic rate is 1 313.35 kcal/day. Men have caloric needs based on their basic metabolic rate of 1 352.95 kcal/day, while female students 1 273.75 kcal/day. The average calorie requirement of male and female students is 2 360 kcal, with details of male students 2 410 kcal and female students 2 311 kcal. Nutrition status of IPB dormitory students has the highest normal status with 64% percentage, 24% status, 4% fat and 2% fat.

The sample of research for the first group point at 8 IPB Experimental Gardens, has 16.28 ha of land area. The percentage of land is obtained only from the total assets of IPB experimental garden of 13.40%. Next, the second point is at the business development location owned by IPB. Of the overall need for sources of animal and animal foods that can be supplied from the research sites point one and two get the percentage of fulfiling requirement of 33.53% for vegetable food sources and 13.25% for animal food sources. That is, the shortage is achieved from Cikarawang Village of 65.84% for vegetable food sources and 86.75% for animal food sources.

References

[1] Creswell J W 2009 Research design: qualitative, quantitative, and mixed method research London (GB): Sage Los Angles
[2] Denzin NK 1978 The research act: a theoretical introduction to sociological methods New York (US) McGraw-Hill Book Company
[3] Douglas AL 2016 Designing agricultural landscape for biodiversity-based ecosystem services basic and applied Ecology journal 1 (12) : 3-10
[4] Foley 2005 Global consequences of land use American Association for the advancement of science New York : (US) 1200 New York Avenue NW, Washington, DC
[5] Gold SM 1980 Recreation planning and design New York (US): Mc Graw-Hill Book Company
[6] Mollison B 1993 The permaculture book of ferment & human nutrition Tyalgum (AU): ATTRA Sustainable agriculture
[7] Morrow R 1993 Earth user's guide to permaculture East Roseville (AU): Kangaroo Press
[8] Pahl-Wostl C and Rabbit H M 2004 Processes of social learning in integrated resources management Journal of community and applied social psychology Osnabriick: 1 (14) 193–206
[9] Priandono A 2006 Perancangan kebun produksi berbasis LEISA untuk pemenuhan kebutuhan gizi sehat: kasus mahasiswa asrama TPB-IPB [skripsi] Bogor (ID): Institut Pertanian Bogor
[10] Raymond MC, Frantzeskaki N, Kabisch N, Pam B, Breil M, Razvan NM, Geneletti D, and Calfapietra C 2017 A framework for assessing and implementing the co-benefits of nature-based solutions in urban areas Environmental Science and Policy Clark University : (US)77 (2017) 15–24
[11] Setiana A 2013 Laporan tahunan University Farm Bogor (ID): University Farm
[12] Soekarno PWB 2011 Pemaparan rektor IPB pada acara rabuan bersama IPB tahun 2008-2010 Bogor (ID): IPB Press
[13] Studer R, Benjamins V, and Fensel D 1998 Knowledge engineering: principles and methods Data and knowledge engineering journal 25 (1) 161–197
[14] [WHO] World Health Organization 2007 Pengkategorian status gizi Genewa (CH): WHO technical report series

IOP Conf. Series: Earth and Environmental Science **179** (2018) 012012 doi:10.1088/1755-1315/179/1/012012

Perception of place attachment between cultural heritage in Yogyakarta City

P Sari[1*], A Munandar[2], And I S Fatimah[2]

[1] Graduate Student of Department of Landscape Architecture, Faculty of Agriculture, Bogor Agricultural University (IPB);

[2] Department of Landscape Architecture, Faculty of Agriculture, Bogor Agricultural University (IPB).

[*]E-mail: rrpawitrasari@gmail.com

Abstract. Yogyakarta City has a very special history, this city has a lot of cultural heritage, either tangible or intangible which are located on the philosophy axis. Along the philosophy axis can be found a lot of cultural heritage, especially inherited objects physically as important elements surrounded by other supporting elements. Tugu, Malioboro, Zero Kilometer, and South Square have public space as basic of taking the objects on the philosophy axis. The aims of this research were to examine the place attachment against the four objects with the perception of natives, newcomers, and tourists as respondents. This research was conducted using qualitative and quantitative methods. Qualitative methods were done by using descriptive approach to literature review and interviews, as well as field observation. The interviews towards informants, local communities, and tourists. Other qualitative methods by distributing of questionnaires involve the perception of local communities and tourists as much as 327 respondents. Quantitative methods by using a non-parametric test of different test with Mann-Whitney U for test whether there is a difference of perception between two samples. The results of this research, the four objects didn't yet an indication of the place attachment very agreement/high and there was no difference in perception between the natives and newcomers. The results of this research were expected to constitute an important aspect of the place attachment towards all the tangible cultural heritage assets for people against public space to appreciate and to love them as a part of the daily life. The hope of this research, Yogyakarta City could achieve as one of World Heritage City about the culture by UNESCO.

Keywords: Cultural heritage, perception, Yogya City

1. Introduction

Yogyakarta is one of the special city of Indonesia that has a "golden asset" in cultural meaning either physical or not physical. The existence of heritage objects is located strategically in the city center which became important elements on the philosophy axis of Yogyakarta City. The concept of philosophy is used as a principle that shows historical obedience and consistency in the spatial that contains various philosophy so as to reflect the privilege of Yogyakarta [1]. The spatial of Yogyakarta City based on the philosophy axis is a straight line between Panggung Krapyak-Kraton-Tugu, each of which stands buildings that have to mean the process of human life, from birth to death [2]. Until now, the philosophy axis has a cosmic power in maintaining Javanese culture. In this research, the selection of heritage objects refers to cultural areas in the corridor of the philosophy axis. The heritage objects also a mental map element of the city because take a long time in forming a city.

The crisis of public space in Yogyakarta City is very important to be developed considering that the existing city is currently losing open space. Especially for the local communities entitled to get urban space to spend time in enjoying a city. Tugu, Malioboro, South Square, and Dutch Colonial buildings at Zero Kilometer are heritage objects providing public open space for gather and social interaction. The heritage objects as well as a cultural tourism for the tourists. The importance of environmental psychology (spiritual) are also involved to know the assessment of local communities and tourists to the four objects of the study. However, in this research only test whether there is a difference of perception between the natives and newcomers. While the tourists are not tested with reasons of routine and destination differences. So, this research is important to do to improve the place attachment between cultural heritage for the quality of life in urban.

Perception is a response to external stimulus sense and activity targeted in a specific phenomenon clearly [3]. Sensory involvement requires some degree of attachment to the stimulus object and vision to make it easy to evoke an emotional response. In this paper, the researcher describes how substantial the local communities (public/society) and the tourist's assessment towards the heritage objects, place belongingness, and place identity based on the perception of the four objects. Various environments can reflect the cultural complexity that distinguishes the group from others [4]. The meaning of identity refers to the terms of place and means a set of the place feature that ensures the place has its uniqueness and continuity, reflecting the meaning of place identity [5]. Place identity comes from the meaning that people give a place, which depends on the level of user engagement and perception of the environment [6]. The identity of a city can be shown in historical places or places that have certain characteristics [7].

The density in urban make from people's lives getting tired and bored thus requiring public space that can be utilized for potential activities. Quasi-public spaces that make it possible to spend free time inexpensively outside the house and build social relationships locally, can hardly be overestimated in relation to this issue [8]. A city born of high historical value brings the culture both tangible and intangible until now. In the cultural aspect, it explains how someone will have a place attachment by identifying elements of a place that reflect a cosmological concept of a particular culture [9]. A satisfying place creates a harmony between the elements of physical, social, and cultural and it is affecting the identity of a place [10]. Reclus intended to achieve by transforming the evacuated historic urban center into a cultural and social core of the rapidly growing city [11]. Prominent features of a unique place, for example, architecture, historical monuments, a cultural community can be the attachment to a person on self-concept, this is also the process of calling "related a place that has a particularity" [12].

2. Methods

2.1. Methodology

This study was conducted in Yogyakarta City, which the philosophy axis is used as the primary for taking the object of this research, there are Tugu, Malioboro, Zero Kilometer, and South Square that has public space. The public spaces are often visited by the local communities and tourists as recreation objects or tourist attraction. This study was started from the fourth week of May 2017 until the fourth week of September 2017. This research uses qualitative and quantitative methods. Qualitative methods of collecting the literature reviews and interviews descriptively, as well as field observation, and distributing of questionnaires. The field survey included observation and interviews with resource persons (stakeholders) and distributed questionnaires of 327 respondents. Besides stakeholder, interviews were conducted with tourist and public/community leaders/figure about heritage objects, cultural, and tourism. The interviews method was used snowball sampling, digging information from sources until the information was saturated. Documentations were done with an approach to the historical documentary investigation (event note) and survey technique as the primary data source. Cross-check was done between the literatures with research location.

The questionnaire character is a combination of the closed and open questionnaire. The goal is to facilitate in selecting the answers that have been provided and can provide opinions/responses from the problem. Closed questionnaires use measurement interval scale that is a Likert model. A Likert scale used is the number of point 5 with comparable grounds, has a better index of reliability, validity, and strength. Questionnaires involve local comunities perceptions of 260 respondents and tourists perceptions of 67 respondents. From local communities, 170 were natives and 90 were newcomers (with a ratio of 65:35), while the tourist, 62 were domestic tourists and 5 were foreign tourists. Comparison of the number of respondents between natives and newcomers were obtained from the information of Central Bureau of Statistics then were calculated from the magnitude of the population in Yogyakarta City, especially the sub-districts passed on the philosophy axis. Size of sample data from tourists based on the number of tourist data in 2016 from Yogyakarta Tourism Office.

Questionnaires are data collection techniques to capture public perceptions towards questions related to the research by giving written questions to a select number of people and can be used as samples to represent the communities in the research area. In this research the distribution of questionnaires using stratified random sampling. How to determine the respondents based on Isaac and Michael's approach [13]. Although the sampling is random the calculation of the sample selection process is proportional stratified random sampling. Population sampling in stages through steps after obtained the number of respondents from each district than in each sub-district based on the number of natives and newcomers. Determination of the number of samples in each sub-district refers to the formula Singarimbun *et al* [14]. The local communities and tourists choose answers of the number 1, 2, 3, 4, and 5, where the higher the number shows the stronger/better weight. The questionnaires collected then enter the numbers of the Likert scale, i.e. 1, 2, 3, 4, and 5 are tabulated into the table. Assessment of the perception of local communities and tourists were conducted in this research with qualitative methods, which then analyzed descriptively by SPSS version 17 software.

The results were obtained from SPSS data processing, then given a rating based on the average score [15]. Hypothesis testing in this research is determined based on prerequisite test results are the test of data normality using Kolmogorov-Smirnov method and data homogeneity using Levene Statistic method. The results of the analysis using Kolmogorov-Smirnov method for each group (natives and newcomers) showed data not normally distributed. A data is said to be normally distributed if the value of Asymp Sig (2-tailed) calculations using Kolmogorov-Smirnov is greater than $1/2\alpha$. After the test of data normality, then the test of data homogeneity can be concluded that the data comes from the same variance (homogeneous). A data is said to be homogeneous if the value of significance is greater than alpha (> 0.05). The level of significance/level of trust used in this research is 95% (0.05), which is generally conducted in social research. Therefore the data comes from the same variance (homogeneous), then it is done by using non-parametric test of a different test "Mann-Whitney U", which is included in quantitative methods. The Mann-Whitney U test is used for two independent samples, there are natives and newcomers. Data analysis using Mann-Whitney U test is the final step in the quantitative method.

2.2. The Study Area

Cultural heritage is endless to learn because it is the wealth of priceless and irreplaceable for the future a region/city. The existence of the philosophy axis has a harmonious meaning in the inheritance both tangible and intangible. In spatial, the philosophy axis is manifested in the form of a path that connects the heritage objects physically symbolizing the journey of life. The philosophy axis has a position in the heart of the Yogyakarta City with a lot of historical objects (physical) as cultural heritage. In addition, the taking of the four heritage objects based on a minimum age of 50 years, a special meaning in history and culture, an important value to form the image of privilege, a value of authenticity, a value of expediency, being a historical and cultural object and related to tourism, and provide the public space.

How precious the public space which is nearby to the relics of heritage objects have the attachment in Yogyakarta City to improve spiritual welfare of the urban. Emphasis on local contexts reassert subjective roles sense of place, and place experience as a legitimate form of knowledge and meaning related to landscape governance and provides the foundation for advancing approach a place based on environmental planning [16]. Individuals need to express the sense of belonging to a collective entity or place, and an individual's identity, which can be achieved by physical separation or uniqueness, and a sense of belonging to a particular area [17]. It is also related to the importance of place symbols as emotional repositories and relationships that give meaning and purpose to life and reflect a sense of belonging [18].

Currently, Tugu has a physical building that is not the same as the beginning was built by Sri Sultan Hamengku Buwana I. Tugu was rebuilt on October 3, 1889, its means Tugu has an age 128 years. Based on the field observation and literature, Tugu has a special meaning in history and culture, such as an event of a festival (e.g. birthday celebration of Yogyakarta City), performing arts (e.g. sculpture), and traditional culinary arts (e.g. 'angkringan sego kucing'). Until now, Tugu has a value of authenticity and expediency because the concept and form of Tugu have different from other. Tugu has an identity of Yogyakarta City formed the image of privilege because its existence which has hundreds of years standing right in the heart of the city (Figure 1). The public space with the small size made Tugu as tourism objects for tourists and recreation objects for local communities.

Figure 1. (left) Tugu Yogyakarta monuments; (middle) surrounding of Tugu Yogyakarta; (right) small public space

Malioboro Street development explain that in 1765-1945 used for ceremonial, politic, and economic, then in 1945 after the Independence of Indonesia changes into politics, culture, and economic, then in 1950's until 1980's changes into economic and tourism, and in 1980's until this day changes into economic and tourism center of Yogyakarta [19]. Its means that Malioboro has an age more than 250 years despite changes in function and style of the building, Malioboro still has a special meaning in history and culture. Based on the field observation and interviews with the tourists, Malioboro is heritage objects as an image of traditional shopping especially souvenirs, such as T-shirts, antique goods, as well as typical Yogyakarta culinary such as 'Sate Kere', 'Pecel', 'Bakmi', etc. Malioboro provides the public space in the form of the corridor to enjoy for local communities and tourists because it is not found in other cities (Figure 2). With a combination of building style and function area, Malioboro also provides economic value for people as a souvenir center.

Figure 2. (left) Malioboro Street; (right) Malioboro Pedestrian

In Zero Kilometer many found of Dutch Colonial buildings among other of them is the Indonesian Post Office, BNI 46, Central Bank, Vredeburg Fortress, Presidential Palace, and others. For example, BNI 46 (in the past Nillmij office) was built in 1921 by Ghijsels with Colonial Modern Architectural which has the symmetric pattern of the building make a neat impression [20]. Other example based on the same source, Indonesia Post Office (in the past Post, Telegraaf en Telefoon Office) designed by Burgelijke Openbare Werken (now as Public Works Agency) in 1910 and was built 1912 with Transitional Architectural which has a construction system 'a bear wall with giddy gevels'. More than 100 years, buildings in Zero Kilometer has an old age and special meaning in history. In this the district can be said has an important value to form the image as Dutch Colonial Architectural because designed and built by Europe architect especially Dutch. Zero Kilometer is located right at the node has medium public space (Figure 3).

Figure 3. (left) BNI 46 Bank building; (middle) Indonesian Post Office and medium public space; (right) Central Bank

In recent years, South Square function as a public space for people to gather and social interaction of urban heritage (Figure 4). Squares to be a successful public space, because must be liveable and sociable, and able to provide a setting that can offer a variety of experience, sensory and activities for people [21, 22, 23]. In the past, a square has a central function where the whole the core elements forming the city will meet, both symbolically and real [24]. Therefore, South Square is open space has hundreds of years with a special meaning in history and culture as a model of authenticity from Javanese culture has from the image of privilege. The authenticity of South Square never undergoes physical changes from year to year. The South Square has large public spaces and equipped with playground area. It's just that functionality has changed for the local communities and tourists as the historical and cultural object and related to recreation and tourism. It is also an effort to make South Square has a more optimum benefit can be felt and enjoyed by the local communities and tourists.

Figure 4. (left) two Banyan Trees; (middle) large public space; (right) playground area

3. Results

3.1. The Respondents

Based on the results of 327 respondents, obtained data that natives 54.1% were male and 45.9% were female while that newcomers 52.2% were male and 47.8% were female. Different with tourist that

64.2% were female. Based on the field observation, also found that female tourists were visiting heritage objects more often than male. Many possibilities that female tourists prefer tourism in the city who spend more time at the place of tourism than travel, for example, nature tourism. Natives age range 20-30 years as much 40.6% and age range 31-40 years as much 29.4%. While newcomers and tourists, the majority of the respondents with age range 20-30 years as 67.8% are newcomers and 62.7% are tourists. The last education from native respondents as 56.5% and new comers respondents as 68.9% are senior high school. The number of respondents who have a background undergraduate education was 76.1%. More than 50%, domestic tourists who travel to the Yogyakarta City is not the first but has 5 to 10 times. Meanwhile, 5 foreign tourists from India, Japan, Singapore, and Spain.

3.2. *Discussion*

The results of questionnaires distribution were found the highest average value on the variable of place belongingness occurred at Malioboro (see Table 1). The tourists gave the highest average value is 3.73 or agree at Malioboro, while the lowest average value is 2.81 or doubt given tourists at Zero Kilometer. The most obvious low value at Zero Kilometer object has an average value of ≤ 3.00, while the other objects have an average value of ≥ 3.00. The highest score 4.04 or agree by tourists, that Malioboro not only as a place/historical and cultural attractions but also has a meaningful symbol. Symbols as a relic of objects that have a traditional shopping image (especially souvenirs) and become an economic wheel that can be enjoyed by all walks of tourists and community. For tourists themselves, Malioboro is a favorite destination object of the historical and cultural tourism which should not be missed. The presence of Malioboro is the pulse in the urban and tourism for helped turn on the city at night. Unfortunate, Malioboro still has not obtained a very optimal value as a heritage object that has a public space on its pedestrian. The feel of calm and relaxed, and togetherness (familiarity) did not give a big influence, notably for the natives.

Table 1. Variable of Place Belongingness based on Perception Score of Natives, Newcomers, and Tourists

The Perception Score of Natives (NT), Newcomers (NC), and Tourists (TR)[a]												
Variable of Place Belongingness	Tugu (TG)			Malioboro (MB)			Zero Kilometer (ZK)			South Square (SS)		
	NT	NC	TR	NT	NC	TR	NT	NC	TR	NT	NC	TR
I have belongingness at TG/MB/ZK/SS.	3.75	3.68	3.58	3.68	3.67	3.79	3.49	3.33	2.85	3.67	3.53	3.34
TG/MB/ZK/SS gives meaningful symbols.	3.87	3.94	3.90	3.62	3.74	4.04	3.43	3.26	2.81	3.58	3.34	3.66
I feel "something" different at TG/MB/ZK/SS.	3.42	3.41	3.16	3.45	3.53	3.45	3.23	3.19	2.75	3.45	3.29	3.30
I feel calm and relaxed at TG/MB/ZK/SS.	3.11	3.18	3.06	3.34	3.42	3.63	3.11	3.30	2.78	3.41	3.47	3.39
I feel to have intimacy, togetherness when I'm at TG/MB/ZK/SS.	3.27	3.30	3.16	3.37	3.39	3.73	3.25	3.28	2.90	3.41	3.46	3.37
Mean value	3.49	3.50	3.37	3.49	3.55	3.73	3.30	3.27	2.81	3.49	3.42	3.41

[a]Mean value;
1.00 – 1.80: Strongly Disagree, 1.80 – 2.60: Disagree, 2.60 – 3.40: Doubt, 3.40 – 4.20: Agree, and 4.20 – 5.00: Strongly Agree. [19]

Other high scores on the questions, Tugu gives meaningful symbols given by the newcomers with a score of 3.94 or agree and the natives with a score of 3.87 or agree. That is, Tugu for local communities provide a symbol that has many meanings in history until this day. Tugu Yogyakarta cannot be replaced with other symbols despite the change of shape caused by the earthquake before. This indicates that local communities have place belongingness Tugu above the average standard. Also, tourists assessing the Tugu with a score of 3.90 or agree with the same of questions. The lowest score or doubt has given at Zero Kilometer, especially the tourists. Two questions, feel "something" different with a score of 2.75 and feel calm and relaxed with a score of 2.78 at Zero Kilometer. There are still has many places/other objects that are considered calm and relaxed for tourists. This is considering when in the daytime is very uncomfortable because of lack of vegetation. In addition, for tourists at Zero Kilometer as well as in other big cities that have buildings of Dutch Colonial Architecture Style. At the same question, also found a low score or doubt given the natives in two of research objects. The question is feeling calm and relaxed at Tugu and Zero Kilometer with a score of 3.11 or doubt. That is, Tugu and Zero Kilometer is not a place/object that can provide tranquility and a sense of relaxed to spend time on vacation for the natives. Found that Tugu and Zero Kilometer does not have adequate facilities as public space to enjoy the place/object, such as the lack of seats for lounging, garbage dump, and toilets.

The results of questionnaires distribution were found the highest average value with the variable of place identity occurred at Malioboro (see Table 2). The value is given by travelers with an average score of 3.80 or agrees, while the lowest average value is 2.78 or doubt given tourists at Zero Kilometer. The highest score of 3.97 or agree given the tourists with a question, going to Malioboro is one of the most satisfying things. The meaning is satisfying because Malioboro is a shopping center (especially souvenirs) that cannot be missed for tourists. Shopping centers at Malioboro are located hallway or terrace along the Malioboro, gangways of Malioboro and Beringharjo Market is known cheap because it can be done bargain. Tourists are very happy and always want to Malioboro when traveling to Yogyakarta. For tourists, a visit to Malioboro is a special satisfaction and a favorite destination. Another high score of 3.90 or agree is contained in two questions at Malioboro. The question, Malioboro is a favorite place to visit and very meaningful. That is, Malioboro is a favorite destination of historical and cultural tourism that should not be missed for tourists. When tourists did not visit Malioboro will have a different impression. In this year, Yogyakarta Government has been finished revitalization of Malioboro pedestrian more feasible. Tourists can enjoy Malioboro on foot, sitting around looking at heritage building, and a different atmosphere from Malioboro. For tourists, Malioboro has a meaning is different from another city.

Table 2. Variable of Place Identity based on Perception Score of Natives, Newcomers, and Tourists

The Perception Score of Natives (NT), Newcomers (NC), and Tourists (TR)[1]												
Variable of Place Identity	Tugu (TG)			Malioboro (MB)			Zero Kilometer (ZK)			South Square (SS)		
	NT	NC	TR	NT	NC	TR	NT	NC	TR	NT	NC	TR
TG/MB/ZK/SS very meaningful.	3.80	3.70	3.67	3.66	3.78	3.90	3.36	3.44	2.87	3.64	3.56	3.63
Going to TG/MB/ZK/SS is one of the most satisfying things.	3.32	3.34	3.33	3.48	3.77	3.97	3.16	3.31	2.88	3.89	3.58	3.40
TG/MB/ZK/SS is the best place to do the activity.	3.05	3.09	2.96	3.44	3.53	3.69	3.17	3.29	2.88	3.48	3.57	3.66
TG/MB/ZK/SS is a physical object that reflects my lifestyle.	2.96	2.79	2.70	3.12	2.97	3.55	2.88	2.94	2.51	3.14	3.07	2.70
TG/MB/ZK/SS is a favorite place to visit.	3.28	3.29	3.12	3.58	3.67	3.90	3.22	3.41	2.78	3.43	3.37	3.51
Mean value	3.28	3.24	3.16	3.46	3.54	3.80	3.16	3.28	2.78	3.41	3.43	3.38

[1]Mean value;
1.00 – 1.80: Strongly Disagree, 1.80 – 2.60: Disagree, 2.60 – 3.40: Doubt, 3.40 – 4.20: Agree, and 4.20 – 5.00: Strongly Agree. [19]

The highest score of 3.89 or agree is given by natives with a question, going to the South Square is one of the most satisfying things. That is, South Square can provide satisfaction to fill the time either ordinary days or holidays. Visiting the South Square is enough to spend money on parking and then can enjoy the public space in urban (the context of heritage). Squares gain more meanings that make activities a significant place with strong social and cultural values rather than merely an open space [25]. Another high score 3.80 or agrees is given the natives with the question, Tugu very meaningful. The existence of Tugu has created the image of the city on its identity and become its own pride for the natives in Yogyakarta. The lowest score or doubt is given by tourists with a question, Zero Kilometer is a physical object that reflects lifestyle with a score of 2.51 or disagrees. A deeper search of the lifestyle can be conveyed through "a different message" or has features that contain ethnic or traditional elements. Dutch Colonial Architecture Style is still found in many big cities in Indonesia. "Singularity or dissimilarity" can be said as the place's identity in the region. The lowest score or doubt is given to the newcomers with a question, Tugu and Zero Kilometer are a physical object that reflects lifestyle with a score of 2.79 is given at Tugu and 2.88 is given at Zero Kilometer. Related to this, the newcomers still have not yet fully possessed a sense of identity for his city.

The used nonparametric test of the Mann-Whitney U test was the final process in this results and discussion, which previously obtained statistical data. In this research using non-parametric test of the Mann-Whitney U test for two independent samples. Based on the results, the variable of place identity obtained Z value of -0,632 with a significance value of 0.527, then the value of significance greater than alpha (0.527 > 0.05). While based on the results, the variable of place belongingness obtained Z value of -0.338 with a significance value of 0.735, then the value of significance greater than alpha (0.735 > 0.05). So, there was no difference between the perception of natives and newcomers to the variable of place identity and place belongingness (see Table 3). Understanding of the place on the physical, psychological, and social dimensions was inseparable from the culture of human experience. It was developed as a result of the meaning and significance of the place for users. The diversity of user backgrounds, such as gender, marital status, age, last educational, income levels, and others are not a benchmark in the environmental psychology dimension.

Table 3. The Different Test Results of the *Mann-Whitney U*

The Different Test Results of the Mann-Whitney U[b]			
Indicator of question	*Mann-Whitney U*	Z	*Asymp. Sig. (2-tailed)*
Place Identity	7290.500	-.632	.527
Place Belongingness	7457.500	-.338	.735
[b]Group Variable: Natives dan Newcomers			

4. Conclusion

It is very interesting for tourists, Malioboro not found in other cities. Malioboro not only as a historical places/objects and cultural attractions, more than that, has a meaningful symbol. The presence public space of Malioboro is the heart of the city for helped turn on the city and created tourist attractions. In spatial, Malioboro has positioned itself as the image of shopping center especially souvenir, in the form of paths, districts, and landmarks. Malioboro as a historical and cultural tourism offers a free of heritage object that can be enjoyed by all local communities and tourists. Contrast with Zero Kilometer has a weakness in showing its identity. This can be found in the other cities that are former Dutch Colonial that leaves a same typical building heritage. Thus, create the image of the place at Zero Kilometer which has been formed is considered weak. It is unfortunate, Tugu, Malioboro, Zero Kilometer, and South Square didn't have yet an indication of the place identity and belongingness very optimal/strong (Figure 5). The perception between the natives and newcomers was no difference, while the tourist perception has more strength than the natives and newcomers. The spiritual values must be further optimized especially for those born and raised in the cities with many inheritances. The tangible cultural heritage assets should be a part of the daily life, especially for the natives and newcomers.

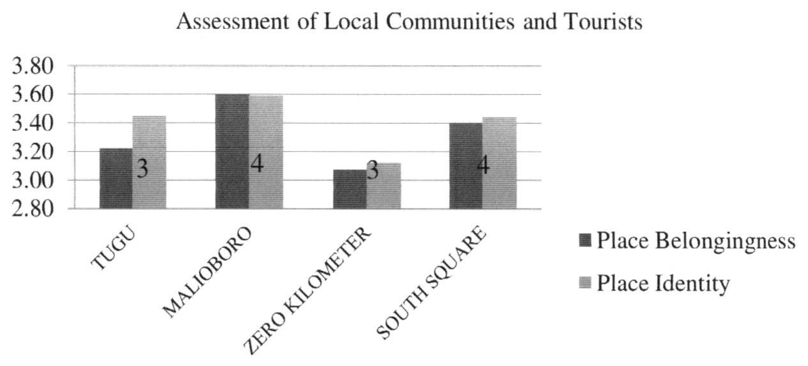

Ranking the Average Value of Public Space with Indicator of Place Attachment:
Malioboro > South Square > Tugu > Zero Kilometer

Figure 5. Ranking the Average Value from Assessment of Local Communities and Tourists

References

[1] [DPRD DIY] Dewan Perwakilan Rakyat Daerah 2017 *Renaisans dalam Perdais Tata Ruang - DPRD DIY* (Yogyakarta: DPRD DIY)
[2] Dinas Kebudayaan Daerah Istimewa Yogyakarta 2015 *Yogyakarta City of Philosophy* (Yogyakarta: Dinas Kebudayaan)
[3] Tuan YF 1974 *Topophilia: A Study of Environmental Perception, Attitudes, and Values* (New Jersey: Prentice-Hall)
[4] Harun NZ, Mansor M and Said I 2013 The Experience of Diversity in Open Spaces of two Historical Towns in Malaysia *Procedia Social and Behavioral Sciences* 85 pp 582-591
[5] Lewicka M 2008 Place Attachment, Place Identity, and Place Memory: Restoring the Forgotten City Past *Journal of Environmental Psychology* 28 pp 209-231
[6] Ujang N 2012 Place Attachment and Continuity of Urban Place Identity *Procedia Social and Behaviour Sciences* 49 pp 156-167
[7] Wahyuni E, Munandar A 2010 Studi Elemen Mental Map Lanskap Kota Pekanbaru *Jurnal Lanskap Indonesia* 2(1) pp 53-58
[8] Mantey D 2017 The 'publicness' of suburban gathering places: The example of Podkowa Leśna (Warsaw urban region, Poland) *Journal of Cities* 60 pp 1-12
[9] Low SM 1992 *Symbolic ties that bind. Place attachment in the plaza* (New York and London: Plenum Pr)
[10] Ginting N, Wahid J 2015 Exploring Identity's Aspect of Continuity of Urban Heritage Tourism *Procedia Social and Behavioral Sciences* 202 pp 234-241
[11] Reclus E 1985 The Evolution of Cities Contemporary *Review* 67 pp 246-264
[12] Twigger-Ross CL, Uzzell DL 1996 Place and identity process Journal *of Environmental Psychology* 16 pp 205-220
[13] Isaac S Michael WB 1981 *Handbook in Research and Evaluation* (California: Edits Publisher)
[14] Singarimbun Masri and Sofian Effendi 1987 *Metode Penelitian Survei* (Jakarta: LP3ES)
[15] Umar H 2005 *Riset Sumber Daya Manusia dalam Organisasi* (Jakarta: Gramedia)
[16] Stewart WP Williams DR Kruger LE 2013 *Place-based Conservation: Perspectives from the Social Sciences* (Dordrecht: Springer)
[17] Carmona M Heath T Oc T and Tiesdell S 2003 *Public Spaces-Urban Places: The Dimensions of Urban Design* (London: Architecture Pr)
[18] Proshansky HM, Fabian AK and Kaminoff R 1983 Place Identity: Physical Worls Socialisation

of the Self *Journal of Environmental Psychology* 3 pp 57-83

[19] Septirina SN, Takeo O and Satoru K 2016 Conservation of historical architecture in Malioboro street, Yogyakarta city, Indonesia *Procedia Social and Behavioral Sciences* 225 pp 259-269

[20] Handinoto 2010 *Arsitektur dan Kota-kota di Jawa pada Masa Kolonial* (Yogyakarta: Graha Ilmu)

[21] Balsas CJL 2007 City Centre Revitalization in Portugal: A Study of Lisbon and Porto *Journal of Urban Design* 12 pp 231-259

[22] Montgomery J 1998 Making a city: Urbanity, vitality and urban design *Journal of Urban Design* 3 pp 93-116

[23] Whyte WH 1980 *The Social Life of Small Urban Spaces* (Washington: Conservation Foundation)

[24] Koesmartadi 1995 *Perubahan Alun-alun dan sekitarnya di Kota Pantai Utara Jawa Tengah, ditinjau melalui Tapak Kasus: Kota Brebes, Tegal, Pemalang* Thesis (Yogyakarta: Universitas Gadjah Mada, Magister Design Kawasan Binaan)

[25] Zakariya K, Harun NZ and Mansor M 2015 Place Meaning of the Historic Square as Tourism Attraction and Community Leisure Space *Procedia Social and Behavioral Sciences* 202 pp 477-486

IOP Conf. Series: Earth and Environmental Science **179** (2018) 012013 doi:10.1088/1755-1315/179/1/012013

Plants production of agroforestry system in Ciliwung riparian landscape, Bogor Municipality

Y B Prastiyo[1,*], R L Kaswanto[2], and H S Arifin[2]

[1] Graduate Student of Landscape Architecture Department, the Graduate School of Bogor Agricultural University (IPB), Jl. Meranti, Darmaga Campus, Bogor 16680,Indonesia.
[2] Landscape Management Laboratory, Landscape Architecture Department, Faculty of Agriculture, Bogor Agricultural University (IPB), Jl. Meranti, Darmaga Campus, Bogor 16680, Indonesia.

[*]Email: yuliusprastyo93@gmail.com

Abstract. Settlement occupation in Ciliwung riparian zone along Bogor Municipality has reduced vegetation formations of agroforestry, such as *kebun campuran* (mixed gardens), *talun* (forest gardens), and *pekarangan* (home gardens). These processes have reduced plants production, one of riparian landscape services. The purpose of this paper is to know the plant production capacity of agroforestry land use, as a basis for arranging the management of agroforestry landscape in Ciliwung riparian, Bogor Municipality. The study used 14 sample plots (4 *taluns*, 5 mixed gardens, 5 *pekarangans*) with purposive sampling method, to plants production analysis of agroforestry land use. Furthermore, the calculation of Land Equivalent Ratio (LER) is conducted to know land productivity. The agroforestry system in Ciliwung riparian, Bogor Municipality has highly potential plant production. The *talun* has arrange area of 38.77 ha and the average plant production 49.20 Mg/ha, total plant production potency of 1907.48 Mg. The mixed garden has arrange area of 9.44 ha and the average plant production 65.41 Mg/ha, total plant production potency of 617.47 Mg. Meanwhile, the *pekarangan* has arrange area of 17.53 ha and the average plant production 48.77 Mg/ha, total potential plant production is 854.94 Mg. The *talun* and mixed garden have high productivity with LER of 1.40 and 1.81, respectively. Where as, the *pekarangan* has low productivity with LER of 0.96. Agroforestry practices are able to provide better plants production, so that to ensure the stability and sustainability of landowners' income.

Keyword: *kebun campuran*, landscape services, *pekarangan*, productivity, *talun*

1. Introduction

Settlement occupation in riparian zones has become an important issue in urban spatial implementation in Indonesia. This process has reduced vegetation formations and ecological functions of riparian landscapes [1]. Vegetation formation that was often discovered in riparian landscapes, such as *kebun campuran* (mixed gardens), *talun* (forest gardens), and *pekarangan* (home gardens). These types of vegetation formations are forms of agroforestry system practices in riparian landscapes.

Agroforestry is an integrated land use system that combines trees, perennial plant with agricultural crops, annual plant and or livestock to produce integrated products [2]. The agroforestry landscape with diversity of components can provide some landscape services, one of which is increases plant production [3]. The plants production capacity of agroforestry landscape is high, as in pekarangan that provides an additional source of income from crop production of 12.9% [4]. However, reduced area of agroforestry system in riparian landscapes due to settlement occupation, has decreased plant production. One of the rivers that experienced the process is Ciliwung River, especially in Bogor Municipality which is one of the cities with a fairly rapid development in Indonesia [5].

The built up area on Ciliwung riparian in Bogor Municipality, increasing by 14.08 ha (0.8%) in seven years (2006-2013) and by 2016, the built up area on middle part of Ciliwung riparian, including in Bogor Municipality is already reached 312.41 ha (37.11%) [6, 7]. This indicates that there is reduction area of agroforestry system which impact to decrease of plant production of Ciliwung riparian in Bogor Municipality. Based on these problems, this research aim to determine the plant production capacity and potency of agroforestry system, as a basis to arrange management of agroforestry landscape on Ciliwung riparian in Bogor Municipality.

2. Method

2.1. Study area

The research was conducted in Ciliwung riparian that crossed Bogor Municipality area. The Ciliwung riparian under study was riparian of main river along 14.43 km, from Katulampa, East Bogor until Kedung Halang, North Bogor (Figure 1). The riparian widths analyzed were determined based on the results of ideal riparian width analysis of Ciliwung River in Municipality area, which is 50 m [7].

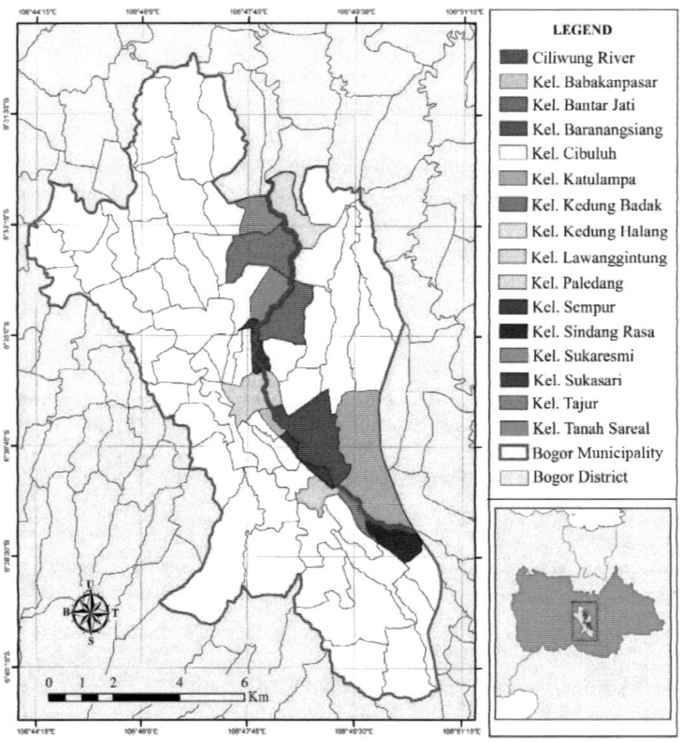

Figure 1. Study area of Ciliwung riparian in Bogor Municipality.

2.2. Sampling technique

The sample plot was made with size of 20 x 20 m, to calculate the structure and composition of riparian vegetation. Determination of plot sample location was done by purposive sampling method which was adjusted to river length, where one sample plot is assumed to represent 1 km of river length (not everyone km one plot). The length of Ciliwung River in Bogor Municipality reaches 14.43 km, so the number of plot sample used there are 14 plots, with the location of alternating (zig-zag) on right and left side of the river. Furthermore, the 14 plots samples were classified based on land use type of agroforestry system, ie 4 plots of *talun*, 5 plots of mixed garden, and 5 plots of *pekarangan* (Figure 2).

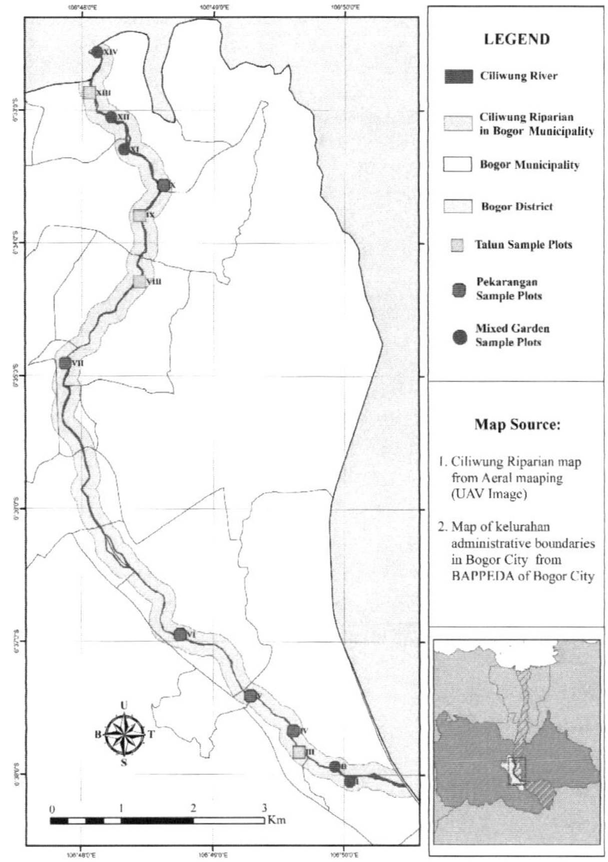

Figure 2. Layout of sample plots of vegetation analysis on Ciliwung riparian in Bogor Municipality.

2.3. Anaysis of agroforestry landuse

Spatial data used is map of land use Ciliwung riparian in Bogor Municipality from classification result using Unmanned Aerial Vehicle (UAV) image with resolution 15 cm/pixel. The agroforestry land useanalyzed in this research is *talun*, mixed garden, and *pekarangan*. The land use classification process was performed using ArcMap 10.4 software with the coordinate system used in geometric correction is UTM with WGS 1984 datum 48S zone. Visual observation of land use is done by digitizing the image of UAV on the screen of monitor (on-screen digitizing) into polygon form based on classification of land use class.

2.4. Analysis of plant production

Plants production calculated only for perennial and annual plant that contained in sample plots. Plants production calculated is harvested plants, including fruits, seeds, vegetables, and tubers. Plant production per hectare is calculated using the following formula:

$$\text{Production (Mg/ha)} = (P/1000)\ K \qquad (1)$$

Where P is production of a plant species (kg), and K is the number of individuals of a plant species in the sample plot area (ind./ha).

Furthermore, the calculation of land productivity based on Land Equivalence Ratio (LER) value. LER is the ratio of plant yield between the agroforestry planted system to the monoculture planted system at same level of management. Plant production data with monoculture system is obtained based on horticultural plants production from Statistics Center Bereau (BPS) of Bogor Municipality, National BPS, and some research on horticultural plant production. LER is one of the ways used to calculate the land productivity from two or more plants grown with agroforestry planted system [8]. Here is the LER equation:

$$\mathrm{LER}=\sum_{i:1,2,3..}^{n}\frac{\mathbf{hi}}{\mathbf{Hi}} \tag{2}$$

Where *hi* is plant production in agroforestry system, *Hi* = plant production in monoculture system, and *i = 1, 2, 3, ...n* is plant species.

3. Results and Disscusion

3.1. Situational analysis of study sites

Ciliwung River flows from Bogor District, Bogor Municipality, Depok Municipality, to Jakarta. Ciliwung Watershed (DAS) is divided into three sections based on topographic characteristics, ie upper-stream, middle-stream, and down-stream. Ciliwung River that crosses Bogor Municipality area is included in the middle-stream Ciliwung watershed. The length of Ciliwung River that crosses Bogor Municipality reaches 14.43 km which belongs to administrative area of 15 villages. The area of Ciliwung riparian in Bogor Municipality when using standard riparian width of 50 m, reaching 148.97 ha. Topography of this area is bumpy and hilly with variations of slope 2-15% and has altitude of 120-350 m above sea level (asl). The soil type found in Ciliwung riparian in Bogor Municipality is red latosol association and rainfall ranged of 3400-4400 mm/year [9].

3.2. Anaysis of agroforestry landuse

Land use of Ciliwung riparian in Bogor Municipality was divided into 9 classes, namely (1) *talun* (forest garden), (2) mixed garden, (3) *pekarangan* (home garden), (4) organic housing, (5) planned housing, (6) bare land, (7) pond, (8) road, and (10) tributary (canal). Agroforestry land use analyzed in this research is *talun*, *kebun campuran*, and *pekarangan* (Figure 3). Based on the results of size area analysis, Ciliwung riparian in Bogor Municipality has been dominated by built area (organic housing and planned housing) compared to agroforestry land use (Table 1).

Table 1. Land use analysis of Ciliwung riparian in Bogor Municipality.

No	Land use class	Number of patch	Area		
			ha	%	Average (ha)
1	*Talun* (forest garden)	93.00	38.77	26.03	0.42
2	*Kebun Campuran* (mixed garden)	52.00	9.44	6.34	0.18
3	*Pekarangan* (home garden)	1464.00	17.53	11.77	0.01
4	Organic housing	467.00	38.71	25.99	0.08
5	Planned housing	259.00	20.67	13.88	0.08
6	Bare land	217.00	8.61	5.78	0.04
7	Pond	95.00	1.39	0.93	0.01
8	Road	153.00	13.32	8.94	0.09
9	Tributary (canal)	18.00	0.53	0.36	0.03
	Total	2818.00	148.97	100.00	-

Based on patch number of land use, *pekarangan* has largest number with 1464 patches, followed by organic housing and planned housing with 467 and 259 patches. The data indicates that number of *pekarangan* is closely related to organic housing and planned housing. *Pekarangan* is part of organic housing and planned housing which is one of owner's incomesource [10]. Agroforestry land use which has largest area in Ciliwung riparian in Bogor Municipality is *talun* with area of 38.77 ha (26.03%), followed by *pekarangan* with area of 17.53 ha (11.77%) and mixed garden with area of 9.44 ha (6.34).

IOP Conf. Series: Earth and Environmental Science **179** (2018) 012013 doi:10.1088/1755-1315/179/1/012013

This is because at study site, there is Bogor Botanical Garden (classified as *talun*) which is an ex-situ conservation area that should not be converted, so that the natural vegetation condition of riparian rivers is still maintained [11] [12]. Average area of *pekarangan* in Ciliwung riparian in Bogor Municipality is relatively small (0.01 ha), which causes land productivity to be lower [13].

Figure 3. Conditions of agroforestry land use of Ciliwung riparian in Bogor Municipality (a) *talun*, (b) mixed garden, and (c) *pekarangan*.

3.3. Analysis of plant production

Plants production of agroforestry land use in Ciliwung riparian in Bogor Municipality is high (Table 2). *Kebun campuran* has the highest production with the average of 65.41 Mg/ha. This is because topography of *kebun campuran* is relatively flat (slope <15%), which makes plant cultivation easier. In addition, with more number of species and populations of productive perennial plant (vegetable and fruit) (Table 2), makes plant production of *kebun campuran* higher, so that their production runs year-round regardless of season [10]. *Talun* has average plants production slightly lower than talun, which is 49.20 Mg/ha. This is because the slope of *talun* is more steep (> 15%), so with high rainfall of Bogor Municipality (3600 - 4400 mm/year), causes velocity of surface flow is greater and land is easily landslide [14]. Therefore, *talun*was dominated by conservative plant of big trees with deep roots (not productive plant). *Pekarangan* has the lowest average plant productionwhich is 48.77 Mg/ha. This is because *pekarangan*in Ciliwung riparian in Bogor Municipality is urban *pekarangan* that has experienced modernity, so plant selection more for aesthetic functions (*ornamental plants*), not for production [15, 16, 17].

Average value of LER of mixed garden is 1.81 which belongs to high category. LER value of *talun* slightly lower, which is 1.40 that still belongs to high category, while LER value of *pekarangan* which is 0.96 that includes low category. The data shows that cultivated plants with agroforestry systemin Ciliwung riparian in Bogor Municipality, in the form of *kebun campuran*, *talun*, and *pekarangan* still have high land productivity. Agroforestry system practices that have optimal land productivity are able to produce balanced outcomes throughout land management, thus ensuring the stability and sustainability of landowners' incomes [4].

Table 2. Plants production of agroforestry land use in Ciliwung riparian in Bogor Municipality.

Plot number	Land use type	Village	Number of harvest species	Production (Mg/ha)	LER
I	*Kebun campuran*	Sindang Rasa	7	112.95	2.25
II	*Kebun campuran*	Katulampa	10	47.92	2.09
XI	*Kebun campuran*	Kedung Badak	5	43.06	1.37
XII	*Kebun campuran*	Kedung Halang	6	65.33	1.67
XIV	*Kebun campuran*	Kedung Halang	9	57.79	1.65
	Average		7	65.41	1.81
III	*Talun*	Tajur	7	44.75	1.07
VIII	*Talun*	Bantar Jati	4	38.41	0.93
IX	*Talun*	Kedung Badak	7	41.42	1.04
XIII	*Talun*	Sukaresmi	10	72.21	2.56
	Average		7	49.20	1.40
IV	*Pekarangan*	Katulampa	7	80.56	1.48
V	*Pekarangan*	Lawang Gintung	1	106.67	0.84
VI	*Pekarangan*	Baranangsiang	6	15.58	1.01
VII	*Pekarangan*	Sempur	9	38.70	1.26
X	*Pekarangan*	Kedung Halang	1	2.32	0.19
	Average		5	48.77	0.96

Notes: low (LER<1.0), moderate (1.0<LER<1.2), andhigh (LER>1.2)

3.4. Plant production potency of agroforestry land use

The landscape services potency of plants production produced by agroforestry system in Ciliwung riparian in Bogor Municipality is high (Table 3). *Talun* with area of 38.77 ha and average plant production of 49.20 Mg/ha, has total plant production potency reaches 1907.48 Mg. *Kebun campuran* with area of 9.44 ha and average plant production of 65.41 Mg/ha, has total plant production potency reaches 617.47 Mg. While *pekarangan* with areaof 17.53 ha and average plant production of 48.77 Mg/ha, has total plant production potency reaches 854.94 Mg. Types of harvested plant commodities are dominated by fruit plant commodities, such as papaya (*Carica papaya* L.), mango (*Mangifera indica*), rambutan (*Nephelium lappaceum*), and so on. It proves that agroforestry system can be the basis for management of Ciliwung riparian landscape in Bogor Municipality that sustainable [13, 18].

Table 3. Plant production potency of agroforestry land use in Ciliwung riparian, Bogor Municipality.

No	Land use type	Area (ha)	Plant production (Mg/ha)	Plant production potency (Mg)
1	*Talun*	38.77	49.20	1907.48
2	*Kebun campuran*	9.44	65.41	617.47
3	*Pekarangan*	17.53	48.77	854.94

4. Conclusion

The agroforestry system in Ciliwung riparian in Bogor Municipality has high potential plant production. *Talun* has arrange area of 38.77 ha and the average plant production 49.20 Mg/ha, total plant production potency of 1907.48 Mg. *Kebun campuran* has arrange area of 9.44 ha and the average plant production 65.41 Mg/ha, total plant production potency of 617.47 Mg. Meanwhile, *pekaranga*n has arrange area of 17.53 ha and the average plant production 48.77 Mg/ha, total potential plant production of 854.94 Mg. *Talun* and *kebun campuran* have high land productivity with LER of 1.40 and 1.81, whereas *pekarangan* has low productivity with LER of 0.96. Agroforestry practices are able to provide better plants production, so that to ensure the stability and sustainability of landowners' income.

References

[1] Gordon M 2004 Stream ecology: an introduction to ecologists ed ke-2 Chichester: John Wiley & Sons
[2] Mbow C, Van Noordwijk M, Luedeling E, Neufeldt H, Minang PA, and Kowero G 2014 Agroforestry solutions to address food security and climate change challenges in Africa Environment sustainable 6: 61–67
[3] Arifin HS and Nakagoshi N 2011 Landscape ecology and urban biodiversity in tropical Indonesian cities J Landscape and ecological engineering 7(1):33-43
[4] Kaswanto RL, Baihaqi M, and Hadi AA 2013 Desain lanskap agroforestri menuju masyarakat rendah karbon Prosiding Forum Komunikasi Perguruan Tinggi Indonesia 2:418-429
[5] Mubarok A 2012 Karakteristik dan permasalahan pedagang kaki lima (PKL) serta strategi penataan dan pemberdayaannya dalam kaitan dengan pembangunan ekonomi wilayah Kota Bogor [Disertasi] Institut Pertanian Bogor Bogor
[6] Susetyo B, Widiatmaka, Arifin HS, Machfud, and Nurhayati 2014 Analisis spasial kemampuan dan kesesuaian lahan untuk mendukung model perumusan kebijakan manajemen lanskap di sempadan Ciliwung, Kota Bogor Majalah ilmiah Globë 6: 51-58
[7] Noviandi TUZ, Kaswanto RL, Arifin HS 2017 Riparian landscape management in the midstream of Ciliwung River as supporting Water Sensitive Cities program with priority of productive landscape. IOP Conf Series: Earth and Environmental Science 91(1):012033
[8] Sardjono MA, Djogo T, Arifin HS, and Wijayanto N 2003 Klasifikasi dan pola kombinasi komponen agroforestri Bogor: World Agroforestry Centre (ICRAF)
[9] [BPDAS] Balai Pengelolaan DAS Citarum-Ciliwung 2013 Rencana pengelolaan DAS terpadu DAS Ciliwung Bogor (ID): Departemen Kehutanan
[10] Arifin HS, Sardjono MA, Sundawati L, Djogo T, Wattimena GA, and Widianto 2003 Agroforestri di Indonesia bahan latihan Bogor: World Agroforestry Centre (ICRAF)
[11] Republik Indonesia 2011 Peraturan Presiden Nomor 93 Tahun 2011 tentang Kebun Raya Jakarta
[12] Purnomo DW, Helmanto H, and Yudaputra A 2015 Peran Kebun Raya Indonesia dalam upaya konservasi tumbuhan dan penurunan emisi karbon Pros sem nas masy biodiv Indon 1:66-70
[13] Nyaga J, Barrios E, Muthuri CW, Öborn I, Matiru V, and Sinclair FL 2015 Evaluating factors influencing heterogeneity in agroforestry adoption and practices within smallholder farms in Rift Valley, Kenya Agriculture, Ecosystems and environment 212: 106–118
[14] Martono 2004 Pengaruh intensitas hujan dan kemiringan lereng terhadap laju kehilangan tanah pada tanah regosol kelabu [Tesis] Universitas Diponegoro, Semarang
[15] Karyono 1990 Homegardens in Jawa: their structure and function eds Landauer K, Brazil M tropical homegardens, United Nations University Press [Tokyo]
[16] Filqisthi TA and Kaswanto RL 2017 Carbon stock and plants biodiversity of pekarangan in Cisadane watershed West Java IOP Conf Series: Earth and Environmental Science 54:012-024
[17] Choliq MBS, Kaswanto RL 2017 Correlation of Carbon Stock and Biodiversity Index at the Small Scale Agroforestry Landscape in Ciliwung Watershed. IOP Conf Series: Earth and Environmental Science 91:012007
[18] Arifin HS, Wulandari C, Pramukanto Q, and Kaswanto RL 2009 Analisis lanskap agroforestri: konsep, metode, dan pengelolaan agroforestri skala lanskap IPB Press Bogor, Indonesia

Attitudinal differences toward riversides based on the distance between local resident and users with the Arakawa River, Japan

C Liu[1*] and K Furuya[1]

[1] Graduate School of Horticulture, Chiba University, 648 Matsudo, Matsudo-shi Chiba 271-8510 Japan.

[*]Email: chengyuliu0809@gmail.com

Abstract. The role of rivers includes flood control and irrigation, as well as offering a place for people to interact with nature in daily life, especially for physical exercises and sports in Japan. However, in Japan, many people do not view the riversides as places for recreation; the rules for use, and the process of getting permission is also very complicated. This research argues that partial revision of existing rules regarding the use of riversides will increase the diversity of activities. The objective of this research is to clarify attitudinal differences towards riversides based on the distance between local resident and users, with the Arakawa River in Japan. In May 2017, questionnaire survey items were derived from the Arakawa Future Image Plan of 2010. This survey was administered from the end of May until the middle of July in 2017. A chi-square test was applied for the result analysis. An attitudinal difference between local resident (n=121) and users (n=63) was examined in two parts: regional understanding, and the utilization of riversides. According to the study results, the perceptions towards riversides are influenced by the distance from the river and the river's utility pattern. Particularly, differences were seen in people's viewpoints concerning riverside usage patterns

Keywords: Arakawa Riversides, local resident, perception, usage patterns, users

1. Introduction

It is generally acknowledged that green open spaces can provide many benefits in terms of amenity, recreational opportunities, improved health, child development, social interaction, ecosystem services, and community identity [1]. In Japan, riverside areas play important roles in social interaction and child development. The reason that riversides play an irreplaceable role in child development is that many student activities are held around them.

The construction of landscapes and environments psychologically reflects how humans understand nature, as well as how they affect it. In turn, these activities reflect how humans are affected by nature [2, 3]. As special public open spaces, the effective attraction distance of riversides is relatively broader. The usage of riversides is therefore diverse. It is essential to understand people`s perception of the river, as well as their preferences for using the riverside during the planning and management stages [4, 5]. It is a basic acknowledgement that different effective attraction distances exist according to the location and acreage of a green open space. Distance is an important factor that not only affects how people use the open space, but also affects how they perceive it. Most articles covering the subject focus on the perception towards riversides without considering factors of distance.

This research intends to clarify the attitudinal differences towards riversides based on their distances to local residents, as well as the attitude of users of the riverside along the Arakawa River in Japan. In addition, this study will explore the relationship between riverside usage patterns and cognition of the riverside.

2. Study Areas

The Arakawa River is 173 km long, and has a 2,940 square kilometer basin. The river flows through the Saitama Prefecture and Tokyo. For this study, we selected the section of the Arakawa River that flows through Tokyo's Katsushika Ward (Figure 1). Katsushika Ward covers an area of 34.80 square kilometers. The ward contains 227,222 households, and has a population density of 12,770 per people per square kilometer. The open space available for public use covers 21.9% of the total area in Katsushika Ward. However, the river accounts for 60% of the public open space. What is more serious is the uneven distribution of parks and green open spaces in Katsushika Ward. The riverside is expected to play a role as an open space for residents to interact with nature in daily life, as well as an evacuation site for times of emergency. Katsushika Ward is also devoted the creation of a blue-green network [6]. For that reason, we selected Katsushika Ward's section of the Arakawa River to perform a survey among local residents.

The user survey area is located downstream of the Arakawa River in Katsushika Ward, and covers an area that is 6.9 to 10.4 km from the estuary. In this area, there are six baseball fields, one football pitch, one parking lot, and one waterside park. The area to the left of town is all lawn space. Since Metropolitan Expressway No. 6 lies between the riverside and the residence area of the Ayase River, access to the riverside is limited. The Ayase River has no surveyed riverside area for people to use. To reach the Arakawa riverside, users must cross bridges that span the Ayase River. Due to the overhead structure of Metropolitan Expressway No. 6, it does not impede access to the riverside.

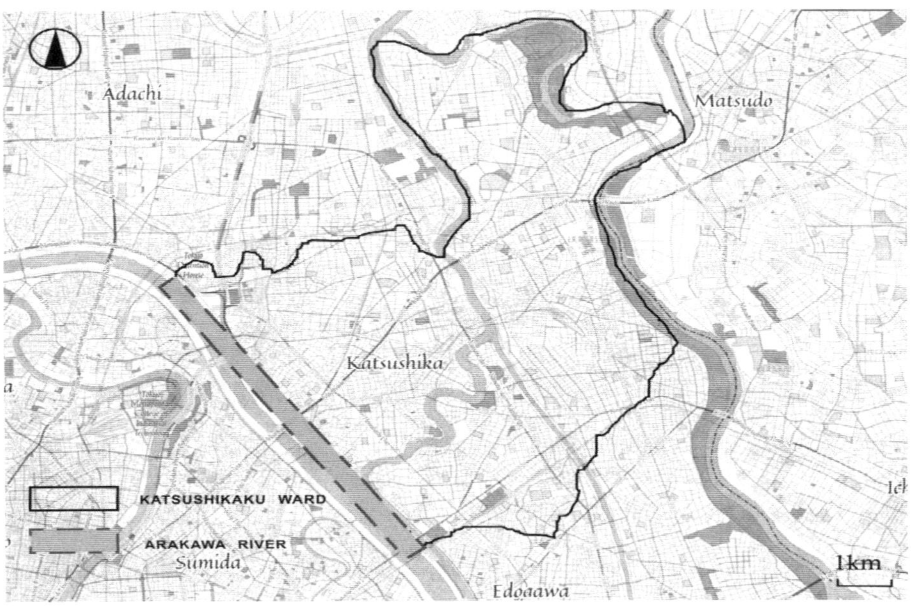

Figure 1. Study area Katsushika Ward and Arakawa River.

3. Methods

3.1. Local resident survey
In May 2017, we divided Katsushika Ward into three zones according to their vertical distance between the residential zone and the riverside. Area one (Z1) is one kilometer from the Arakawa River, while area two (Z2) is one to three kilometers away, and area three (Z3) is three to five kilometers away. A marked questionnaire was distributed to one resident per household in each of the three zones. We randomly distributed questionnaires numbering 300, 300, and 400 to the respective areas of Z1, Z2, and Z3. A total of 121 surveys were returned from the three residential areas of the study. The valid response percentages were 14.3% (Z1), 21% (Z2), and 4% (Z3). The local respondent users numbered 44 (Z1), 63 (Z2), and 15 (Z3). In this survey, respondents were first asked to answer questions about their perception of the riversides from the standpoints of their attitude toward and preference for the structure. The answer was set on a scale of 4 levels, where 4 represented "strongly agree", 3 represented "agree a little", 2 represented "disagree a little", and 1 represented "strongly disagree". Second, residents were asked if they had used the Arakawa riverside. If a resident had never used the Arakawa riverside, the reason was reviewed. If a resident had used the Arakawa riverside, they were asked to answer basic information about riverside usage. Finally, the questionnaire collected background data from local residents, including age, gender, and occupation.

3.2. Users survey
The survey was conducted along the Arakawa riverside in Katsushika Ward, and 65 responses were collected from June to July in 2017. In this survey, questionnaire survey items were derived from the Arakawa Future Image Plan of 2010. The questionnaire was divided into two parts. The first involved regional understanding, while the second involved riverside usage. Questions about riverside usage included the reasons and purpose for doing so, as well as an evaluation. Regional understanding covered perceptions toward the riverside, and preferences for its structure. As for riverside usage patterns [7], immobile use and mobile use were set. Immobile use refers to the activity of the user in a defined place, such as a football pitch. Mobile use indicates that the user is moving along the riverside for activities such as walking or cycling.

3.3. Analysis
Local residents` perception data were analysed in two steps. First, all of the local residents` perception data were analysed without the consideration of any other factor. Then, only the perception data from residents who had used the Arakawa riverside were analysed alongside the distance data. The perception data of users along the Arakawa riverside were compared with local resident users. Simple aggregation and chi-square testing was applied for the result analysis. Furthermore, the mean and deviation of the answers were determined. A Wilcoxon/Kruskal-Wallis test was also used for data analysis.

4. Result
A total of 121 valid responses were gathered from the local residents of Katsushika Ward. The responses were from 49 men, and 72 women. The largest age group was people over fifty (66.1%). There were 86 (71.1%) respondents among the 121 responses that indicated use of the Arakawa riverside, while 35 (28.9%) respondents had never done so. The local respondent users traveled to the Arakawa riverside predominantly by walking (44.2%) or cycling (51.2%).

We collected 65 responses from users along the riverside, and recorded 63 valid responses. User residences were concentrated in Katsushika Ward and four other surrounding wards in Tokyo. The respondents were 34 men, and 29 women. Most respondents were in their forties (50.8%). Riverside users chose a variety of transportation methods to get to the Arakawa riverside, such as private car (27.0%), train (4.8%), bus (3.2%), cycling (52.4%), and walking (9.5%).

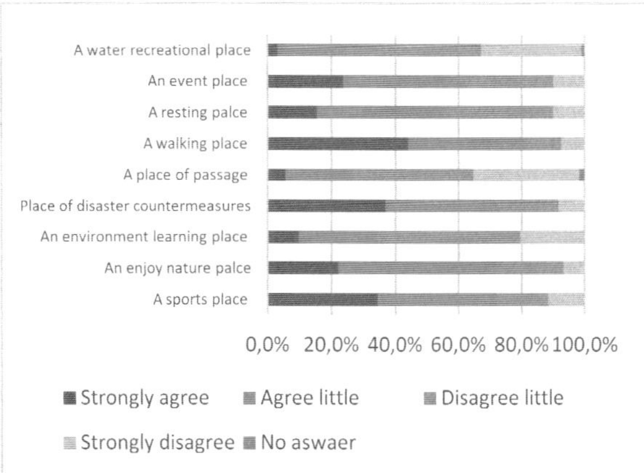

Figure 2. The perception of all local resident on riverside.

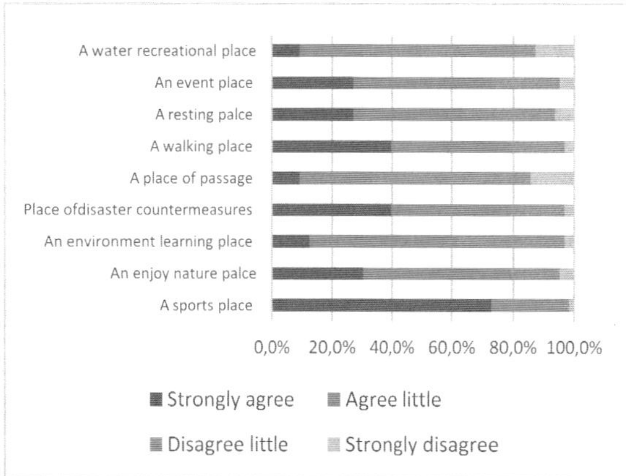

Figure 3. The perception of users on riverside.

4.1. Local resident perception
Local residents mainly viewed the riverside (Figure 2) as a place for walking (81.8%), while users along the riverside (Figure 3) viewed it as a place for exercise or sports (96.8%). A total of 73% of users strongly agreed that the riverside was a place for sports. On the whole, both local residents and users viewed the riverside as a place for sports, an evacuation area, a place for walking, and a place to enjoy nature. The agreement percentage for those four items was over 70.0%. With the exception of a place for rest, every item of data from users had a higher agreement rate than it did for local residents. This was especially true for users who agreed that the riverside was a place for environmental learning, which was two times higher than that of local residents. Finally, the riverside is highly accepted as a place in which events can be held.

Table 1. The perception of all local resident based on distance.

Perception items	Z1 Means	Z2 Means	Z3 Means
A sport place	**3.02**	2.86	**3.07**
An enjoy nature place	**3.02**	2.79	2.80
An environment learning place	2.35	2.17	2.20
Place of disaster countermeasures	**3.07**	2.94	**3.13**
A place of passage*	1.65	2.10	2.13
A walking place* **	**3.51**	**3.02**	2.93
A resting place*	2.81	2.46	2.80
An event place	2.88	2.65	2.93
A water recreational place	1.94	1.97	1.87

Note: Means are based on a four-point rating scale. Body type is means over 3.00.
* Significant different between Z1 and Z2 ($P<0.05$)
** Significant difference between Z1 and Z3($p<0.05$)

The local resident users` perception data (Table1) was analyzed by using a t-test. The agreement rate for the riverside as a place for walking decreases gradually in the three areas (Z1, Z2, and Z3 ($P<.05$)). When comparing Z1 and Z2, significant differences were found in the items "a walking place ($p <. 0017$)", "a place of passage ($p <. 0294$)", and "a resting place ($p <.0366$)". Then, in areas Z1 and Z3, there was a significant difference for " a walking place ($p <. 0252$)". Comparing immobile use ($n = 26$) and mobile use ($n = 60$) among local residents, there were results for " a walking place (p

<. 0406)", "an event place (p <.0161)", "good sense of openness (p <.0372)", and " a reservation for ground (p <.0041) ".

4.2. The users on riverside perception

While the immobile use category contained 20 males and 27 females among users on the riverside, the mobile use category for that area contained 14 males and 2 females. Here, significant differences can be seen. Between the immobile use and mobile use categories, significant differences for the items "an enjoy nature place (p<0.0001) ","a walking place (p<0.0001)", and "a resting place (p<0.0001)" can be seen. The average values of immobile use are as follows: exercise area (3.64), place of Disaster Countermeasures (3.28). The average value of immobile use is as follows: exercise area (3.81), a walk place (3.75), a place can enjoy the views (3.44), a rest place (3.31), and an event place (3.0). From these average values, immobile users have a high evaluation of riversides function. There are also significant differences in "the entrances are not enough (p<0.0001)", "difficult to approach (p<0.0001)" that could be relevant (Figure 6). Immobile users attach more importance to the entrance of riverside than mobile users.

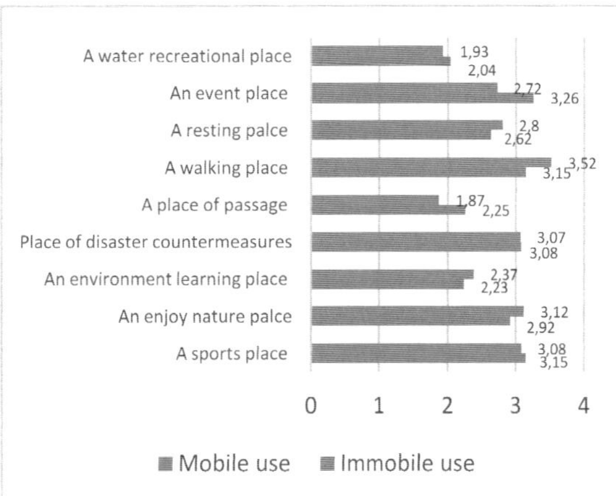

Figure 4. The perception of local resident users based on usage pattern.

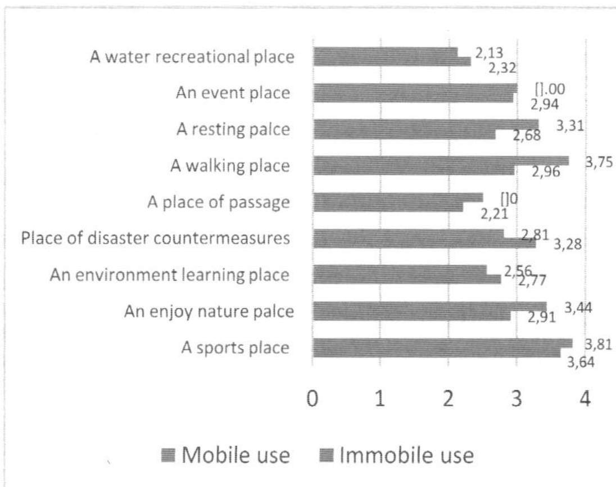

Figure 5. The perception of users on riverside based on usage pattern.

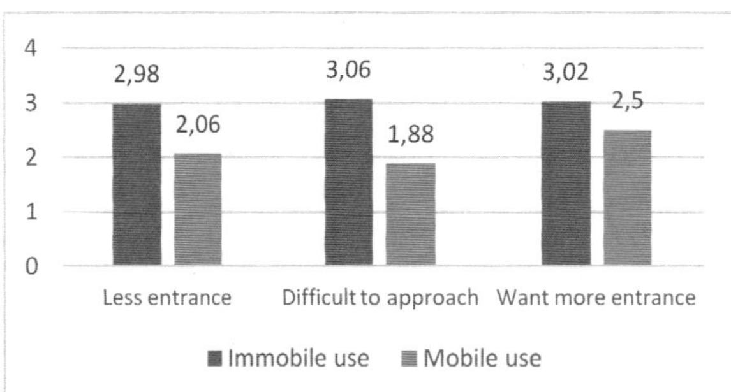

Figure 6. The perception towards riverside entrance based on usage pattern.

5. Discussion

It is well known that human behavior and environment affect each other. Yoshinomi's research shows waterfront experiences and consciousness have positive correlation [8]. Sampels research indicates resident's evaluation of the river and the frequency using of the river were influenced by resident's river experiences [9] and resident's experiences or socialites effect on their images of the river environment [10]. In this research, the relationships between human's waterfront activities and perception also can be found. Users along the riverside show differences between mobile use and immobile use. In particular, few women are involved in mobile use activities, such as walking or jogging, but women are highly involved in immobile use activities, such as watching sports. With regard to mobile use, the riverside was recognized as a walking place, a resting place, and a place to enjoy the view. Riverside access exerts a significant impact on its perception. For immobile use, it was viewed as a place to exercise and a place to deal with disasters. In summary, emphasis was placed on riverside facilities for immobile use, while emphasis was placed on using the software aspects to enjoy scenery while using the riverside in the case of mobile use.

Next, local residents use the riverside for many mobile use activities, as well as taking walks there and holding events. Inside of a 1 km distance, the riverside was especially recognized as a place to take a walk, to find passage, and to rest. Differences in consciousness were also observed at the 1 km boundary. Moreover, since the Z3 effective response percent was 4%, it was considered that local residents' perception of the riverside was low at distances exceeding 3 km.

6. Conclusions

Landscape planning for the open space of the river has been completed. However, as can be seen from this survey, the consciousness and perception among local residents differed in areas 1 km and over from areas that were within 1 km of the riverside. Through this research, we were able to clarify the difference in perception between users and local residents, as well as the attitudinal differences towards the riverside based on the distance of local residents from the riverside.

Reference

[1] Arakawa future image promotion plan 2010
[2] Environment master plan of Katsushika Ward Second Vision 2011
[3] Masaru Y and Shinichi K 1988 A study on the relationship between behaviour and settings in the river, Committee of Infrastructure Planning and Management 6, 113-120
[4] McHarg J A 1969 Design with nature. Natural History Press, Doubleday, Garden City, NY
[5] Niamh M and Mark S 2005 Renewing Urban Communities: Environment, citizenship and sustainability in Ireland pp 481
[6] Sampei Y, Hide M, Koji S, Muneo H and Toyoko A 1989 The process of producing the image of the river environment and the characteristics of the recreational uses of riverside, Annual journal of Hydraulic Engineering, JSCE 33, 31-36
[7] Sampei Y, Hide M, Koji S, Muneo H and Toyoko A 1990 Analyses of resident's evaluation of the urban river and the frequency of their visits to the river space based on their histories of relationship to the river, Annual journal of Hydraulic Engineering, JSCE 34, 31-36
[8] Sampei Y 2002 Perception and evaluation of water in landscape: use of Photo-Projective Method to compare child and adult residents' perception of a Japanese river environment, Landscape and Urban Planning 62, 3-17
[9] Thomas C, Terry B and Raymond D Y 2006 The Riverside and Berwyn experience: Contrasts in landscape structure, perceception of the urban landscape, and their effects on people, landscape and Urban Planning 75, 282-299
[10] Yoshinomi O and Toshiya A 2014 A relational assessment between water experiences and urban river consciousness by association analysis, Committee of Infrastructure Planning and Management 70, 365-372

Bio-Retaining wall as an adaptive design of constructed riverbank into sustainable urban riparian landscape management

A Mosyaftiani[1,*], Kaswanto[2], HS Arifin[2]

[1]Graduate Student of Landscape Architecture Department, Bogor Agricultural University (IPB), Indonesia
[2]Landscape Architecture Department, Faculty of Agriculture, Bogor Agricultural University (IPB), Indonesia

* Email : amarizni_29@apps.ipb.ac.id

Abstract: Indonesia has been struggling in managing erosion and existed adjacent settlement in urban riparian landscape. Retaining wall construction has been done to handle the problems. It has protected the riverbank, but has not considered the ecological issues. Moreover, some of the existing retaining wall in Indonesian urban riparian has been built without any standard and proper management by the community who lives surrounding it, include in Ciliwung Riparian along Bogor City. The retaining wall may potentially risk and clearly not have any environmental benefit. This research was conducted to discover the design of retaining wall that has more adaptive to environment and household grey water biofiltering, called as Bio-Retaining Wall. The case study was carried out in Ciliwung Riparian along Bogor City. Research data such as Retaining Wall Information Collection and Assessment System (WICAS) and design processes by adaptive model approach were combined as the methods for generating design in this research. The assessment result showed that most of retaining walls in the riparian were in risky condition of erosion, landslide, and poor household drainage. Bio-Retaining Wall is designed as adaptive retaining wall by having particular components for wastewater separation, water filtration, and plant cultivation. Despite the fact that Bio-Retaining Wall need further development, it may have prospect for the future of constructed riverbank into more sustainable by considering ecological aspect.

Keywords: Bio-Retaining Wall, Ciliwung Riparian, biofiltering, grey water, Bogor City

1. Introduction

Riparian are a narrow green area right adjacent to the water bodies. It has unique plant communities with soil and topography formed from the availability and dynamics of river water. Riparian landscapes are preferred to natural areas (forest vegetation, grasses, etc.), but many urban riparianlandscapehas changed for settlement use.

Based on the Ministerial Regulation of PUPR No.28 / PRT / M / 2015, river banks are "the space between the banks of the river basin and the inner embankment legs located on the left and / or right of

IOP Conf. Series: Earth and Environmental Science **179** (2018) 012015 doi:10.1088/1755-1315/179/1/012015

the river bank". The boundary of the riverbank is the virtual line on the left and right of the riverbed which is defined as a river protection boundary."

Riverbanks in urban areas, such as Ciliwung River in Bogor City, are not only affected the quality of river water, but also affected by the community's behavior on their environment. Improving the quality of riparian and riverbed landscapes can support scale-change changes [1]. In addition, the riverbank facilitates lateral and vertical connectivity along the river. People in developing countries make the river as a place to do various activities such as fishing, washing clothes, supplying raw water, recreation, or farming. The riverside farm provides local crop production for the surrounding settlements. Such riparian and river utilization activities are very useful and serve to maintain and improve its quality [2].

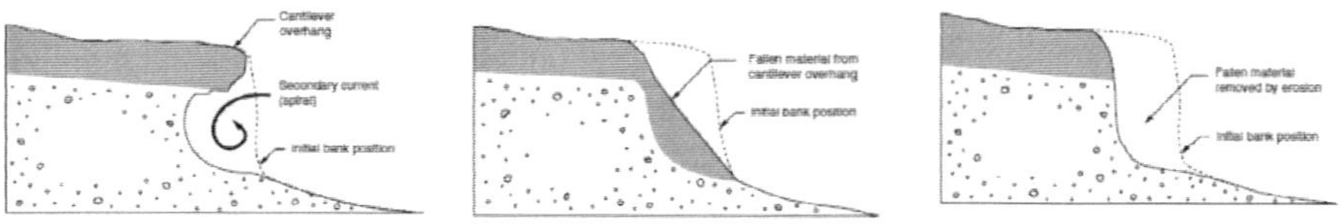

Figure 1. Riverbank destruction process [3]

However, the smaller spaces in the riverside make it difficult to develop the river as a riverfront. In addition, urban riverbank tend to be constructed with rigid and high retaining walls that prevent people from interacting with water. In fact, based on history, access to the river is so easy that it used to be a busy human activity center for washing, fishing and transportation because of the relatively better water quality. However, at the end of the 20th century, the riverbank has undergone many changes from publicly accessible open green spaces to private property, restaurants, police stations, including highways that disrupt the environmental quality and comfort for all residents [2]. Whereas the vegetated riverbankhas ecosystem functions as climate amelioration regulator, erosion control and flood wave [4].

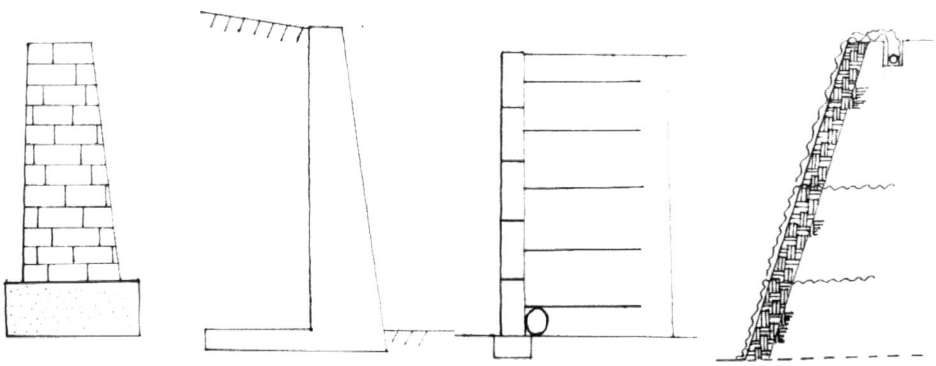

Gravity Wall Cantilever Metal Strip Wall Geotextile

Figure 2. The types of retaining wall [5]

Retaining walls with rigid structures were used if the slope of the ground is above 60% because the more natural retaining wall of the slope is not able to withstand the soil mass and may cause erosion. If conventional structures can be stronger in protecting soil and water, additional environmental considerations should be made more ecologically by the ecological engineering approach [6]. The use of ecological engineering can be done with the use of vegetation and materials that can support the quality of the environment in the river (e.g. riparian tree planting, geotextile use, etc.). Thus, adaptation and modification are required to produce adaptable, high-quality ecological engineering design model.This research was conducted to discover the design of retaining wall that has more adaptive to environment and could potentially support household grey water bio-filteredto the river, called as Bio-Retaining Wall.

2. Materials and Methods

2.1. StudySite

The research was conducted in December 2016 until August 2017. The location of this research is in riparian landscape of Ciliwung River in Bogor City. In this research, Ciliwung Riparian studied were divided into two segments: North Segment (NS) and South Segment (SS) based on Bogor Botanical Gardens (KRB) as the center of Bogor City. Each segment has two research locations. The four locations are location 1 and location 2 were in NS, location 3 and location 4 were in SS. A retaining wall was selected on two side of each sample location, east side and west side. The retaining wall samples were measured for visual assessment and further observation.

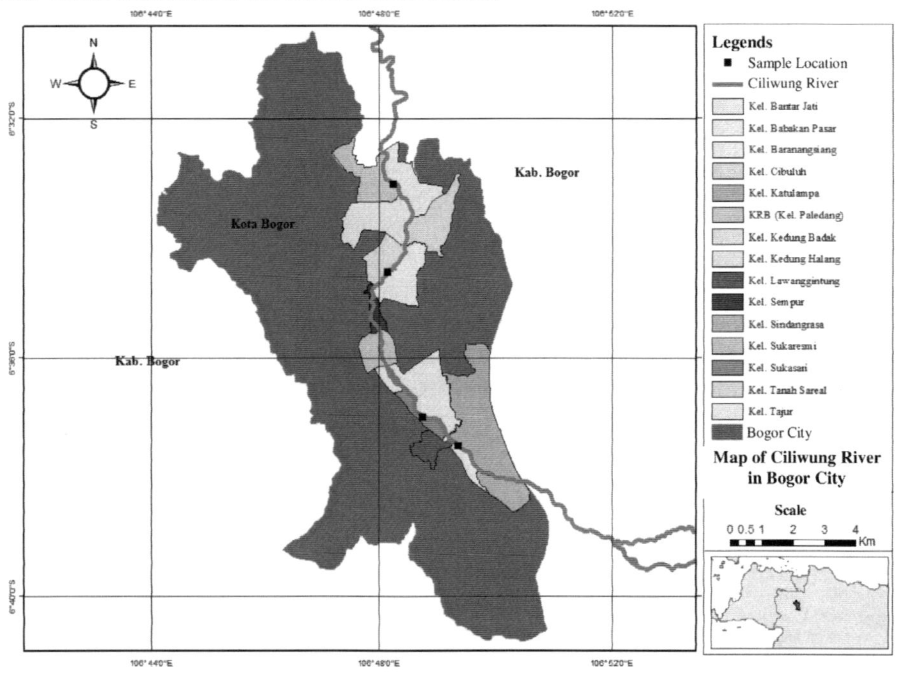

Figure 3. Site location

2.2. Retaining Wall Assessment Method

The strength of the constructed riverbank on two sides of each location was tested in the field using the Retaining Wall Information Collection and Assessment System (WICAS) method. WICAS is an important approach to prioritize management and improve the function of riparian edges. Retaining walls on the river banks are the focus of this visual analysis. The fault can be detected early. The assessment was started by evaluating each character from the criteria through a value of 1-4 by identifying a sample. Then all the information was written into a form that was developed by Butler [7]. The criteria identified included six sections, namely: 1. General wall data; 2. Assessment of wall elements; 3. Assessment of backfill elements; 4. Assessment of drainage element; 5. Assessment of other observations (optional); 6. The rating process by synthesizing those elements.

After all criteria of each parts (include: wall, backfill and drainage elements) were rated with scale of 1 that shows a "good" value to a value of 4 indicating a "bad" value, then all values are calculated and averaged. Accordingly, walls with composite values of less than 1.5 are considered as a good condition, walls with composite scores of 1.5 to 2.4 are considered as moderate / moderate conditions, walls with composite scores of 2.5 to 3.4 are considered as bad shape, and walls with composite scores of 3.5-4.0 are considered as having severe state conditions. The overall assessment method that is done is to consider the average and the worst value. Therefore, if any element value received a rating of 3 or 4 which are "bad", then the overall rating of the wall is 3 or 4 or worse. This decision based on the predetermined criterion which is a safety standard on a retaining wall [7].

2.3. Design Analysis

The next part of this research is design analysis. This study was conducted on the riparian of the Ciliwung River, especially on riverbank. The study in design analysis was using integration method approach between design and research, which is integrating adaptive design model with design model [8].

According to Milburn and Brown [8], the adaptive model is very responsive to the site or research location. Therefore, the design as a landscape product was based on research results that were implemented as general concept in the design. Therefore, the resulting design was inspired by research conducted in the field so that the concept was adapted to be a design.

In this case, research was carried out as an element that was absorbed as a concept in design. Subsequently, the concept containing the information was processed and inspired to be adapted to site design in response to the target site[8].

Both sections are carried out as an integral part in the preparation of ecological engineering design recommendations on the banks of the Ciliwung River. The integration of research in design can lead to an objective process while still considering creativity. This is a challenge and complexity for research in the field of landscape architecture to be able to offer solutions to complex problems based on research done [8].

Field observations and expert or stakeholder engagements are essential for learning in designing a future product design or landscape concept. Experiments conducted in landscape approaches tolerate uncertainty, acknowledging alternatives for conceptualizing and generating innovations. [9,10]

3. Result and Discussion

3.1. Retaining Wall Assessment Analysis

The Changes of land cover and land use along the watershed, including the riparian, affected the hydrological conditions of watershed. All matters relating to rivers and streams are interconnected so that instability of landscape caused morphological conditions and river functions unstable. The river and its riparian are need to be improved physically, rehabilitated and even restored. If it has be done, the landscape services will automatically greatly benefit humans and the environment [11].

In this study, physical condition of the constructed riverbanks of Ciliwung River was possibly assessed by visual assessment [7] of sample locations. It is important to do assessment because the constructed riverbank influenced by intensive human activity, it holds as a settlement barrier, the road or flood retaining structures. After all the sample locations were assessed, the result can be seen that only one structure of a retaining wall on the riverbank has 'good' criteria, i.e. in the 4th location of the West side (L4 West) with "*bronjong*"or gabion structures. In addition, five structures of 'fair' retaining wall were in the 1st East side location (L1 East), 2nd East side (L2 East), 3rd West side location (L3 West), 3rd East side location (L3 East) and 4th West side location (L4 West). While the retaining wall on the location of the West side and the location 2(L2 West) were 'poor'.

Table 1. The assessment rating of the constructed riverbank of Ciliwung riparian in Bogor City

Location	Type of retaining wall	Element Assessment			Rating	Status
		Wall	Backfill	Drainage		
L1 West	*Bronjong*(Gabion)	2.8	3	2	3	Poor
L1 East	Riprap stone, Masonry	1.7	1.5	NA	1.6	Fair
L2 West	Riprap stone	2.63	2.67	2.5	2.6	Poor
L2 East	Riprap stone	2	1.6	2.5	2.03	Fair
L3 West	Riprap stone	2.5	2.1	2.5	2.38	Fair
L3 East	Riprap stone, Masonry	2.2	2.25	2	2.15	Fair
L4 West	*Bronjong*(Gabion)	1.3	1	1	1.1	Good
L4 East	Riprap stone, Dry Laid stone	1.9	1	2	1.63	Fair

Note: L(number) East/Westmeanslocation in the riverbank, west or east side

The riparian condition of the sample locations represented the overall condition of the Ciliwung Riverbank. Based on the observation along the riverbanks of the Ciliwung River, the researchers focused at the condition of the retaining wall structures as the result of structural condition analysis (Table 1).In fact, the Ciliwung River riparian region was the territory of the Ciliwung Cisadane River Basin Authority (BBWS Cilicis). However, based on BBWS Cilicis planning staff, there was no environmental planning for the construction of the Ciliwung River bank structures in Bogor City. The construction of the Ciliwung Riverbank structures should be used by the government in issuing regulations to make riparian landscape conserved from developments such as settlements and buildings that caused destruction to river.

3.2. Bio-retaining Wall Model Design Analysis

From field observations and previously collected data, we found that riparian conditions with different characteristics needed different approaches in ecological engineering design and management. The watershed conditions which are discussed in this analysis of ecological engineering designs construct the riverbank [12,13].

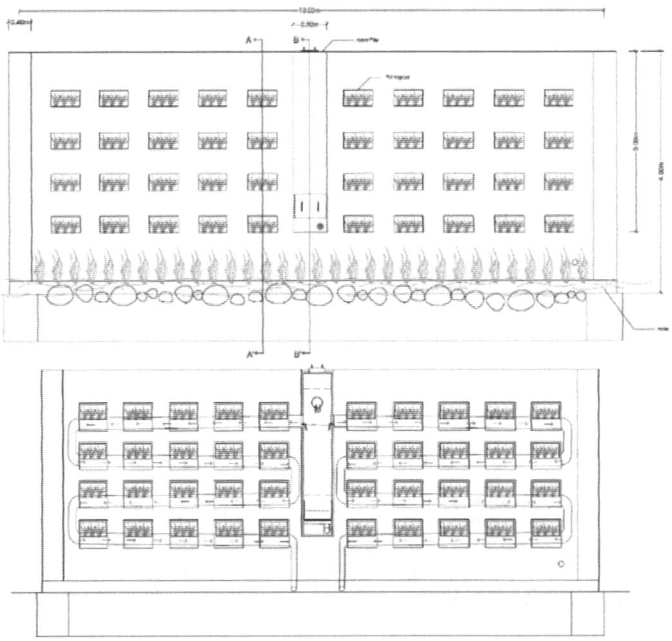

A. Front view B. Cross-section view

Figure 4. Illustration of Bio-Retaining Wall design on the constructed riverbank of Ciliwung Riparian in Bogor City (not scaled).

One of the efforts to establish ecological engineering of riverbank is adding ecological function of existing construction by revitalizing the existing retaining wall. The existing retaining wall only has the function of preserving riverbank fromlandslide, but there is no additional benefit to conserve sustainability of Ciliwung River. Whereas existing retaining walls have the potential to play a significant role in improving the services of riparian landscapes.

Thus, we approached the modification of gravity wall [14] using a more ecological concept of construction. Modificationwas done by making Bio-Retaining Wall. The concept of Bio-Retaining Wall was a gravity wall modification that has a water filter column component from the road drainage and residents' wastewater channel and the media box planting component.

Bio-Retaining Wall work system was to drain wastewater and drainage to the planting medium along the Bio-Retaining Wall. The wastewater stream from the settlement is not directly discharged into the river but can be flowed first to the filtration column to be filtrated and then the resultant water from the column was flown into the planting media box connected by the pipe.

The term of Bio-Retaining Wall was not widely found in the literature. In principle, Bio-Retaining Wall was similar to the living wall [15], eco-retaining wall, green-wall [16] and other terms. It refers to the vegetated wall to improve the more ecological function (multifunction) of construction, in the context of this study is the modification of retaining wall [16]. Bio-Retaining Wall was to make water flow into the river will be cleaner. In addition, various plants can grow on retaining walls to improve landscape services in absorbing pollutants or as a provider of other organisms' habitats.

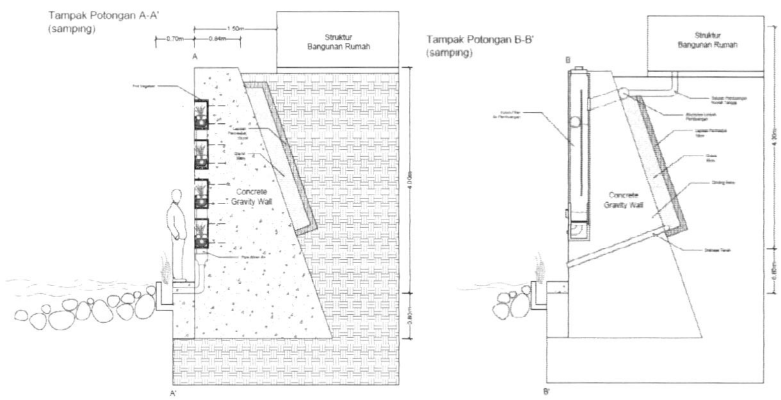

Figure 5. Bio-RetainingWall design (side pieces look) on the constructed riverbank of Ciliwung Riparian in Bogor City (not scaled).

The system was a part of an effort of ecological engineering in construction by restoring morphology and ecology. Restoration experts argue that this construction (Figure 6) can be applied to riparian that have been built by RW to be reconstructed so that Bio-Retaining Wall also plays a role in helping to improve river quality by filtering the water [13, 17, 18].

Alternative management that can be implemented was the delay/stoppage and supervision of construction of riverbank wall construction, whether done by self-help, private and local government; improving the function of existing retaining walls by reconstructing them to be multifunctional such the concept of Bio-RetainingWall. Upgrading theexisting retaining wall by revitalization using Bio-Retaining Wall design is important and this concept could be developed in the future. Many urban areas were concerned to be able to handle using Bio-retaining Wall, such as Katulampa, Tajur, Sukasari and Babakan Pasar areas.

4. Conclusion

The condition of the Ciliwung River in Bogor City was dominated by steep slope above 75°. The existence of constructed riverbank development poses a considerable threat to the ecological quality and diversity of riparian floor vegetation and has an impact on the degradation of river ecosystem. Bio-retaining wall design model is a way of improving environment of riparian. Bio-Retaining Wall becomes important alternative solution while the building removal strategy from the riparian is not yet a solution. In addition to considering its strength, ecological functions also become the main concern to support the quality of the river by this Bio-Retaining Wall.

References

[1] Grêt-Regamey A, Weibel B, Vollmer D, Burlando P and Girot C 2016 River rehabilitation as an opportunity for ecological landscape design *Sust. Cit.& Soc.* 20 pp 142–146.
[2] Kondolf G and Pinto P J 2016 The social connectivity of urban rivers. *Geomorp.*277 pp 182-196
[3] Johnson A W and Stypula J M 1993 Guidelines for bank stabilization projects in the riverine environments of King County (Seattle: King County Department of Management Division)
[4] Vollmer D and Gret-Regamey A 2013 Rivers as municipal infrastructure: Demand for environmental services in informal settlements along an Indonesia river *Glo. Enviro. Ch.*23 pp 1542-1555.
[5] Khan A J and Sikder M 2004 Design and economic aspects of different types of retaining walls *Civ. Eng. J.*Institute of Engineers BangladeshDhakapp 17-34
[6] Wu H L and Feng Z 2006 Ecological engineering methods for soil and water conservation in Taiwan *Ecol.Eng.* 28pp 333–344
[7] Butler C J, Gabr M A, Rasdorf W, Findley D J, Chang J C and Hammit B E 2015 Retaining wall field condition inspection, rating analysis and condition assessment *J. Perform. Constr.Facil.* 04015039doi: 10.1061/(ASCE)CF.1943-5509.0000785
[8] Milburn L A S and Brown R D 2003 The relationship between research and design in landscape architecture*Lands. & Urb.Plan.*64 pp 47–66.
[9] Hulse D W, Branscomb A and Payne S G 2004Envisioning alternatives : using citizen guidance to map future land and water use *Ecol. App.*14 pp 325–341
[10] Nassauer J I 2012 Landscape as medium and method for synthesis in urban ecological design. *Lands.& Urb. Plan.*106 pp 221–229
[11] Matlock M D and Morgan R A 2011Ecological engineering design: restoring and conserving ecosystem services (New Jersey: John Wiley & Sons Inc.)
[12] Cavaillé P, Dommanget F, Daumergue N, Loucougaray G, Spiegelberger T, Tabacchi E and Evette A 2013 Biodiversity assessment following a naturality gradient of riverbank protection structures in French prealps rivers*Ecol. Eng.*53 pp 23–30.
[13] Maryono A 2017 Pengelolaan kawasan sempadan sungai (Yogyakarta: Gadjah Mada University Press)
[14] Harris C W and Dines N T 1988Time-saver standards for landscape architecture (New York:Mc-Graw Hill)
[15] Fowdar H, Hatt B E, Breen P, Cook P L M and Deletic A 2017 Designing living walls for greywater treatment *Wat. Res.* doi: 10.1016/j.watres.2016.12.018
[16] Prodanovic V, Hatt B E, McCarthy D, Zhang K and Deletic A 2017 Green walls for greywater reuse: Understanding the role of media on pollutant removal. *Ecol. Eng.*102 pp 625–635
[17] Noviandi T U Z, Kaswanto R L, Arifin H S 2017 Riparian landscape management in the midstream of Ciliwung River as supporting Water Sensitive Cities program with priority of productive landscape. IOP Conf Series: Earth and Environmental Science 91(1):012033
[18] Karima A, Kaswanto R L 2017 Land use cover changes and water quality of Cipunten Agung Watershed Banten. IOP Conf Series: Earth and Environmental Science 54(1):012025

A set of sustainable urban landscape indicators and parameters to evaluate urban green open space in Bandung City

Firmansyah[1*], AR Soeriaatmadja[1], and R Wulanningsih[1]

[1] Magister of Landscape Architecture Study Program, Institut Teknologi Bandung, Jl. Ganesha 10 Bandung 40132, Indonesia.

[*]Email: firmansyah@ar.itb.ac.id

Abstract. To improve the environmental quality and to increase citizens' index of happiness, Bandung City government has built or either rebuilt and revitalized some urban green open spaces, or in popular are known as public park. Until now the government has built hundreds of parks scattered throughout the city of Bandung, ranging from neighbourhood park to city parks. With the availability of many public parks in Bandung, this attract the citizens' attention, both Bandung's citizen and surrounding citizen, to having activities at public park as an alternative destination. Based on the current conditions, the development of green open space in Bandung more likely to focus on meet user's satisfaction and put aside the fulfilment of other function (ecological, health, socio cultural, economy, and aesthetic) as part of the efforts to increase Bandung city urban green open space sustainability. Since the development of green open spaces only expected to accommodate more visitors, this will result in expanding pavement area which potentially reducing the park's ability to absorb water and reducing existing vegetation. In order to prove that, it is necessary to evaluate the performance of urban green open spaces in Bandung. The evaluation's aims to value the balance between its aesthetical and social function, with its ecological function. This paper will propose a set of indicators which can be used as a tool to evaluate performance of urban green open space as an effort to increase Bandung city urban green open space sustainability. These indicators were obtained through in-depth research and review of some related literatures, sub indicators which obtained at that stage will combined and modified to appropriate with the Bandung City urban green open space characteristics. There are four main evaluation indicator categories, which are ecology, socio-cultural, health, and economy indicator. These categories of indicator would be developed into sustainable urban landscape sub-indicators and parameters based on sustainable landscape design principles. In the end, these indicators were expected not only to be applied in Bandung City area but also in other cities as well with appropriate adjustment to the city's character and condition. In the end, these indicators are expected not only can be applied in Bandung City area, but also in other cities, with appropriate adjustments base on the character and condition of the city.

Keywords: sustainable landscape, urban green open space

1. Introduction

In the last three years, Bandung City has gained a significant benefit by the introduction of urban green open space as one of the city's image, it can increase the visitors both from Bandung citizen and tourists.

IOP Conf. Series: Earth and Environmental Science **179** (2018) 012016 doi:10.1088/1755-1315/179/1/012016

Bandung City Government has made redesign, renovation and rebuilding of the urban green open space. This became one of success' parameters in building the city.

The enthusiasm of city residents in utilizing urban parks (especially those that have undergone the process of revitalization/rearrangement) as a place to indulge in public space, has become an indicator of increasing happiness index of Bandung citizens. City Government in this case managed to increase awareness of the citizens to have activities in urban parks that have been rarely done by the public. The City Government also managed to make the activity in the park as a regular need of Bandung city residents and tourists, either as a daily or periodic activity, especially on weekend [5].

The success of Bandung City Government in attracting Bandung citizen and tourist to utilize and have activities in the city park should be appreciated, which different from the activities of other cities' residents, who spend their time by visiting shopping areas and other commercial areas. but the success should be reviewed critically and thoroughly related to the impact of the existence and sustainability of Bandung's city park as part of Bandung's green open space system. In that perspective, it is necessary to review the success of urban park utilization with the necessity of maintaining balance with other aspects that should exist in a green open space of the city. Micro aspects related to the existence of the main elements that support economic factors, socio-cultural and including its role in macro in maintaining and increasing the ecological value of a city. The existing urban green open space rearrangement aims to accommodate new activities and facilities, which often pay less attention to ecological values, including the increasing pavement area that reduces and even minimizes the catchment area within the city's green open spaces. Therefore, it is necessary to find the indicators needed in the process of evaluating green open space in Bandung. Indicators that can cover all aspects of planning and designing sustainable urban green open spaces.

The revitalization project of the park that tends to be people-oriented are potentially disregard the ecological function of the green infrastructure itself. The new parks were built on former park, so it was not significantly added green public space area in Bandung. These new parks development was not spreading distribution of urban public space. This development of new green open spaces was expected to accommodate more visitors, results in expanding pavement area and potentially reducing the park's ability to absorb water and reduce existing vegetation.

Based on data from Bandung City Government Parks Service in 2016, the amount of green open space recorded as a park in the city of Bandung there are 430 parks with a total area of 216.195 hectares. The data includes parks of various shapes and types with varying extents from the smallest of roadside gardens and 16 m^2 road islands to large-scale city parks such as Tegallega Park with an area of 190,011 m^2. Distribution of urban parks in the city of Bandung by its location almost mostly located in two areas of Cibeunying Region and Bojonagara Region whose position is in the northern part of Bandung. This pattern of distribution occurs because most of the green parks/ open spaces in Bandung are built during the Dutch East Indies colonial period. A total of 33 existing parks in the city of Bandung is then revitalized and reorganized into thematic parks by the city council. While in other parts of the City was the parks and green open space that most of them spread in the area of new housing and green areas in road corridors.

2. Revitalized Bandung Urban Park Condition

Based on the results of field observations, especially on 16 parks from 33 parks and green spaces of Bandung city that has been revitalized/ redesigned, can be presented various recent conditions. The term "Thematic park" were introduced: Thematic park were given as one of the ways the municipality provides specificity of structuring in every revitalized park by trying to give meaning and function to the theme chosen on the park, including new landscape furniture that suitable to the theme [1]. In searching for themes for individual parks, it is a theme that is linked to specific community activities in the city community (football, photography, skate boarding, music, movies, flower lovers) [4], or tailored to a specific age category of user; children, teenagers, elderly) or themes based on it's unique landscape character.

Planning and design only focused on each park with it's special themes thus, there is no interconnection plan between parks, especially in viewing the function of the park as part of the city's green open space system. With the inclusion of activities and facilities within the parks there is a change of park conditions that reduce the fulfillment of ecological aspects such as the function of maintaining microclimate, biodiversity, hydrology, and functions as a recharge area.

Aspects designed according to the results of the design and the design document: focus on fulfilling the facilities of potential users/communities, selection of vegetation types for visual/ aesthetic needs, artificial water utilization (fulfillment of recreation and beauty aspects, pavement and material use) making icons in accordance with the theme park [1]. From the aspect of urban park utilization by the public (resident and visitors/tourists) then the thematic park is mostly used for sports activities, recreation and education, public activities/events (special events, meeting point), economic activity of street vendors and CFD effects around Dago-Gasibu of surrounding parks, socializing (social media influences).

In order to conduct the evaluation, it is necessary to look for the indicators which are the basis for evaluating. So that all aspects and function of the park are represented and not also make an evaluation that is inconsistent with the context when the park has been designed with a specific purpose and facilities according to the developed theme of each thematic park.

3. Research Framework and Method

The research background is an issue regarding sustainability of revitalized Bandung urban park, which need to be evaluated its performance. There are some criteria and parameter that needs to be concerned in evaluating urban park performance. To reach those criteria, evaluations were done by assessing the park components and elements, so it has to be a quantitative assessment. The indicators were combined through selected landscape indicator, sustainable green open space indicator, and successful public open space indicator, which results in a proposed set of urban park indicator, adjusted with Bandung urban park's characteristics.

Method used in this research are literature review and site observation. A literature review is conducted on topics ranging from landscape indicators, successful public space indicators, to the qualitative and quantitative parameters of each component. At this stage the general indicators of the landscape are collected and defined. After the initial proposed indicators are established, a preliminary site observation to the 30 parks in Bandung city conducted to check whether the indicators are suitable with the park's condition and its context. After these preliminary site observations were conducted, further refinement of the initial indicators was carried out, and resulting a more comprehensive indicator set, with scoring method and parameter for each component (see Figure 1).

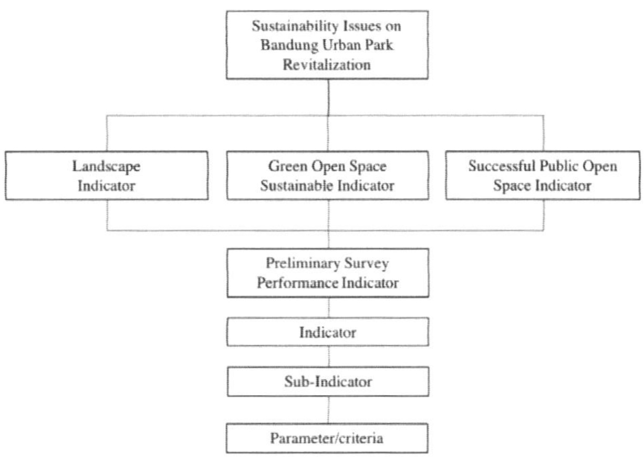

Figure 1. Research framework and method.

4. Sustainable Landscape and Landscape indicator

Urban green open space has three benefits that are *ecological and environmental benefits (*a habitat to support biodiversity conservation, improve micro-climate, decrease carbon emission levels, filters out other particles and dust, controls water regime and reduces run-off, absorb and reduce the noise generated by human activities, and also act like wind breaker), *social benefits* (offer citizens various activity choices which promote social cohesion, designed and serve to everyone in the community, serve the community's leisure needs and socialize, and environmental education), and *health benefits* (to protect from exposed to pollution and infectious diseases, against an urban passive lifestyles lead to some serious health problems, urban citizens have less contact opportunities with nature which is also linked to health and well being) [7].

A conceptual framework has proposed (Figure 2) which links green infrastructure performance into ecosystem services, ecosystem health and human health and wellbeing. This framework provides a conceptual basis to establish a composite indicator-based model for assessing green infrastructure sustainability performance [6].

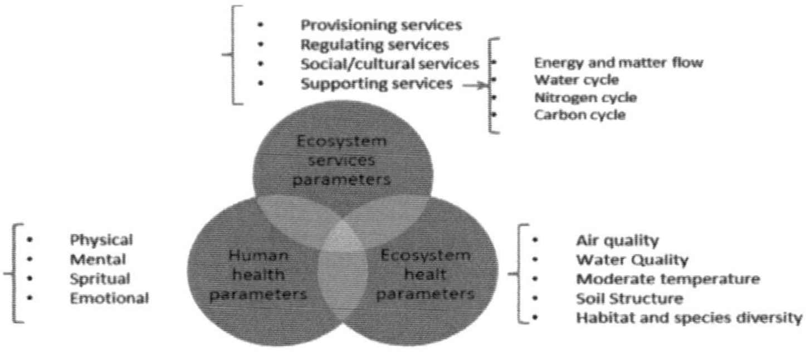

Figure 2. Conceptual framework of green infrastructure [6].

4.1. Environmental indicator

The indicator is a parameter associated with an environmental phenomenon, which can provide information on the characteristics of the event in its global form (OECD 2003). Then, an indicator is an element of the environment or characteristic of it able to represent, individually or in combination with other parameters, environmental characteristics not directly measurable through physical units (Malcevschi, 1991). By using the indicators we can obtain targeted information, in order to concisely represent the problems studied while maintaining the informative content of the analysis intact (Schmidt 1986). Selected indicators must have unique, strategic, computable, and statistically monitorable characteristics (Carollo, 2008). Other useful features to consider when formulating and selecting indicators are that they be representative, accessible, and reliable (Vismara, 1990). We distinguish, therefore, "descriptive" indicators and "performance" indicators: the first kind is used to characterise the environmental situation and monitor the planning process; the second kind enables measuring of the degree to which objectives are achieved [2].

One consolidated instrument for the integrated analysis of the social-economic and environmental aspects in the field of sustainability assessment is the system of environmental indicators known as the DPSIR model (Driving forces, Pressures, State, Impacts and Responses), established by the Organisation for Economic Co-operation and Development in the early 1990s (OECD 1993), and acknowledged by the European Environment Agency (EEA 1995) [2].

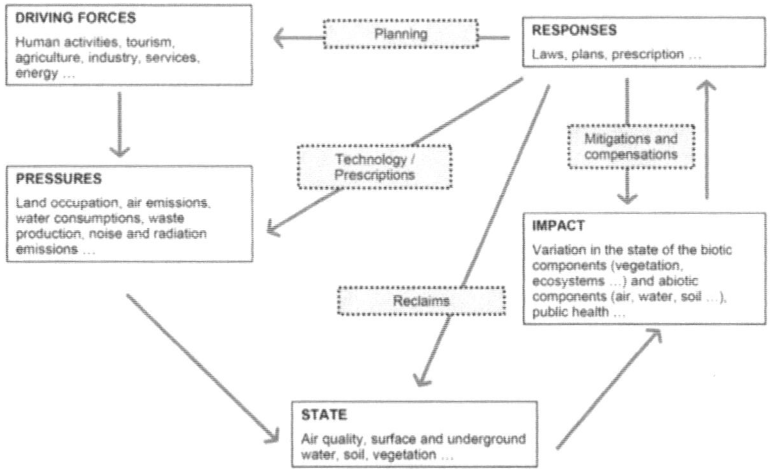

Figure 3. DPSIR model [4].

4.2. Landscape indicator

Vallega (2008) first establishes the functions of the indicators as recognition (monitoring and measuring), evaluation (judgement of the value), and orientation (supplying indications) function. Landscape indicators can be identify at least in three categories: "structural", for which we can make objective measurements that, when changes occur, affect more subjective cognitive aspects; "functional", or physical–biological indicators, mainly attributable to the aspects of ecosystems that occur at different scales; and "cognitive-functional", relating to the nature of the first two categories of indicator, seeking to address the possible subjectivity of other indicators.

Landscape indicators are very important as they identify resources, qualities, and criticalities of a particular area and act within the four (ecological, social, cultural, and institutional) contexts of sustainability: they should not be used alone but in a broader context [3].

Table 1. Landscape indicators, its characteristics and function [3].

chain	indicators	characteristics	function
biological quality	1. loss of species 2. richness of species 3. endangered species 4. protected species 5. environmentally sensitive areas	environmental indicators to use in introductory phase	risk evaluation
environmental quality	1. transparency air 2. transparency of water bodies 3. protection of the caves 4. forest fires	quality estimation	risk evaluation
urban quality	1. wellness acoustic 2. spaces rural pedestrian 3. revitalization historic spaces 4. enhancement urban green	indicators to evaluate how urban sprawl has contributed to the transformation of the territory	risk evaluation
Culture tangible	1. protection of the Archaeological Heritage 2. enhancement of industrial 3. creating cultural trails 4. protection UNESCO sites	include indicators that relate to the material signs that culture imprints on the territory connoting the landscape	• richness • valuing • development
intangible culture	1. viewpoints 2. places of Taste 3. places heterotopic 4. places of events 5. places of personality	indicators that lead to the symbolic value of places	• membership • richness • enhancement • development
aesthetic quality	1. value of the skyline landscape 2. safeguarding the terraced landscape 3. landscapes injured 4. pressure of parking	indicators to monitor the extent to which the infrastructure as a pressure factor	risk evaluation
institutional action	1. efficiency of the measures referred to landscape planning	indicators to monitor the extent to which	• planning • development

According to DSE (2007) sustainability assessment is 'a generic term for a methodology that aims to assist decision making by identifying, measuring and comparing the social, economic and environmental implications of a project, program, or policy option' (DSE, 2007,pp1). Green infrastructure performance indicators play an important role in successfully achieving the urban sustainability targets. An Attilia Peano (Turin Polytechnic) study group conducted the following, final example of research surrounding landscape indicators in which categories were grouped as ecology, perception, land uses, cultural heritage, and economy [4].

Figure 4. Landscape indicator category [4].

5. Development of Indicators and Criteria/Parameter to Evaluate Bandung Public Green Open Space Performance

Performance evaluation on urban park, in which its function as the city's green infrastructure, has many different methods depends on what indicators and parameter used. As physical and nature feature of the park represents the perceived environment, the assessments are often based on visual perception. Measuring the perception of the landscape is not easy and involves many critical assumption, in other words, the quality assessment based on perception may be considered subjective. Landscape is considered a resource used and consumed by people, so the quality of this resource is important. Landscape can be modified and managed to provide services for society and based on the society's needs. To obtain a better-quality assessment than the subjective one, further steps have to be made for a more objective result. To assess performance and success of green infrastructure, a selection of indicators has to be combined and adjusted with the city's characteristics in order to generate a reasonable result [8]. These 28 indicators have been selected based on previous literature review of landscape indicator, sustainable green open space indicator. The indicators that considered appropriate to the green open spaces context in Bandung will be selected as an indicator in this study. The proposed variables or the indicators are both qualitative and quantitative. Preliminary survey within 16 new revitalised park in Bandung has been conducted to identify which indicator be in those parks. The preliminary survey of the 16 Bandung urban parks was conducted in Babakan Siliwangi Park,

Cikapayang Park, Ganesha Park, Maluku Park, Tongkeng Park, Pustaka Bunga Park, Saparua Park, Cilaki Park, Pet Park, Inklusi Park, Persib Park, Karang Taruna Park, Lansia Park, Karta Park, Superhero Park, and Seram Park.

Based on these preliminary survey, the following results are obtained: The park that has the most indicators is Maluku park. The following are the maximum and minimum score of each performance indicator. The maximum number of ecological indicators were found on Maluku Park with total score were 0.69 out of 1. The minimum number of ecological indicators were in Karta Park with score 0.31. The maximum number of health indicators were in Saparua Park with score 0.72 while the minimum number of health indicatros were in Cilaki Park is 0.28. The maximum number of ecological indicators were found also in Saparua Park is 1 and the minimum number of ecological indicators were found in Pet Park is 0. The maximum number of Socio-Cultural indicators were found in Maluku Park is 0.70, while the minimum number of socio-cultural indicators were found in Cilaki Park is 0.16.

Based on initial result of indicator mostly found in the preliminary survey, a more comprehensive proposed indicators set was obtained in order to carry out evaluation of Bandung city thematic parks performance. Table 2 below shows a proposed indicators set, including 18 performance indicators and 73 assessment components. Each assessment components also completed with parameters and criteria to make it quantitative.

Table 2. Proposed indicators set to evaluate Bandung urban park performance.

PERFORMANCE INDICATORS			Assessment Components
ECOLOGICAL INDICATOR	C1	Climate & microclimatic modifications	Regulation of solar radiation
			Lowering air temperature through evapotranspiration and shading
			Wind breaking/wind speed modification
	C2	Reduced building energy use	shading by trees
			covering building by green roof and green wall
	C3	Hydrological regulation	Biopore holes
			Natural Swale
			Drainage system
			Permeable pavement
			Retention pond
			Detention pond
			Water reuse and recycling
	C4	Waste decomposition and cycling	Waste sorting
			Waste container
			Waste treatment facility
			Waste treatment system
			Garbage collector
	C5	Biodiversity enhancement	Vegetation and Planting
			Wild animal
HEALTH INDICATOR	H1	Improving physical well-being	seating area
			sport/ physical exercise facility
			Jogging/running track
			playground
			play equipment
			stairways
	H2	Improving social well-being	social interaction
			social integration and community cohesion
	H3	Improving mental well-being	reducing mental fatigue
			healthier feeling
			emotional benefit
			spiritual benefit
ECONOMIC INDICATOR	E1	Increased property values	Land and property price per m^2
	E2	Supporting local economic activity	Local visitor (from Bandung)
			non-Bandung visitor
			street vendor

	PERFORMANCE INDICATORS		Assessment Components
SOCIO-CULTURAL INDICATOR	S1	Urban farming	urban farming
			kitchen or medicinal garden
			community garden
	S2	Opportunities for recreation, tourism, and social interaction	public toilet
			Shelter/gazebo
			Plaza/stage
			art installation
			pets area
			water fountain
			interactive water play
			pond/lake
			colorful planting/ flowery landscape
			entertainment
			aromatic plants
			landscaping beds
	S3	Improving pedestrian ways and connectivity	walkway inside park
			sidewalk
	S4	Improving accessibility	entrance/gate
			bike rack/shelter
			vehicle parking space
			bus and *angkot* shelter
			map
	S5	Supporting education and research	historical monument/ historical feature
			information board
			plant identification board
			park library
			thematic garden
	S6	Reduction of crimes and fear of crime	lamp at walkway, park area, sitting area, shelter, entrance
			park border
			visibility and openness within the entire park, to the street and surrounding
			regulation board
			park/security officer
	S7	Attachment to place and sense of belonging	historical mark
			landmark
			historical story of the park/city
	S8	Enhancing attractiveness of cities	enhancing attractive view
			restricting undesirable view
			vantage point

To assess each parameters of indicators, likert scale method was used to value parameters, ranged from 1 to 5, in which the minimum score/ worst value is 1, and the maximum/best value is 5 (Table 3). To find the suitable parameter and criteria, each assessment components was reviewed and through literature reference, all performance indicators could be valued quantitatively by the likert chart. After preliminary survey and further discussion through the result, the survey was re-conduct using criteria and parameters to give score to each assessment component. But the assessment was done not only by scoring but also be given description to each component based on surveyor's perspective.

Table 3. Likert scale to score parameters

Item	Item Description	Score range
5	Very good/ highly suitable to parameter	4.21 – 5.00
4	Good/ partly suitable to parameter	3.41 – 4.20
3	Moderate/ fair	2.61 – 3.40
2	Bad/ low suitability	1.81 – 2.60
1	Very bad/ component is not existed	1.00 – 1.80

The second observation was conducted in *Lansia* Park, located on Jalan Diponegoro Bandung. This park was selected because it has complete data, especially climate data. Beside of that, this park was considered as one of the most popular park in Bandung, widely known not only by the citizen but also by visitors from other regions. On this phase, the assessment was conducted to value the assessment components. The parameters assigned to each component, however can be quantitatively assessed and thus, can be compared with other park performance evaluation results.

Table 4. Evaluation of *Lansia* Park based on ecological indicator category C3-C5.

PERFORMANCE INDICATORS		ASSESSMENT COMPONENT	CRITERIA/PARAMETERS	score	OBSERVATION RESULT
C3 Hydrological regulation	regulation of water quality, increased rainwater retention, flow control and flood reduction	Biopore holes	availability of biopore holes with n amount. n= (rain intesity mm/day x impermeable area m^2): velocity of water infiltration per hole.	1	no biopore holes
		Natural Swale	natural/bioswale that infiltrate run-off, slope 1:4, channel width 0.5-2.5 m	1	no natural swales
		Permeable pavement	all pavement area using porous/ permeable material	3	pedestrian way pavement is impermeable. Plaza and seating area using paving block which is porous material (but the lower layer is unknown)
		Retention pond	retention pond with sedimentation pond, water plants on slope, depth >1.8 m	2	2 retention pond, 1 was leaked, 1 was functioning with sluice-gate to control waterflow
C4 Waste decomposition and cycling	nutrient cycling, promoting good habit on waste control and waste treatment	Waste sorting	waste sorting container based on waste type, at least 2 containers for organic and non-organic waste	2	some types of bin had 2 separated compartments
		Waste container	1 waste bin per radius <25m, garbage bin at certain area; sitting area, dining area, smoking area, entrance, parking, plaza; Durable and strong material. Harmony with park visual quality, Aesthetic and Minimize vandalism.	3	20 waste bins in entire park. Some were broken. No unifying design (many different types and material)
		Waste treatment facility	waste collecting and treatment facility to compost/treat waste without disturbing visual quality of the park and visitor comfort	1	no waste treatment facility
		Garbage collector	at least 1 garbage collector or waste manager, for 100 m^2 park area	3	garbage collector also as caretaker
C5 Biodiversity protection and enhancement	promoting conservation, harboring wildlife	vegetation and planting	enhancing local vegetation, non-toxic, not dangerous to visitors (thorn, big fruit/branches falling), easy to maintain, enhancing visual quality	4	local plants species, various height and ages. Tree branches should be trimmed to let sunlight in
		wild animal	wild animal found living/transit in the park without threaten visitor or animal itself	4	species of birds, reptiles, insects, etc. were found. Park and trees also is a green corridor

6. Conclusion

In order to evaluate the sustainability performance of the revitalized Bandung's Urban Green Open Space, Indicators and Parameters are needed. Those indicators and parameters make the evaluation process can fulfil various aspects of sustainability, so that the success or failures of the revitalization result are no longer seen partially.

From the result of various literature studies, that used to explore the aspects and indicators of park's sustainability performance evaluation, verified by observing the presence of those indicators in revitalized parks. There are 28 performance indicators, assessment components, and parameters for each of assessment component.

Further and deeper researches are still needed for more measurable parameters development by involving interdisciplinary science fields in both engineering and social fields against each element directly related to the science fields.

References

[1] Ari MM, Zulkaidy D, Pratiwi WD 2016 Evaluasi dampak penyediaan taman-taman tematik kota bandung berdasarkan persepsi masyarakat sekitar Prosiding temu ilmiah IPLBI Bandung

[2] Bottero M 2011 Indicators assessment systems Landscape indicators: assessing and monitoring landscape quality Springer 15-29

[3] Bruni D 2016 Landscape quality and sustainability indicators Agriculture and agricultural science procedia 8: 698–705

[4] Cassatella C and Voghera A 2011 Indicators used for landscape Landscape indicators: assessing and monitoring landscape quality Springer 31-46

[5] Ilmiajayanti F, Dewi DIK 2015 Persepsi pengguna taman tematik kota bandung terhadap aksesibilitas dan pemanfaatannya RUANG Volume 1 Nomor 1: 21-30 Universitas Diponegoro

[6] Pakzad P and Osmond P 2016 Developing a sustainability indicator set for measuring green infrastructure performance Urban planning and architecture design for sustainable development UPADSD 14-16 October 2015 Procedia - social and behavioral sciences 216: 68 – 79

[7] Ozyavuz M 2012 Landscape planning Urban landscape design Memlük Murat Z Ankara University Turkey

[8] Gavrilidis A et.al 2016 Urban landscape quality index – planning tool for evaluating urban landscapes and improving the quality of life International Conference – Environment at a Crossroads: SMART approaches for a sustainable future Procedia environmental sciences 32 (2016) 155 – 167.

[9] Dipa A 2017 Teenagers happiest people in Bandung: survey (http://www.thejakartapost.com/news/2017/09/09/teenagers-happiest-people-in-bandung-survey.html)

Towards sustainable cultural landscape: A challenge in developing ecomuseum in Minahasa Region

C E V Wuisang[1,4*], J Rengkung[2,4], and D M Rondonuwu[3,4]

[1] Department of Architecture, Landscape Architecture Laboratory, Faculty of Engineering, Unsrat, Manado, Indonesia
[2] Department of Architecture, Architectural Design Laboratory, Faculty of Engineering, Unsrat, Manado, Indonesia
[3] Department of Architecture, Urban Design Laboratory, Faculty of Engineering, Unsrat Manado, Indonesia
[4] Indonesian Built Environment Researchers Association (IBERI)

[*] Email : cynthiawuisang@gmail.com

Abstract. Recently, numbers of ecomuseums are developed around the world, particularly in Europe and Asia.Notably in Indonesia, ecomuseums are provided in various forms, covering large areas to small areas, and isolated sites in order to reach sustainable cultural landscape. This by means protecting the natural resources and cultural heritage. Minahasa region has considerable potential to develop ecomuseum model, encouraging community's local sustainable development. The purpose of this paper is analyzing the potential aspects of ecomuseums in selected rural villages in Minahasa, include; Leilem, Pulutan,Watumea and Kayu Uwi, as case studies. Current development at these villages are evaluated by comparing the local community involvement with the basic principles of ecomuseum concepts, by exploring local community involvement. A qualitative methods was carried out, such in-depth survey between April and September 2017. Field observations include several activities; documentation, also interviewing local participants for their persepective on the ecomuseum developments.In order to reach sustainable landscape development in the region, Ecomuseum is also pivotal and practical model as functional tools for protecting the cultural heritage and developing of local areas. This research proves that Minahasa has got many cultural heritage sources, which are using productively for local sustainable development. The villages are samples of natural and cultural heritage areas of Minahasa for a sustainable cultural landscape development in the region.

Keywords: **Ecomuseum, Minahasa, local community, cultural landscape, sustainable**

1. Introduction

The concept of sustainable cultural landscape development is now widely accepted as a means of protecting natural resources and cultural heritage [20, 21]. Ecomuseum is one of approach towards sustainability, in relation to cultural tourism [9]. Indonesia possesses considerable potential ecomuseum ecological models to encourage sustainable local development. There were no ecomuseums developed in the North Sulawesi Province, so the aim of this paper was to explore the potential for concept development at the provincial level and at self-identified demonstration sites. Ecomuseums are locally-driven, place-based organizations that encourage sustainable community development, based on *in situ* heritage conservation and interpretation.

This paper examines the ecomuseum model from Minahasan perspective and illustrating how an ecomuseum compares to a park, a heritage district, a traditional museum, other shape of cultural preservation.

1.1. Understanding of Ecomuseum

Ecomuseum is identified as a dynamic way in which a community preserves, interprets, and manages its heritage for a sustainable local development.An ecomuseumgrows from below, rather than being imposed from above: it begins with people, not with objects. Its philosophy and practice are based on the intimate relationship between local people and their heritage. Thus, an ecomuseum is "a place where people live which is interpreted (or represented) in the manner of a museum, where traditional museums, natural areas, and historic sites are an integral part of it [3,7].Ecomuseum represents a relevant phenomenon for the local community development. The ecomuseum model was developed first in France in the early 1970s, with aimed to foster holistic interpretations of cultural heritage, combining tangible objects, sites, and built structures with the aspects of traditions, practices, and customs associated with intangible or living heritage". The ecomuseum model then was broadened to include natural heritage, including local flora and fauna, important wildlife habitats and geomorphology sites[3, 6, 7].Ecomuseums provide local people with an inclusive process for rescuing fragments of heritage – such as a vernacular building, a redundant factory, a woodland habitat or an intangible heritage - from loss or destruction, and ultimately lead to the development of a tangible expression of their sense of place, a means of celebrating their heritage[4].Ecomuseum preserves and interprets heritage assets, also work to apply this knowledge to address local development issues, with a focus on sustainability[8, 20]. Ultimately, ecomuseums provide a valuable conduit between the heritage community, the private sector, and elected officials. Ecomuseums matter because it focus on sustainable forms of community development [9, 20, 21, 22].There are many factors which communities contribute and respond to globalized economies, changing and chaotic environments, and a host of societal changes, while heritage is an important consideration, since the community need to appreciate the history of their communities and landscapes, if they are going to understand their potential for adaptation [20].

2. The Case of Minahasa Region

Cultural landscapes of Minahasa, represent an important part of the cultural landscape and ethnographic heritage of a community, both at the local and at the regional level. The sustainability of the qualitative and quantitative characteristics of traditional agrarian practices mostly is based on the use of the traditional landscape and of the environment as resources.The living rural landscapes present high potential for the maintenance of the environmental stability and for the sustainable development of a given territory.In particular, the natural and cultural landscape of Minahasa presents many traits of outstanding relevance at the environmental, aesthetic and historical levels. The area bears a great value and potential for local sustainable development not only for its ecologic, historic and cultural significance, but also in respect of its broader territorial context and of the socio-culture and historical identity of the local population.

3. Objectives and Methodology

In this paper the Cultural Landscape of the Minahasans is examined to reposition their cultural attributes in Indonesian Cultural Heritage schemes, by viewing ancient cultures and social, economic, environmental and spiritual systems as interdependent. This approach thereupon can make a valuable contribution to better understanding landscapes as to their incorporation into regional development.

This research was conducted in Indonesia in the Minahasa Region, North Sulawesi Province. Minahasa is a region on the north-eastern peninsula of the Sulawesi Island (formerly called Celebes Island). The inhabitants are called 'Suku Minahasa' or the Minahasa Ethnic group 1, 2, 11, 23]. The

Minahasa has a significant cultural inheritance which is currently being researched for scientific and educational purposes. However, Minahasan cultural landscape, ecological landscapes or either Indigenous architecture is rarely investigated. Bibliographic survey and field research were primarily done relevant to socio-economic and cultural aspects of Minahasa region inhabitants. This research was qualitative study-based, and involved observing and collecting data and exploring the bodies of knowledge relevant to local traditions and wisdoms on the management of the Minahasan landscapes. Multi-methods were used against the framing of cultural landscape dimensions, and included participant phenomenon observation, in-depth interviews, group discussions and survey strategies.An ethnographic approach was used to analyze and compare local knowledge and land-relationships developed by Minahasan people to curate their unique characteristics as well as ensuring adoption their cultural, social and economic values. This study relied on survey research methods, key informant field interviews, and examination of environmental variables such as varieties of crops and plants. Primary data collection has been taken during the 5 months fieldwork in April-September 2017. Data collection methods included interview, and participant phenomenon observation of which the latter was based upon observing human-land interactions. The researcher was involved as a participant observer. [16].The site selection-criteria was based upon bio-cultural characteristic. To seek variability between people and nature, the researcher invited participant individuals living in settlements located in the Minahasa highland that offer a diversity of features. Qualitative data was obtained during the fieldwork sessions. An in-depth interview and group discussions was applied to 22 respondents, that were selected using purposive sampling and snowball techniques [8, 18]. Interviews were divided into different classifications of elderly (9 respondents), head of village (3 respondents), and cultural subjects (6 respondents).Consistent interview questions were applied to all respondents and key informants to gain local knowledge, perceptions, connections and interactions with the landscapes and places. Four factors that are considered important in establishing the trustworthiness of findings are credibility, transferability, dependability, and confirmability [8, 16].

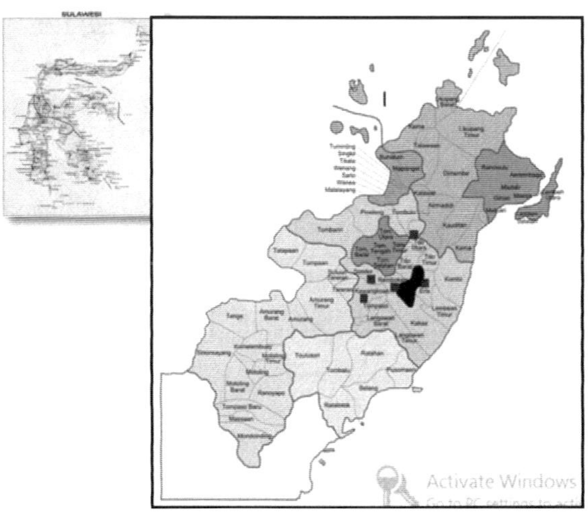

Figure 1. Map of Sulawesi and Map of Research Locations

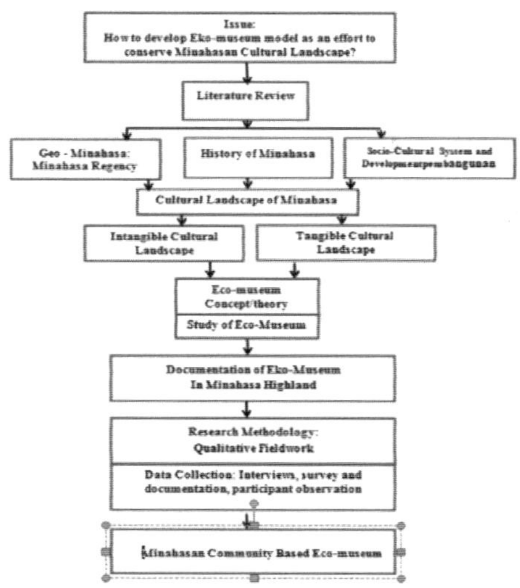

Figure 2. Research methodology Scheme

In contrast to phenomenology, grounded theory, ethnographic, or narrative studies that are based upon specific methodological frameworks can emerge from distinctive disciplinary traditions [8]. In transferring data into research results, various methods are attained to make it understandable. The processes of the analysis started with the transcription, description and inscription documentary data, interview and field notes. This paper profound the tangible and intangible cultural landscape values of

Minahasans from four aspects: (1) traditional Biocultural knowledge, (2) cultural and art , (3) social traditions, and (4) Settlement characteristics, (5) cultural materials/artifacts and, (6) Historical Significant.

4. Results

4.1 The Context of Minahasa: People and Their Land

The Minahasa regency is distinguished by two lowland areas encompassing the northwest to the southeast side of the Sulawesi Island peninsula. These lowlands divide Minahasa into four parts, with the Central Minahasa as its historical heartland (Figure 1). The region presents an interesting case study to this current discussion. The region has a patchy landscape and environment consisting of mountainous and hilly terrain, dominated by an agricultural landscape and a lake ecosystem in the heartland [15, 24, 25]. Specifically, the local ethnic groups, who have inhabited the region for generations, represent one of the oldest regional ethnic groups in Indonesia whom were originally identified during the Portuguese and Dutch colonization and missionaries periods [11, 14, 17].Historically, the Minahasa ethnic groups have been progressively integrated or assimilated into the greater Indonesian population, and after European colonization experienced an extensive Westernization of their socio-cultural systems. Other study about Indonesian culture, claim that the Minahasans have become less traditional compared to other Indonesian ethnic community groups, resulting in a major adoption of Christianity and the loss of Indigenous cultures in contemporary Minahasa [2, 11, 12, 15, 21, 25]. In a contemporary context, the Minahasan people are today re-engaging with their cultural traditions, with their traditions of life and meaning. Cultural values that links their past with their present and give insights into what their ancestors undertook in the past are rekindling the remaining tangible and intangible Minahasan cultural heritage including their occupation sites or places, where Minahasans continue to live and use the natural resources, and sites that represent substantial natural landscapes (for example, Lake Tondano, rivers and spring water sites, remain burial grounds and sites, and cultural artefacts) (Table 1).

Table 1. Cultural Landscape Components in Sellected Territory (Villages) in Minahasa Highland Region

No	Sellection Criteria of Cultural Landscape Potentials							
	Villages (*Wanua*)	**Landform**	**Natural Resources**	**Art and Culture**	**Customs/ Traditions**	**Settlem ent Charac teristic**	**Cultural artifacts**	**Historic al significa nt**
1	Pulutan - Romboken	Lakesides - Hills	Clay and Agriculture	Traditiona l Music and dance	Social Traditions in death and wedding, "Sarikat/Clan gathering, thanksgiving	Highlan d village betwee n two hills	Historic stone, Traditional Houses	Village History, Folk story/ narrative
2	Watumea	Lakesides	Lake/fisher y, Agriculture	Landscape narratives and dance	Social Traditions in death and wedding, Clan Gathering, thanksgiving	Village at the edge of rural lake	The oldest church building in Minahasa, Traditional Houses	Village History, Folk story/ narrative
3	Leilem	Hills	Wooden furnitures, Geother-mal, agriculture	Kolintang music and dance	Social Traditions in death and wedding, Clan Gathering, thanksgiving	Highlan d village surround ed by hills	Historical Stones Traditional Houses	Village History, Folk story/ narrative

4	KayuUwi	Hills	Traditional agriculture	Kolintang, and maengket	Social Traditions in death and wedding, clan gathering, thanksgiving	Highland village surrounded by hills	Traditional Houses	Village History, Folk story/ narrative

Source; Analysis 2017

Discussions and debates with the Minahasa community about land involve environmental, socio-cultural, economic, and their history in being able to comprehend the historical and contemporary interplay between their people and their environment [10, 24, 25]. Landscapes studies in the Minahasa region can reveal multi-diverse local wisdomsand traditions. This is because the Minahasan people's worldview has changed overtime, which the use of landscape retains their ultimate expression in the earliest pre-histories before European colonialization. Over the last few years a massive development has occurred across much of this landscape, aided by Indonesian policy, resulting in many Minahasan people parcelling up their farmland and selling it for development. The value of Minahasa landscape is accordingly possess now more economic values than cultural and emotional values. This shift prompts any to ask about balance; how can the historic and cultural values of Minahasans be conserved while at the same time allowing the community evolve into the future. Such were denied in the past, resulting in the fact that the cultural dimensions of the Minahasa were not integrated into the regional development strategies. There are some records of development failures that were due to ignorance of the cultural potentials and barriers. Such ignorance led to the implementation of development projects which often distorted cultural knowledge and also created an attitude of rejection towards further development planning.In managing the Minahasan's land and environment, local traditional knowledge is seen as a link to their Indigenous thoughts that converts knowledge of their spiritual insights into sustainable living concepts. The word 'knowledge' is applied to Minahasan Indigenous understandings of nature, interpretation of natural processes, knowledge of resources and so forth. In the following analysis, evidence of the substantive properties of this Indigenous knowledge are applied to Minahasan resource management, in particular agricultural or farming practices, hunting and gathering, and ethno botanical plant usage. Evidence of Biocultural knowledge of selected village show that the Minahasans are continuing to exploit their natural resources according to their cultural rules [23, 24]. Through biological adaptation and adoption of local knowledge derived from their ancestors, the Minahasans continue to use the landscape and the forest in sustainable ways. What is considered as spirit of the land is reflected through the lessons learned from Minahasan's traditional agricultural practices, thoughts, and from their mediation of different cultures (traditional and modern) based upon self-experience. The knowledge of the Minahasans has been gained through in-depth interviews with elders in the case study area are presented in this paper as examples. The majority of the Pulutan community rely on exploiting "land" that contains clay at upper layer structure that can be found around the village environment to make ceramics or other types of objects. The Pulutan community manage their production in traditional and semi-traditional ways. The respondents inherited the tradition of ceramic making from their families, and also have their own knowledge and experience, ranging from 10 to 45 years. Watumea community sustain their everyday life and rely on the lake and agriculture, particularly cloves and, red beans and corns. Community in Lailem Raya for years has been producing wooden furnitures/meuble and agriculture in sustaining community's life. Community of KayuUwi possesses their traditional agriculture and social cultural traditions.

4.2 The Ecomuseum Model

Basically, the ecomuseum model is characterized by the community participation in facing the current socio-economic challenges. The key elements of ecomuseum are the collection, human activities (material culture and trades), the environment (geographical and ecological sense), and the community (Figure 2). The Ecomuseum Models for Minahasa Region is consists of four elements (Figure 3) and the proposed of ecomuseum model has three strategies; Community Participation, Local Enhancement and In-Situ Conservation (Open Air Museum), whilst the actions of those strategies are constructing platform for public-private collaboration, initiating tourism community empowerent and legally defining cultural landscape towards sustainable regional development and sustainable landscape development (Figure 4).

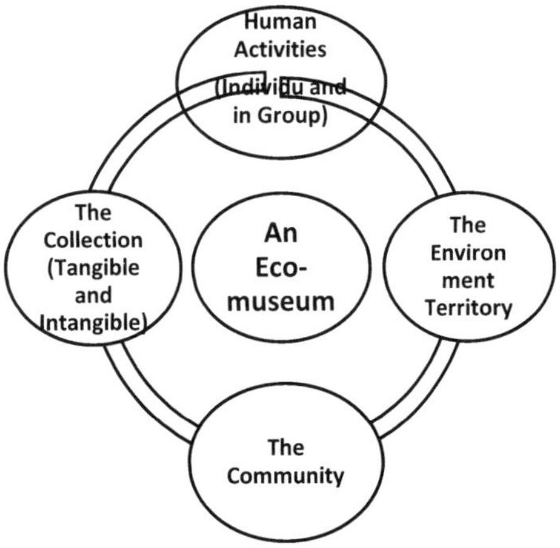

Figure.3 The Key Elements of Ecomuseum. Source: Author 2017

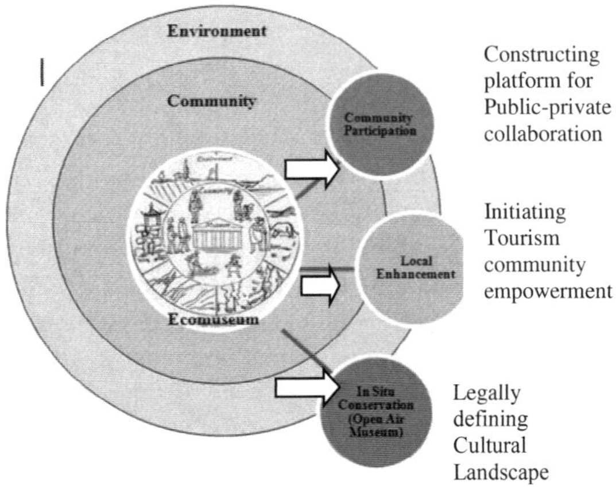

Figure 4. Proposed Ecomuseum Model for Minahasa Region Source; Author2017

5. Discussion

The discussion gives an overview on how Cultural Landscape can be sounded in Cultural Heritage Conservation scheme, as there have been a lot of recent debate and research about issues pertinent to an ethnic community in Indonesia, including Minahasan people. While this is not the first research inquiry about this people, the majority of this research to date has been concentrated on socio-cultural and environmental issues. In this kind of research, ideas, beliefs and theories about the social world as what social sciences are based upon are applied into practical realities of human environment in cultural landscape realm. Therefore, communities' based eco-museum should be utilized in any planning and development that concerns their living environment in order to conserve their living culture and natural resources. The discussion summarizes and provides description and understanding on past cultural attributes to the present and the positive roles of local knowledge of Minahasans and the important of historical information in informing conservation planning and management strategies for an Ecomuseum concept. This research finding recommends and assists Minahasans comunity to continue their traditional cultural knowledge and build their positive roles, subject to their strong connection to the Minahasa land as their "home land",as for recognition to self-identity and in maintaining the local traditions and its values. Moreover, the value of cultural landscape knowledge of the Minahasans reflects their relationship to the current environment characteristics. The cultural attributes is more attuned to the Minahasan's living quality than Western or Indonesian generic concepts. From a holistic perspective, environmental knowledge benefits the locals and could be recommended for developing Ecomuseum and integration into local/state (and potentially national) place and landscape planning initiatives both in schemes and plans (rural or urban).

6. Conclusion

The challenge of increased population and development through modern and global influences in the Minahasa region will continue and will occupy more landscape spaces around and within Minahasan living places. These influences lead to landscape change and threaten to bring about the disappearance of local traditions amongst the community. Therefore, conserving the Minahasan landscape and their traditional ethnoecological knowledge through Ecomuseum are necessary for the Minahasans to sustain their land and environment. This research recommends both local and regional government in terms of reaching sustainable landscape development, which the opportunity of environmental knowledge may help to achieve a sustainable development in which many natural resources have been cultivated and used in a sustainable way for thousands of years. The skills and techniques of local communities provide valuable information to the global community and can be a useful model for environmental policies. The international community has recognized this close and traditional dependence of the indigenous and local communities on biological resources in the preamble, and more concretely. There is a growing international awareness of the links between cultural diversity and natural diversity, and the vulnerability of both outside processes. For future the Safeguarding of the Tangible and Intangible Cultural Landscape heritage it would be recommended to legally bound by the Convention to designate and empower organisations to document intangible cultural heritage and create inventories thereof, and also to encourage the presentation, preservation, protection, and transmission of intangible cultural heritage by working closely and cooperatively with the relevant communities. Reconnection to land will provide a reconnection to Spirit; rejoining a long-time cultural history with the land that made it. The landscape formed the relationships, language, dance, music and symbols which held a people's identity for millennia and still holds it despite recent disconnection arising from violent European land acquisition, displacement policies and language loss. The Spirit of the land is there in both tangible and intangible cultural heritage but it is the intangible which remains the conduit for that reconnection.

References

[1] Blumtritt P 2002*North Sulawesi Tourism Promotion Board-NSTPB*. Retrieved September 28 2010 from http://www.north-sulawesi.org/index.html

[2] Buchholt H 1994 The impact of the Christian Mission on the Processes of Social Differentiation, In: Schefold, Reimar (eds.), *Minahasa Past and Present*, Research School CNWS Leiden.

[3] Corsane G 2005 From 'outreach' to 'inreach': how ecomuseum principles encourage community participation in museum processes. Communication and Exploration - International Ecomuseum Forum, Zijin City Publishing House, Guizhou, China

[4] Corsane G 2006 *Using Ecomuseum Indicators to Evaluate the Robben Island Museum and World Heritage Site*, Landscape Research, t. 31 nr 4, ss. 399-416

[5] Corsane G Davis P Elliot S Maggi M Murtas D and Rogers S 2007a, Ecomuseum evaluation: experiences in Piemonte and Liguria, Italy, International Journal of Heritage Studies, t. 13, nr 2, ss. 101-116

[6] Corsane G Davis P Elliot S Maggi M Murtas D and Rogers S 2007b Ecomuseum performance in Piemonte and Liguria, Italy: the significance of capital, International Journal of Heritage Studies t. 13, nr 3, ss. 223-239

[7] Davis P 2011 Ecomuseums: A Sense of Place, Continuum, London

[8] Denzin N K and Lincoln Y S eds 1994*Handbook of Qualitative Research* Thousand Oaks California 1–17

[9] Donghai 2008 *The Concept of the ecomuseum and its Practice in China*Journal Museum International Vol.60 no.1-2 pp. 33

[10] Moningka B H 1985*Beberapa Bahan Obat dan Ritus dalam pengobatan Tradisional di Tonsea* [Medicinal resources and rituals of traditional healing in Tonsea]. *Makalah Seminar Penelitian Indonesia Bagian Timur, Lembaga Ekonomi dan Kemasyarakatan Nasional* (LIPI dan UNSRAT, Manado 23-29 Juli 1985

[11] Montolalu L R 1991*Minahasa: Negeri, Rakyat dan Budayanya* [Minahasa: country, people and their culture], Pustaka Utama Grafiti, Jakarta (translation of Graafland's book De Minahasa, Haar verleden en tegenwoordige toestand)

[12] Renwarin P R 2007 *Matuari wo Tonaas: Mawanua, Jilid I*, Jakarta, JKT: Cahaya Pineleng.

[13] Riviere G H 1985*Ecomuseum: an evolutive definition*. Museum International 148: 182-183

[14] Schouten M 1983 Minahasan Metamorphoses: Leadership and social mobility in a Southeast Asian society, *c*.1680-1983 (PhD Vrije Universiteit, Amsterdam, The Netherlands)

[15] Schwarz J A T 1908b *Ethnograpica uit de Minahassa* [Ethnography of the Minahasa]. IAE 18, 44-63.
[16] Strauss A L and Corbin, J (1990). *Basic of qualitative research*, Sage, Newbury Park, California.
[17] Supit B 1986 Minahasa dari Amanat Watu Pinawetengan Sampai Gelora Minawanua, Jakarta: Penerbit Sinar Harapan
[18] Tongco D C 2007 Purposive Sampling as a tool for informant selection. Ethnobotany Research and Applications, 5, 147-158
[19] UNESCO 2003 the state of world heritage in asia-pacific region, *World Heritage Reports* No.12
[20] UNESCO 2009 World heritage cultural landscapes,a handbook for conservation and management, *World Heritage Papers* No.26
[21] UNESCO 2003 Linking universal and local values: managing a sustainable future for world heritage, *World Heritage Papers* No.26
[22] Walker B H and D Salt 2006 Resilience Thinking: Sustaining Ecosystems and People in a Changing World.
[23] Wenas J 2007*Sejarah dan Kebudayaan Minahasa* [Minahasa history and culture], Tompaso, Minahasa, Institut Seni Budaya Sulawaei Utara
[24] Wuisang C E V and Jones D S (2011) Challenges in Conserving Indigenous Culture in Minahasa: Culture, Genius Loci and the Environmental Planning System, Conference paper on Track 16 (Special Session on Indigenous and First/Nations Communities and Planning) at the 3rd World Planning Schools Congress, Perth (WA), 4-8 July 2011.

Public perspective on conservation value of Malay rural traditional village landscape in Langkawi Geopark

A A Rahman[1*], S M Ismail[2],and N F M Ariffin[3]

[1] Centre of Studies for Park and Amenity Management, Faculty of Architecture, Planning & Surveying, UniversitiTeknologiMARA, 40450 Shah Alam, Selangor.
[2] Institute for Environment and Development (LESTARI), Universiti Kebangsaan Malaysia, 43650 Bangi, Selangor.
[3] Department of Landscape Architecture, Faculty of Design and Architecture, University Putra Malaysia, 43400 UPM Serdang, Selangor.

[*]Email: azmanar62@gmail.com

Abstracts: The Malay rural traditional village landscape in under tremendous pressure being transformed into other uses to fulfil current socio-economic needs and urbanisation. Due to various anthropogenic activities, they are being abandoned and neglected. Furthermore, people less depended on the rural landscape for the provision of food and other requirements. The conservation and preservation of traditional rural village are crucial not only to protect cultural values but also as a national heritage for future generation. In this study, we estimate the value of rural landscape through their willingness to pay (WTP) for the conservation of the Malay rural traditional landscape. This study was conducted to explore the public perspectives on the conservation value of the Malay rural traditional village landscapes, cultural value, and their willingness to pay. The findings suggested that 15.3% of respondents were willing to contribute between RM 50.01 to RM 200.00 on the conservation of the Malay traditional village landscape. The highest contribution was RM 500 000. The results also show that 30% of the respondents, were satisfied with the present situation of traditional village environment and 50% of the respondents stated that the government should be responsible for the cost of conservation of the Malay traditional village landscape.

Keywords: conservation, malay rural, traditional village landscape

1. Introduction

With rapid urbanisation, rural landscape is under constant threats to converted to more valuable economic activities. Due to various anthropogenic activities, they are being abandoned and neglected. Furthermore, people were less dependent on the rural landscape for the provision of food and other requirements. However, over the past few decades, concerns about disappearing rural landscapes and encroachments of new and emerging features in the landscapes has become a severe issue in both advanced and developing countries. The conservation of rural landscape required in-depth analysis to enable the preservation of the rural setting as a traditional heritage for future generation.

The rural landscape made up where people have developed an area from a wilderness to a small settlement area. They built their settlement from the natural resources such as timber, and building material founded from their surrounding environment. The rural landscape is very rich in flora and fauna which provide the need of people for medicinal and food. According to Benson [1], the rural landscape is the land which is populated by human beings and worked on by them, exploiting its natural resources, primarily by agriculture, aquaculture and silviculture, but excluding areas densely built over and housing human beings living and working nearby. In many ways, the rural landscapes are also very rich in people custom and beliefs that represent the culture of the community in that area which includes religion, language, mannerism, dressing, rituals, music, festival, food, norms, among others. In Malaysia, for example, culture is being practised and expressed by its multi-ethnic groups.

The traditional land use systems in Peninsular Malaysia have changed over time, as a function of the interplay of socioeconomic and technological factors [2]. In the early human settlements were located along the major rivers and coastal region. The village was mainly fisher, hunter and gatherer of forest produce. The Malay settlements concentrated along the coastal and estuarine fishing villages followed by wet paddy growing including the establishment of the fruit orchard.

As the rural resources entail all natural and human-made elements that depict the character of rural areas. In the most common landscape character can be seen until today are the paddy fields, vegetable farms and or orchards [3]. Since the early 1970's, palm oil plantation has become the dominant new look for the traditional landscape of the country, while, paddy fields were found in the north-western region of the peninsula, as well as along the foothills in the east coast. Meanwhile, rubber and palm oil cultivated areas mainly founded in the foothills scattered along the Peninsular Malaysia, besides these areas, the natural landscape was primarily covered by forest or some on the coastal areas [4]. In the most common landscape character can be seen until today are the paddy fields, vegetable farms and orchards.

The Malay rural traditional village landscape is under pressure to being transformed into other uses to fulfil current socio-economic needs and rapid urbanisation. However, some development continues to erode the natural environment, and cultural values that embodied in the landscape that has maintained the traditions are under threat. Due to it's the villages, and *kampong* house also has lost much of its heritage value as the Malays become indifferent to and or ignorant of its cultural importance [5]. The aim of the study examines the fundamental knowledge on landscape characteristics of conservation in the rural area by enquiring the publics' perspectives on aesthetics value of landscape characteristics and their willingness to contribute to the preservation. Also, public knowledge and recognition of ecosystem services generated by Malay rural traditional landscape and their expectation of the traditional rural landscape, and others influencing factors explored. One of the most precious parts of Langkawi that attract every public in their visits is the Malay rural landscape, cultural and geological heritage. In Langkawi, tourism activities have significantly influenced on land use land cover. Although this island attracts more than 2 million tourists from local or abroad yearly, not much changes of its land use land cover had occurred in the last 30 years. However, some of the areas undergoing significant changes mostly centred around Kuah area and near existing population centres such as in Padang Mat Sirat and Pantai Chenang. The result of this study will contribute towards to protection and conservation of natural resources, the rural landscape assets and heritage. In this regard, the Government is committed to the stewardship and preservation of our natural environment and non-renewable resources.

1.1. Malay Rural Traditional Village Landscape

The traditional Malay landscape are invariable could be found in the rural setting, and the *kampong* (Malay village) house represents the conventional form of the dwelling of the Malay people in Peninsular Malaysia. According to Lee & Lau [5] to in the last decade of the 20th century, the landscape of Malay villages in the rural areas of Peninsular Malaysia seems to paint an assuring picture that the Malays have managed to preserve their non-urban socio-cultural character.

According to Yuan[6], the *kampong* or village is a rural settlement sustained traditionally by subsistence activities like paddy-growing, fishing and other agricultural practices. The *kampong* randomly distributed with Malay houses, trees, compounds and paths which blends harmoniously with the environment. There are usually not many main roads in the *kampong* except occasionally for the access road leading into the *kampong*, instead of the paths of a link to the village that lead from one house to another, winding through the houses and leading to other parts of the villages. Due to the preference for community intimacy over personal privacy, house compounds are often open and unfenced[6]. The villages have no clear geometric boundaries, the bunds and irrigation canals more clearly define the boundaries in paddy fields. They also do some farming, fishing, chicken raring and raising livestock activities for daily consume. Fishing villages are found close to the sea while the inland village usually located near the paddy fields, rubber or oil palm plantations [7].

The most distinctive features of Malay vernacular houses built from timber. The houses constructs on the traditional Malay architecture contributes to the shaping of the Malay aesthetical such as the carving design of the roof, wall, door, and window.The numerous features in the traditional Malay house are to providing sufficient ventilation; the house is raised on stilts as to catch winds of a higher velocity. Meanwhile, the elongated structure of the traditional Malay house with minimal partitions in the interior as to allows smooth passage of air and cross-ventilation. The traditional Malay house used ventilation, and solar radiation control devices as to provide climatic comfort in the house atmosphere and these are the most effective means for climatic support in a house in the warm and humid Malaysian climate[6].

The characters of Malay rural landscape influenced mainly by neighbouring regions of the Malay Archipelago, which they brought the traditional cultural values and lifestyles. Furthermore, the arrivals of Islam in the 13th century in the Peninsular, have made the Malays were strongly animistic in their beliefs and the spirit on the dependent decision-making became the foundation in their daily routines [8]. According to Maliki[9], the *kampung* environment largely depends on the principles of the Malay socio-cultural practice, where Malay culture is closely related to Islamic traditions. In the *kampong* architecture the most common and prominent structures in *kampong* are the village mosque (*masjid*) often stands out in a *kampong* landscape, and every village has a *masjid* or a smaller prayer house (*madrasah* or *musola*) which is the Malay going to performs their prayers [10].

1.2. Valuing Rural Landscape

Landscape character is a product of an interaction between ecological and socioeconomic processes made by the human. Nevertheless, human activities are always dominating the natural processes resulting in landscape change and habitat loss of the environment. Human has made modifications to the landscape as to extend their settlements, creating the fields and pastures for agriculture, clearing the forests and draining the marshes and wetlands as to fulfil their socioeconomic needs. Indeed, due to various anthropogenic activities, global climate change and dwindling natural resources, the landscapes change rapidly as well as changes in lifestyles and the competing demands for land for local and global food and fuel security [11].

As almost people valuing landscape as about places and locations that reflect land use, however rapid changes on landscape dynamics make people feel the urgency of preserving (traditional) landscape values, or at least of managing and developing landscapes sustainably. Since placing monetary values on the landscape has never been an easy task due to the complexity of representing landscape features in economic models. In interpreting many environmental goods, a landscape is often considered to be a public good as well no market price exists to indicate their economic value and consequent impacts on the landscape.

In interpreting the public good of landscapes, landscape elements, such as hedgerows, trees, grass strips, ditches, or walls, are interpreted as publicly, commonly, or privately owned. As far as they are publicly owned, the government need the decision making proses in designing the landscape for best to the society. Due to the public good character of the landscape, the effective governance of landscape management is complex and challenging. As it is a fundamental part of this process is to establish the 'value' of the landscape conservation assets within an area. This value can then be compared with data for other ecosystem services to identify management priorities under different future scenarios. However, there is little consensus on how to perform such an evaluation. In this study, we explore the public perspectives on the conservation value of the Malay rural traditional village landscapes, their knowledge and recognition of ecosystem services generated by the rural landscape, their expectation on the cultural value, and their willingness to pay. To calculate the monetary value of such ecosystem services by using contingent valuation method (CVM) as to value the stewardship and preservation of our natural environment and non-renewable resources. Thus, this paper aims to estimate their willingness to pay (WTP) for the conservation of the Malay rural traditional landscape on the existing area. Under certain circumstances, consequences of policy changes towards the rural landscape as to support for conservation, intervention in rural areas, as well as changing economic demands and policies towards the rural people.

2. Materials and methods

2.1. Study Area

Langkawi Geopark is Malaysia's and Southeast Asia's first established Geopark in this region. Covering an area of about 478 square kilometres, the rocky tropical legendary islands are rich in geodiversity, many of which have a scientific value of national and regionally significant. It is a tropical island located off the north-west coast of Peninsular Malaysia between 6^0 10'N and 6^0 30'N latitude and 99^0 35'E and 100^0 0'E longitude. It is unique in the sense that it formed of 104 islands that made up the legendary Langkawi Archipelago. The topography of this island is mountainous covered by forest reserved area of 26,266 ha that is 54.6% of the total land area (Langkawi Municipal Council).

2.2. Sampling Method and Data Analysis

There were several methods used in the data collections in this research such as observations, interviews and questionnaire surveys. The instrument of this survey was designed to measure the following: i) the public perspectives on the conservation value of the Malay rural traditional village landscapes, their knowledge and recognition of ecosystem services generated by the rural landscape, and their willingness to pay towards the conservation of the scenes. The survey consisted primarily of categorical choice questions and Likert-type survey items that provided a list of statements or characteristics of interest (e.g., visitors understanding towards the traditional village landscape and character for conservation elements) and asked respondents to tick the space associated with their level of agreement or concern about that particular statement or characteristic.

The primary data for this study gathered through interviewing by selecting the visitors randomly. A total of 456 respondents were interview into six sub-districts namely is Kuah, Ulu Melaka, Bohor, Kedawang, Padang Matsirat and Ayer Hangat to elicit their preference for traditional rural development, their perceptions of rural character, and their attitudes towards rural living in general. Most of the visitors gave full cooperation during the field survey. The data were analysed using the SPSS software.

3. Result and Finding

Respondents were asked to determine the value of their willingness to contribute to the conservation of Malay traditional village landscape, the result of the values of WTP study showed that 23% of respondents indicated their willingness to contribute and 77% of respondents did not want to contribute as shown in Table 1. This studies also show thatthere are only 8.1% of respondents able to contribute as much as RM100.00 and 5% of respondents able to donate as much as RM200.00.

Meanwhile, the high value of donations was RM500 000, and the lowest amount is RM1.00 among the respondent who wish to contribute or the purpose of conservation of Malay traditional kampong landscape. Based on descriptive analysis shows that the mean and median scores. The study also found that a higher mean score valued at RM 4,927.00, while the median score valued at RM 150.00, this indicates that this means score cannot state the majority characteristics of the other scores in the value of an ability to pay by respondent.

Table 1. Willingness to Pay (WTP) Value

No.	Value (RM)	Frequency	Percentage (%)	No	Value (RM)	Frequency	Percentage (%)
1.	1	1	0.2	12.	250	7	1.5
2.	2	1	0.2	13.	300	9	2.0
3.	3	1	0.2	14.	350	2	0.4
4.	10	2	0.4	15.	400	1	0.2
5.	20	3	0.7	16.	500	1	0.2

6.	50	4	0.9	17.	1 000	1	0.2
7.	100	37	8.1	18.	500 000	1	0.2
8.	120	1	0.2	19.	No Information	351	77.0
9.	150	9	2.0	20.	Total	456	100.0
10.	200	23	5.0	21.	Mean	4,927.01	
11.	230	1	0.2	22.	Median	150.00	

The analysis on the willingness to contribute to the conservation of Malay traditional village landscapes found that only 8.1% of respondents were willing to commit from RM 50.01 to RM 100.00. Meanwhile, there are 2.7% of respondents were willing to contribute between RM 0 to RM 50.00 while 12.2% of respondents willing to pay between RM 100.01 to RM 500 000.00 as shown in Table 2.

Table 2. WTP for the Malay Rural Traditional Landscape Conservation

Fund Range	Frequency	Percentage (%)
Below RM 25.00	8	1.8
RM 25.01 to RM 50.00	4	0.9
RM 50.01 to RM 100.00	37	8.1
RM 100.01 to RM 200.00	33	7.2
Higher than RM 200.00	23	5.0
No information	351	77.0
Total	456	100.0

On the other hand, the analysis showed that 30% of respondents or 137 respondents, were satisfied with the present situation of the traditional village environment as shown in Table 3. Meanwhile, 50% or 228 respondents stated that the government should be responsible for the cost of conservation of the Malay traditional village landscape. In this study also indicate that 11% of respondents or 50 respondents gave their view that allowing others to contribute to the conservation and conservation of the traditional Malay village landscape. Meanwhile, only 2.6% of respondents did not provide any reason for their views on the traditional Malay village landscape.

Table 3. Respondent's View of Willingness to Contribute for the Funds

Visitor Perception Malay Rural Traditional Landscape	Frequency	Percentage (%)
a. I am satisfied with the current situation of traditional village environment.	137	30.0
b. Let others donate.	50	11.0
c. The government is responsible for the costs of conservation	228	50.0
d. Other reasons.	12	2.6
f. No information.	29	6.4
Total	456	100.0

The selection of this placement makes the residence of the Malay or kampong community unique and distinctive that is a combination of physical, artistic and environmental aspects. Given the identity of the traditional Malay traditions and its surroundings at the beginning of the settlement of the Malay community it was described by Moorhead (1965), that the river was the only means of communication and transport to the interior; while its output is the perfect place for placement.

The analysis showed that 26.2% of respondents placed a high priority on the determination of the traditional character of traditional Malay landscape and 57.7% of the respondents rank a moderate of priority of the character of the traditional Malay landscape. This shows that the physical environment, topography, drainage, land and sampling capabilities to the settlement area still have the uniqueness of the Malay identity as shown in Table 4.

This traditional Malay environment also means that there are several elements such as home, town, fort, chart, base, granary, madrasah, mosque or musola or agricultural activity. However, 16.1% of respondents stated that the identity of the characteristic of the traditional village landscape did not depend heavily on the physical environment of the village.

Majority of the respondent or 53.5% of respondents have chosen the architectural aspect as one of the factors that contributed to the retention of the identity of a traditional village, at a moderate level of preference, while 35.3% of respondents placed at high priority levels in determining the character identity of the landscape. These also means the uniqueness of the traditional Malays' identity based on the architectural characteristics of the Malay identity that still present in the structures of houses, towns, fortifications, stubs, bases, granaries, madrasahs, mosques or musola.

The analysis also found that 51.8% of respondents' views of putting a level of moderate priority on the land use environment in agricultural activities such as rice fields, fields, hamlets, gardens, farms and open space still exists in most traditional Malay villages. Meanwhile, 14.1% of respondents set the land use environment is not a priority in determining the character of traditional Malay village landscape.

Meanwhile, 31.2% of the respondents have set high levels of priority in determining complement activities which it contributes to the retention of traditional Malay village identity. The complementary activities can be explained as the socio-cultural values such as culture, handicraft, processing and sales of forest products while socio-economic activities were family gatherings such as feast or banquet and religious activities. The analysis suggested that the identity of the traditional Malay village in the aspect of architecture is the main priority of maintaining Malay character, which is at the mean 1.76 range which was high priority ranges as in Table 4.

Table 4: Uniqueness of Malay Identity

Malay Identity Aspect	Highly Priority	Moderate	Low Priority	Mean
1) Built Environment (houses, cities, forts, jetty, paddy fields, religious schools, mosques and musola)	35.3%	53.5%	11.2%	1.76
2) Agricultural Land Use (paddy fields, orchards, gardens, fields and open spaces)	34.2%	51.8%	14.1%	1.80
3) Complementary activity (cultural, handy craft, selling forest product)	31.2%	49.1%	19.6%	1.88
4) Physical Environment (topography, drainage, soil and accessibility)	26.2%	57.7%	16.1%	1.90

The analysis showed that 38% of respondents stated that the thought factor was at a moderate level of preference. These means that the thought of the present generation has changed dramatically because of the modernisation of science and technology. Meanwhile, 46% of respondents stated that confidence factor was at moderate-high priority. These imply that the beliefs inherent in their parents from previous ancestors can be the basis of a change in the development of traditional Malay landscape at present.

Table 5. Influencing Factor Towards the Development of Malay Traditional Village Landscape

	Factor	Very low Priority	Low Priority	Moderate	High Priority	Very High Priority	Mean
1)	Customs	34.5%	37.9%	15.5%	9.8%	2.3%	2.08
2)	Practices	35.5%	23.2%	21.1%	17.2%	3.0%	2.29
3)	Thought	18.5%	22.7%	38.4%	19.4%	0.9%	2.62
4)	Believes	13.4%	15.3%	20.9%	46.0%	4.5%	3.13
5)	Others	4.1%	5.8%	6.6%	7.9%	75.6%	4.45

The analysis also found that 75.6% of respondents stated that apart from the above factors were at high priority. These means that the thinking of the present generation has changed dramatically due to the modernization of science and technology. Meanwhile, 46% of respondents stated that confidence factor was at moderate-high priority. This implies that the beliefs inherent in their parents from previous ancestors can be the basis of a change in the development of traditional Malay landscape at present. The result shows that the factors affecting the traditional Malay village landscape are traditional factors, practices, ideas, beliefs that are a modest priority in influencing the development of the traditional Malay village landscape, which at the mean 2.08 to 3.13 which is within simple priorities as in Table 5.

4. Conclusion

The study found that the public views on the importance of preserving and the value of Malay rural traditional landscape value and the WTP value for the Langkawi Geopark conservation. The public hold on the government should be responsible for the cost of preservation and protection of Malay traditional village landscapes. These shows that public also provides support to the conservation of the rural landscape through their willingness to pay for the preservation of the Malay rural traditional landscape.Almost less than one-fourth of respondents indicated their willingness to contribute to the conservation of Malay traditional landscape.

The Malay rural traditional landscape alteration in the form of modernism of the rural Malay village it has given the most significant impact toward the cultural landscape of the Malay traditional village in Malaysia into the level of loss it peculiar identity in every place in the country. The study of rural landscapes identifies the appearance of areas to understand processes that are affecting a scene and meanings in the human features of the landscape. Moreover, the cultural sustainability is vital to create a sustainable rural development and preserve the cultural and traditional values for future generations.

Acknowledgement

Centre of Studies for Park and Amenity Management, Faculty of Architecture, Planning & Surveying, Universiti Teknologi MARA, 40450 Shah Alam, Selangor, Institute for Environment and Development (LESTARI), Universiti Kebangsaan Malaysia, 43650 Bangi, Selangor. Department of Landscape Architecture, Faculty of Design and Architecture, University Putra Malaysia, 43400 UPM Serdang, Selangor, are greatly acknowledged.

References

[1] J. Benson, "Aesthetic and Other Values in the Rural Landscape," Environ. Values, vol. 17, no. 2, pp. 221–238, 2008.
[2] B. M. Kumar and K. Takeuchi, "Agroforestry in the Western Ghats of peninsular India and the satoyama landscapes of Japan: a comparison of two sustainable land use systems," Sustain. Sci., vol. 4, no. 2, pp. 215–232, 2009.
[3] A. Saiful Arif, "The Characteristics of the Cultural Landscape in Malaysia: Concept and Perspective Landscape Ecology in Asian Cultures," S.-K. Hong, J.-E. Kim, J. Wu, and N. Nakagoshi, Eds. Springer Japan, 2011, pp. 41–53.
[4] N. H. M. Hussain and H. Byrd, "Towards a Compatible Landscape in Malaysia: An Idea, Challenge and Imperatives," Procedia - Soc. Behav. Sci., vol. 35, no. 0, pp. 275–283, 2012.
[5] H. Lee and S. S. Lau, "Transformation of Malay Vernacular Architecture in 20th-Century Peninsular Malaysia," in ACSA International Conference 1998, 1998, pp. 13–18.
[6] L. J. Yuan, "The Malay house: Rediscovering Malaysia's indigenous shelter system. Malaysia: Institute Masyarakat." Phoenix press, 1991.
[7] N. A. Ismail and N. F. Mohd Ariffin, "Longing for Culture and Nature : the Malay Rural Cultural Landscape 'Desa Tercinta,'" J. Teknol., vol. 9, no. August 2015, pp. 21–24, 2015.
[8] A. Ani, N. Mohamed, and N. A. Rahman, "Socio-Cultural Influences in The Composition of Traditional Malay House Compounds in Rural Melaka," ALAM CIPTA J., vol. 5, no. 1, 2012.
[9] N. Z. Maliki, "Rural-urban Migrants ' Interpretations of Their Home," pp. 1–259, 2008.
[10] H. M. Zaharah, Traditional Landscapes of the Malays of the Peninsula. Universiti Pendidikan Sultan Idris, 2004.
[11] C. M. Van Der Heide and W. J. M. Heijman, The Economic Value of Landscapes. 2012.

Storm water management approach to increase the quality of urban streetscape in Sudirman Street, Bogor City

C N Ainy[1*], D Harlan[2], and H Hanan[3]

[1] Landscape Architecture Department, Bandung Institute of Technology, Jln. Ganesha No 10 Bandung 40132, Indonesia.
[2] Research Division of Water Resources Engineering, Bandung Institute of Technology, Jln. Ganesha No 10 Bandung 40132, Indonesia.
[3] Architecture Department, Bandung Institute of Technology, Jln. Ganesha No 10 Bandung 40132, Indonesia.

[*]Email: nurulainyherish@gmail.com

Abstract. The storm water management approach is still scarce in developing countries. Streetscape is one of the city's elements that suitable for storm water management because of the dominance of impervious surface. Storm water runoff will cause damage of streetscape facility and river biodiversity in the long term due to the pollutant consist on it. This study focuses on the use of hydrological analysis to determine the main problem regarding the drainage capacity and the design solution with storm water management approach. The case study takes place in Sudirman Street, Bogor City. There are two types of drainage basins in Sudirman Street and delineated to two different catchment areas. The result of study shows that both of existing channels have enough capability to accommodate peak runoff until 5-year storm, yet the flash flood still occurs during the heavy rain. This is leading to conclusion that the existing inflow design is not effective to convey the runoff into existing drainage. So, the design solution is utilizing green area in the streetscape as a green infrastructure that give a multiple benefit to the environment such as filter, infiltrate, and convey storm water runoff slowly to the main drainage.

Keywords: drainage capacity, hydrological analysis, runoff, storm water, streetscape

1. Introduction

The International Geosphere-Biosphere Programme (IGBP) identifies three main components of global change: land-use/cover change, changes in atmospheric composition, and climate change [1]. Current environmental problems as we face them because of climate change are encouraged to have green spaces in the city [2]. Streetscape is one of the city's elements that plays important role I the environmental quality because it has 25% area of the city [3], yet it has the dominance of impervious surface. Directly Connected Impervious (DCI) areas are areas where a runoff is directed into a storm sewer, Swale, or creek with no opportunity for the runoff to encounter absorptive soil or vegetation. These areas are the most damaging to stream health and flooding because all precipitation is translated immediately into a runoff in the stream system [4]. To prevent water pollution, site designers must now address storm water management practices not only for the finished development, but for the construction site as well, from the time the site is first disturbed [5].

Hydrological analysis is very critical step to ensure the sustainability of site towards storm water runoff. Storm water management approach in Indonesia still unpopular while we need more responsive design to encounter the climate changes effect. Widening the drainage capacity more and more was the quickest solution and ended up failing to solve the problem regarding storm water runoff in the streetscape. It is essential to conduct the computation of the existing drainage capacity to determine the main problem. There are multiple benefits to enhance our hydrological analysis with the calculation of storm water runoff discharge: first, it will help to find the cause of storm water runoff that occurs due to the lack of capacity or not; second, it will be easier to design the drainage system with the proper capacity. This paper will focus on calculation practice on both benefits. Therefore, it will be very beneficial for designer to design the environment in a more sustainable way. Sudirman Street, in Bogor City, has been selected as the study area. As a one of the main road in the Bogor City, it faces so many problems due to the development of the city and the effect of climate change. Storm water runoff affects the damage of streetscape's facility and decrease user comfortability. Thus, with this study it will help to determine the main problem and the design solution to reduce storm water runoff in the Sudirman Street.

2. Methods

2.1. Study area

This study was conducted in Sudirman Street, Bogor City (approximately 6°24'49" S- 6°35'39.24" S, 106°47'44" E- 106°47'46" E). Bogor has high annual precipitation compared with other city in the same region about 3500-4000 mm/year. It has historical street which is part of road development in the Colonial Era (1808) called Groote Post Weg now the name is changed to Jalan Sudirman and classified into Secondary Artery Road (Figure 1). In the South of Sudirman Street, there is Bogor Botanical Garden, the largest green open space that classified as a City Forest [6].

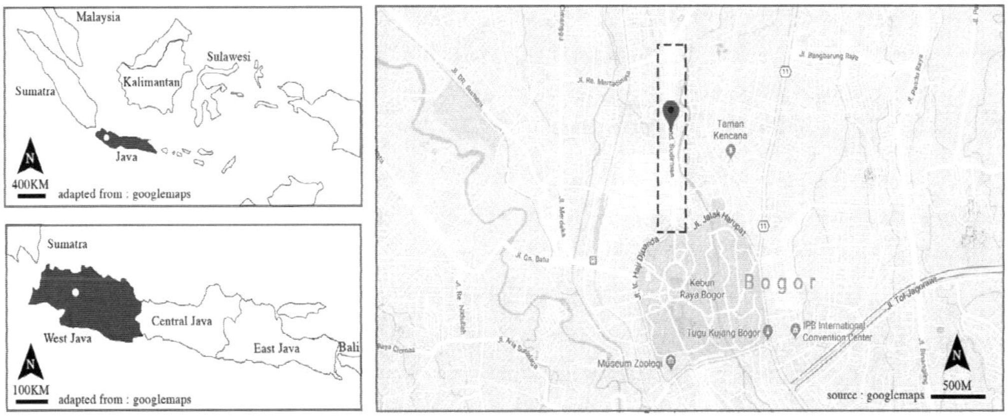

Figure 1. Map of study area.

2.2. Data collection

The main data for this study is the maximum precipitation (Figure 2) for over the past 10 years (2006 - 2015). These data were collected by Department of Meteorology, Climatology, and Geophysics located in Bogor Botanical Garden (06°35'55.7" S- 106°47'44.1" E). Beside the maximum precipitation, the other data were land cover, topography, flow direction, and the size of existing drainage. Land cover data were obtained from Google Earth to determine the runoff coefficient. Topography, flow direction, and the size of existing drainage were obtained from the secondary data and field observation were used to determine the catchment area and computed the runoff discharge.

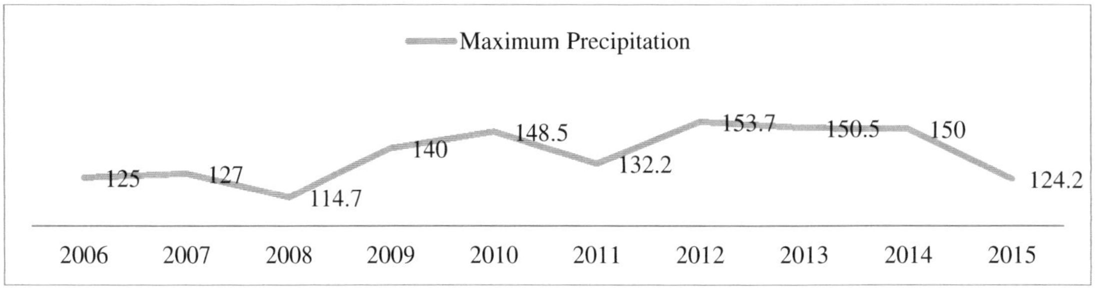

Figure 2. Maximum precipitation/month (mm) from year 2006-2015.

2.3. Runoff discharge

2.3.1. Manning equation

The flow in open channels is a function of a velocity and the cross-sectional area of flow. Velocity is a function of slope, surface roughness, and cross-sectional shape. The Manning Formula is used to compute velocity [7]:

$$R = \frac{A}{P} \tag{1}$$

$$v = \frac{1}{n} R^{2/3} S_o^{1/2} \tag{2}$$

Two other arrangements of the Manning Formula are convenient in conveyance design. Discharge (***Q***) can be expressed in terms of the Manning Formula using the following formula [7]:

$$Q = \frac{1}{n} R^{2/3} S_o^{1/2} A \tag{3}$$

Remarks:
R = the hydraulic radius (m)
A = the cross-sectional area (m^2)
V = velocity of flow (m/s)
S_o = longitudinal slope of the hydraulic grade line (m/m)
N = Manning's coefficient for friction

2.3.2. Gumbel method

The Gumbel method is used to determine predicted 5-years rainfall in the study area based on the maximum rainfall data in the last 10 years. The formulas used are as follows [8]:

$$X_T = \overline{X}(1 + KC_v) \tag{4}$$

$$K = (y_T - y_n) / \sigma_n \tag{5}$$

Remarks:
X_T= the probable maximum precipitation with a return period of T years
C_v = coefficient of variation
Xbar = mean maximum precipitation
K = frequency factor

σ_n = standard deviation of data
y_T = -lnln(T/T-1)
y_n, σ_n = expected mean and standard deviations of reduced extremes to be found from Gumbel's table.

2.3.3. Mononobe rainfall intensity

The intensity of rainfall is defined as the mean of the rainfall intensity assumed to fall uniformly over the catchment area; to determine the duration and frequency of the units, the commonly used unit is mm/hr. The Mononobe formula is used to calculate rainfall intensity each time based on daily rainfall. Here is the formula used [9]:

$$I = \frac{R_{24}}{24}\left(\frac{24}{t}\right)^{2/3} \qquad (6)$$

Remarks:
I = Rainfall intensity (mm/jam)
R_{24} = Rainfall in 24 hours (mm)
t = Rainfall duration (jam)

2.3.4. Rational method

Many methods to compute runoff has been developed over the years and the first and most enduring of these are the Rational Method. The Rational Method is used to compute the peak runoff (***Qp***) following a rainfall event. It makes no attempt to estimate runoff before or after the peak, but simply estimates the one quantity of flow that is greatest [10]. The formulas are:

$$Q = kCIA \qquad (7)$$

Remarks:
Q = peak runoff (m^3/s)
K = coefficient (0.278)
C = runoff coefficient
I = rainfall intensity (mm/jam)
A = drainage area (m^2)

3. Results and Discussion

3.1. Hydrological analysis

3.1.1. Drainage capacity

Before computing the capacity of the drainage, field observations were conducted to know the size of the drainage and catchment area delineation. There are two types of drainage in the study area, both are open channel drainage, but different in material and size. Drainage Type 1 lies in the west of road while Drainage Type 2 lies in the east of the road. The catchment areas were determined based on the area where the flow of water entering the drainage; so, it is divided into two catchment areas. One is for the catchment area in Drainage Type 1 (180717 m^2) and the other one is for catchment area Drainage Type 2 (56775 m^2). The highest point is in the south with elevation 258 m above sea level (asl) and the lowest point is in the north with elevation 233 m asl. The length of Drainage Type 1 is 1439 m and Drainage Type 2 is 1404 m (Figure 3).

The result of field observation about the size of existing drainage and the boundary of the catchment areas could be seen in Figure 4. The material of the channel will affect Manning Coefficient of Friction. Drainage Type 1 has 0.023 coefficients because it made from gravel bottom, sides dry rubble. Drainage Type 2 has 0.012 coefficient because the material is made from concrete, formed and trowel finish [7]. For the slope of Drainage Type 1 is 0.017 m/m and Drainage Type 2 is 0.018 m/m. Based on that, the result of computation using Equation (1), (2), and (3) is the capacity of Drainage Type 1 is 6.56 m^3/s and Drainage Type 2 is 0.85 m^3/s.

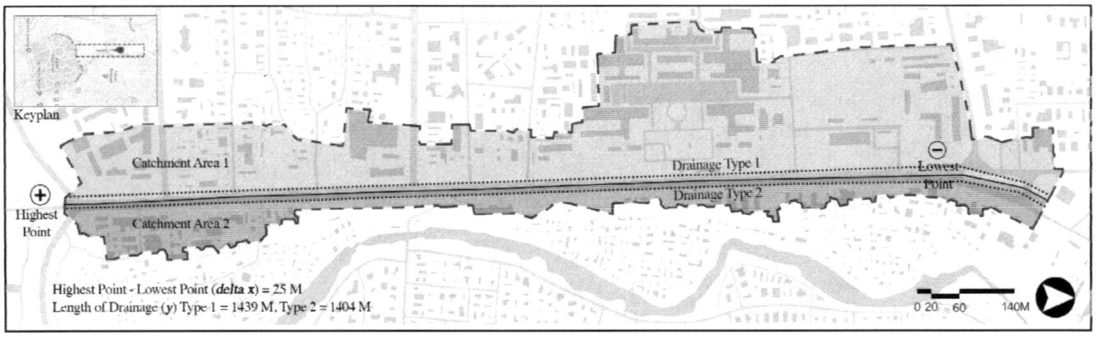

Figure 3. Catchment area and drainage map.

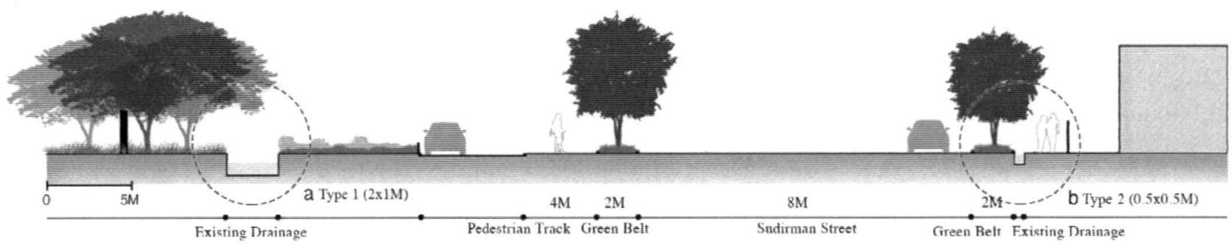

Figure 4. Typical existing street section.

3.1.2. Runoff discharge

Rainfall data processing is needed before computing the runoff discharge. Gumbel Method (4) is one of the methods to predict the rainfall. In this study, the trend of rainfall is predicted in 5-years rainfall. The result is rainfall in the study area (**Xt**) will reach 151.03 mm/day in the 5-years rainfall. Then, this data will be processed on Mononobe Formula (6) to determine the rainfall intensity per hour. The result is rainfall intensity (**I**) in the study area is 2.65×10^{-5} mm/s. From the result of land cover analysis, there are seven types of land cover in the study area: roof, green open space, green belt, pedestrian track, paving, road, and parking area.

The results are the runoff discharge of 5-years rainfall on Catchment Area 1 is 0.94 m^3/s while in the Catchment Area 2 is 0.37 m^3/s. Compared to the existing drainage capacity, the runoff discharge of 5-years rainfall in both catchment areas could be well accommodated by both drainage existing. So, this leads to the conclusion that the main problem of the local flooding in the street during the rain is not because the existing drainage is not having enough capacity to accommodate the stormwater runoff, instead the design of the drainage system is very poor; therefore, it will stagnate in somewhere before reach the channels.

IOP Conf. Series: Earth and Environmental Science **179** (2018) 012019 doi:10.1088/1755-1315/179/1/012019

3.2. Design recommendation

At the local and site scales, it has been defined as a storm water management approach that mimics natural hydrologic processes [11]. Traditionally, the country relies on a network of concrete canals and rivers to quickly channel water into the reservoirs and the sea to prevent flooding. These rivers and canals are designed to quickly and efficiently convey storm water runoff to the sea or to the nearest waterbody [12]. When new developments are planned, appropriate management interventions, including infrastructure and development controls are required to manage storm water and prevent flooding of properties [13]. Based on study there are two important design solutions to effectively reduce stormwater runoff and create a more sustainable environment on the streetscape:

3.2.1. A multifunctional green belt

Storm water runoff on the streetscape is very difficult to reach the drainage due to inadequate design and maintenance of conventional drainage. On the roads that have a green belt as in the study area, it can be utilized as a bioretention swale before it enters the existing drainage. This will help shorten the flow of water and reduce the runoff significantly. In essence, the aims of bio-swale approaches are two folds: to reduce the amount of impervious surface in order to reduce storm water runoff, to utilize the landscape and soils to move naturally, store and filter storm water runoff before it leaves the development site [14]. Bio-swale is one of the types of green infrastructure that suitable for green belt. Figure 5 is a sample of the bio-swale system that can accommodate the storm water runoff. Use of local plant species is very recommended to enhance the biodiversity and easier maintenance. Plants such as Cymbopogon citratus, Pennisetum alopecuroides, Sanchezia nobilis, Ipomea pes-caprae, and Widelia biflora are proven to help reduce water pollutants in the runoff [15].

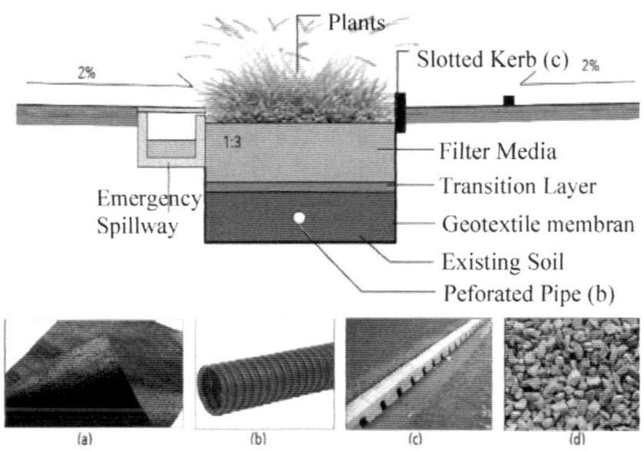

Figure 5. Cross section of bioswale (adapted from PUB 2011).

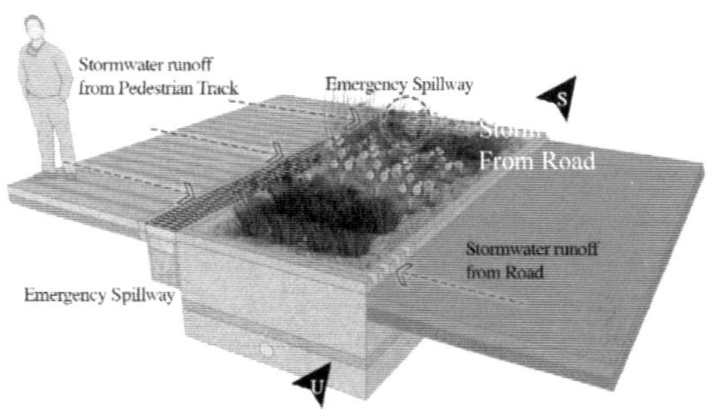

Figure 6. The design flow of water in the bioswale.

3.2.2. Capacity of bio-swale

To create more efficient and sustainable infrastructure, the capacity of bio-swale should properly accommodate storm water runoff within the study area. At first, the design of flow direction must be clear and efficient to convey the water to the bio-swale, as well as the boundary of the catchment area. Figure 6 shows the design of flow direction in the streetscape. Once catchment area is determined, the size of bio-swale could be designed by calculating the capacity with Manning Equation (3) and Rational Method (7). The results are the runoff discharge within the catchment area is 0.07 m^3/s and the capacity of bio-swale is 0.5 m^3/s (Figure 7). Thus, the bio-swale has more than enough capacity to accommodate up to 5-years storm.

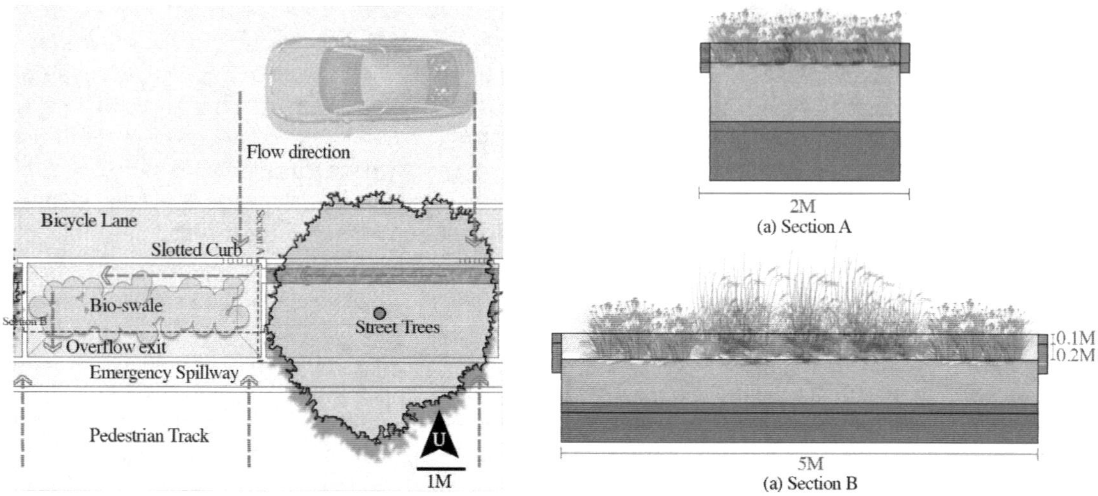

Figure 7. Design solution for utilizing green belt as bio-swale to reduce storm water runoff in the streetscape; Plan (left) and Section (right).

4. Conclusion

The existing drainage has enough capacity to accommodate storm water runoff up to 5-years rainfall, yet the runoff in the streetscape still occurs. Therefore, this leads to conclusion that the existing drainage system needs to be reevaluated to decrease storm water runoff significantly. The design solutions are to utilize green belt as bio-swale and calculate proper capacity towards the site's runoff condition. It will help to reduce storm water runoff up to 100%. The local plants will absorb the pollutant and will contribute to increase the quality of water within the area.

This paper is trying to enhance the hydrological analysis in the landscape design process with storm water management approach. So, hopefully the design product will effectively work and contribute to environmental sustainability. It will be much related to city planners, engineers, and landscape designer to design our environment in more sustainable ways.

References

[1] Head L 1969 Key issues in environmental change Cultural Landscapes and Environmental Change (New York: Oxford University Press Inc.) p 44
[2] Dan M B 2014 Planning and designing sustainable and resilient landscapes Green Walls (New York: Springer) p 180
[3] UACDC 2010 Low impact development A Design Manual for Urban Areas (Fayetteville: University of Arkansas Press) p 92
[4] Donald W and Adams M 1937 Design for flooding Architecture, Landscape, and Urban Design for Resilience to Flooding and Climate Change (New Jersey: John Wiley & Sons.Inc) p 92
[5] Russ T H 2002 Site Planning and Design Handbook (New York: McGraw-Hill) p 211
[6] RTRW Kota Bogor 2011-2031
[7] Harris W C, Dines T N 1998 Time Saver Standards for Landscape Architecture(USA: McGraw-Hill.Inc) p 330-27
[8] Al M G 1978 A Simple Version of Gumbel's Method for Flood Estimation Hydrological Sciences Bulletin Taylor and Francis Group p 374-380 http://dx.doi.org/10.1080/02626667809491810
[9] Sulung G 2012 Probabilitas Kejadian Hujan Maksimum untuk Perencanaan Saluran Air pada Tambang Terbuka Studi Kasus: PT Adaro Indonesia Online Publication (Bandung: Program Studi Meteorologi Institut Teknologi Bandung)

[10] Gribbin J E 2014 Introduction to Hydraulics and Hydrology with Applications for Stormwater Management (Delamr: Cengage Learning) p 219
[11] Rousse D and Ossa I 2013 Green infrastructure A Landscape Approach (Chicago: American Planning Association) p 1
[12] PUB 2011 ABC Waters Design Guidelines (Singapore: Public Utilities Board)
[13] Melbourne Water 2013 Stormwater strategy A Melbourne Water Strategy for Managing Rural and Urban Runoff (Australia: State Government of Victoria)
[14] Nigel D and Clayden A 2006 Rain gardens Sustainable Water Management for the Garden and Designed Landscape (Portland: Timber Press.Inc)
[15] Hunt W F 2015 Plant Selection for Bioretention Systems and Stormwater Treatment Practices SpringerBriefs in Water Science and Technology DOI 10.1007/978-981-287-245-6_2

The concept of Betawinese decorative in culinary zone planning at Setu Babakan cultural village

D Radnawati[1], D A Fatayati[2], D F Makhmud[1], P T Putra[1], and R M Syahadat[1]

[1] Department of Landscape Architecture, National Institute of Science and Technology
[2] PT Nindya Karya

*E-mail: daisy.arl@istn.ac.id

Abstract. Setu Babakan Cultural Village is a settlement area established by the Government of Jakarta as a place for the conservation and development of Betawinese culture. It is located in Srengseng Sawah, South Jakarta. It has a natural lake, called as Setu Babakan, as the focal point. It is one of the interesting tourist attractions for those want to enjoy the typical rural atmosphere or see the indigenous Betawinese culture directly. The lack point of Setu Babakan Cultural Village is the character as the village area based on Betawinese culture. Culinary zone as one of tourist destination has a dirty look with the number of carts and food stalls that are not well organized. Thus the visitors feel uncomfortable with the situation in culinary zone of Setu Babakan Cultural Village. To reinforce the impression as Betawinese Culture Village, the culinary zone needs to be strengthening by design. This research use planning approach to interpret the Betawinese decorative concept such as ornament, materials, and color based on Betawinese architectural philosophy. The result of this study is not only to create an interesting, comfortable, and aesthetic atmosphere of culinary zone in SetuBabakan Cultural Village but also strengthen the cultural identity that must be preserved.

Keywords: character, design, planning, settlement, tourist.

1. Introduction

Jakarta consists of many tribes from all around Indonesia archipelago. It has a local tribe called as Betawinese. Betawinese is a new tribe made from mix cultures such as Malays, Sundanese, Javanese, Buginese, Makassar, Balinese, Ambonese, Arabian, Chinese, and Portuguese [1,2]. Betawinese people mostly living in periphery of greater Jakarta. Islam the main beliefs of Betawinese people and affect their daily life and culture.

One of the area in Jakarta that inhabited by the majority of Betawinese people located on SetuBabakan in SrengsengSawah, Jagakarsa, South Jakarta. The land use of this area dominated with green open space and settlements [3]. Thus this area designated as Betawinese cultural center by the government namely SetuBabakan Cultural Village (SBCV). SBCV is the embryo of Betawinese cultural center, a place of growth of natural beauty of Betawinese tradition which includes religious, culture, and arts. The main entrance of SBCV located on M. Kahfi II Street and the side entrance is on Srengseng Sawah Street located in the southern part of the SBCV area. Visitor activities around the Setu Babakan area include sitting (relaxing), sports, enjoying views of Setu Babakan, playing, selling food (culinary area), and boating. Besides enjoying the scenery and relaxation, Betawinese culinary is also one of the tourist destinations. The problem is in the culinary area that is currently still looks slum with the number of carts and stalls that not arranged. It makes visitors feel uncomfortable. The objectives of this research is to create a concept for culinary zone in SBCV through landscape planning. Landscape planning in the culinary zone of the SBCV is basically one of the efforts so that visitors can enjoy the culinary of various regions but while maintaining the nuances of Betawi.

The planning approach used in this research is to interpret Betawi decoration concept such as ornament, material, and color as Betawi architectural philosophy so that the result of this research is not only to create an interesting, comfortable and aesthetic atmosphere but also to have a cultural identity that can be enjoyed and preserved the Betawinese culture.

2. Methods

Research was conducted on March until April 2015 at culinary zone of SBCV which are Zone B1 (3700 m^2) and Zone B2 (1800 m^2), Srengseng Sawah, South Jakarta. Figure 1 shows the site location. The data was collected through physical, biophysical, and socio-cultural observation. The landscape planning method used Gold's planning process which are preparation, site inventory, analysis and synthesis stages. At the last stage, the language of landscape theory from Spirn used as an approach to create an attractive and impressive tourism landscape [4].

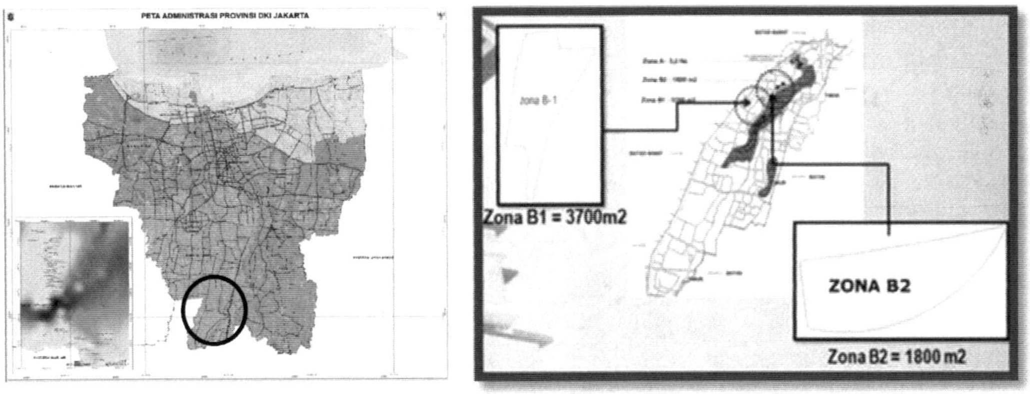

Figure 1. Culinary zone of SBCV, South Jakarta

3. Results and Discussions

3.1. Betawinese ornamental variety

Betawinese wealth of art and culture supports the creation of ornaments that characterize Betawinese architecture. The ornaments are not only as a decoration of landscapes and buildings but also have a philosophy in the life of Betawinese people.

- Lisplang

The fascia board of Betawinese, called as *lisplang*, is an ornament of teeth or *gigibalang* made from wooden boards. It has lined up triangular form to resemble the teeth of grasshoppers symbolize that life must always be honest, diligent, persistent and patient [5]. The grasshoppers can only break wood if it done continuously and usually in the long period. It also means as 'strong defense'.

- Banji

Banji is a rectangular pattern. This pattern developed from the basic ornaments of *swastika* which is the influence of Hinduism culture, means dynamic. The pattern of *banji* similar with stars or solar circulation.Thisornaments come from Chinese culture which means 'many' [6]. The philosophy of *banji* is a cheerful or the spirit of life. *Banji* ornaments are often combined with elements of plants. The most widely chosen is *bunga tapak dara* or *Vinca rosea.*

- Jasmine

Jasmine (*Jasminum sambac*) symbolizes the happiness, fragrance, and friendliness towards anyone. No wonder that Betawinese are always open to anyone who wants to visit their village. Just like jasmine flowers, cempaka (*Michelia champaca*) flower ornament is a symbol that the life of the home owner must always be fragrant and harmonious.

- Sunflower

Sunflower (*Helianthus annuus*)ornaments in the form of a translucent engraving that is usually located at the top of the door of the living room as a symbol that the life of the home owner should be an inspiration for the surrounding community. Sunflower also symbolize a source of life and light [6].

- Langkan

Langkan is a guardrail that placed on the terrace of the yard made from wood, symbolized like a human statue that also has a moral message that is good ethics in the visit must pass from the front page of the house. In Betawinese belief, a visit through the back or side of the house is an unfavorable ethic.

- Tumpal

Tumpal is a triangle form symbolizes the natural power which is consists of macrocosms, microcosms, and metacosms [6].

3.2. Site analysis

- Structure and landscape elements

The food stalls are not well managed. Figure 2 shows the actual condition of culinary zone. The food stalls owner occupied the road and not well organized so that it created a filthy environment and unattractive area. The culinary zones need to be improved as the solution in tourism planning development of SBCV. The form of the food stalls will be adjusted with the need of the owner. The Betawinese decorative concept will be applied in landscape elements in culinary zones that strengthen the Betawinese character and aesthetics landscape of SBCV.

Figure 2. The food stalls and its furniture

- Material analysis

Mostly, the material that used in the site are paving block, both pedestrian pathway and vehicle circulation. Figure 3 shows the information board and traditional gate made from natural materials.

Wood is the dominant material used in Betawinese traditional house, gazebo, and food stalls. The materials are eco-friendly and influence the visitors' perception of SBCV.

Figure 3. Materials in the site

- Vegetation analysis

In the whole area of SBCV vegetation are planted for ecology and production purposes. There are so many shades trees and fruit trees that scattered around the area. There are also no aesthetic trees for view amelioration so that makes the area quite monotone. Vegetation selection are needed to fit with the site. Figure 4 shows some existing vegetation in the site.

(a) (b) (c)

(d) (e) (f)

Figure 4. Existing vegetation in SBCV (a) tanjung (*Mimusops elengi*); (b) glodokan tiang (*Polyalthia longifolia*); (c) matoa (*Pometia pinnata*); (d) durian sitokong (*Durio zibethinus* var. Sitokong); (e) bunga kupu-kupu (*Bauhinia purpurea*); and (f) sawo duren (*Chrysophyllum cainito*)

3.3. Theoretic approach

Betawinese ornaments are widely known by the community is a tooth ornament, called as *gigibalang*, that is located in the roof. This ornament is widely known because it is often applied to building public facilities such as schools, bus stops, gates, or government offices in Jakarta. The form of Betawinese ornaments comes from plants because the Betawinese people are close to nature. Betawinese people are also known as a society that is obedient to the religion it embraces (Islam). This can be seen in many buildings that have a green and yellow color that characterizes religious home owners of religion [5].

Planning in this research use approach of using theory of the language of landscape [4]. The theory is a theory that can be used in planning and design to create an attractive environment and have the impression and message so memorable and liked by the users of the site. As a communication, the contents of the theory include emphasis, framing, exaggeration, distortion, alliteration, echoism, assonance, rhythm, metaphor, euphemism, antithesis and oxymoron, and aposiopesis.

3.4. Spatial concept

Decorative variety used in the concept of space is a variety of ornamental *banjil* which is defined as the sun that means joy and spirit of life. The form of "half-sun", shows in Figure 5, in the Betawinese philosophy symbolizes "Life is coming forward with Full of Spirit". The shape of the half-sun is chosen to adjust the identity on the Setu Babakan site which is a tourist area of Betawinese culture located in Jakarta.

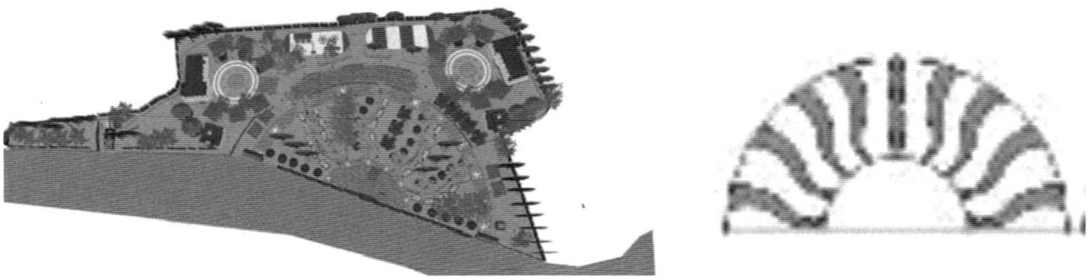

Figure 5. The form of half-sun in Zona B1

3.5. Vegetation concept

The vegetation concept using a typical local Betawinese vegetation that provides identity and education to the user. The vegetation concept also should have architectural, social, cultural, and aesthetic functions. Figure 6 shows some scented flower used in Betawinese decorative concept such as *Jasminum sambac, Rosa sinensis,* and *Michelia champaca*. Based on Jakarta's Governor Decree No. 2359/1987 about Betawinese unique plants, there are seven types that are still many found in the yard of SBCC such as *jambu* (*Psidium guajava*), *jambubol* (*Syzygium malaccensis*), *matoa* (*Pometia pinnata*), *mengkudu* (*Morinda citrifolia*), *rukem* (*Falcourtia rukam*), *sawo manila* (*Manilkara zapota*), and *sirsak* (*Annona muricata*). Therefore, the selection of the identity-enhancing vegetation that will be used in SBCC include *tanjung* (*Mimusops elengi*), *buni* (*Antidesma bunius*), *belimbing* (*Averrhoa carambola*), mengkudu (*Morinda citrifolia*), *mangga* (*Mangifera indica*), *sawokecik* (*Manilkara kauki*), and *menteng* (*Baccaurea racemosa*).

(a) (b) (c)

Figure 6. Scented flowers used as decorative concept (a) melati (*Jasminum sambac*); (b) mawar (*Rosa sinensis*); and kenanga (*Cananga odorata*)

Figure 7 described the spatial concept of zone B2 derived from the form of jasmine. Jasmine flowers symbolize sanctity or it can be interpreted that the SBCV region has a mission to preserve the authenticity of Betawi culture in Jakarta. In addition, the form of jasmine flowers selected to adjust identity on the footprint of SBCV.

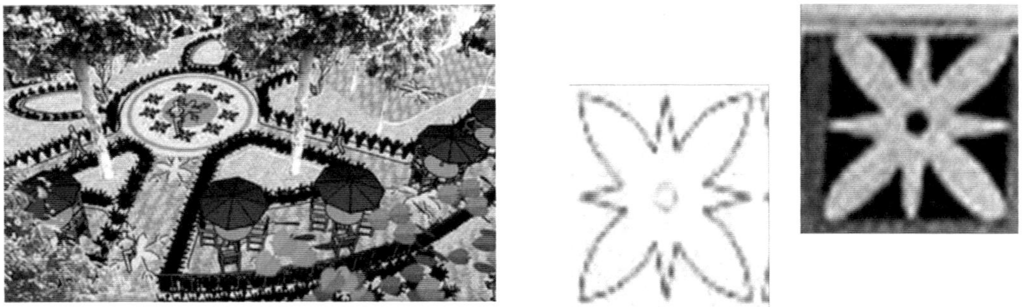

Figure 7. The spatial concept of zone B2 from the form of jasmine

The concept of vegetation that is presented also provides aesthetic value and provides a special characteristic to give identity, create elements of beauty, and soften the atmosphere on the site. The use of crop ping pattern in cultivation landscape planning of SBCV uses several cropping patterns such as linear pattern, distribution, hierarchy, and concentric. The shape of the jasmine (*Jasminum sambac*) ornament is applied to Zone B2. Figure 8 showed the vegetation concept both B1 and B2.

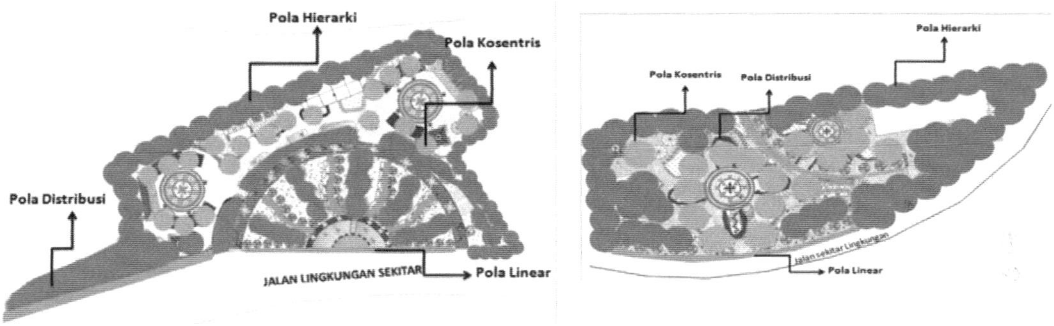

Figure 8. Vegetation concept at zone B1 (left) and B2 (right)

3.6. Colors concept

The color selection is an important part of the planning process. The use of colors also helps the nuance of space [7]. Figure 9 showed the colors concept. In this study, color selection is based on site characteristics. Colors to be planned using the colors typical Betawi include soft colors such as brown and old-brown that are more modern-nuance while the colors are dominantly bright using the color green and yellow which is the color of Betawinese people who live on the outskirts of Jakarta. The acculturation element of Chinese culture in the form of the use of red color will also be applied in the site. Red symbolize a leadership, energy, love, and warmth, yellow symbolize prosperity and wealth, green symbolize calmness and harmony with the nature [6, 7]. Different color combinations will create harmonization on the site planning that results in a culinary landscape identity at SBCV.

Figure 9. The colors concept

3.7. Materials concept and landscape buildings

The concept of material and landscape buildings in the culinary zone of SBCV interprets the Betawinese culture which makes an identity on the site. Material concept on B1 and B2 zone treads includes carts, gazebos, dining chairs, stalls, lamps, lanterns, garbage bins, and others. These buildings and appendages will be displayed using materials appropriate to environmental conditions and providing symbolism as a culinary area. Materials or materials to be used include wood, iron, concrete, bricks, ornamental stone, and stone times. While the image will be displayed using many organic patterns (arch) and some in the form of a line (geometric).

- Cart

The cart has a length of 120 cm and a width of 80 cm. Figure 10 showed the proposed idea for cart. Ornaments used are ornaments of Betawi form philosophy such as the form of half the sun, jasmine flowers, and *lisplang gigibalang* material. In addition, the wood is also resistant to pressure and flexible. In terms of aesthetics of wood has a good texture and beautiful for various types. And in terms of physical properties of wood has a fairly light weight that can float and resonance properties.

- Gate

Figure 11 showed the proposed gate for culinary zone. In the welcome area will be planned a gate at the entrance with traditional accents using Betawinese philosophy ornaments such as *lisplanggigibalang*and formation of sun patterns on the gate column. The size of the gate is 3.5 mlong, 2.8 m high and 0.8 m wide. The material used in the form of concrete because it can last long and strong in the structure.

- Gazebo

The planned gazebo on the site has a size of 3 m x3 m. The material used is wood with the intention of providing ease in staining and anti-rust material. There are half sunflower ornaments, and *lisplang gigibalang* with the intention to characterize the identity of the site. The illustration of gazebo can be seen on Figure 12.

- Kiosk

The kiosk available on the site consists of Betawinese and non-Betawinese food stalls, kiosks of equipment, and souvenir kiosks. Besides offering the local food of Betawinese, the kiosks also serve non-Betawinese food. The kiosks are managed by local people in order to improve the local income and welfare. Figure 13 showed the illustration of the proposed kiosk.

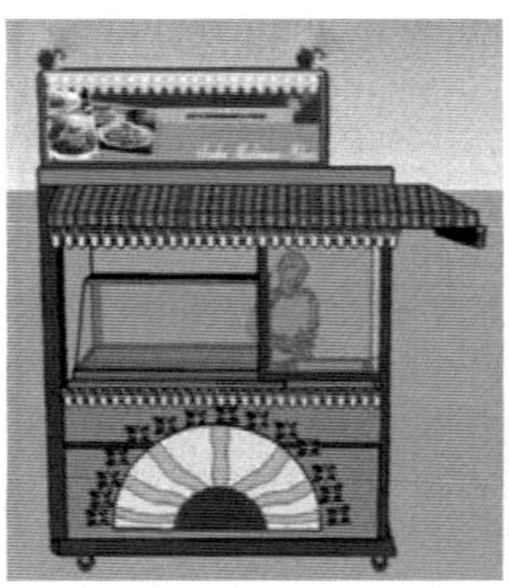

Figure 10. The Cart

Figure 11. The gate

IOP Conf. Series: Earth and Environmental Science **179** (2018) 012020 doi:10.1088/1755-1315/179/1/012020

Figure 12. The gazebo

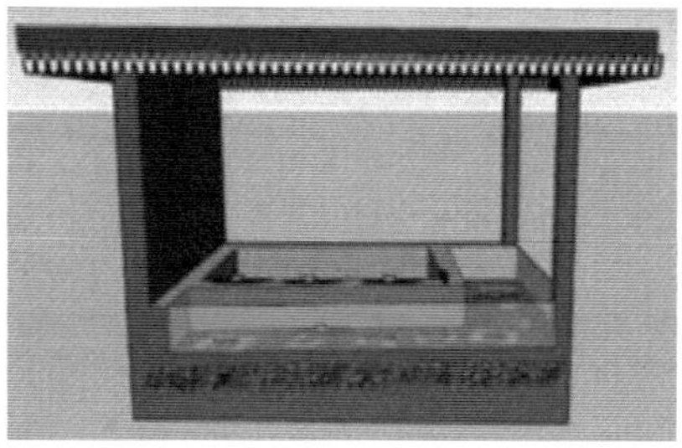

Figure 13. The kiosk

4. Conclusions

The culinary landscape planning of SBCV is basically one of the efforts to make SetuBabakan as a recreational area that has a special culinary zone where visitors can enjoy the culinary from various regions but while maintaining the nuance of Betawi and culture contained in SetuBabakan. The culinary zone B1 and Zone B2 planned as Betawinese culinary landscape of SBCV interpret the concept and forms of ornaments, materials, and colors from the characteristic or Betawinese philosophy. The concept is not only creating an interesting atmosphere, comfortable, and aesthetic but also have a cultural identity that can be maintained. Thus an integrated planning needs to be done to create an exciting culinary landscape and can provide benefits to visitors. Figure 14 shows the concept illustrations of Betawinese decorative in culinary zone B1 and B2.

To compensate for the increasing demands of the needs and desires of visitors who want to enjoy the tour at the SBCV required the addition and repair of some facilities either in the form of social facilities, sports, and recreation that has a nuance of Betawi with the intention not to override the original local culture. So hopefully in the development of SBCV itself becomes one of the interesting tourist attraction for tourists who want to enjoy the typical rural atmosphere or see the original Betawinese culture directly.

(a)

(b)

Figure 14. Concept illustrations of culinary zone in B1 (a) and B2 (b)

IOP Conf. Series: Earth and Environmental Science **179** (2018) 012020 doi:10.1088/1755-1315/179/1/012020

References

[1] Blackburn S 2011 Jakarta Sejarah 400 Tahun (Jakarta: Masup Jakarta)
[2] Hidayah Z 2015 Ensiklopedi Suku Bangsa di Indonesia (Jakarta: Yayasan Pustaka Obor Indonesia)
[3] Zaenuddin H M 2012 212 Asal-usul Djakarta Tempo Doeloe Disertai Fakta, Foto, dan Kesaksian Unik Yang Membawa Anda ke Masa Lalu (Jakarta: Ufuk Press)
[4] Spirn A W 1998 The Language of Landscape (London: Yale University Press)
[5] Chaer A 2012 Folklor Betawi Kehidupan dan Kebudayaan Orang Betawi (Jakarta: Masup Jakarta)
[6] Swadarma D, Aryanto Y 2013 Rumah Etnik Betawi (Jakarta: Griya Kreasi)
[7] Susanta G 2007 Lantai (Jakarta: Penebar Swadaya)

IOP Conf. Series: Earth and Environmental Science **179** (2018) 012021 doi:10.1088/1755-1315/179/1/012021

Sustainable development concept of rain harvesting for public flat in Tamansari Village, Bandung

F F Devitama[1*], B Paramita[1]

[1]Dept. of Architectural Education, Universitas Pendidikan Indonesia

[*]E-mail: fahmifauzandevitama97@student.upi.edu

Abstract. Bandung is the capital city of West Java province in Indonesia. The city is located on 768 meters (2,520 feet) above sea level and lies on a river basin surrounded by volcanic mountains. The annual rainfall is considered as high level. The most rain happen on January with total precipitation is 973.0 millimetres, while the less rain happen on August with total precipitation is 70.0 millimetres. In addition, La Nina phenomenon occurs lately to trigger high rainfall level in Bandung city. Based on Meteoblue Bandung City, the average of precipitation in Bandung is 2164mm / year. The potential of this large rainfall later can be utilized as water reserves, in urban settlements with high population density. In line with the Mayor of Bandung city Regulation (Perwal Bandung) no. 1023 year2016 about the Green Building mentioned that any building with area more than 5,000 m^2 requires to preserve balance of environment, including concerning social, cultural and ecosystem. Later, this paper aims to explore the sustainable development concept of rainwater harvesting for a public flat in Taman Sari, Bandung. The need for clean water for the residents is as much as 102,200,000 litters/year for 2800 occupants. The design of Tamansari Village public flat then able to accommodate 23,700,993.6 litters/year, which means 23% of the total needs of clean water.

Keywords: Concept, Rain Harvesting, Sustainable

1. Introduction

West Java province is the most population province in Indonesia. The capital city of West java province, Bandung city, is a fast growing metropolitan city. The city is recently facing environmental problem and rapid urbanization. The most critical environmental issues is related to water resources, including flood, pure water supply crisis, water pollution and ground water level. Flood disaster in Bandung city is a common news topics whenever rainy season. [1]. Conversely, in dry season, the city has problems related to pure water scarcity faced by the residents. Overall, the supply of pure water by Water Supply Company (PDAM) is only available to serve36.5 million m^3 per year, including 40% of surface water and 60% ground water (BPS, 2015). As consequences, the new groundwater companies were established because the PDAM cannot provide public demand of water. The decrease of groundwater level has been happened since 1980s. This is related with the rapid development of industrial and residential settlements. The most severe ground water levelling occurred in industrial areas and high density of settlements [2].Meanwhile, rainwater utilization technology as a source of water has not been explored optimally yet. The study and implementation of rainwater harvesting in Indonesia is still less number compared to the drought area. The study in Petra University shows that building able to harvest 100% of rainwater based on WAC (Water Conservation) in GBCI Greenship methods [3]. Rainwater harvesting also conducted in IPB Dramaga Campus, which potentially amounted to 319.734 m^3 per year or equal to 52% of the average annual needs.[4]. The study in Novotel Hotel, Yogyakarta, during the rainy season, rain water can supply up to 21% of total water consumption in a month. Annually, the rain water can supply the water consumption of the hotel for about 8.6%.[5]. Meanwhile, rainwater harvesting technology as a source of clean water in households are still not optimal. This paper then explores the concept of sustainable in high density areas such as Tamansari Village for rainwater harvesting of public flat.

The location has been selected based on the Urban Slums Development Plan (RKP-KP) 2015.Those study shows that Tamansari Village is one of urban villages identified as having severe slum conditions (PU, 2015).Public flat here is a plan for the development of vertical housing in densely populated areas.

2. Public Flat in Tamansari Village, a Densely Populated Area in Bandung

Bandung was a high density of population in West Java and second city after Jakarta. In line with the strategic plan of Ministry of Public Work and Housing [6], Bandung City Government has planned the location of vertical housing development including on high density population, Tamansari Village.

Bandung was located in 6°54S and 107°3E with elevation on 768m above sea level. Bandung city is famous as a mountainous city with moderate average temperature 26 °C on 2017. The precipitation average was 2.164 mm/year. The rate of precipitation was consider decreased 184.74 mm on 2015 198.8 mm on 2016.

2.1. Study Area

The area of public flat is 36.508 $m^2$with elevation is 725 to747 m above sea level. The geographical coordinate of area is 6°53'47.01"S and 107°36'24.71"E. The boundary of the site as follow: Cikapundung riverside, Jl. Plesiran, Jl. Kebon Bibit and local settlement (Figure 1). The existing area consist of building (90%) with few open space. The rainfall precipitation average is 10 – 20 mm per days.

Figure 1. Area of Tamansari Village in Bandung City
(Source :http://static.panoramio.com/photos/large/41790853.jpg)

Population in this region is about 2.400 people with 1.080 households. The average population aged around 40-60 years, with the number of children in each family is about 2 children. Beside that age, there are also people whose age 60 years old and over.

Cikapundung River is the main river of Bandung City and cross Tamansari Village. The problems in this area is number of people who still throw waste into the river. The water supply for residents who rely on river water cannot be fulfilled. The Water level is reduced due to siltation of the river caused by piles of garbage, other than the potential of flood disaster on rainy season that was also a problem for Tamansari Village district.

3. Rainwater Harvesting Concept

Rainwater Harvesting is a rainwater collection system for reuse in daily activities, such as for watering plants, flushing water, drinking water for farm animals, irrigation water, washing, and others. Rainwater is also very suitable to be utilized for drink water resources. The amount of rainwater is abundant and the quality is better than river water. It takes a bit of processing to be able to use it as drinking water.

Figure 2. a. Retention pool. Source: images.google.com b. Gutter. (Source: images.google.com)

Rain water harvesting is aimed to take advantage of rain water. There are various methods to do this rainwater harvesting. The most simple and most applicable method is to utilize the roof as a medium of falling rainwater, then channeled through a horizontal gutter which channeled down and combined with retention pool (Figure 2b) [7].

Juliana has studied the rainwater harvesting (RWH) system based on tank capacity and catchment area [8]. The RWH system is capable of replacing up to occupant and 72% for 4 occupants, especially for areas with annual rainfall above 2500 mm / year. The effectiveness of tank storage is 2 m^3, but for more efficiency, the capacity can be reduced to 0.5 – 1.0 m^3. The extensive catchment area that is effective for capturing rainwater is at least 70 m^2.

The buildings such as flats with roof to land area can accommodate large water discharge. The water harvesting system is more appropriate to use for the following reasons (Chris Reardon, 2013):

a. Water harvesting uses a simpler system than recycling water. For example, to utilize waste water, necessary process and more complex systems such as cleaning and sterilizing water.
b. Water recycling can damage / contaminate groundwater reserves because waste water contains chemicals that may be harmful to health. Thus, in addition to the substance is damaging the soil, the substance can also contaminate the wells in the vicinity.

Rainwater collected can be move into the soil so that water reserves in the soil can be maintained. In addition, it also aims to maintain soil fertility so as not to decrease soil. Given the design area in the middle of the city that has a little water absorption

Based on Green Building Council Indonesia (GBCI), the score for Water Conservation (WAC) mention about rainwater harvesting as follow [9]:

1 star = 50% of surface roof area can harvest the water
2 star = 75% of surface roof area can harvest the water
3 star = 100% of surface roof area can harvest the water

The Water conservation method was then adopted by the Government of Bandung City which stated in Mayor of Bandung Regulation (*Perwal Bandung*) no. 1023 of 2016 about the Green Building. [10]

Water Storage can be accommodated in the upper water tank and bottom water tank, depending on design of water tower. The tank will be stored on top so that water distribution can be carried out using gravity (Figure 3). In addition to the above, there is also a tank below which will be discussed at the next point.

Figure 3. Left: The example of placing Rainwater storage on a flats. Right: The water storage must be placed in service area. Source: housing-estate.com.b The tank for harvesting water (Source: http://storagetangkindonesia.co.id)

In order to maintain the aquifer (groundwater reserves), the system must consider water percolation so that ground water reserves are maintained. In addition, percolation is aimed to maintain the soil surface height so as not to decrease soil.

Water that was collected, some of which must be collected into the soil so as to maintain soil fertility on the land. In order to take rain water into soil, geo textile was used on shallow soil surface so that water run off can be provoked to enter underground tank storage. The underground tanks storage were intended to add water reserves, as well as to maintain soil fertility because water can be absorbed directly when the soil is dry. The scheme of rainwater harvesting method is explained in Figure 4.

That scheme in Figure 5 shows that the rainwater harvesting is a simple idea. First, the rainwater precipitation which fall on the roof, is collected to the water tank through gutter. Water will be distribute to the tower. Once it filtered, it may consume as a fresh water.

The other method for rain harvesting is collecting from the water runoff which through grill and store at the ground reservoir. After filtered and pumped it is able to consume as a fresh water. This scheme at Figure 6 shows how the rainwater which fall on the asphalt can be collected.

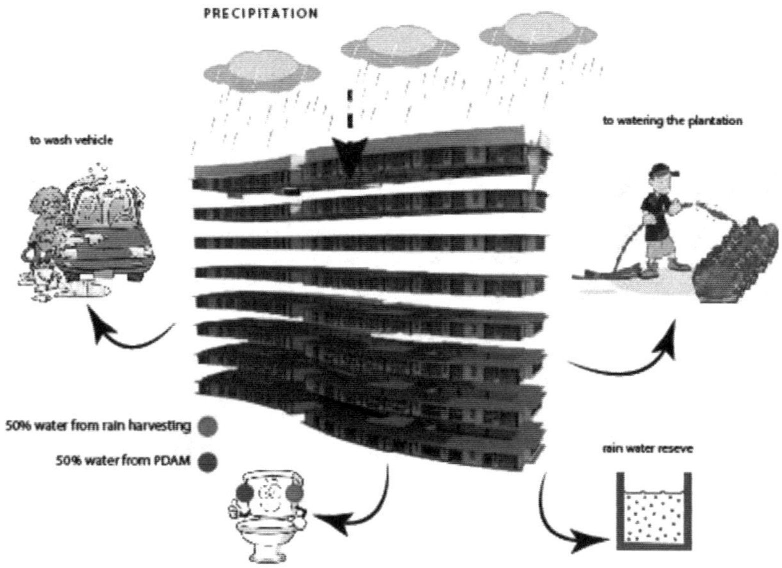

Figure 4. Rain harvesting concept in Tamansari Village Flats

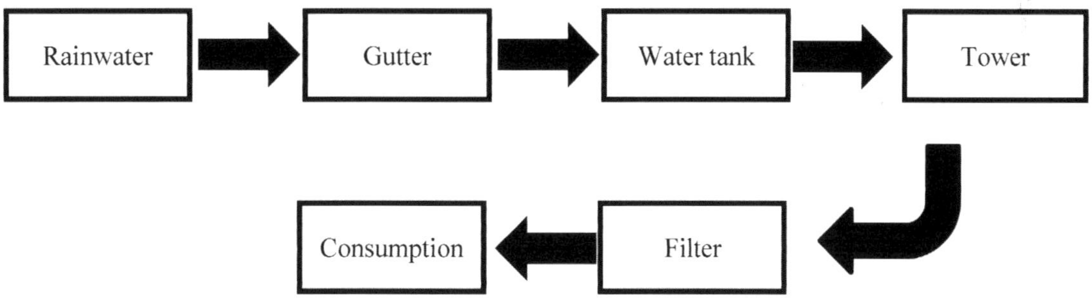

Figure 5. Rain harvesting method through gutter and vertical water tank in Tamansari Village Flats

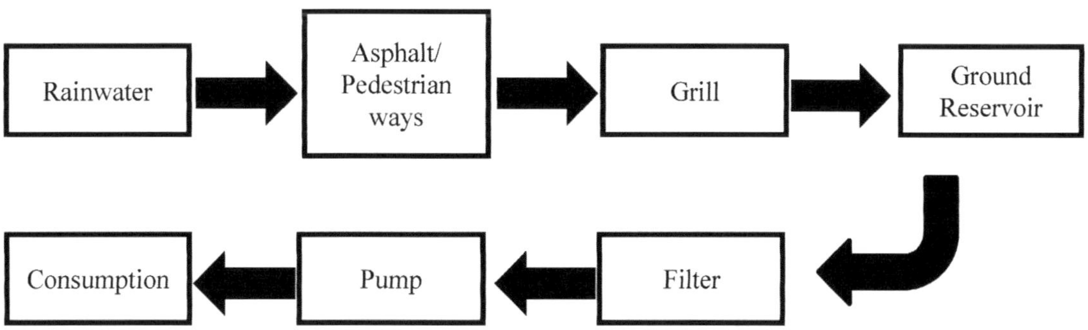

Figure 6. Rain harvesting method through grill and stored at ground reservoir in Tamansari Village Flats

Placement of a water tank, this should consider the height of the building. The rate of water velocity due to the force of gravity needs to be considered the slope of the gutter. In addition, design considerations according to dimensions and volume for aesthetic purposes are also important. Thus, as a flat with 8 stories, the flow of rainwater from the roof, will be distributed into 2 directions, the first gutter will serve 5 – 8 stories, while on second gutter will serve 1-4 stories. The scheme of this distribution as seen at Figure 7.

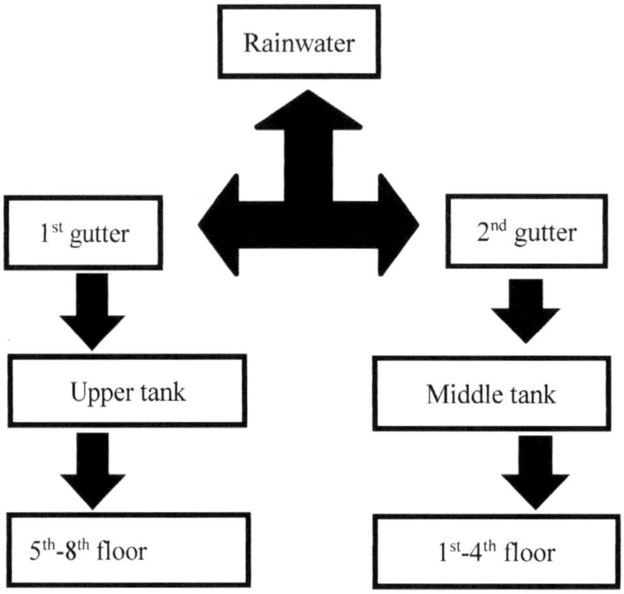

Figure 7. Rain harvesting distribution through gutter and floor distribution

The rainwater contents the chemical substance such as SO_4 and H_2S, especially in urban region which is directed from traffic and industrial pollution. The chemical substance very dangerous and may harm for human health. The layer in the water storage with minimum 20 cm gravel at least able to hold back precipitates containing impurities. For a better filtration, make the addition of fiber, sand, coconut charcoal, fiber and on the top layer covered with brick collision. The layer of water filtration as seen at Figure 8.

Figure 8. The filter to remove chemical substance in rainwater

IOP Conf. Series: Earth and Environmental Science **179** (2018) 012021 doi:10.1088/1755-1315/179/1/012021

4. Rain Water Harvesting Design and Calculation

4.1. Flat Design

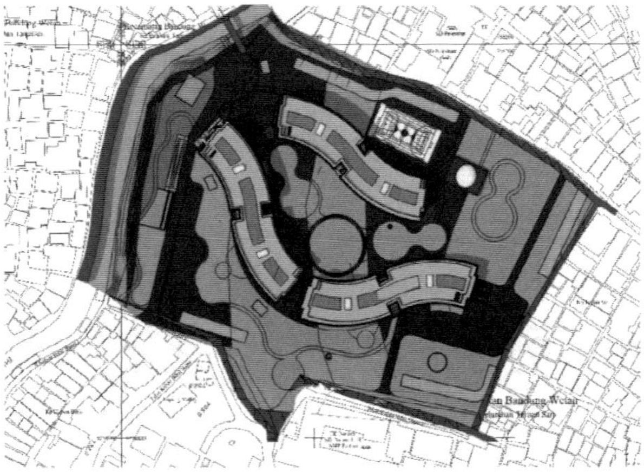

Figure 9. Site Plan of Public Flat

Figure 10 shows the rooftop plan that able to utilize rainwater harvesting, the roof has designed in flat design to collect the water easily than diagonal roof, and meanwhile the system of utilization from the water tank to the building is seen at the flat perspective at Figure 11 and Figure 12.

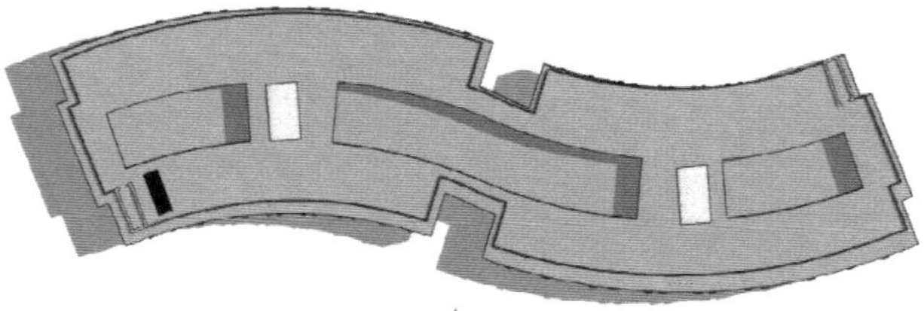

Figure 10. Roof Top Plan

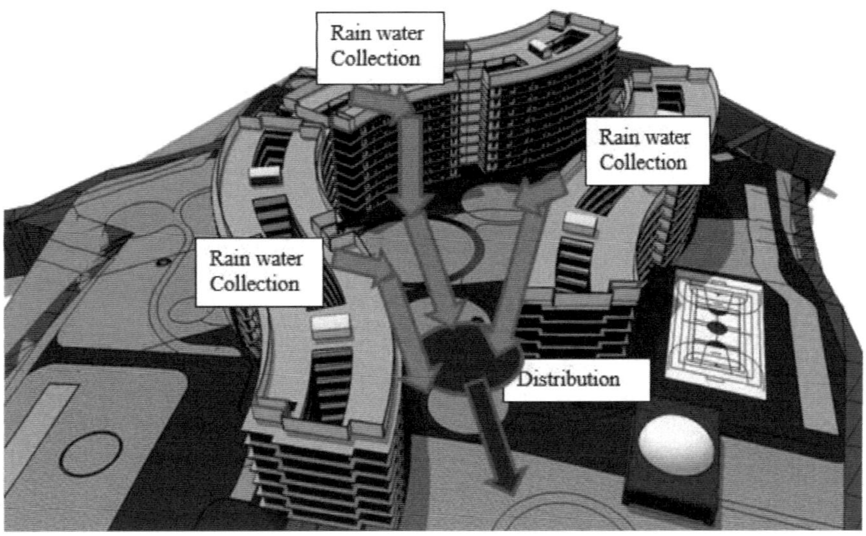

Figure 11. Water storage system and distribution

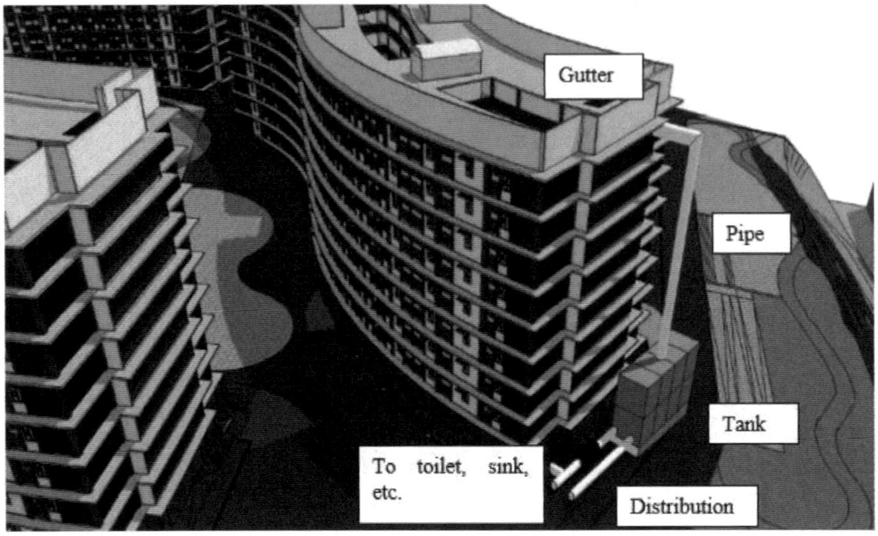

Figure 12. Flat RWH system

4.2. Rain Water Harvesting Calculation
The rainwater harvesting will be calculated based on the precipitation data, and the following is the amount of discharge water that can be accommodated:

Roof area that can harvest the water and distributed to the tank:
= 3,650.8 m^{2}/ roof
= 3,650.8 m^{2}/ roof x 3 towers
= 10,952.4 m^2
Rainfall precipitation: 2,164 mm / year

The amount of water that falls to the surface of roof:
= 10,952.4 x 2,164
= 23,700,994 liters of water per year
Which means the maximum (in average) of rainwater in 1 day will be: 64,935 liters

The total needs of clean water for residents: 100 liters/person/day, then in 1 year:
= 100 x 2,800 people x 365 days
= 3,894,200 litters
= 102,200,000 litters
Based on the calculation above, by accommodating the rain water it is about 23% of total water needs will be supplied from the rainwater harvesting.

The number of water tanks required to accommodate this rainwater harvesting will be:
64,935 litters/11,000 liters ≈ 6 water tank.
@tank capacity is 5,500 litters because in one spot, we must to divide the tank into 2 tank.
Figure 6 shows the place of the tank in the site, there are 6 water tank spot that utilized in the site.

The rainwater not only fall on the roof, but it fall on site (asphalt) too. It will be calculated based on the precipitation data, and the following is the amount of discharge water that can be accommodated:

Site area is 36800 m^2. Asphalt presentation area that can harvest the water and distributed to the tank:
25% x 36,800 m^2 = 9,200 m^2
Rainfall precipitation: 2,164 mm / year

The amount of water that falls to the surface of roof:
= 9,200 x 2,164 mm / year
= 19,908,800 liters of water per year

Based on the calculation above, by accommodating the rain water it is about 19% of total water needs will be supplied from the rainwater harvesting.

5. Conclusion

The ability of Tamansari Village flat to accommodate rainwater harvesting has performed the possibility to use water storage concept and method that was in line with design and planning. The roof top design is significantly able to influence the method of rainwater harvesting. The flat roof is able to accommodate 80% if the total area of the rooftop. The roof design in Tamansari Village flat shows that the rainwater harvesting is possible to obtain 23% (23,700,993.6 litters) amount of water supply, and by the ground cover (asphalt) is possible to obtain 19% (19,908,800 litters) amount of water supply. Based on Bandung city data, the limitations of PDAM in providing drink water supply in the drought months is possible to be reduced by as much as 60%. The implementation of Concepts and methods of rainwater harvesting will overcome resource of water supply during dry season. The water footprint capacity of more than 50% of the area will also be one of the requirements of the Green Building Council of Indonesia (GBCI).

References

[1] BNPB, "Data Bencana Kota Bandung," 2014.
[2] E. Maharani, "Muka Air Tanah di Bandung Turun 75 Meter," *Republika*, Jakarta, Sep-2015.
[3] T. Steviani Dewi, P. Jimmy, and S. Nugroho, "Kajian Penerapan Prinsip Water Conservation Sesuai Standar Greenship New Building Versi 1.2 Studi Kasus: Gedung P1 Dan P2 Universitas Kristen Petra Surabaya," *J. Dimens. Utama Tek. Sipil*, vol. 2, no. 1, 2015.
[4] T. Hamonangan and Y. Sugiarto, "Analisis pemanenan hujan dari atap bangunan: studi kasus gedung-gedung dikampus IPB Dramaga Bogor," Bogor, 2011.
[5] A. S. Fathi, S. S. Utami, and R. Budiarto, "Perancangan Sistem Rain Water Harvesting, Studi Yogyakarta, Kasus: Hotel Novotel," *Teknofisika*, vol. 3, no. 2, pp. 35–45, 2014.
[6] KemenPUPR, "Renstra KemenPUPR 2015-2019," Jakarta, 2015.
[7] N. I. Said and W. Widayat, "Bab 6 Pemanenan Air Hujan," in *Pengisian Air Tanah Buatan, Pemanenan Air Hujan Dan Teknologi Pengolahan Air Hujan "Studi Kasus Kota Depok*, 1st ed., Jakarta: BPPT - Kelair, 2014.
[8] I. C. Juliana, M. S. B. Kusuma, M. Cahyono, H. Kardhana, and W. Martokusumo, "Performance of rainwater harvesting system based on roof catchment area and storage tank capacity," in *MATEC Web of Conferences*, 2017.
[9] GBCI, "Green Ship Untuk Bangunan Baru," no. April, 2013.
[10] Pemkot-Bandung, *Perwal Bandung No 1023 Tahun 2016 - Bangunan Gedung Hijau*. 2016.

IOP Conf. Series: Earth and Environmental Science **179** (2018) 012022 doi:10.1088/1755-1315/179/1/012022

Green space planning based on biophysical and local wisdom of Luhak Agam in Lubuk Basung, Agam Regency, West Sumatra Province

S K Apriliani [1*] and Q Pramukanto [2]
[1] Alumni of Department of Landscape Architecture IPB, Indonesia
[2] Department of Landscape Architecture IPB, Indonesia

*E-mail: simplisittii@gmail.com

Abstract: Lubuk Basung is the capital city of this regency which is currently being developing and increasing in the population. By increasing those development activities in the city, attention to protect and anticipate the problem lead to environmental degradations is needed. Moreover, any development activities in the urban area almost creating the homogeneity in spatial development and ignored the local characteristics. One of the preventing action of these environmental problems and creating local characteristics is to conduct spatial planning for the existence of green space. In order to be able to reduce the homogeneity in creating of green space planning, integrating the local wisdom in this research is needed. The objective of the research is to understand the characteristic of existing green space, identify and analyze the green space needs, identify the value of local wisdom, and to plan the green space in Lubuk Basung city. The research consists of preparation, inventory, analysis, synthesis, and planning. The analysis was conducted to determine the green space (biophysical-based green space) toward to protected area, green space along the river corridor, and green space for urban community needs (social-based green space). It was evaluated by comparing those with the existing green space. The result of analysis was used as a reference in directing green space planning to be developed by integrating with Luhak Agam local wisdom value (cultural-based green space). The final result of this research is arranged as green space planning in Lubuk Basung City. Those green space planning consist of green spaces for protected forests, limited production forests, agricultural cultivation areas, on building areas, greenway for the river, road power line corridor, and parks.

Keywords: biophysical-based, cultural-based, green space planning, local wisdom, social-based

1. Introduction

Lubuk Basung city is the capital of Agam Regency which is in a developing condition. In the past, Agam Regency was included in the Luhak Agam (one of the three indigenous Minangkabau areas). It was recorded that population growth rate increased per year, due to an increase of development activities in this city. This needs to be considered and anticipated in order not to cause further problems such as degradation of environmental quality. One form of anticipation of this is to plan the existence of Green Space well as a form of urban spatial arrangement.

Green Space (GS) in the city are open spaces in various places in an urban area that are optimally used as a reforestation area and function both directly and indirectly for the sustainability and beauty of the environment [4].The condition of the city of Lubuk Basung as the central government area of Agam regency as well as the Hinterland area (the area of food staple supplies) shows the availability of green areas such as protected areas and agricultural areas but lessen of Green Spaces. Therefore, it is necessary to do the planning of Green Space in this city.

Each region has distinctive characteristics and identities that are different from other areas as well as the area of Lubuk Basung City. The life of local people who always interact with each other and with the environment produce local values of local wisdom based on customs and religion. Local wisdom is a legacy of ancestors that blend in the form of religion, culture, and customs. In its developed people to be adapted to their environment [6].

Local characteristics and local identity of Lubuk Basung can be valued as Identity and uniqueness that will enhance and strengthen the value of Lubuk Basung area. The value of local wisdom to the natural history, customs, and culture of Minangkabau born in Luhak Agam area can be expressed not only with symbols and ornaments but also with the spatial arrangement of territory. One of them is the arrangement of Green Space. In addition, the GS plan can provide a local identity that can reduce the value of homogeneity of similarity of existing green space structure in Indonesia.

The objectives of this research are: (1) to identify the characteristics and distribution of existing Green Space in Lubuk Basung, (2) to identify and analyze the needs and distribution of Green Space in Lubuk Basung, (3) to identify local wisdom of Luhak Agam community in Lubuk Basung, 4) to arranging the GS planning by integrating GS with local wisdom of Luhak Agam community in Lubuk Basung City, Capital of Agam Regency.

2. Methodology

2.1. Location, Time Equipment and Material

This research was conducted in Lubuk Basung City, the Capital of Agam Regency. The city occupied an area about 123.1 km^2 or 12.311 ha. The research activities begin from April to August 2016. Equipment used in site inventory include laptops, digital cameras, questionnaires as well as drawing tools and stationery. While for data processing using some software like MS Word, MS Excel, Arc GIS 9.3, and Adobe Photoshop. The materials used are satellite imagery, base map, thematic maps, biophysical data, socio-cultural data, and interview data.

2.2. Research methods

The research was conducted in five stages, namely preparation, inventory, analysis, synthesis, and Green Space planning (Figure 1). Preparation of research includes the preparation of research proposals, making detailed research activities, preparing letters and administrative supplementary documents, details and tracking of required data sources, and preparation of tools and materials for research.

Inventory consists of data collection, data processing, and checking data to the field. The primary data obtained through interviews with resource persons and direct field observation while the secondary data obtained through literature study and documents from relevant local government agencies. Data used in this study include administrative data, biophysical data, and socio-cultural data. Administrative data obtained in the form of administrative area and regulations related to green space planning. Biophysical data was obtained in the form of topographic data and slope, geology and soil, climate, natural infrastructure, city infrastructure (road and electricity network/power line), and land cover. While socio-cultural data obtained in the form of the demographic population, toponymic aspects (origin of the name of the region), aspects of custom and religion, and ethno botany (local plants and plants that interact with humans since the past until today).

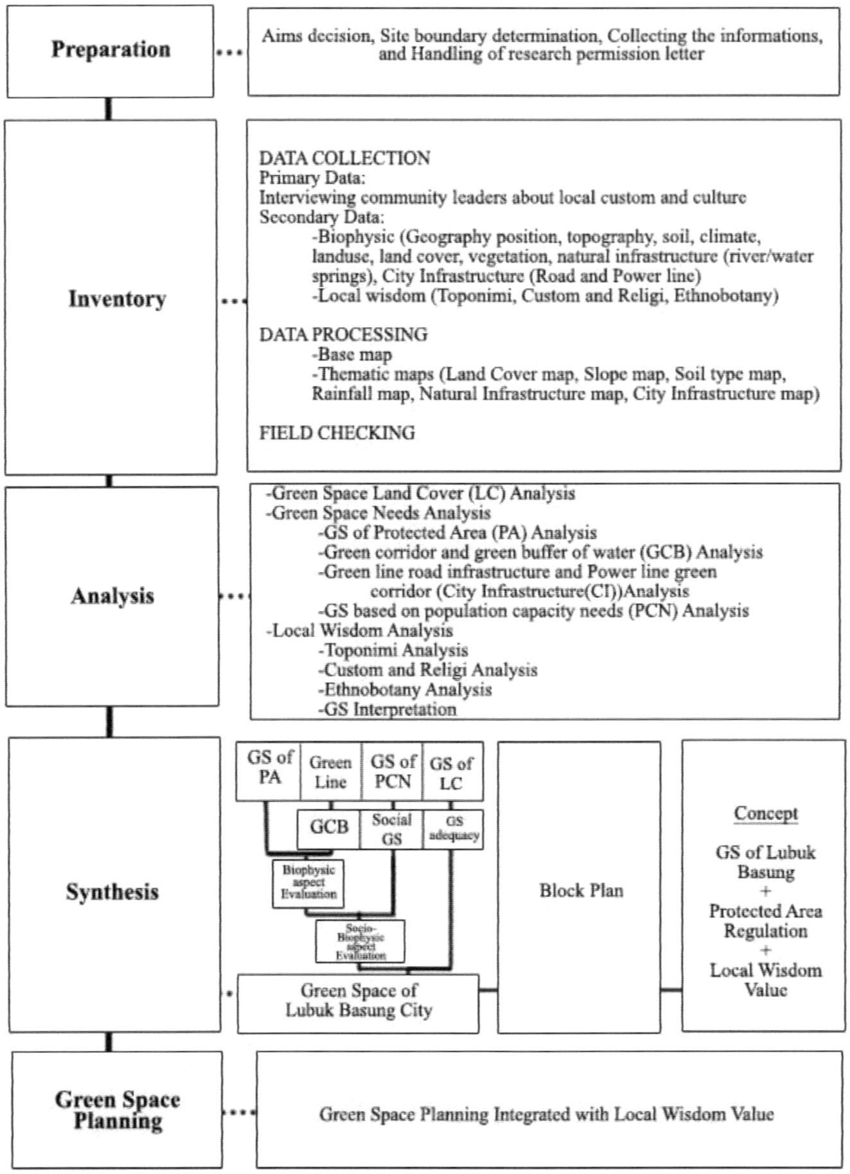

Figure 1. Stages of the Research Flow

The analysis was conducted descriptively and spatially. The analysis consists of GS analysis and local wisdom value analysis. The green space analysis was conducted to determine the availability of existing green space, biophysical function (protected area, local protection area, and urban green infrastructure), and green space based on per capita population needs for social space.

GS analysis was divided into 4 (four) part of the analysis that is: (1) Analysis of green space dissemination based on land cover, (2) Analysis of Protected Area based on Decree of Minister of Agriculture no. 837 / Kpts / Um / 11/1980, (3) Local Protected Area Analysis under Law no. 26 of 2007 on spatial planning and Decree of Minister of Public Works. 28 / PRT // M / 2015 on the determination of river corridor and water body border area, (4) GS analysis based on population needs to GS which is regulated by Minister of Public Works No. 05 / PRT / M / 2008 (For social space). Data of population used is based on the projected population growth in 2020.

Furthermore, the analysis of local wisdom values consists of: (1) Toponymic analysis based on place name, origin, and regional typology, (2) Analysis of custom and religion aspects based on custom institutional analysis, values, procedures, and local special provisions, and (3) Ethnobotany analysis based on community perception of some local plants.

The synthesis was performed base on Green Space evaluations. Those evaluations consist of green space for protected area, green space for river corridor and setback (buffer) of water body and infrastructure corridor, green space for community (requirement of green population per capita) and existing Green Space. Evaluation is also conducted to determine the adequacy of existing GS of 30% as a minimum requirement of urban green space based on Decree of Minister of Public Works no. 05 / PRT / M / 2008 on guidelines for the provision and utilization of green space in urban areas. The result of synthesis above is followed by creating a block plan and the formulation of planning concept.

Green Space plan was conducted by integrating blok plan with local wisdom value of Luhak Agam region in Lubuk Basung capital of Agam Regency. This GS planning product will illustrakte the distribution of green spaces, local wisdom elements, and the alternative plan of plant species selection.

3. Results and Discussion

3.1. Biophysical Aspects

3.1.1. *Green Space of Land Cover*

Based on land cover data provided by Local Development Planning Agency at Sub-National Level of Agam Regency 2010, Lubuk Basung has ten types of land cover, i.e. forest, rice field, open field, shrubs, mixed plantation, coconut plantation, oil palm plantation, moor, settlement, and body water. Lubuk Basung is dominated by forest which is 27% or 3,329.6 ha. In addition, based on the Decree of the Minister of Forestry No.35 of 2013, protected areas have been defined in Protected Forest, Production Forest, Nature Conservation Areas, and Other Use Areas. Forest Area Map in Lubuk Basung City is used by government of Agam Regency as a base of Spatial Plan in Spatial Planning of Agam Regency.

Based on data of land cover above can be known the distribution pattern of GS. The calculation results obtained that the presence of GS in the city of Lubuk Basung area is 95.23% or 12,216 ha. Green Space is in the form of Forest, Rice Field, Farm, Moor, and Open Field. The results of this calculation indicate that Lubuk Basung has fulfilled (beyond the minimum GS requirement of 30%) according to the criteria established in Government Law No. 26 of 2007.

3.1.2. *Green Space of Function Protected Area*

GS Analysis of Protected Areas (Decree of Minister of Agriculture No. 837 / Kpts / Um / II / 1980) was assessed based on three determinants factor, i.e. slope, soil type, and rainfall, 79.35% of the total area is a flat slope class (0-8%). Meanwhile, the type of land in the city of Lubuk Basung based on data processing obtained from Soil Research Center is dominated by Alluvial soil (94.72%) and slightly Latosol land area (5.27%). Based on the criteria for the determination of protected areas, according to their sensitivity to erosion, alluvial soils are classified as non-sensitive criteria so that no erosion is possible. While the latosol soil belongs to the category rather sensitive and the possibility of erosion is very small.

Based on rainfall data of Class II Climatology Station, Sicincin, Padang Pariaman year 2010-2015 obtained the average rainfall of 3,247 mm / year with average rainy day as much as 160 days/year. The results of interpolation of data between stations surround Lubuk Basung was obtained by the division of the region with rainfall of 23.4 (mm/day) and 22.3 (mm/day) are classified into the middle class and the area with rainfall 15,23 (mm/day) belong to the lower class.

The overall result of the above three-component analysis had been overlain for the Protected Area. The result of the analysis shows that there is no area that belongs to the protected forest area. The highest

IOP Conf. Series: Earth and Environmental Science **179** (2018) 012022 doi:10.1088/1755-1315/179/1/012022

percentage is obtained for agricultural cultivation area or another usage area with value 97,9% or area of 12,561,72 ha. Furthermore, suitable areas for production forests have a percentage of 2.06% or an area of 264.69 ha.

3.1.3. *Green Corridor and Green Buffer of Water*
Based on the provisions of Law no. 26 of 2007 on Spatial Arrangement Article 5 paragraph (2), natural infrastructure can be categorized into local (corridor) protected areas. Lubuk Basung City has a natural infrastructure in the form of a river that is better known by the community as *batang*. For smaller rivers are usually also called *Banda*. Based on Agam regency's spatial plan, the required of corridor along the river flow is 15 meters along the left and right of the river. It also refers to the assumption that the river is not placed in urban areas with a depth of 3 meters to 20 meters. Meanwhile, based on The Decree of Minister of Forestry No. P.21 / menhut-II / 2011 on environmental management and monitoring of forestry activities for river border in protected areas shall be stipulated ± 100 m from the river.

3.1.4. *Green Line Road Infrastructure and Power Line Corridor*
Based on the regional spatial plan of Lubuk Basung (2011-2031), the information on road infrastructure and High Voltage Transmission (RPLC-150 kV) was obtained. The provisions on the Ministry of Public Works Ministerial Decree No.05 / PRT / M / 2008 state that a green road can be provided by planting between 20-30% of the Right of Way (ROW) in accordance with the road class. The city of Lubuk Basung has the main road segment consisting of the National Road (major road), Provincial Road (medium road), and District Road (small road). In green space analysis on the road obtained a green path area of 41.6 ha.

The green line of the power line is determined in the same regulation as the electric power line borderline is 64 m set from the midpoint of the power grid. In addition, there is a minimum clearance provided between the power line to the ground and other objects of 20 m from the network. Analysis of green space of the power line green corridor space obtained area of 74.7 ha.

3.2. Social Aspects

3.2.1. *Green Space Based on Population Capacity Needs*
Based on the provisions of Regulation of Minister of Public Works No. 05 / PRT / M / 2008 concerning the provision of green space based on the total population, can be determined the area for the park based on the projected population growth in 2020. Based on the current data obtained from the Central Bureau of Statistics Agam regency in 2014, GS for social space (Table 1).

Table 1. Green space area for social space based on per capita population by 2020

Nagari (Village)	Population by 2014	Population growth rate	Projected population by 2020	Park Area (Population *0.3m) (m^2)	Allocation
Lubuk Basung	36,347	1.82	40,501	12,150	Village Park
Geragahan	6,884	-0.19	6,805	2,041.5	
Kampung Pinang	3,684	-0.22	3,636	1,090.8	
			50, 942	15.282,6	City Park

Source: The result of population projection calculation using geometric method

3.2.2. Cultural Aspects (Local Wisdom)
The value of local wisdom is obtained through the interpretation of interviews of custom leaders (Datuk) in three villages (Nagari) in Lubuk Basung City. The data obtained include toponymic, custom, and

religious aspects, as well as ethnobotany which then each will be interpreted to give values on the concept of Green Space.

3.2.2.1 Toponymic

Toponymic is a term about the name of the place, its origin, meaning, use, and typology. The data obtained in each *Nagari* (village) has its own toponymic value. This can provide the identity of each villages with the existence of a distinctive ornament that became the origin of a *Nagari*.

There are plants as the source of the name of a *Nagari* such as Pulai Basung better known by the name of the Basung tree (*Alstonia spatulata*) as for the toponymic of Lubuk Basung and Areca nut (*Areca catechu*) as for the toponymic name of Kampung Pinang. And there is a term of *agahan* (term residents of the sound of the river) that describes the existence of the river (Batang antokan) in Garagahan village. This term can strengthen the existence of rivers by providing local protection areas of river border.

3.2.2.2. Custom and Religion

3.2.2.2.1. Customary Institution

Traditional institutions in Minangkabau are generally the same at every level of government. It is just special for the district in the Province of West Sumatra using the concept of territoriality with the *Nagari* system (village level but the boundaries determined by local customs). A *Nagari* will be lead by a *Wali Nagari* chosen based on a customary decision of *Nagari* (customs of a *Nagari*) on certain conditions. The *Nagari* administration will also be assisted by the secretary of the *Nagari* and the Civil Servant, as well as traditional leaders comprised of scholars, intellectuals and leaders of indigenous tribes in a *Nagari* (Also called *Tungku Tigo Sajarangan*-Three Custom leaders).

The *Nagari* district structure where the *Nagari* is quite clear is listed in the Minangkabau *Tambo* (historical literature of Minangkabau tribe). The element of the formation of a *Nagari* as contained in the *Tambo* consists of the inter-sub village boundary (*Jorong*) occupied by at least 4 sub-tribes of Minangkabau, there are roads, there are rice fields and fields, there are houses, there are courtyards and terrains (community gatherings), and there are graves (either in groups as custom or individual).

3.2.2.2.2. Value system

There is a foundation of custom (Minangkabau) and religion (Islam) which underlies the regulation of life in West Sumatra as a whole. In addition, there are customary systems in general (all areas of West Sumatra) and specifically (per *Nagari*). The value system produces the stipulation that each *Nagari* has different customary provisions. But still, have in common in common as Minangkabau custom unity. There is a characteristic of community habits both in interacting with other human beings as well as to nature.

3.2.2.2.3. Procedures

In the ordinary course of community life, there is a societal way of life closely related to the agricultural system (planting rice together-the *Julo-julo* community's terms, planting coconut or banana before marrying in the yard, and giving the produce to the newlyweds). That is, there is the value of the daily landscape of the people (Ordinary Landscape) in interacting with human beings as well as to nature. For example, formerly coconut and banana are required to be planted with a certain amount. Currently, the yields of these plants can be utilized by the community as the next generation. Until finally people cultivate these plants to be utilized continuously. In addition, there is also a community gathering location for religious or customs activities, customs such as traditional tribal chief election (*datuk*), customary marriage, and traditional art show area such as mosque courts, markets, and open fields.

3.2.2.2.4. Special provisions

Lubuk Basung area has several areas that are local special provisions namely *Rimbo Larangan* (Prohibited Forest) and *Banda Larangan* (Prohibited River). *Rimbo larangan* is a forest area (*rimbo*)

IOP Conf. Series: Earth and Environmental Science **179** (2018) 012022 doi:10.1088/1755-1315/179/1/012022

that should not be disturbed by the community or anyone. If any citizen wishing to take the wood, it must obtain permission from the *Nagari* apparatus. Determination of prohibited forest areas that serve as forest conservation efforts.

While the provision of Banda ban is a river flow that is kept so as not polluted from materials or objects that can destroy the biota in the river. In addition, in the provision of the community should not catch fish by using electricity and so forth. The opening time of the *Banda larangan* is determined by the *Nagari* apparatus.

3.2.2.3. Ethnobotany

Lubuk Basung is famous for its Coconut and Pinang plants (Lubuk Basung is the highest producer of Coconut and Pinang in Agam Regency). In addition, there are also plants that are often used as indigenous food and symbols for traditional ceremonial activities such as Coconut (*Cocos nucifera*), Fern (*Pteridiun aquilinum*), Jackfruit (*Artocarpus heterophyllus*), Banana (*Musa* sp.), Betel (*Piper betle*), Areca nut (*Areca catechu*), and Gambir (*Uncaria gambir*). It can be interpreted that the need to cultivate crops that are often consumed by the community and used for such customary activities.

3.3. Synthesis

3.3.1. Evaluation

Based on the analysis of GS needs biophysically obtained GS of Protected Area, GS of River corridor Greenway of Road corridor and Greenway of Power Line. The result of the analysis is in overlay with Social Green Space (Population Needs per capita and Cultural Green Space) to get GS of Protected Area, GS of River corridor, Greenway of Road corridor, and Greenway of Power Line, City Park, and *Nagari* (Village) Parks. The results of the biophysical and social analysis are revaluated with existing land cover. The evaluation resulted in several GSs, namely GS of Protected Area, GS of River Corridor, GS of Forest, GS of Rice Field, Open Field GS, GS of Bushy Woodland, GS of Mixed Garden, GS of Coconut Plantation, GS of Palm Oil Plantation, Green Space of Dry land, GS at Settlement, Greenway of Road corridor, Greenway of Power line corridor, as well as City Park and *Nagari* park.

GS analysis based on land cover data can be seen that the existing green space has an area of 95.23% (12,216 ha) of the total area of Lubuk Basung city. This indicates that the city of Lubuk Basung has met even exceeding the minimum extent of GS area of urban areas based on Law no. 26 of 2007 covering 30% of the total area of urban areas, so there is no need to increase the proportion of green space so that this study only needs to make efforts to structuring green space.

3.3.2. Block Plan

The block plan consists of GS of protected areas, GS of river corridor, forest GS (Limited Production), GS of agricultural cultivation area, GS in the built area, GS of Urban Infrastructure, and Parks Space (Table 2). It is obtained from the generalization of GS result above.

Table 2. Matrix of RTH Block Plan of Lubuk Basung City.

Land Cover	GS Block Plan	Area (ha)
Protected Area	Protected forest	256
River Border	River Border	1,226
Forest	Limited production forest	175
Rice Field Mixed Garden Coconut Garden Palm Oil Garden Moor	Agricultural cultivation area	7,211

IOP Conf. Series: Earth and Environmental Science **179** (2018) 012022 doi:10.1088/1755-1315/179/1/012022

Settlement Open field Shrubs	Built Area	541
Green road line	City Infrastructures	42
Green power line		75
City Park	Park	3
Nagari parks		

3.3.3. Concept

The concept of GS planning in Lubuk Basung City was developed based on the block plan (Table 2). The integration of the block plans with local wisdom values generates the concept of the GS development plan as presented in Table 3.

Table 3. Concept Matrix of GS Development Plans

Green Space Block Plan	Integration with Local Wisdom Value
Nature reserve area*	There is a valuable area of *Rimbo Larangan*
Protected forest*	
Limited production forest*	
River Corridor*	There is a valuable area of *Banda Larangan* on the river located in the cultivation area.
Agricultural cultivated area	There are ordinary landscape values for: • Rice Planting Activities "Julo-julo" (the social gathering system) • "*Padi Sumpik*" activity (Giving crops to newlyweds) at Kampung Pinang *Nagari* There are several areas worth as *Rimbo Larangan*
Built area	The existence of the home yard that has several plants that are used for daily needs (Medicinal plants and spices, Coconut, Jackfruit, and Banana) already existed since the ancient time.
City infrastructures*	Provision of typical local plant elements at several locations on the Green Line Road (e.g. on pocket garden on the street) such as Pulai Basung *(Alstonia spatulata)* or Areca nut (*Areca catechu*)
Park	City Park: Potential for community gathering areas and local traditional art performances *Nagari* Parks: A fixed area that is used as a community gathering area for customary and social activities on a *Nagari* scale Can be given the typical local plant elements for each location

* Fixed status for RTH (not subject to change) based on (in the provisions of pre-defined rules)

3.4. Green Space Planning

Urban Planning of the City of Lubuk Basung was created to create a green space order that still pay attention to its functions ecologically, economically, aesthetically, socially, and culturally. The cultural element of local wisdom values into the identity and characteristic of GS of Lubuk Basung City. The green space plan based on the concept in Table 3 above is presented in Figure 2.

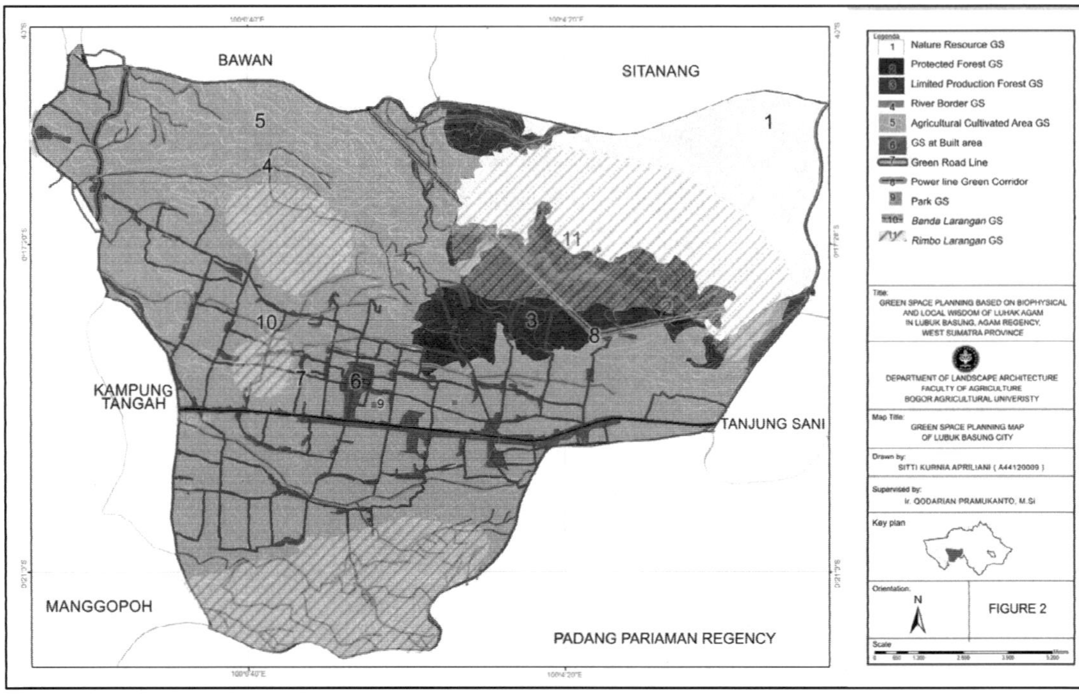

Figure 2. Green Space Planning Map

4. Conclusions and suggestions

4.1. Conclusion

1. The result of analysis and distribution of existing Green Space based on land cover is about 95.23% (12.216 ha) of the total area. The existing GS characteristics consist of forest land cover, rice field, open field, shrub land, mixed garden, coconut plantation, dryland and oil palm plantation. Based on the characteristics of this existing GS quantitatively, the city of Lubuk Basung has the GS that exceeds the minimum requirement of 30% urban Green Space,
2. Based on the analysis of the distribution of demand for green space, it could be determined the needs of Protected GS, Local (corridor) Protection GS/ Green Corridor and Green Buffer of Water, Greenway of Road corridor, Power Line Green Corridor and Social Green Space, respectively 264.69 ha, 1,226 ha, 41. 6 ha, 74.7 ha, and 3 ha,
3. Based on the interpretation of local wisdom which includes toponymic, custom, and religious analysis, and ethnobotany can be determined the green space based on customary provisions, namely *Rimbo Larangan* GS, *Banda Larangan* GS, and Cultural Green Space, as well as information of typical local plants and indigenous plants,
4. The GS planning in Lubuk Basung city can be arranged by integrating biophysical and local wisdom into each functional space namely: GS of protected forest (255,76 ha), GS of limited production forest (175,09 ha), GS of river corridor (1,226 ha) GS of agricultural cultivation area (7,211 ha), GS of open field (108 ha), greenway of road corridor (41.6 ha) and Greenway of Power line (74.7 ha)), and The Parks (City Park (1.5 ha), Lubuk Basung Nagari Park (1.2 ha), Geragahan Nagari Park (0.2 ha), and Kampung Pinang Nagari Park (0.1 ha)).

4.2. Suggestion

Green Space Planning in Lubuk Basung City, can be used as an input to the government of Agam Regency in the planning of city development.

References

[1] Agency for Regional Development Planning Agam Regency 2010 Regional Spatial Plan of Agam Regency (2010-2030) (Agam(ID): Agency for Regional Development Planning Agam Regency) (In Indonesian)
[2] Ministry of Forestry of the Republic of Indonesia 2010 Regulation of the Minister of Forestry of the Republic of Indonesia Number: P.32 / Menhut-II / 2010 on Changing the Forest Area (Jakarta(ID): Ministry of Forestry of the Republic of Indonesia) (In Indonesian)
[3] Ministry of Public Works of the Republic of Indonesia 2008 Regulation of the Minister of Public Works No. 05 / PRT / M / 2008 on the Guidelines for Preparation and Utilization of RTH in Urban Areas (Jakarta (ID): Ministry of Public Works of the Republic of Indonesia) (In Indonesian)
[4] Nurisyah S 1996 Strategies to Enhance and Preserve the Diversity of Flora and Fauna of Urban Areas. (Class papers PSL 702) Graduate School. Management of Natural Resources and the Environment. (Bogor (ID): IPB) 116 (In Indonesian)
[5] Soil Research Center of Republic of Indonesia 2016 Land Unit of Lubuk Basung Sub-district (Bogor(ID): Soil Research Center) (In Indonesian)
[6] Suhartini 2009 Community Review of Community Wisdom in the Management of Natural Resources and the Environment. In Proceedings of the National Seminar on Research, Education and Application of MIPA, Faculty of MIPA, State University of Yogyakarta, May 16, 2009 (Yogyakarta (ID): State University of Yogyakarta) (In Indonesian)

Redesign of Karapan Sapi Park as an urban green open space in Jakarta

T C Julian[1*] and B Paramita[2]

[1] Dept. of Architectural Education, Universitas Pendidikan Indonesia
2 Dept. of Architectural Education, Universitas Pendidikan Indonesia

[*]E-mail: trezachandra@gmail.com

Abstract: Volker Park, an abandoned park located in North Jakarta which potentially develop as green open space in urban area. The green open areas of DKI Jakarta itself currently only 9.98% of the total area, which is planned to be extended into 30%. This expansion in accordance with UU No.26 of 2007 that Municipal government have to provide green open space to improve the quality of the environment and as a means of environmental protection. It is also creating harmony of the natural environment and the built environment that is useful for the benefit of society. The planning and design concept then needed to implement this vision. This paper brings the idea of urban harvesting and farming theme park to encourage the collaboration within the community. The area then divided into six stages, which are 1870 m^2 for the first phase, followed by 3094 m^2 in second phase, then 3094 m^2 in the third phase, the 2808 m^2 in the fourth phase, the 700 m^2 in the fifth phase and finally the sixth phase is 752 m^2.

Keywords: environment, green open space, urban area

1. Introduction

1.1. Background

The main problems in the field of the population these days are population growth and urbanization. Urban population is predicted to reach 60% of the total world population by 2025, with growth reaching three times faster than the growth of villagers [1].

The city as a dynamic entity that develops continuously, demanding continue to build facilities and infrastructure to serve its citizens. Physical development is done by changing the natural landscape and spatial area which resulted in the decrease of environmental supporting capacity and environmental services [1, 3]. As a result, city residents faced environmental and social stress that negatively impact the physical and psychological of their citizens [4].

Along with the development of a city, population, and physical development continue to increase resulting in the availability of land that is currently less and less. Attention to the role of green open space that plays a role in sustainability and urban comfort increases with the increasing impact of urbanization [5].

The importance of the city's green open spaces, especially in the neighborhoods, has been proven in several studies. Urban green open space is an important component that affects the quality of human life, both ecologically and socially-psychologically. However, the current proportion is diminishing as a result of the high population density due to the growing human population [6].

Therefore, this research is based on the results of research that examines the needs of green open spaces, especially in the capital city of DKI Jakarta and establishes a green open space system in accordance with the needs of the community, and can contribute to the system on a larger scale, with case studies of the region Karapan Sapi Park (Taman Volker) on RE Martadinata street, Tanjung Priok Sub-district, North Jakarta. The findings of the research are a description of the characteristics of the case study case as the green open space is misused and it becomes a negative impact on the surrounding environment. From these findings, a model of empowerment and a design framework that can be developed in the development, utilization, and maintenance of green open spaces by the local community in the form of redesigning design that can revive an unproductive area to be productive.

1.2. Green Open Space in Jakarta

Green Open Space under the Instruction of the Minister of Home Affairs No. 14 of 1988 on the arrangement of green open spaces in urban areas is the spaces within the city or wider area, either in the form of area or in the form of elongated areas/paths where in its use is more basic without the building. Green Open Space itself is normatively expected to have multifunctional for city life, that is ecological function, socio-cultural and economic function.

UU No. 26 of 2007 on Spatial Planning mandates spatial planning of urban areas should contain a plan for the provision and utilization of the Green Open Space is a minimum of 30% of the city area. Green Open Space in the urban area consists of public and private where the proportion of green open space in the urban area is at least 30% consisting of 20% public green open space and 10% consists of private green open space [7].

The proportion of 30% is a minimum measure to ensure the balance of the ecosystem of the city, both the balance of the hydrological system and the balance of microclimates, as well as other ecological systems that can increase the availability of clean air that is needed by the community, as well as to increase the aesthetic value of the city. The broad target of 30% of the city area can be achieved gradually through the typical urban land allocation (Ministry of Public Works Regulation No. 5 Year 2008 on Guidelines for the Provision and Utilization of Green Open Space in Urban Areas).

Development of an increasingly developed city in Indonesia, especially in DKI Jakarta has an impact on the wide change of land use including the extent of green open space. The existence of green open spaces in big cities including DKI Jakarta has become a demand of modern life that wants a more fresh and natural atmosphere and the aesthetic demands of the city. DKI Jakarta as the capital city of Indonesia has a total land area of 661.52 km2. However, the proportion of green open spaces of DKI Jakarta currently does not meet spatial policy standards (based on Law No.26 Year 2007 on Spatial Planning), which is 30% of the total urban area. DKI Jakarta Provincial Government keeps trying to make Jakarta temperature cooler. One of them is by building more green parks and open spaces, amid the limited land in Jakarta. But the number of parks currently considered not enough to sustain the needs of green space in the capital.

Based on data from Jakarta Parks and Cemetery Agency, the number of Green Open Space in Jakarta reached 3,131. This green open space in the form of city parks, environmental parks, interactive parks and also the path of the green road. Although the number has been widely spread throughout the DKI Jakarta, the area of green open space in Jakarta only amounted to 9.98% of the total area. This figure is still far from 30% that should be owned by DKI Jakarta. During this time the development of green open space constrained land acquisition and too many problems in the process of land acquisition and land purchase.

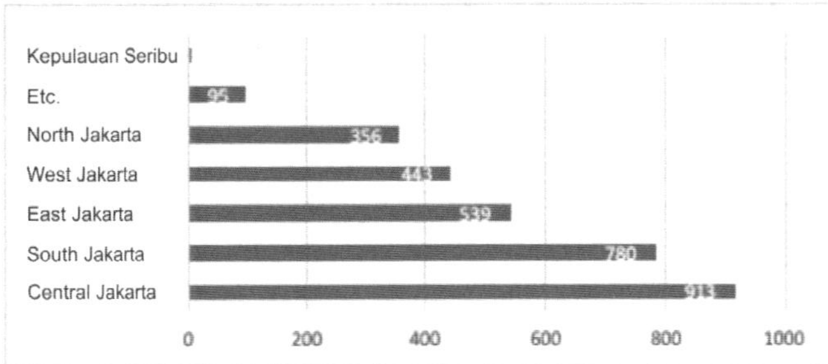

Figure 1. Number of Green Open Space by Region in Jakarta 2017

Based on Figure 1 Central Jakarta became the region with the largest green open space, as many as 913 green open spaces. Although the number has been widely spread throughout the DKI Jakarta, the area of green open space in Jakarta only amounted to 9.98% of the total area. Followed by 780 parks, East Jakarta as many as 539 parks, West Jakarta as many as 443 parks, North Jakarta as many as 356 parks, others as many as 95 parks, and Kepulauan Seribu only 5 parks. This figure is still far from 30% that should be owned by DKI Jakarta because so far green open space development is constrained land acquisition and too many problems in the process of land acquisition and land purchase.

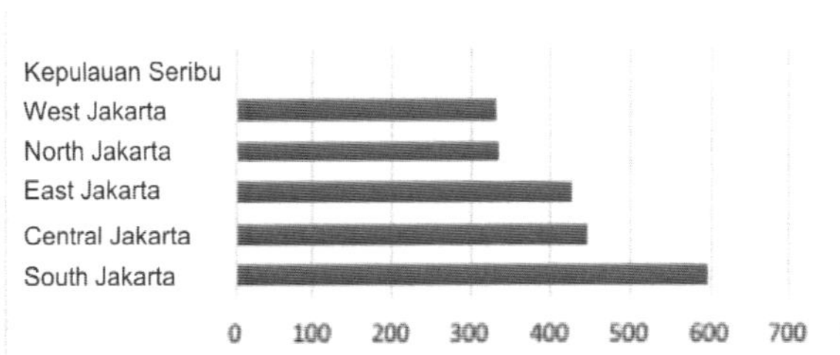

Figure 2. Number of Park City by region in DKI Jakarta 2017

Based on data from Jakarta Central Bureau of Statistics (BPS) until 2013, the number of urban parks spread in the capital amounted to 2,166. Table 2. Explains that South Jakarta is the region with the urbanest parks, which is 597 parks, followed by Central Jakarta and East Jakarta as many as 446 and 426 parks. The areas of North Jakarta and West Jakarta have urban parks of 336 and 331.

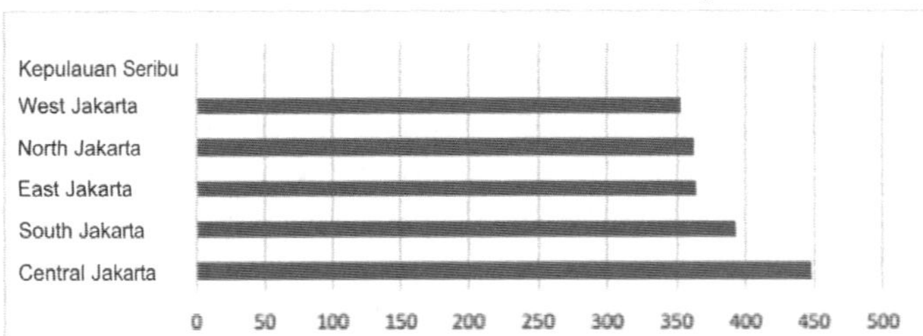

Figure 3. Area of city park by area in DKI Jakarta 2017

Based on Central Bureau of Statistics (BPS) data of DKI Jakarta Province until 2013, the urban park area in all Capital City area is only 1,912.1 Ha. Table 3. Explains that Central Jakarta is the region with the largest city park, which is 446.4 Ha, followed by South and East Jakarta area of 391.6 and 362 Ha. The area of North Jakarta and West Jakarta has a city park area of 361.3 and 350.7 Ha.

1.3. Karapan Sapi Park (Taman Volker)

Karapan Sapi Park (Taman Volker) is a park that is categorized as an environmental park located on RE Martadinata street, Tanjung Priok Sub-district, North Jakarta. The condition of Karapan Sapi Park (Taman Volker) as a missed green open space is actually a negative impact on the surrounding environment, especially for Tanjung Priok Sub-district residents. The balance of functional and aesthetic aspects is important in every green open space, taking into consideration the proportion of functional aspect and ideal aesthetic aspect in describing its development, it is expected that Karapan Sapi Park (Taman Volker) can become a landscape of harmony as the identity of Tanjung Priok Sub-district of North Jakarta from redesign process and increasing green open space productivity in the capital city of DKI Jakarta.

2. Methods

This research is done by field survey method, then the data is analyzed descriptively. The type of data used is primary data and secondary data.

2.1. Tools and materials

Tools and materials used in this study are as follows:

Meter measure, Compass, Wood stakes, Ruler, Pen / pencil, Camera, Handphone, Printer, Paper, and Ink.

2.2. Field survey

Activities undertaken during the field survey are as follows:

1. Contour, topographic, and boundary data retrieval
2. Weather test and citizen activity
3. Documentations

2.3. Preparation of reports

Preparation of reports in the form of papers and image panels using Autodesk AutoCad 2016, Sketchup Pro 2016, Adobe Photoshop CC 2017, Adobe Illustrator CC 2017, and Microsoft Excel 2016.

2.4. Details of the design stage as follows

2.4.1. Preliminary stage

- Initial survey
- Initial consultation

2.4.2. Survey stage and site analysis

- Survey of existing site conditions
- Topography
- Contour
- Site Limit
- Circulation and Achievement
- Water Resources
- Site Drainage System
- Power Source
- Site Rating
- Potential Assessment
- Looking for Potential View
- Consultation of Survey Results

2.4.3. Conceptual Planning Stage
- Site Planning Concepts
- Concept Consultation

2.4.4. Redesign stage
- Programming
- Schematic Design
- Design consultation

3. Result and Discussion

3.1. Existing Condition of Karapan Sapi Park

Figure 4. Case study sites

(Source: Personal Documentation, 2017)

Established since 1997 on the ownership of PT KAI, Karapan Sapi Park (Taman Volker) was originally created with the aim of creating a public area that can be an absorption area and a place for people to engage in social activities and interactions. In Figure 1. The 1870m^2 park is located in Tanjung Priok sub-district close to the residential area. Name Karapan Sapi is made because of the majority of residents Kec. Tanjungpriok who lived in the surrounding area is a native of Madura. There are also monuments shaped Karapan Sapi at the entrance of this park.

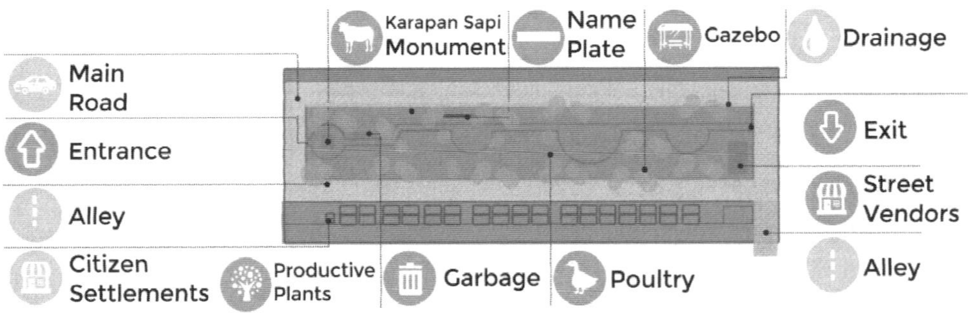

Figure 5. Existing Conditions of Karapan Sapi Park
(Source: Personal Documentation, 2017)

This park has been misused far from the real green open space function, which is a place of sale of used goods and storage of heavy equipment. From heavy machine wreckage to poultry cages scattered at several points makes the area a slum such as iron, tin, wood, even anchor ships, ship propellers and ship turbines. Some of the illegal parks along the park are also impressed, some of which are also used as illegal parking lots for trucks or expedition vehicles that will unload at the port of Tanjung Priok. It

adds to the chaos of the park which seems to have never been touched by the Municipality government.

Figure 6. Existing Conditions of Karapan Sapi Park
(Source: Personal Documentation, 2017)

3.2. Residents of Tanjung Priok

Based on the results of field surveys and interviews with one of the community leaders, the majority of citizens based on the age in Tanjung Priok around Karapan Sapi Park (Taman Volker) is 45% of parents, 40% of youth, and 15% of children. As for the community profession of traders by 70%, factory workers 25%, and civil servants of 5%.

The majority of people who use the park are children. Despite the very worrying condition of the existing, children who are residents of the Tanjung Priok desperately needs a Green Open Space around their residence. Based on the results of interviews with several children in the field, children usually do activities such as playing and relaxing around the park.

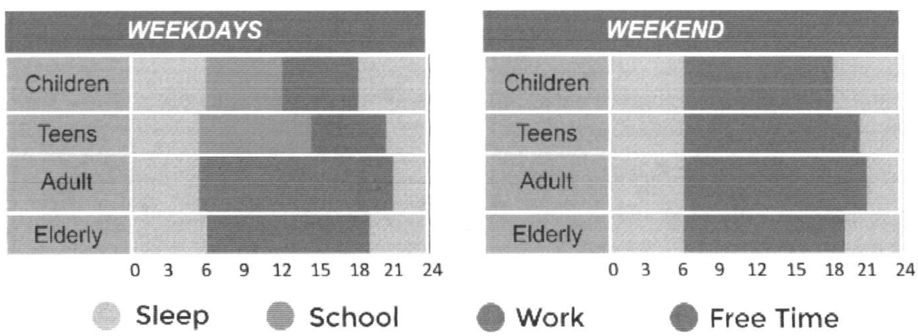

Figure 7. Residents of Tanjung Priok Activities
(Source: Personal Documentation, 2017)

Based on fig. 4 the layers of residents around the park are children, adolescents, adults, and the elderly. In the spare time (yellow in the figure) all walks of life can use the Karapan Sapi Park (Taman Volker) as a public green open space during weekdays and weekend.

IOP Conf. Series: Earth and Environmental Science **179** (2018) 012023 doi:10.1088/1755-1315/179/1/012023

3.3. Redesign Karapan Sapi park (Taman Volker)

Redesign is an activity of planning and redesigning a physical change without changing its function either through expansion or location transfer (Jalal, 2010). Karapan Sapi Park (Volker Park) is located between R.E. Martadinata street with residents of Tanjungpriok Sub-district. This means that the park's position as an environmental park for the surrounding community should be a buffer system that separates the highway from the settlement. The planned area is generally a lowland with a very low slope (ramps). The type of soil in the area is alluvial gleik and belongs to clay loam texture category. This area is located at a lower altitude than the highway and parallel to the residential area. This makes the area often affected by flooding when heavy rain falls because the existing drainage system in the area can't accommodate enough water and eventually overflowed into the area even up to the settlement. The climate inside the area is very hot at around 39 ° C at midday, which is good for plant growth as a process of photosynthesis, but there is a need for shade for the users of the area.

Karapan Sapi Park (Taman Volker) is part of Jakarta Green Open Space. In RUTRK Jakarta stated that Karapan Sapi Park (Taman Volker) is included in the environmental park area which prioritized its development as part of North Jakarta City. The Jakarta City Government already has the RPTRA (Child-Friendly Public Open Space) program and has run more than 100 parks spread across Jakarta area. To achieve that objective, the researcher made some special design concept for Kencana Sapi Park (Taman Volker) by considering a design to be able to restore the portrait of real green open space and able to revive the area around the park into a productive area.

Karapan Sapi Park (Taman Volker) is planned as an active park. Tread developed with Urban Harvesting and Farming theme. Many changes to the grading, the surface of the tread is made higher than the surface of the highway and residents settlements. The concept of Urban Harvesting and Farming is used as a way to answer the problems of the people in Tanjung Priok Sub-district. On the social aspect, people need a place for self-expression, interaction space, and as a container of citizen activity. In the cultural aspect, the majority of Tanjung Priok Sub-district enjoys planting and raising. This is evidenced in the existing state of the area that is planted with many productive plants and there is some domestic poultry. In the economic aspect, the majority of citizens are middle-lowers, citizens need areas that can provide benefits as well as turning capital for them.

Urban Farming Activity is done in the area around the tread area of fruit, vegetables, and flowers with the division of schedules and alternatives based on the results of studies and analysis according to the season in Indonesia on each month.

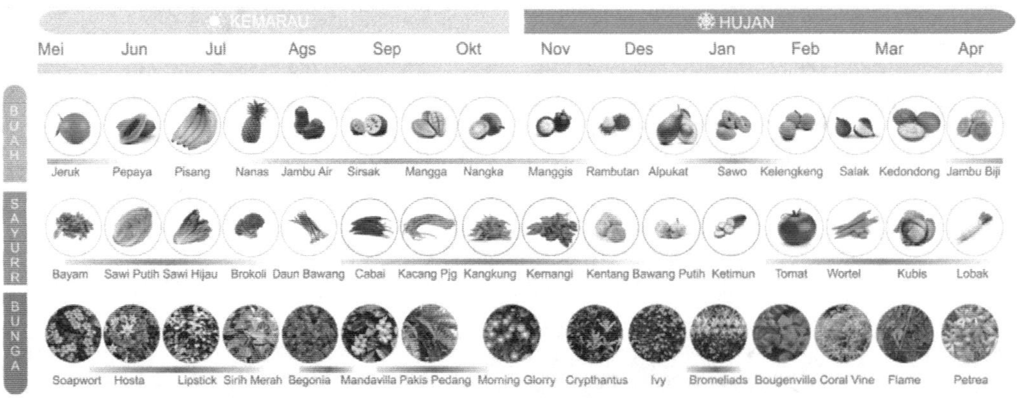

Figure 8. Schedule Analysis of Urban Farming in Indonesia

(Source: Personal Documentation, 2017)

Based on the fig.5 urban farming activities carried out in accordance with the schedule of growing plants every month. The table is the result of the seasonal analysis in Indonesia, it is expected that the residents of Tanjungpriok Sub-district have land to grow fruit, vegetables, and flowers in the park area that can make a magnet for residents come to the park every day.

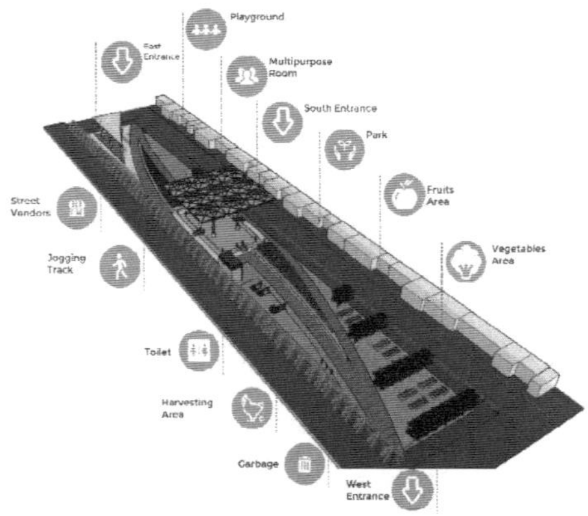

Figure 9. Planning Concept
(Source: Personal Documentation, 2017)

This concept is then developed and continued at the design stage because it is regarded as the most capable concept to showcase the distinctive features of Karapan Sapi Park (Taman Volker). Based on fig.6 Redesign of Karapan Sapi Park (Taman Volker) is done by adding some facilities in the park that is playground (playground of children), multipurpose room, garden, urban farming area (fruit area, vegetable, and flower) urban harvesting, toilets, jogging tracks, and street vendors to sell.

Entrance in the region consists of 3 access to access to the west, east, and south. It is chosen to make it easier for users to enter the park area, as well as avoid public transportation or stop vehicles in the north (front of the park) which can cause congestion in the direction of RE. Martadinata street. The zone on the site consists of 3 play zones, interaction zones, and production zones each representing several park facilities. Production zones are located in the east because urban farms need more light during the day

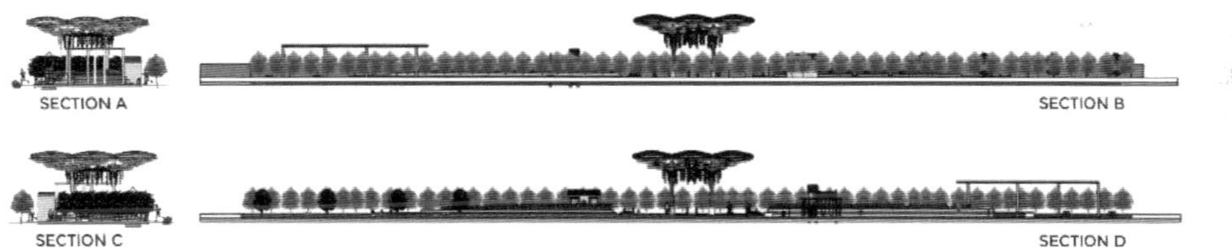

Figure 10. Section of Plan
(Source: Personal Documentation, 2017)

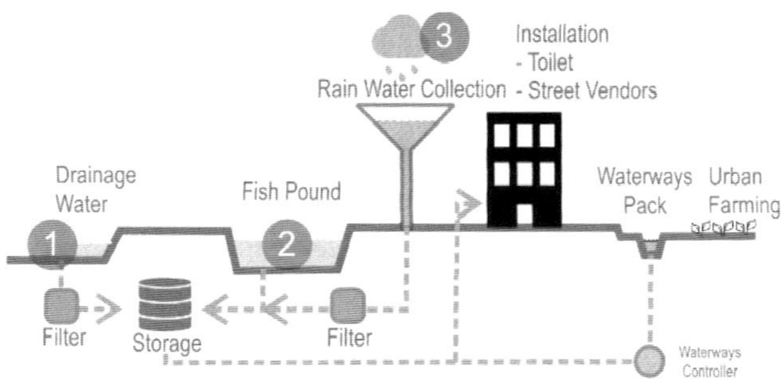

Figure 11. Water storage process
(Source: Personal Documentation, 2017)

To answer the problem of water that often overflows from RE. Martadinata street and drainage around the area carried out a water storage process that is accommodated into a storage or storage tub. This water storage process is divided into 3 water sources which will be water that will be re-used (secondary use water) for the benefit of the park for example to irrigate urban farming and garden water source.

Based on Figure 11 the source of water is water from drainage, fish ponds stored around the park, and rain water collection that is stored on the roof of the multipurpose room in the park area.

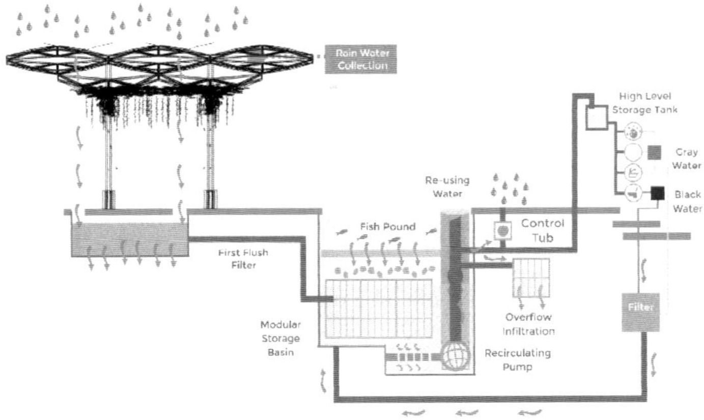

Figure 12. Water storage process
(Source: Personal Documentation, 2017)

Based on Figure 12 explained how the rain water collection process is accommodated and channeled back to the needs of Karapan Sapi Park (Taman Volker). The rain water collection process is obtained from rainwater which is then accommodated into a modular storage basin through a process of clarification assisted by a control tub as a regulator of the water circulation. Water that is in storage then distributed to some point through the help of the pump.

Figure 13. Redesign Karapan Sapi Park situation
(Source: Personal Documentation, 2017)

Planning and designing green open spaces based on community needs allows for a more productive and environmentally conscious community [8]. Based on Figure 14 Planning and facility design on redesigning Karapan Sapi Park (Taman Volker) enables the sub-district of Tanjungpriok to become an active and productive area through the presence of existing environmental parks in their area of residence.

3.4. Green Open Space Vision for Tanjung Priok Sub-district

Figure 14. Green Open Space Vision for Tanjung Priok Sub-district
(Source: Personal Documentation, 2017)

After creating a design framework with Karapan Sapi Park (Taman Volker) case study, the author tried to make a vision for the future. This idea aims to develop, utilize, and maintain green open spaces in the area of DKI Jakarta. The first phase is its own Volker park of 1870 m^2. Followed by a second phase of 3094 m^2, the third phase of 4698 m^2, the fourth phase of 2808 m^2, the fifth phase of 700 m^2, and the sixth phase of 752 m^2. Based on the calculation of this open space planning, it is expected to develop the green open space in DKI Jakarta to affect the quality of human life, both ecologically and socially-psychologically.

4. Conclusion

1) Green Open Space is an important component of people's quality of life, the presence of green open space can revive society into a productive society.
2) The existence of green open space in the city of DKI Jakarta that is less than standard requires the government must move actively to fix open green spaces and revitalize the existing ones.
3) Redesign of Karapan Sapi Park (Taman Volker) is done by adding various elements to the site, that is how to apply urban harvesting and urban farming can be the hallmark of the park in accordance with the needs and potentials of Tanjung Priok Sub-district.
4) With an area of 1870 m^2 of Karapan Sapi Park (Taman Volker) can be used as a place to express themselves to the people around Tanjung Priok Sub-district with playground facilities, multipurpose room, garden, urban farming area (fruit area, vegetables, and flowers), urban harvesting areas, toilets, jogging tracks, and street vendors to sell.

References

[1] J. Wu, "Ecology, Planning, and Management of Urban Forests," *Ecol. Plan. Manag. Urban For.*, no. November, pp. 10–28, 2008.
[2] P. Bolund and S. Hunhammar, "Ecosystem services in urban areas," *Ecol. Econ.*, vol. 29, no. 2, pp. 293–301, 1999.
[3] D. J. Nowak, R. Hoehn, and D. E. Crane, "Oxygen Production by Urban Trees in the United States," vol. 33, no. 3, pp. 220–226, 2007.
[4] M. M. Carreiro, "Introduction : The Growth of Cities and Urban Forestry," pp. 3–9, 2008.
[5] C. C. Konijnendijk, R. M. Ricard, A. Kenney, and T. B. Randrup, "Defining urban forestry – A comparative perspective of North America and Europe," *Urban For. Urban Green.*, vol. 4, no. 3–4, pp. 93–103, 2006.
[6] W. A. Rahmy, B. Faisal, and A. R. Soeriaatmadja, "Kebutuhan Ruang Terbuka Hijau Kota pada Kawasan," *Lingkung. Binaan Indones.*, vol. 1, no. 1, pp. 27–38, 2012.
[7] R. Dwihatmojo, "Ruang Terbuka Hijau Yang Semakin Terpinggirkan," *Badan Inf. Geospasial*, 2010.
[8] L. Widaningsih, T. Busono, and E. Krisnanto, "Community Architecture Dan Community Based Development Dalam Pengelolaan Ruang Publik Kampung Kota," pp. 73–83, 2008.

Formulation of design concept of urban park using butterflies as a good urban environment bio-indicator

D Ariyani [1*], B Sulistyantara [2], and T Budiarti[2]

[1]Graduate Student of Landscape Architecture, Department of Landscape Architecture, Faculty of Agriculture, Bogor Agricultural University (IPB), Indonesia
[2]Department of Landscape Architecture, Faculty of Agriculture, Bogor Agricultural University (IPB), Indonesia

[*]Email: shevyri@gmail.com

Abstract: Creating an urban eco-park an efforts to restore habitat in urban areas. The Urban Eco-Park is useful to be a wildlife refuge, as well as functioning the urban park to become an indicator of environmental quality by introducing butterflies. The butterfly is one of the animals that can be observed more easily and closely than other species of insects. Butterflies have a significant ecological role, which can be found in several habitats of urban areas. The benefit of butterfly is acting as a good urban environment bio-indicator. Some butterflies' species mentioned in a case study as a bio indicator but the abundance and diversity of butterflies have more important role to indicate a good urban environment. The objective of this study is creating model to formulate a concept of urban park where the park is become an indicator of a good environment showing by butterflies appearance. The Site Plan was made with concept to invite butterflies by using their proper plants preferences (food plants and host plants). The method in this study using descriptive method comprised literature review from case studies and field observations. The result showed that a well landscape management was able to provide suitable food plants and host plants which were important to maintain the existence of butterflies in an urban park.

Keywords: urban park, butterfly, food plant, host plant, bio-indicator.

1. Introduction

The appearance of butterfly is an aesthetic and ecological value. The existence of butterfly in the environment are mentioned to be pollinators of plants, food sources for other animals and also has an important role in scientific discoveries. Butterfly can live in many different habitats and their populations undergo normal fluctuations form year to year. There can be large differences in butterfly's abundance and diversity of their habitat in any given landscape. Unfortunately, environmental disturbance overcome decrease of the butterfly population. Environmental disturbance is a change in environmental components forming the habitat thereby affecting the ecosystem imbalance and biotic communities inside [4].

In order to reduce the impacts of disturbance, the thematic garden is needed, such as the ecological park (eco-park), wildflower gardening, and the butterfly park, which were useful in attracting butterflies and other pollinating insects. Currently, the appreciation of butterfly has evolved from consumptive use as collectibles and souvenirs to conservation measures. Increased knowledge and understanding of the importance of butterflies and ecological balance are getting [1].

Butterfly is one of the animal species that responds to environmental disturbances and is easy observed. Almost all butterfly species are herbivores, as well as its larvae or caterpillars are plants, flowers, or fruits' leaf-eaters, which are often species-specific. The caterpillar of *Papilio polytes* and *Papilio demoleus* eats citrus leaf (*Citrus aurantifolia, Citrus grandis). Graphium Agamemnon* lays their eggs on *Annona muricata* (soursop) and *Polyalthia longifolia* (asoka tree). This shows that the diversity of butterfly species is associated with the diversity of plant species. The existence butterflies in a park will enhance the aesthetic value along with blooming flowers. In order to maintain these butterflies in the park, it necessary to ensure the feeding is sufficient for caterpillar and adult butterflies. Providing proper trees as host plants and flowers for food plants in the right place of the park. The abundance and diversity of butterflies indicate a good quality of urban environment.

The study aims to formulate the concept of an urban park as an indicator of good environment, with the approach of using butterflies. This study also made a concept of garden model that possible to attract butterflies. The concept consist of arranging suitable plant of butterflies' preferred (food-plants and host plants), in order to live and survive in urban park.

2. Methods

The study located in Cibinong, Bogor, West Java, Indonesia. The urban park area in this district has a potential habitat of butterfly which is near water bodies and have a patch area that consist trees and herbaceous plants that would provide food plants for butterflies. The elevation is about 141-151 meter above the sea level.

The analysis method was descriptive qualitative by analyzing case studies and adapting information proposed to the butterfly and preferred plants [1,5]. In addition, field observations around the site was conducted to study butterflies behavior along the influence factors.

The scope of the study is determine butterflies behavior in some habitats in urban areas, identifying the type of preferred plants (food-plants for adult butterflies and host plants for caterpillars) and adding the preferred plants in urban park concept.

3. Discussion

3.1. Butterfly Diversity on Various Habitat and The Affecting Factors

Habitat is the environment in which an animal or plant normally lives or grows. Habitat degradation refers to a reduction of habitat quality. Habitat quality is reduced when the breeding success of a particular species is compromised, or when the number of individuals of a species that the habitat can support and the diversity of organisms supported by the habitat is reduced. Habitat may be degraded via the alteration of community components (species composition and diversity), structural components (physical habitat structure), or environmental components (climatic, chemical, and other abiotic variables) [18].

The existence of a butterfly in a habitat affected by the characteristics of the habitat itself. Forest reserve is one of the least distracted urban forest, which will be the most effective and affluence conservation for butterfly species [4,10,16].

Based on the level of sensitivity, tolerance, and the ability to adapt, species of organisms are divided into specialist species and generalist species, which require specific habitat requirements and conditions to live and breed.

A study was conducted to understand the value of conservation in forest reserves, forest fragments, and urban parks in Singapore. Forest reserve is part of the Central Catchment Nature Reserve of Singapore, which consists of secondary forests, tropical lowland rain forest, and freshwater swamp forest [10].

The forest reserves location has the highest diversity of butterflies than the other habitats. An urban park located near forest fragments, which has a similar composition and diversity, may increase the diversity of butterfly species. This can occur due to the varied microclimatic conditions and ecological niches in both of the habitats.

The research results in Singapore indicate that the butterfly cosmopolitan species or generalist species is able to survive in an urban park rather than the border areas. This may be due to the wide and easy adaptable distribution of this species than the butterfly specialist species. Differences in ecological treatment (host plants specificity) and environmental factors (such as canopy cover) in butterfly species may affect the distribution between different habitats. The number of survived butterfly species in an urban park depends on the number of available potential host plants.

Unmanaged areas are an area of semi-natural, which often in the early stages of ecological succession with a high diversity, and have plants that fit the needs of a butterfly [14].Microhabitat components are consisting of, namely (a) shrub; b) *hedges;* c) nectar plants, and d) unmanaged areas. The diversity of microhabitat components determines the diversity of butterflies in the urban parks.

Butterfly communities respond and predisposes to its habitat conditions according to its needs and preferences. Butterfly is one of the organisms that have a sensitivity and specificity to a particular environmental condition that is indicated to be negatively affected, but also reflect the condition of its habitats [3].

Abiotic factors that affect the characteristics of the fourth urban forests are sunlight intensity, temperature, and humidity. Leaf area index was calculated to determine the flatness of canopy crown and the characteristics of canopy closure. The humidity required for the butterfly ranges between 50 to 75%, and the temperature ranges in 20 to 40°C [17]. While tolerance limit of light intensity ranges from 500 to 7500 lux [12].

3.2. Correlation Between Various Plants Preference and Butterfly Abundance

A study of four-type urban forest in East Jakarta, Indonesia had identified 22 species of butterflies. Urban Forest Rawa Dongkal, which has the settlement type of urban forest, reached the highest species diversity with as many as 18 species; followed by Urban Forest Kopassus Cijantung located in an office area with as many as 15 species, Urban Forest PT. JIEP located in an industrial area with 11 species; and the lowest was in Urban Forest Cawang UKI, located in the highway area, with only as many as 7 species of butterflies. The differences in observation periods resulted in differences in the number of butterflies. The highest number of species is found in the dry season. While in the transition between the dry season to rainy season, 20 species were found. And in rainy season, only found a total of 17 species.

Four species of butterflies found in overall locations and observation periods, which are *Graphium Agamemnon, Deliashyparete, Euremahecabe,* and *Leptosianina*. These four types of butterflies found in a large number in overall locations and observation periods. These species can be categorized as generalist species type [9].

Some butterfly species are found only in one location (site specific), among others, *Ideopsis juventa* and *NeptisHylas* at Urban Forest Rawa Dongkal, *Papilio demoleus* at Urban Forest PT. JIEP, and *Floss aniella* at Urban Forest UKI Cawang. In addition, there are two species of butterflies found only at certain locations and periods, which are *Polyurahebe* at Urban Forest Rawa Dongkal and *Ypthimahor sfieldii* at Urban Forest Koppasus Cijantung, both only during the dry season. The low frequency of attendance indicates that these species have a strong attachment to particular habitat components that only exist in locations that provide these habitat components and are sensitive to certain environmental conditions [2].

The results of correlation analysis showed a significant positive correlation between plant species richness with the species richness and diversity of butterflies. This supported the theory that abundance food plants and host plants species can be used as predictors of the variety of butterfly and is the most dependent factors in influencing population size and butterfly's diversity [15]. Further confirmed that the size of the butterfly population in an urban environment is related to the availability of various types of food plants, not by their abundance[16].

With abundance food and host plants and the most diverse plant species make Urban Forest Rawa Dongkal the highest variety of butterflies. Caterpillars have specificity to certain food plants while on the other hand, adult butterflies have no specifications for their food plants. Based on the observations described in this study, in addition to nectar plants, adult butterflies are also often found to consume tree sap, decomposing fruit, animal feces, or animal carcasses.

Throughout the study, butterflies are often found perched on a variety of ground cover habitats. For example, *Mimosa pudica* was the most cover ground plants that was found in the four research sites and as food for the *Eureka hecabe* butterflies. The *Imperata cylindrica* (reed) is often visited by the *Nymphalidae* family, while *Lantana camara* (big-sage) due to its attractive colors and scents, is often visited by the genus *Graphium*.

These observations confirmed the theory that most butterflies are found in *herbaceous* feeding plants [11]. Various types of plants of the family *Fabaceae* found in the urban forests provide diverse host plants for the caterpillar. The legume is a plant family that is most widely used by the caterpillar as host plants. The presence of shrubs in overall locations, in addition to functioning as host plants, also serves as a shelter so that its presence supports the existence of butterflies. The areas with bush vegetation will affect the high diversity of butterflies in those areas[11].

Figure 1. Nectar plants

Those butterflies' preferences and requirements of the components of a certain habitat, where the resources availability of the habitat will affect the existence of butterflies[10].

Wildlife Habitat Management Institute and Wildlife Habitat Council of America (2000) mentioned various factors of butterflies' requirements which to be considered in developing a suitable habitat for the survival of butterflies. These factors are:

- Cover Plants
  Tall trees that have wide canopy will protect butterflies from strong gusts of wind and rain [13]. Butterflies will choose the areas with less compact canopy, because less dense areas will allow sunlight to enter and create warmer micro-climate that can maximize the growth of caterpillar [7]. Butterflies also favor these areas because the sunlight maintains their temperature and metabolism.
- Water
  In the life cycle of butterflies, water requirements are fulfilled mostly in the caterpillar stage by its consumption of green vegetation. Whereas for the adult butterflies, the water requirements are fulfilled by the nectars. Nevertheless. Adult butterflies will gather in small mud puddles as well as wet sand areas to absorb minerals and salts contained in the puddles known as mud puddling. Puddling is more of nutrient source than water.

Figure 2. Host plants for caterpillar

- Food – butterflies (food plants)
  Adult butterflies depend largely on the availability of nectar as food source. While other species take nutrition and minerals from rotting fruits, tree sap, animal feces and urine, and animal carcasses. Nectar plants are the main food for adult butterflies. The sufficient provision of light-colored nectar plants needs to be considered because butterflies can quickly recognize the plants. Planting in areas with an abundance of sun intensity is important for plants and butterflies. Nectar plants are flowering plants that require sun exposure. Butterflies also demand a high intensity of sunlight as butterflies use solar heat to increase metabolism and help them fly.
- Food - Caterpillars (host plants)
  Each species of butterflies laid their eggs on certain plants as host plants. Caterpillar's host plants can be shaped as leaves on trees, shrubs, vegetables, legumes, wild flowers, grasses, and weeds in Figure 2. Female butterfly is quite selective in choosing host plant to lay their eggs to ensure the growth of caterpillars that fit the feed. The existence of host plants will also invite the female butterflies to lay their eggs. There are several important factors that affect the pattern of oviposition (egg-laying) on butterfly, namely microclimate, vegetation structure, the number of host plants available, and the availability of nectar-producing plants. When the number of plants that can be used by butterflies as oviposition sites increases, the abundance of caterpillar found will also increases [7].
- Interspersion of habitat component
  In ensuring the reproduction of butterflies goes well and their ability to survive, all the necessary habitat components must be available. The ideal butterfly habitat component consists of a complex of closely spaced vegetation, comprising with diverse sources of host plants and food plants (trees, shrubs, wildflowers, legume and grasses), open water, and puddling areas.
- Minimum area for Habitat
  Although butterflies can search for food in large areas, there is no minimum size for a butterfly habitat. Butterfly species has potential and can be found anywhere as long as the closure of plants required by caterpillars and adult butterflies are available. In general, minimum habitat size is not a limiting factor for each species, as the territory of each species varies.

3.3. Landscape Design Concepts for Inviting Butterflies into The Urban Park

The beauty of butterflies has been admired so far because it appears attractive with various patterns and colors. Butterflies roam the sunny wildflower gardens; lawns and garden fields, searching for feed on the nectar of flowering plants. The correct selection of plants will help to influence the visit of interested butterflies, which then settle and reproduce there [8].

Environmental consideration becoming an important part in creating a concept planning of a landscape [5]. Analyzing site's micro and macroclimates, ecosystem, hydrology, vegetation and soil condition will conclude a valuable consideration.

The organizing of a landscape concept has to be well designed for the sustainability of an urban park. Base on various factors purposed by Wildlife Habitat Management Institute and Wildlife Habitat Council of America [1], the criteria that establish the landscape space concept consist of :

- Butterfly habitat components
  Six components of butterfly life requirements that have described previously are priority factors in preparing the concept of an urban park as a butterfly park. Furthermore, the next consideration is the abiotic factors and environmental disturbances that exist around the site. Urban park area that has been qualified according to these criteria will determine the survival of the butterfly.
- Butterfly Maintenance Area
  This area was created with the aim of providing life support for the butterfly. The live support component may consist of a caterpillar conservation area, cocoon (chrysalis) space, and adaptation area for butterflies before being released into the wild.
- Vegetation arrangement by function
  In essence, there are four categories of vegetation areas based on their function, namely the shelter plants (canopy cover), the caterpillars' host plants, the butterfly's food plants, and the nursery plants. This vegetation placement is arranged in zones according to its function. Plants are arranged by the composition of plants' strata (ground cover, shrubs, trees) to look aesthetic.
- Butterfly display and collection room
  This is an enclosed area, which designed according to the life requirements of a butterfly. The sunlight should sufficiently penetrate through paranet. Plants display in this area is only for food plants for butterfly due to the presence of caterpillars may cause inconvenience for visitors. Butterflies are supplied from the butterfly maintenance area.
- Butterfly Collections Introduction
  The determination of the butterfly species that will be displayed is related to food plants and abiotic factors that influence it. Butterflies that will be introduced are local butterfly species that already exist on the site. To maintain its existence, the plant feed should always be available and assisted with well treatment in the butterfly maintenance area.
- Management and service area
  The park management requires an area that serves as management space, office, and visitor services. The management and office area are private, while the visitor service area is semi-private.
- Visitor management program
  The urban park is a green open space that accessible and can be enjoy by visitors as part of recreation and tourism. The visitor management serves to provide visitors comfort, as in determining the visit time and also serving the group visits (school and community), route visits, and educational tourism program packages.

The green open spaces in urban park should be optimally functioned to maintain biodiversity, including the butterflies' habitat. Furthermore, the selection of plants located in green open spaces should include the caterpillar host plants and nectar sources for the butterflies [7]. This statement is validthat food plants and host plants have the important role in attracting butterflies. Concept in arranging butterflies habitat in urban park should be focusing on a proper placement for food plants and host plants in Figure 3.

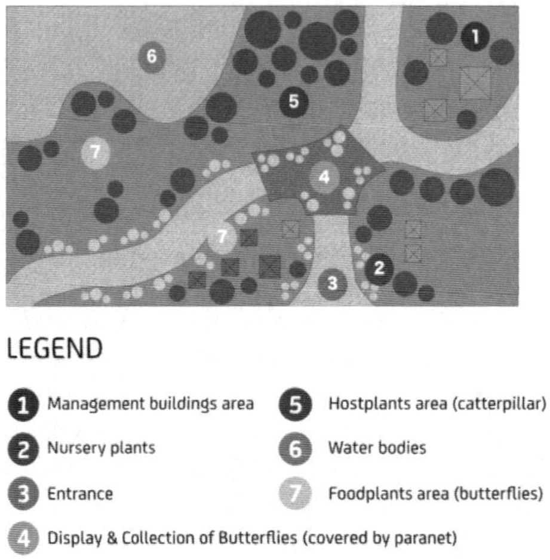

Figure 3. Zoning and developed concept

Nursery plants area have to be well managed in order to maintain sustainability of butterflies of the urban park. Since the caterpillar needs plenty of particular leafs while adult butterflies will search plants with more flowers. These plants in the nursery area should have always been ready to support food plants and host plants. Not only concern in plants but also periodically, the caretaker should check butterflies' eggs and caterpillar. And then bring them to the breeding house to keep away from predators such as aphids, spider and mantis. When the chrysalis is hatch and the young butterfly ready to fly, it means they are ready to be moved to the display area. A good management will support the life of butterflies in the urban park.

4. Conclusions

Some result study mentioned that several species of butterfly has sensitive responses to disturbance of environment called as bio indicator. However, the higher diversity of butterflies'occupied particular areas shows that the environment is in good condition.

Urban parks ecological function could be enhance to be a habitat for butterflies in urban areas, hence it becomes a patch *(stepping stones)* or even as a small habitat for the butterflies. The diversity of butterfly species and population that exist in an urban park indicates that the environment is in a good quality and suitable as butterflies' habitat. Food plants and host plants are the important factors to maintain the existence of butterflies in an urban park. Butterflies will seek another area if the food plants and host plants are insufficient.

Further research was necessary to explore more detail information in providing sufficient food plants and host plants in the urban parks, as it has become the important factor to create a sustainable environment for the butterflies in urban parks.

References

[1] [WHMI;WHC]Wildlife Habitat Management Institute;Wildlife Habitat Council (US).2000 *Butterflies* (Madison: Natural Resources Conservation Service)

[2] Azahra S D 2016 *Perbandingan Komunitas Kupu-Kupu di Berbagai Tipe,Karakteristik dan Gangguan Lingkungan Hutan Kota* Media Konservasi Vol. 21 No. 2 Agustus 2016: 108-115

[3] Brown K S and Freitas A V L 2002 *Butterfly communities of urban forest fragments in Campinas Sao Paulo, Brazil: structure, instability, environmental correlates, and conservation* Journal

ofInsect Conservation 6(4): 217-231.

[4] Collinge S K, Kathleen L P, Jeffrey C O 2003 *Effects of Local Habitat Characteristics and LandscapeContext on Grassland Butterfly Diversity* Conservation Biology Volume 17, No. 1, February2003

[5] Chiara J D and Koppelman L E 1978 *Site Planning Standard*, McGraw-Hill,Inc.

[6] Connor EF, Hafernik J, Levy J, Moore VL, Rickman JK. 2003. Insect conservation in an urban biodiversity hotspot: the San Francisco Bay area. *Journal of Insect Conservation* 6(4): 247-259.

[7] Diana P, Djamahar R, Isfaeni H 2016 *Hubungan Karakteristik Lanskap denganPreferensi Situs Oviposisi Kupu-Kupu di Ruang Terbuka Hijau Kawasan Jakarta Timur* Bioma 12 (1) (Biologi UNJ Press)

[8] Daniel J C, Schaefer J, Huegel C N, Mazzotti F J 2014 *Butterfly Gardening inFlorida* WEC 22 (IFAS Extention, University of Florida)

[9] Kitahara M, Sei K, Fuji K 2000 *Patterns in the structure of grassland butterflycommunities along a gradient of human disturbance: further analysis based on the generalist/specialist concept* Population Ecology 42(2): 135-144.

[10] Koh L P and Sodhi N S 2004 *Importance of reserves, fragments, and park for butterflyconservation in a tropical urban landscape*. Ecological Applicaton 14 (6): 1695-1708.

[11] Koneri R and Saroyo 2012 *Distribusi dan keanekaragaman kupu-kupu (Lepidoptera) diGunung Manado Tua, Kawasan Taman Nasional Laut Bunaken, Sulawesi Utara*. Jurnal Bumi Lestari 12(2): 357-365.

[12] Nurjanah S T 2010 *Biologi Troides helena helena dan Troides helenahephaestus(Papilionidae) di penangkaran* [tesis] (Bogor (ID): Institut Pertanian Bogor)

[13] Royer T 2015 *Landscaping to Attract Butterflies, Moth and Skippers* HLA-6430Oklahoma Cooperative Extension Service, Oklahoma State Unversity.

[14] Sing K W, Jusoh W F A, Hashim N R, Wilson J J 2016 *Urban parks: refuges fortropical butterflies in Southeast Asia?* Urban Ecosist (DOI 10.1007/s11252-016-0542-4, Springer)

[15] Smart P 1989 *The Illustrated Encyclopedia of The Butterfly World* (SalamanderBooks,UK)

[16] Tiple A D, Padwad S V, Dapporto L, Dennis R L H 2010 *Male mate location behaviourand encounter sites in a community of tropical butterflies: the influence of taxonomy, biotopes, landscape structures, resources, morphology, and population variables* Journal of Biosciences 35(4): 629–646

[17] Tresnawati E 2010 *Siklus Hidup dan Pertumbuhan Kupu-Kupu Graphiumagamemnon L. dan Graphium doson C (Papilionidae: Lepidoptera) denganpakan daun cempaka dan daun sirsak* [tesis] (Bogor (ID): Institut Pertanian Bogor)

[18] Venhaus H 2012 *Designing the Sustainable Site* (New Jersey: John Wiley Sons, Inc.)

Urban farming: people preference towards *verticulture* model in small housing type-settlements in Malang as sustainable landscape movement

N S S Giriwati[1*], A Citraningrum[1], and I Setyabudi[3]

[1] Department of Architecture, Brawijaya University, Jl. MT. Haryono 167, Malang 65145, Indonesia.

[2] Department of Landscape, Tribhuwana Tunggadewi University, Jl. Telaga Warna, Tlogomas, Kec. Lowokwaru, Kota Malang, Jawa Timur 65144, Indonesia.

*Email: novie_gieriwati@yahoo.com

Abstract. The rapid growths of Malang city as education and tourism city makes a higher demand for housing and increases land prices. The number of small housing type-settlement development with no adequate garden or open space is always increase. Housing that lack of greenery has led to many problems such as bad quality of fresh air, lack of playing area for kids and mental problem. On the other side, the widespread of city development has resulted in reduced agricultural space which has led to inadequate domestic food supplies and the declining quality of food crops availability. In the wake of various issues on food and energy security, urban farming by Food Oriented Development (FOD) is a concept of urban development that can make the city as a food provider for its own citizens on an ongoing basis. This concept considers aspects of food security as well as socio-economic considerations in urban physical development. The objective of this research is to get the information of people preference of *verticulture* model as urban farming method in small housing type-settlement. The questionnaire survey using Likert scale is conducted to measure people perception and preference. This study explores factor analysis for decision-making process. The result indicate that people tend to choose simple and smart system of *verticulture* model as a vertical garden in private residential. Giving recommendation of the *verticulture* model, hopefully this study can be implemented in small housing-type settlement in Malang city in order to be self-sufficient in food supply.

Keywords: small housing, sustainable landscape movement, urban farming, verticulture model

1. Research background

1.1. The Rapid growth of population in Malang and small housing type settlement

The rapid growths of Malang city as education and tourism city makes a higher demand of housing and increases land prices. According to the head of Demography information of Malang city in 2016, the population of Malang always grow every month as Malang is education city. The population growth is about 1,58 % [1]. The increasing number of the population mainly because of many newcomers. With the land size of 252.1 km2, the population of Malang city in the last five years has increased. By 2015 there are 50.116 people that increase from 2012 which is only 845.271 people [2].

The number of small housing type-settlement development with no adequate garden or open space is always increasing. Housing that lack of greenery has led to many problems such as bad quality of fresh air, lack of playing area for kids and mental problem.

1.2. Agriculture issues and food resilience

On the other side, the widespread of city development has resulted in reduced agricultural space which has led to inadequate domestic food supplies and the declining quality of food crops availability. In 2000-2012 the agricultural land in Malang reduced and left about 1300 ha, and in 2015 the Head of Department of Agriculture of Malang City Hadi Santoso said productive agricultural land in the area currently only left 865 Ha spread in District Blimbing, Kedungkandang, Sukun, and Lowokwaru [3]; [5]. From year to year agricultural land continues to shrink. The agriculture land reduces about 68 Ha Every year [4]. even in the District Klojen now there is no agricultural area at all because it has changed functions into public facilities, offices, trade and industrial centers and housing. It causes an effect on the availability of regional food. Because of these conditions, the productivity of rice produced by farmers has not been able to meet the food needs of residents of Malang, although the productivity per hectare is quite high, reaching 7.25 tons of rice equivalent. The demand of rice in Malang residents reaches 96,600 tons per year, while the production is only about 13,500 tons, so every year the average rice shortage reaches 83,000 tons. Hence, the innovation of development is needed to protect the food security.

1.3. The important of the vertical urban farming

In the wake of various issues on food and energy security, urban farming by Food Oriented Development (FOD) is a concept of urban development that can make the city as a food provider for its own citizens on an ongoing basis. This concept considers aspects of food security as well as socio-economic considerations in urban physical development. Recently, hydroponic culture technology started to gain favor in the developing because of population growth in urban areas represented an opportunity to grow food near consumers. The 100-200 million urban farmers worldwide providing the city markets with fresh agricultural products are the evidence of how food security can be achieved by urban agriculture [1].

The objective of this research is to get the information of factors that become people preference of vertical agriculture model as an urban farming method in small housing type settlements.

2. Theoretical Approach

According to Druckman & Lupia (2000) about the nature of Human preference, a preference is a comparative evaluation of (i.e. a ranking over) a set of objects. A preference serves as a cognitive marker that reminds people how to interact with various aspects of their environment. Preferences are stored in memory and drawn on when people make decisions. The objects of preference are aspects of the environment that are evaluated relative to one another. They can include observable, physically continuous phenomena (such as bowling balls) and unobservable, physically discontinuous phenomena [2].

The objects within a preference are those that a person can imagine as substitutable. On the other side, the predominant view of human cognition for nearly 2000 years has been that the objects of preference (alternatively, the categories of phenomena over which preferences can be held) are strictly external. In recent years, the evidence against this view of cognition has been piling high. For example, architecture objects. Most people prefer some architecture objects to others, and people have a favorite. People treat architecture objects as basic attributes of other objects.

Consumer preference might change since the experiences change. The preference of customers changes over time because of changes in demographics and lifestyle or more attractive competitors product, a target of marketing after a certain time. Consumer preference according to Kotler is like or dislike choice by someone to one product (goods and service) that consumed. Customer preference analysis is an analysis to decide product important hierarchy/order of important which is important or mostly preferred [3].

One of the fundamental issues in consumer behavior is the way consumers develop, adapt and use decision-making strategies. Consumer decision making could be defined as the "behavior patterns of consumers, that precede, determine and follow the decision process for the acquisition of need-satisfying products, ideas or services". Consumer decision-making has long been of great interest to researchers. Early decision-making studies concentrated on the purchase action. It was only after the 1950's those modern concepts of marketing were incorporated into studies of consumer decision-making, including a wider range of activities [4]. The contemporary research indicates that more activities are involved than the purchase itself. Many other factors influence the consumer decision-making than the final outcome. Vast numbers of studies have investigated this issue and many models have been developed accordingly. Models aim to depict the purchase decision-making process and its influential factors.

Factors that influence for the consumer in decision-making process are categorized into psychological and personal factors. The primary Psychological factors that influences on consumer behavior: 1). Personality and self-concept, 2). Motivation, 3) Learning, 4) Perception, and 5) the impact of attitudes. While the personal factors include demographic and situational variables such as sex, ages, race, origin, income, family life cycle, and occupation. Situational variables as external conditions like the amount of time for the consumer to make a decision [5].

The concept of vertical farming as the way for the citizen to do farming activity in an urban area especially in their neighborhood is a sustainable solution for the rapid growth in Malang city with the high demand of settlements and no adequate land. Vertical farming as a component of urban agriculture is the practice of producing food in vertically-stacked layers, vertically-inclined surfaces and/or integrated into other structures. Vertical farming is not a new idea. In 1915, Bailey coined the term "vertical farming". Since then, architects and scientists, especially towards the end of the twentieth century, have repeatedly looked into the idea of producing food in urban environments because of constant human population growth and the pressures exerted on resources for food production. Denmark was the first country to attempt to implement the concept of agricultural integration in a built environment in a house in the 1950s; they tried to grow watercress (*Nasturtium officinale*) on a large scale. Today a more evolved urban agriculture, where the product is grown in a totally controlled urban environment, in closed vertical structures, is attracting more attention in several countries. In the past two decades, scientists in the United States, Europe, and several Asian countries have been conducting research and development to bring this concept into reality [1].

Asia countries such as South Korea, Japan, China, Singapore, and Europe such as Italy, Holland United Kingdom, also Middle East areas such as Jordan, Saudi Arabia, United Arab Emirates and Canada, are moving ahead in the development of vertical farming projects. Vertical farming technology has been seen as a solution to the problems of limited land area suitable for agriculture, as well as a more rational use of water resources, thus providing better opportunities for a sustainable food supply in both developed and developing countries. Because of advances in hydroponic and aeroponic technology, lighting through LEDs and energy provided using solar cells, it is now possible to have agriculture in cities and possibly even in individual households to create centers of production and consumption integrated with urban and suburban communities. One can grow crops inside multi-story city buildings, using the very little land to produce food that would not need to be shipped far to the end of a consumer. Moreover, the vertical farming technology could contribute to a reduction in some of the following social, economic and environmental issues faced in the country [1].

3. Typology of vertical greenery

There are two typologies of vertical greenery that we can modify and adopt the construction to be implemented in the *verticulture* model for the vertical urban farming in settlements such as Green facade and living wall like in the table below:

Table 1. Vertical greenery classification.

Classification	Construction on the building facade	Vertical greenery type	
Green façade	Direct	*Traditional wall climbers*	
	Not direct	*Modular trellis panel*	
		Mesh structure	
		Wire structure	
		Perimeter flower pots	
Living wall	Not direct	*Modular living wall*	
		Vegetated mat wall	

4. Research methodology

4.1. Study area

The study area consists of 5 settlements in Malang City. Several case study was used to determine how the respondent/public evaluate *verticulture* model and how they perceive them. The settlements that become case studies are Swarna Housing, Saxophone Housing, de Prima Housing, d Ahsana Regency and Dapenza Housing that located in Malang Region Table 2).

Table 2. Settlements characteristic.

1. Housing I-Swarna	
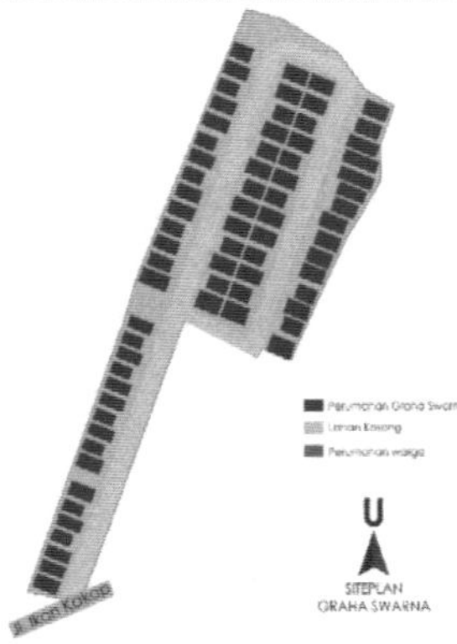	Housing Swarna is located in the Lowokwaru district in Malang City, the total occupancy in this housing is 55 units. It consists of types 45 and 36.  
2. Housing II-Saxophone	
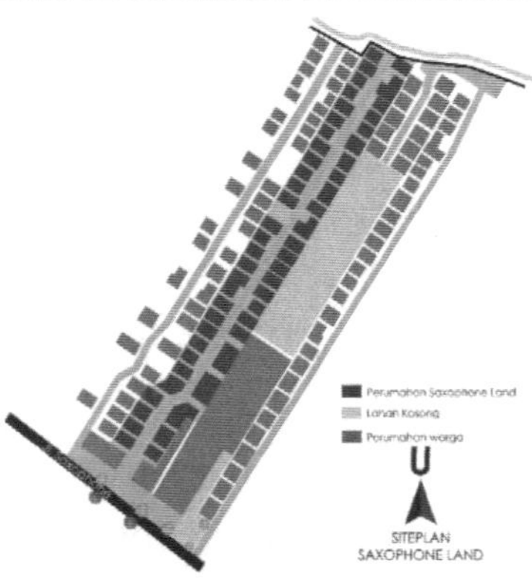	Saxophone Housing located in the Lowokwaru district of Ketawanggede Malang City, the total occupancy of this housing is 65 units. It consists of types 45 and 36.  
3. Housing III-de Prima	
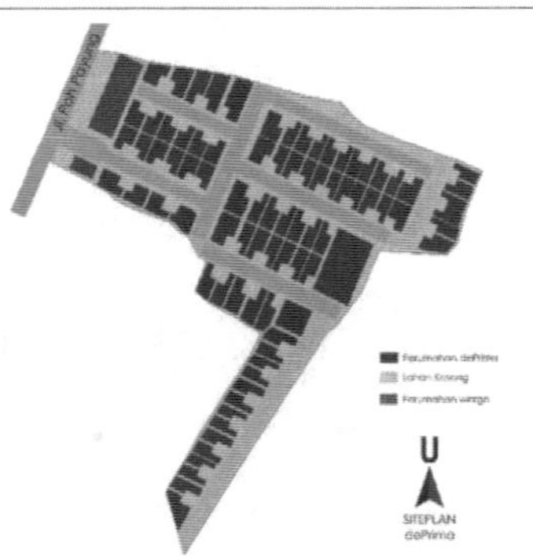	Housing de Prima is located in Lowokwaru district of Malang City, the total occupancy in this housing is 60 units. It consist of type 45 and 36.  

IOP Conf. Series: Earth and Environmental Science **179** (2018) 012025 doi:10.1088/1755-1315/179/1/012025

4. Housing IV-d'Ahsana regency

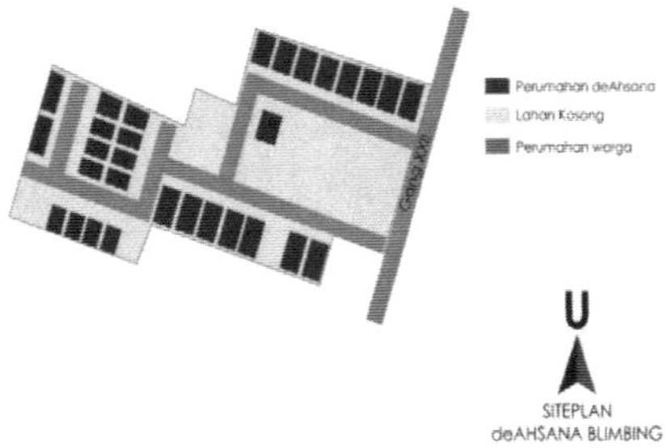

Ahsana housing is located in Pandanwangi district of Malang City, the total occupancy in this housing is 44 units. It consists of type 45 as many as 14 units as the focus of research. While other units of type 65 and 75 are not used as a research focus.

5. Housing V-Dapenza

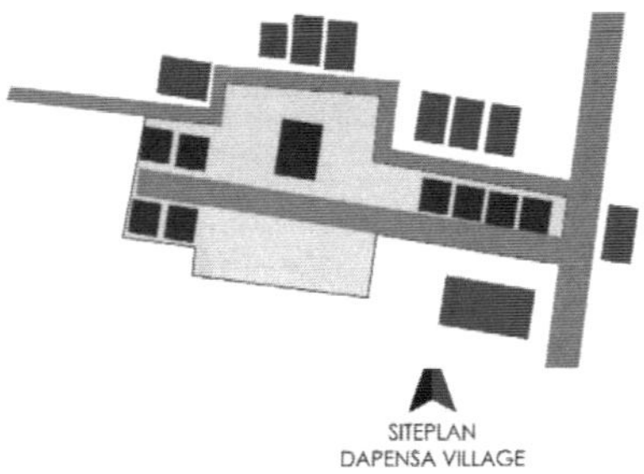

Dapenza Housing located in Batu City, the total occupancy in this housing is 60 units. It consists of types 45 and 36.

1. Housing I-Swarna

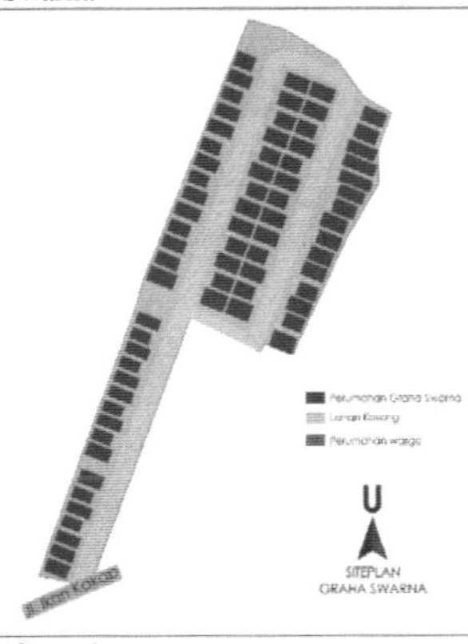

Housing Swarna is located in the Lowokwaru district in Malang City, the total occupancy in this housing is 55 units. It consists of types 45 and 36.

2. Housing II-Saxophone

Saxophone Housing located in the Lowokwaru district of Ketawanggede Malang City, the total occupancy of this housing is 65 units. It consists of types 45 and 36.

3. Housing III-de Prima

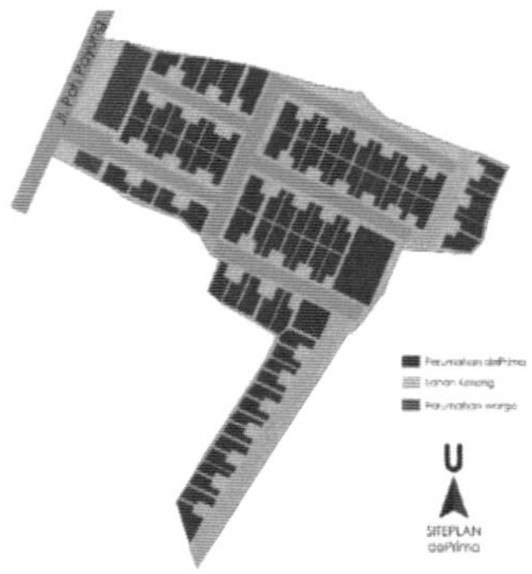

Housing de Prima is located in Lowokwaru district of Malang City, the total occupancy in this housing is 60 units. It consist of type 45 and 36.

4. Housing IV-d'Ahsana regency

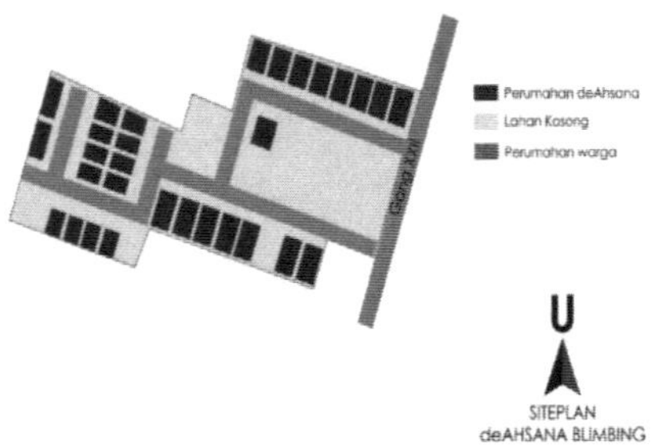

Ahsana housing is located in Pandanwangi district of Malang City, the total occupancy in this housing is 44 units. It consists of type 45 as many as 14 units as the focus of research. While other units of type 65 and 75 are not used as a research focus.

5. Housing V-Dapenza

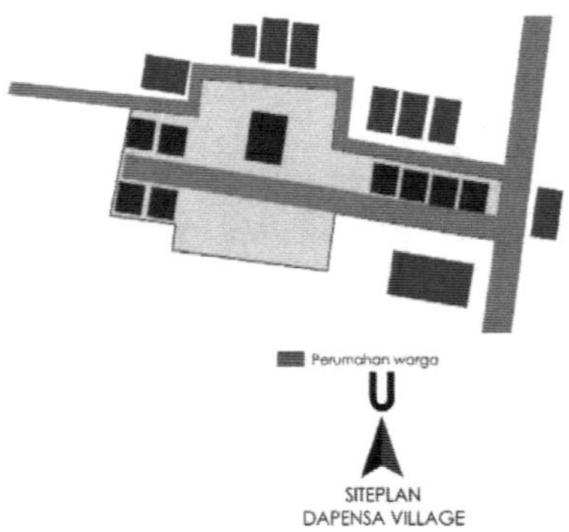

Dapenza Housing located in Batu City, the total occupancy in this housing is 60 units. It consists of types 45 and 36.

4.2. Data collection and research variable

This research using quantitative approach. The population of the research is small housing type settlement in Malang Region. The data collection procedure using field survey and questionnaire in order to get clear description of the research focus and public preference toward the vertical urban farming. In this research, 5 experimental groups were selected from 5 settlements. The respondent limited to productive ages people (17-50 years old) of house's owner and settlement manager. The number of sample determined from Gay and Diehl, that is depending upon the type of research. The experimental research has a minimum sample which is 15 subjects per one group [12].

The questionnaire survey using a Likert scale to measure people perception and preference. The variable that becomes research guideline in this paper are described in Table 3.

Table 3. Research variables.

Vertical agriculture and greenery			
1	Vertical greenery model	a	Green facades-GF
			• Traditional green facade (creeping and hanging)-GF1
			• Double skin green facades with trellises, wires and nets-GF2
			• Green facade with Perimeter flower pot-GF3
		b	Living Wall-LW
			• Living wall with landscape wall-LW1
			• Living wall with vegetated mat wall-LW2
			• Modular living wall-LW3
2	Vertical agriculture application	a	Growing media-GM
			• Recycle/used materials-MT1
			• PVC Pipe-MT2
			• New materials-MT3
			• Perimeter Flower Pot-MT4
			• Vegetated mat wall-MT5
			• Modular living wall-MT6
		b	Types of plants-JT
			• Fruits-JT1
			• Vegetables-JT2
			• Fruits and vegetables-JT3
		c	Plants watering-P
			• Watering plants manually-P1
			• Watering plants automatically-P2
		d	Target of harvesting-TP
			• Depending on the type of plant and planting period-TP1
			• Not targeted-TP2
		e	Space efficiency-EF
		h	Planting media should facilitate maintenance-MTMP
		i	Crop support structures must be sturdy and durable, resistant to rain, heat and cold-SP
3	Farming benefit for the community	a	Social value-NS
			• Increase public awareness of the importance of greening-NS1
			• Increase public awareness that greening can also meet the needs of micro foods crops-NS2
		b	Ecology value-NE
			• The benefit to improve urban green public space-NE1
			• The benefit to reduce the pollution in urban-NE2
			• Improve air quality around the dwelling area-NE3
		c	Aesthetic value-NES
			• Improve building and settlements uniqueness-NES1
			• Increase the interesting view around the dwelling-NES2

4.3. Method of analysis

This study explores factor analysis to get people preference and decision-making process. The analysis consists of several steps including descriptive analysis and to test the hypothesis using Factor Analysis (Bartlett's test of Sphericity (BTS), Kaiser-Meyer-Olkin (KMO), Principal Component Analysis (PCA) and the determination of the number of factors by extraction.

5. The result of the analysis

5.1. Respondent

Table 4. Characteristic of research respondent.

Demography		Swarna	Saxophone	De prima	Ahsana	Dapena
		n	n	n	n	n
Gender	Woman	15	4	1	1	1
	Man	13	6	6	5	6
Domicile	Malang					
	The other city					
Occupation	College student	10	3	2	1	1
	Stay at home mother	8	1			
	Company Employee	5		1	6	6
	Civil servant	3	1			
	others		5	2		1

5.2. People preference on the vertical urban farming

5.2.1. Normality test

According to normality test table, the significance of one sample Kolmogorov-Smirnov test us 0.089, that is bigger than 0.05. It indicates that the distribution of the data is normal.

Table 5. Kolmogorov-Smirnov test result.

One-Sample Kolmogorov-Smirnov Test

		Unstandardized Residual
N		58
Normal Parameters[a]	Mean	.0000000
	Std. Deviation	.86184232
Most Extreme Differences	Absolute	.164
	Positive	.078
	Negative	-.164
Kolmogorov-Smirnov Z		1.246
Asymp. Sig. (2-tailed)		.089

a. Test distribution is Normal.

5.2.2. Descriptive statistic

1. Bartlett Test and KMO

Bartlett test and KMO was conducted to know the appropriateness of all the indicator to analyzed using factor analysis. The Validity using factor analysis is the same principle with correlation analysis. It means if one indicator valid to measure one *latent variable*, that indicator must be correlated significantly and strongly with another *latent variable*. The significance of correlation can be seen in Bartlett's of Sperchity's Sig value. The strength of the correlation is in KMO value (Kaiser-Meyer-Olkin Measure of Sampling Adequacy). In this research, the Bartlett sig value is 0, that is smaller than 0.05 which means there is a correlation between indicators. All Indicator is indicated valid that can be seen from KMO score 0.836 which bigger than 0.05, so the correlation is strong.

Table 6. KMO and Barlett's test result.

Descriptive Statistics

	Mean	Std. Deviation	Analysis N
GF1	3.5345	1.12726	58
GF2	3.6897	1.17289	58
GF3	4.1207	.97473	58
LW1	3.9310	.98874	58
LW2	3.7241	1.12067	58
LW3	3.6897	1.07942	58
MT1	3.3793	1.34852	58
MT2	4.1034	1.11905	58
MT3	3.8276	1.02833	58
MT4	3.8966	1.08724	58
MT5	3.5345	1.11159	58
MT6	3.7414	1.06886	58
JT1	3.3966	1.43807	58
JT2	3.5345	1.41688	58
JT3	4.2586	1.08515	58
P1	3.3793	1.42444	58
P2	3.9828	1.33102	58
TP1	3.8793	1.29882	58
TP2	3.3793	1.37430	58
EF	3.0345	1.99090	58
MTMP	4.2931	1.09238	58
SP	4.3276	1.11431	58
NS1	4.2931	1.07620	58
NS2	4.2414	1.11309	58
NE1	4.2414	1.08110	58
NE2	4.3103	1.09556	58
NE3	4.4138	1.07662	58
NES1	4.1379	.96333	58
NES2	4.3793	1.07324	58

KMO and Bartlett's Test

Kaiser-Meyer-Olkin Measure of Sampling Adequacy.		.836
Bartlett's Test of Sphericity	Approx. Chi-Square	1.639E3
	df	406
	Sig.	.000

2. Anti-Image Matrice

On the MSA scores result, the part which written as 'a' and has shaped a diagonal line indicated MSA score of each variable. The score characterized bigger than 0.5. It showed that the variables have been predicted to be processed appropriately in the future analysis. The MSA scores for indicator GF1 in variable Green Wall is 0.757. It means GF1 have a strong correlation with another indicator. The MSA scores divided into three parts of the result. The first part described the *anti-image correlation result* of several variables including variables such as Double skin green facade (GF2), Perimeter flower pot (GF3), living wall with landscape wall (LW1), living wall with vegetated mat wall (LW2), modular living wall (LW3), Recycle/used materials (MT1), PVC Pipe (MT2), New materials (MT3), Perimeter Flower Pot (MT4).

Table 7. MSA result part 1.

Anti-image Matrices

		GF1	GF2	GF3	LW1	LW2	LW3	MT1	MT2	MT3	MT4
Anti-image Correlation	GF1	.757a	-0.292	-0.034	-0.163	-0.333	0.287	-0.330	0.205	-0.052	0.011
	GF2	-0.292	.772a	-0.335	0.187	0.160	-0.141	-0.248	-0.584	-0.078	-0.075
	GF3	-0.034	-0.335	.871a	-0.392	-0.109	0.225	0.225	0.111	-0.058	-0.224
	LW1	-0.163	0.187	-0.392	.875a	-0.163	-0.167	0.047	-0.099	0.087	-0.386
	LW2	-0.333	0.160	-0.109	-0.163	.839a	-0.155	0.117	-0.274	0.114	0.392
	LW3	0.287	-0.141	0.225	-0.167	-0.155	.885a	0.059	-0.114	-0.005	-0.197
	MT1	-0.330	-0.248	0.225	0.047	0.117	0.059	.715a	-0.051	0.230	-0.047
	MT2	0.205	-0.584	0.111	-0.099	-0.274	-0.114	-0.051	.847a	-0.138	-0.023
	MT3	-0.052	-0.078	-0.058	0.087	0.114	-0.005	0.230	-0.138	.871a	-0.213
	MT4	0.011	-0.075	-0.224	-0.386	0.392	-0.197	-0.047	-0.023	-0.213	.857a
	MT5	0.401	-0.161	-0.131	0.158	-0.449	0.180	-0.146	0.001	-0.016	-0.234
	MT6	-0.150	-0.205	0.338	-0.230	-0.162	-0.127	-0.123	0.323	-0.127	-0.242
	JT1	-0.316	-0.017	0.182	-0.067	0.068	-0.093	0.385	0.008	0.363	-0.054
	JT2	0.385	-0.107	-0.078	-0.021	-0.103	0.011	-0.240	-0.020	-0.263	0.102
	JT3	0.002	0.053	-0.162	-0.116	0.194	-0.372	-0.050	-0.045	-0.323	0.352
	P1	0.042	0.102	0.145	0.072	-0.132	0.214	-0.352	-0.007	-0.248	-0.251
	P2	-0.222	0.243	-0.100	0.098	-0.195	-0.018	-0.129	0.005	-0.083	-0.235
	TP1	0.049	0.152	-0.084	-0.071	-0.102	-0.013	-0.089	-0.121	0.226	0.070
	TP2	-0.128	0.271	-0.485	0.101	0.138	-0.187	-0.128	-0.040	0.100	0.250
	EF	0.199	-0.117	-0.113	-0.171	-0.017	0.088	-0.174	0.127	0.111	0.172
	MTMP	-0.032	0.016	-0.187	-0.097	0.260	0.075	0.182	-0.098	0.063	0.079
	SP	-0.051	0.047	0.114	0.127	-0.115	-0.146	-0.247	0.067	-0.220	-0.014
	NS1	-0.065	0.147	-0.204	0.340	-0.039	0.123	0.104	-0.125	0.120	-0.153
	NS2	-0.085	-0.248	0.266	-0.151	-0.007	-0.044	0.340	0.033	0.255	0.023
	NE1	-0.313	-0.294	0.189	-0.212	0.208	-0.310	0.147	0.364	-0.181	0.137
	NE2	0.191	0.310	-0.323	0.311	-0.033	0.224	-0.161	-0.341	0.169	-0.065
	NE3	0.366	-0.121	0.002	-0.142	-0.166	0.224	0.094	-0.082	0.004	0.056
	NES1	0.111	-0.173	0.027	-0.230	-0.260	-0.243	-0.135	0.303	-0.363	0.230
	NES2	-0.233	0.380	0.058	0.104	0.041	-0.011	-0.272	-0.177	0.038	-0.309

a. Measures of Sampling Adequacy(MSA)

The second part described the *anti-image correlation result* of several variables including variables such as Vegetated mat wall (MT5), Modular living wall (MT6), Fruits (JT1), Vegetables (JT2), Fruits and vegetables (JT3), Watering plants P1 manually, watering plants automatically P2, depending on the type of plant and planting period (TP1), Not targeted (TP2), space efficiency (EF).

Table 8. MSA result part 2.

		MT5	MT6	JT1	JT2	JT3	P1	P2	TP1	TP2	EF
Anti-image Correlation	GF1	0.401	-0.150	-0.316	0.385	0.002	0.042	-0.222	0.049	-0.128	0.199
	GF2	-0.161	-0.205	-0.017	-0.107	0.053	0.102	0.243	0.152	0.271	-0.117
	GF3	-0.131	0.338	0.182	-0.078	-0.162	0.145	-0.100	-0.084	-0.485	-0.113
	LW1	0.158	-0.230	-0.067	-0.021	-0.116	0.072	0.098	-0.071	0.101	-0.171
	LW2	-0.449	-0.162	0.068	-0.103	0.194	-0.132	-0.195	-0.102	0.138	-0.017
	LW3	0.180	-0.127	-0.093	0.011	-0.372	0.214	-0.018	-0.013	-0.187	0.088
	MT1	-0.146	-0.123	0.385	-0.240	-0.050	-0.352	-0.129	-0.089	-0.128	-0.174
	MT2	0.001	0.323	0.008	-0.020	-0.045	-0.007	0.005	-0.121	-0.040	0.127
	MT3	-0.016	-0.127	0.363	-0.263	-0.323	-0.248	-0.083	0.226	0.100	0.111
	MT4	-0.234	-0.242	-0.054	0.102	0.352	-0.251	-0.235	0.070	0.250	0.172
	MT5	.826a	-0.269	-0.250	0.381	-0.132	0.093	-0.095	-0.182	-0.179	0.127
	MT6	-0.269	.827a	0.162	-0.162	-0.081	0.085	0.195	-0.103	-0.269	-0.112
	JT1	-0.250	0.162	.611a	-0.748	-0.368	-0.334	-0.057	-0.006	0.030	0.055
	JT2	0.381	-0.162	-0.748	.589a	0.374	0.070	-0.188	-0.120	-0.357	-0.064
	JT3	-0.132	-0.081	-0.368	0.374	.860a	-0.022	-0.257	-0.092	0.006	-0.071
	P1	0.093	0.085	-0.334	0.070	-0.022	.586a	0.377	-0.292	-0.249	-0.104
	P2	-0.095	0.195	-0.057	-0.188	-0.257	0.377	.895a	-0.086	0.085	-0.141
	TP1	-0.182	-0.103	-0.006	-0.120	-0.092	-0.292	-0.086	.852a	0.415	-0.040
	TP2	-0.179	-0.269	0.030	-0.357	0.006	-0.249	0.085	0.415	.734a	0.175
	EF	0.127	-0.112	0.055	-0.064	-0.071	-0.104	-0.141	-0.040	0.175	.710a
	MTMP	0.020	-0.170	0.009	0.001	0.047	-0.035	0.040	-0.173	0.012	0.142
	SP	-0.113	0.171	-0.037	-0.052	-0.027	0.095	-0.041	0.024	0.045	-0.188
	NS1	-0.034	-0.361	0.087	-0.267	-0.337	-0.049	0.168	0.186	0.239	0.118
	NS2	-0.327	0.275	0.336	-0.206	-0.035	-0.275	-0.005	0.136	-0.059	-0.132
	NE1	-0.253	0.468	0.197	-0.233	0.153	-0.086	-0.037	-0.161	-0.063	-0.059
	NE2	0.253	-0.533	-0.123	0.129	-0.198	0.096	-0.048	0.028	0.187	0.186
	NE3	0.343	-0.263	-0.186	0.411	0.116	0.055	-0.205	-0.007	-0.146	0.040
	NES1	-0.140	0.254	-0.157	0.138	0.422	-0.137	-0.006	0.076	-0.030	-0.037
	NES2	0.160	0.161	-0.030	0.017	-0.181	0.216	0.172	-0.085	-0.240	-0.203

a. Measures of Sampling Adequacy(MSA)

The third part described the *anti-image correlation result* of several variables including planting media should facilitate maintenance (MTMP), Crop support structures must be sturdy and durable belongs to(SP), resistant to rain (NS1), heat and cold (NS2), Increase public awareness of the important of greenery (NE1), Increase public awareness that greening can also meet the needs of micro foods crops (NE2), Benefit for improving urban green public space (NE3), Benefit for reducing pollution, Improving air quality around the dwelling area, Improving building and settlements uniqueness (NES1), Increase the interesting view around the dwelling (NES2).

Table 9. MSA result part 3.

		MTMP	SP	NS1	NS2	NE1	NE2	NE3	NES1	NES2
Anti-image Correlation	GF1	-0.032	-0.051	-0.065	-0.085	-0.313	0.191	0.366	0.111	-0.233
	GF2	0.016	0.047	0.147	-0.248	-0.294	0.310	-0.121	-0.173	0.380
	GF3	-0.187	0.114	-0.204	0.266	0.189	-0.323	0.002	0.027	0.058
	LW1	-0.097	0.127	0.340	-0.151	-0.212	0.311	-0.142	-0.230	0.104
	LW2	0.260	-0.115	-0.039	-0.007	0.208	-0.033	-0.166	-0.260	0.041
	LW3	0.075	-0.146	0.123	-0.044	-0.310	0.224	0.224	-0.243	-0.011
	MT1	0.182	-0.247	0.104	0.340	0.147	-0.161	0.094	-0.135	-0.272
	MT2	-0.098	0.067	-0.125	0.033	0.364	-0.341	-0.082	0.303	-0.177
	MT3	0.063	-0.220	0.120	0.255	-0.181	0.169	0.004	-0.363	0.038
	MT4	0.079	-0.014	-0.153	0.023	0.137	-0.065	0.056	0.230	-0.309
	MT5	0.020	-0.113	-0.034	-0.327	-0.253	0.253	0.343	-0.140	0.160
	MT6	-0.170	0.171	-0.361	0.275	0.468	-0.533	-0.263	0.254	0.161
	JT1	0.009	-0.037	0.087	0.336	0.197	-0.123	-0.186	-0.157	-0.030
	JT2	0.001	-0.052	-0.267	-0.206	-0.233	0.129	0.411	0.138	0.017
	JT3	0.047	-0.027	-0.337	-0.035	0.153	-0.198	0.116	0.422	-0.181
	P1	-0.035	0.095	-0.049	-0.275	-0.086	0.096	0.055	-0.137	0.216
	P2	0.040	-0.041	0.168	-0.005	-0.037	-0.048	-0.205	-0.006	0.172
	TP1	-0.173	0.024	0.186	0.136	-0.161	0.028	-0.007	0.076	-0.085
	TP2	0.012	0.045	0.239	-0.059	-0.063	0.187	-0.146	-0.030	-0.240
	EF	0.142	-0.188	0.118	-0.132	-0.059	0.186	0.040	-0.037	-0.203
	MTMP	.857a	-0.894	0.191	-0.004	0.065	-0.005	0.141	-0.258	-0.230
	SP	-0.894	.855a	-0.134	-0.123	0.026	0.013	-0.279	0.267	0.208
	NS1	0.191	-0.134	.891a	-0.250	-0.253	0.245	-0.339	-0.229	-0.070
	NS2	-0.004	-0.123	-0.250	.886a	0.106	-0.278	-0.039	0.090	-0.338
	NE1	0.065	0.026	-0.253	0.106	.825a	-0.785	-0.304	0.167	-0.017
	NE2	-0.005	0.013	0.245	-0.278	-0.785	.819a	0.123	-0.391	-0.034
	NE3	0.141	-0.279	-0.339	-0.039	-0.304	0.123	.888a	-0.132	-0.311
	NES1	-0.258	0.267	-0.229	0.090	0.167	-0.391	-0.132	.864a	-0.218
	NES2	-0.230	0.208	-0.070	-0.338	-0.017	-0.034	-0.311	-0.218	.891a

a. Measures of Sampling Adequacy(MSA)

5.2.3. Extraction

1. Communalities

The extraction number of the traditional green facade (GF) 1 is 0.742. It means 74.2% variance from GF1 variable can be explained by the factor that has been shaped. The extraction number of double skin green facades (GF 2) is 0.795, GF3 0.792, LW1 0.752, LW2 0.870, and LW3 0.715. The smaller the value of communalities means the weaker the relationship with the factors formed.

Table 10. Communalities score.

Communalities

	Initial	Extraction
GF1	1.000	.742
GF2	1.000	.795
GF3	1.000	.792
LW1	1.000	.752
LW2	1.000	.870
LW3	1.000	.715
MT1	1.000	.805
MT2	1.000	.739
MT3	1.000	.763
MT4	1.000	.758
MT5	1.000	.804
MT6	1.000	.729
JT1	1.000	.888
JT2	1.000	.865
JT3	1.000	.732
P1	1.000	.795
P2	1.000	.782
TP1	1.000	.755
TP2	1.000	.779
EF	1.000	.562
MTMP	1.000	.862
SP	1.000	.894
NS1	1.000	.901
NS2	1.000	.870
NE1	1.000	.873
NE2	1.000	.905
NE3	1.000	.928
NES1	1.000	.825
NES2	1.000	.918

Extraction Method: Principal Component Analysis.

2. The total variance explained

The total variance explained table show that there are 7 new factors have been shaped. The eigenvalue must be >1. According to the table of total variance explained, Factor1 eigenvalue: 13.5 with variance 46.927%, factor2 eigenvalue 2.767 with variance 9.54%, factor 3 eigenvalue 2.136 with variance 7.366%, factor 4 eigenvalue 1.541 with variance 5.314%, factor 5 eigenvalue 1.318 with variance 4.545%, factor6 eigenvalue 1.047 with variance 3.609%, factor 7 eigenvalue 1.010 with variance 3.484%. The total variance of all factors is 80.784.

Table 11. Total variances score.

Total Variance Explained									
	Initial Eigenvalues			Extraction Sums of Squared Loadings			Rotation Sums of Squared Loadings		
Component	Total	% of Variance	Cumulative %	Total	% of Variance	Cumulative %	Total	% of Variance	Cumulative %
1	13.580	46.827	46.827	13.580	46.827	46.827	7.514	25.910	25.910
2	2.767	9.540	56.368	2.767	9.540	56.368	5.292	18.250	44.159
3	2.136	7.366	63.734	2.136	7.366	63.734	2.804	9.668	53.827
4	1.541	5.314	69.047	1.541	5.314	69.047	2.794	9.634	63.461
5	1.318	4.545	73.593	1.318	4.545	73.593	1.834	6.325	69.786
6	1.047	3.609	77.202	1.047	3.609	77.202	1.605	5.534	75.320
7	1.010	3.484	80.686	1.010	3.484	80.686	1.556	5.366	80.686
8	.727	2.508	83.194						
9	.627	2.164	85.357						
10	.558	1.924	87.281						
11	.514	1.774	89.055						
12	.443	1.526	90.581						
13	.392	1.352	91.933						
14	.378	1.304	93.237						
15	.331	1.140	94.377						
16	.293	1.009	95.386						
17	.279	.961	96.346						
18	.187	.646	96.993						
19	.149	.516	97.508						
20	.140	.484	97.992						
21	.126	.436	98.427						
22	.116	.401	98.828						
23	.092	.317	99.145						
24	.073	.252	99.396						
25	.055	.190	99.586						
26	.049	.170	99.756						
27	.032	.112	99.868						
28	.021	.074	99.941						
29	.017	.059	100.000						

Extraction Method: Principal Component Analysis.

3. Rotation component matrix

The value of loading factor are calculated by correlation between factors and variable. From line one in the table of Rotated component matrix, traditional green facade (GF) with loading factor 1 is 0.339, loading factor 2 is 0.129. Traditional green facade with loading factor 3 is 0.194, traditional green facade with loading factor 4 is 0.122, traditional green facade with loading factor 5 is 0.727, traditional green facade with loading factor 6 is 0.069, traditional green facade with loading factor 7 is 0.160. According to the correlation score requirements, strong correlation must be 1 or -1. In this indicator, the biggest one is 0.727, so traditional green facade indicator belongs to factor 5.

Double skin green facade (GF2) belongs to factor 2, Perimeter flower pot belongs to factor 1 (GF3), living wall with landscape wall (LW1) belongs to factor 5, living wall with vegetated mat wall LW2 belongs to factor 6, modular living wall (LW3) belongs to factor 2, recycle/used materials (MT1) belongs to factor 7, PVC Pipe (MT2) belongs to factor 2, new materials (MT3) belongs to factor 2, perimeter flower pot (MT4) belongs to factor 2, vegetated mat wall (MT5) belongs to factor 2, modular living wall (MT6) belongs to factor 2, fruits (JT1) belongs to factor 4, vegetables (JT2) belongs to factor 2, fruits and vegetables (JT3) belongs to factor 7, watering plants P1 manually belongs to factor 7, watering plants automatically P2 belongs to factor 3, depending on the type of plant and planting period belongs to factor 3 (TP1), not targeted (TP2) belongs to factor 4, space efficiency (EF) belongs to factor 3, planting media should facilitate maintenance (MTMP) belongs to factor 3, crop support structures must be sturdy and durable belongs to factor 3 SP, resistant to rain belongs to factor 1 (NS1), heat and cold (NS2) belongs to factor 1, increase public awareness of the important of greenery (NE1) belongs to factor 1, increase public awareness that greening can also meet the needs of micro foods crops (NE2) belongs to factor 1, benefit for improving urban green public space (NE3) belongs to factor 1, benefit for reducing pollution belongs to factor 1, improving air quality around the dwelling area belongs to factor 1, improving building and settlements uniqueness (NES1) belongs to factor 1, increase the interesting view around the dwelling (NES2) belongs to factor 1.

Table 12. Rotated component matrix.

Rotated Component Matrix[a]

	Component						
	1	2	3	4	5	6	7
GF1	.339	.129	.194	.122	.727	.069	.160
GF2	.128	.829	.032	-.014	.096	.175	.223
GF3	.471	.559	.091	.158	.403	.146	-.203
LW1	.293	.485	.195	.210	.513	.229	-.182
LW2	.384	.257	.111	.140	.171	.771	.037
LW3	.322	.602	.176	.371	.046	.123	-.251
MT1	.107	.204	.391	.128	.413	.035	.641
MT2	.293	.719	.220	.024	-.136	.260	.043
MT3	.365	.760	.083	.109	.109	-.140	-.055
MT4	.324	.718	.073	.023	.363	.007	.002
MT5	.259	.646	.184	.064	.024	.528	.052
MT6	.478	.547	.079	.125	.317	.250	.128
JT1	-.004	.059	.220	.891	.046	.195	-.033
JT2	.023	.136	.147	.894	.042	.009	.154
JT3	.480	.461	.369	.182	.098	.129	-.306
P1	.071	.021	.076	.522	-.020	.103	.707
P2	.322	.281	.454	.169	.326	.264	-.435
TP1	.067	.195	.710	.166	.013	.424	.017
TP2	.394	.063	-.149	.668	.326	-.104	.184
EF	.047	-.058	.699	.039	.174	.030	.186
MTMP	.439	.451	.643	.171	.053	-.124	-.063
SP	.438	.478	.663	.153	.007	-.083	-.054
NS1	.864	.349	.048	.086	.067	.138	.019
NS2	.860	.268	.200	.040	-.011	.104	.077
NE1	.875	.251	.078	.079	.159	.078	-.036
NE2	.893	.235	.097	.019	.149	.135	-.036
NE3	.898	.269	.141	-.015	.117	.103	-.067
NES1	.816	.228	.021	.144	.185	.208	.097
NES2	.887	.123	.236	.105	.218	-.031	.033

Extraction Method: Principal Component Analysis.
Rotation Method: Varimax with Kaiser Normalization.

4. Component transformation matrix

Component transformation Matrix shows Varimax rotation result. All variables have been distributed to each factor that has been shaped. Factor I named Benefit of Vertical Agriculture including Green facade with perimeter flower pot (GF3), Fruits and vegetables (JT3), Increase public awareness of the important of greening (NS1), Increase public awareness that greening can also meet the needs of micro foods crops (NS2), Benefit for improving urban green public space (NE1), Benefit for reducing pollution (NE2), Improving air quality around the dwelling area (NE3), Improving building and settlements uniqueness (NES1), Increase the interesting view around the dwelling (NES2). Factor 2 named Modular Vertical Agriculture including Double skin green facades with trellises, wires and nets (GF2), Modular living wall (LW3), PVC Pipe (MT2), New materials (MT3), Perimeter Flower Pot (MT4), Vegetated mat wall (MT5), Modular living wall (MT6). Factor 3 named Technology for Vertical agriculture including watering plants automatically (P2), Depending on the type of plant and planting period (TP1), Space efficiency (EF), Planting media should facilitate maintenance (MTMP), Crop support structures must be sturdy and durable, resistant to rain, heat and cold (SP). Factor 4 named Vertical agriculture plants including Fruits (JT1), Vegetables (JT2), Not targeted harvesting (TP2). Factor 5 named Traditional

Alternative for greenery type. Traditional green facade/creeping and hanging (GF1), the living wall with landscape wall (LW1). Factor 6 named Alternative for living wall including living wall with vegetated mat wall (LW2). Factor 7 named environmentally friendly Vertical agriculture including Recycle/used materials (MT1) Watering plants manually (P1).

Table 13. Component score coefficient matrix.

Component Score Coefficient Matrix

	Component						
	1	2	3	4	5	6	7
GF1	-.042	-.093	.020	-.068	.549	-.020	.049
GF2	-.106	.321	-.100	-.085	-.040	.003	.200
GF3	-.029	.086	-.086	.019	.229	-.009	-.152
LW1	-.093	.043	-.031	.020	.346	.069	-.163
LW2	.011	-.108	-.072	-.018	.021	.609	.012
LW3	-.031	.147	-.037	.158	-.097	-.037	-.181
MT1	-.065	.018	.155	-.110	.241	-.067	.411
MT2	-.023	.221	.011	-.041	-.264	.083	.090
MT3	-.030	.280	-.062	.018	-.037	-.284	-.004
MT4	-.076	.225	-.079	-.062	.201	-.148	.019
MT5	-.055	.136	-.035	-.047	-.120	.332	.068
MT6	-.006	.092	-.090	-.033	.123	.088	.089
JT1	-.046	-.067	.009	.394	-.073	.100	-.136
JT2	-.034	.015	-.024	.386	-.095	-.076	.001
JT3	.020	.034	.095	.052	-.046	-.021	-.202
P1	.026	.000	-.016	.149	-.157	.046	.432
P2	-.043	-.084	.145	.031	.203	.123	-.326
TP1	-.059	-.083	.316	-.021	-.091	.274	.005
TP2	.060	-.039	-.180	.276	.152	-.151	.029
EF	-.030	-.125	.378	-.092	.093	-.033	.113
MTMP	.022	.067	.277	.008	-.109	-.268	-.024
SP	.022	.076	.286	-.002	-.154	-.236	-.010
NS1	.174	-.013	-.074	-.002	-.122	.021	.040
NS2	.195	-.048	.033	-.034	-.193	-.003	.090
NE1	.178	-.064	-.046	-.005	-.026	-.021	-.010
NE2	.185	-.083	-.033	-.037	-.033	.035	-.003
NE3	.185	-.065	-.005	-.050	-.061	-.001	-.014
NES1	.162	-.078	-.090	.007	-.012	.101	.067
NES2	.191	-.125	.060	-.013	.023	-.114	.024

Extraction Method: Principal Component Analysis.
Rotation Method: Varimax with Kaiser Normalization.
Component Scores.

6. Conclusion

The result indicated that people tend to choose the simple and smart system of the vertical urban farming model as a vertical garden in private residence. Based on factor analysis, the public preference of the vertical urban farming model can be classified in 7 important factor: 1. Factor 1 (Benefit of vertical agriculture), 2. Factor 2 (Modular vertical agriculture), 3. Factor 3 (Technology for vertical agriculture), 4. Factor 4 (Vertical agriculture plants), 5. Factor 5 named (Traditional alternative for greenery type), 6. Factor 6 (Alternative for living wall) including living wall, 7. Factor 7 (Environmentally friendly vertical agriculture).

However, this research has a limitation, the respondent could not in minimum amount in one group. It was suggested that the sample should be 15 in one group for the minimum; however, some of the samples was lower than 15 dues to the different occupancies of the house. The other issue is not all housing owner could fulfill the questionnaire. However, due to time constraints and the normality test which indicated normal, this research is finished. Future research can pinpoint this problem and thus could facilitate the improvement of research methodology.

Giving recommendation of the vertical urban farming model based on people preference, hopefully this study can be implemented in a small housing-type settlement in Malang City in order to be self-sufficient in food supply. It is hoped that the evaluation model that we established can serve as a constructive reference for professionals in the design of sign systems and for academicians regarding their further studies. On the other hand, the evaluation model in the research can be used for long-term follow-ups concerning user requirements, and for implementation when the vertical urban farming in a small housing type settlement are designed.

References

[1] Anda JD and Shear H 2017 Potential of vertical hydroponic agriculture in Mexico Sustainability 9:140
[2] Citraningrum A 2016 Persepsi terhadap sistem penghijauan vertical di Indonesia, Universitas Brawijaya
[3] Druckman JN and Lupia A 2000 Preference formation annual review of political science 3:1-24
[4] Embaye A 2004 an empirical investigation into the factors influencing consumer perception of the no name brand food items: special focus-South Africa University of Kwa Zulu Nathal: Durban, Sudrafika
[5] Gay LR and Diehl PL 1992 Research methods for business and publishing company, New York
[6] http://jatim.tribunnews.com/2017/02/14/lima-tahun-penduduk-kota-malang-bertambah-50116-orang
[7] http://suryamalang.tribunnews.com/2016/05/30/penduduk-kota-malang-bertambah-158-persen-tiap-tahun
[8] http://kabar24.bisnis.com/read/20130303/186/138615/alih-fungsi-lahan-pertanian-di-malang-hanya-tersisa-1-dot-300-ha-sawah
[9] https://www.antaranews.com/berita/516608/lahan-pertanian-kota-malang-berkurang-68-hektare-per-tahun
[10] http://surabaya.bisnis.com/read/20160621/9/89686/alih-fungsi-lahan-pertanian-kota-malang-hanya-tersisa-865-ha
[11] Karimi S 2013 A purchase decision-making process model of online consumers and its influential factor a cross sector analysis Manchester Business School
[12] Tsiotsu H and Goldsmith E 2012 Strategic marketing in tourism services Emerald Group publishing: Florida, USA

The development of Bogor Botanic Garden design from 1817 to 2017 base on spatial and functional

D Safarinanugraha[1*], A Gunawan[2], and W Q Mugnisjah[2]

[1] Graduate Student of Landscape Architecture Department, Bogor Agricultural University (IPB), Indonesia
[2] Landscape Architecture Departments, Faculty of Agriculture, Bogor Agricultural University (IPB), Indonesia

[*]Email: dina.safarinanugraha@gmail.com

Abstract: Bogor Botanic Garden (BBG) is one of the ex situ conservation areas with overall site approximately 87 ha. BBG is the oldest botanical garden in Southeast Asia and become an icon of the city of Bogor. From its inception in 1817 to its present age of 200 years, the botanical garden has evolved into an institution that combines diverse activities in science and horticulture with public education, as well as a rich heritage of historic buildings and monuments of living plants. The development of a botanical garden that presents new roles and functions led to a design change in the botanical garden. The objective of this study was to identify the development and function of The Bogor Botanic Garden. This research was conducted using descriptive methods consisting of historical reference search, compare, from initial map and current condition. The results of The Bogor Botanic Garden design established based on taxonomy patterns with the classic English garden concept. From the function of the botanical garden initially only serves as a garden where the collection of plants, then became the center of research, education, and now developed as a place for recreation and educational activities of nature observation.

Keywords: Bogor Botanic Garden, english garden, function, history, conservation

1. Introduction

The Botanical Garden is an ex situ plant conservation area that has a collection of documented and arranged plants based on the taxonomic, bioregion, thematic, or combination pattern of those patterns for conservation, research, education, tourism, and environmental services [1]. Currently, there are about 3.422 botanical gardens in 148 countries worldwide [2]. Most of the Botanical Gardens are in the urban area of about 60%. Botanic gardens in South-East Asia made a significant contribution during the discovery and dissemination of flowering and economic botany throughout the world from the mid-1800s to the mid-1900s became an important supplier of tropical trees, plants and orchids to Europe and North America. The Bogor Botanic Garden is one of significant Botanical Gardens, in Southeast Asia together with the Calcutta Botanic Gardens in India (1787), Peradeniya Botanic Gardens in Sri Lanka (1843), and Singapore Botanical Garden (1859).

Tropical botanic gardens remain important for their records of local floras. This work is the essential basis for all other studies of plants, whether scientific or technological. The gardens also remain important as collections of living plants for botanical study of all kinds and for local education. [2]. The origins, history and functions of modern botanic gardens predate the European Renaissance and that they reflect the social, economic and environmental circumstances of their times [3]. As expansion proceeded, botanic gardens established in the colonial region were designed to function as imperial way stations for the acclimatization and exportation of economically desirable plants [4]. .

The planning and design of Bogor Botanic Gardens have changed over time to reflect changes in purpose, changes in science, and changes in the aesthetic expressions of human cultures over time. As such, the conservation botanic garden is one that evolved in response to the conservation desires and requirements of local cultures; contains or is associated with areas of natural vegetation within the context of local biotic zones; focuses exclusively on plants native to the region; and engages in public education [5]. From its inception in 1817 to its present age of 200 years, The Bogor Botanic Garden has evolved into an institution that combines diverse activities in science and horticulture with public education, as well as a rich heritage of historic buildings and monuments of living plants. The development of a botanical garden that presents new functions and roles led to a design change in the KRB that adopted the function. The objective of this study was to identify the development and function of The Bogor Botanic Garden.

2. Methods

2.1. Study area

The study area is located in Bogor Botanic Gardens, located in the center of Bogor City, West Java (Figure 1). Bogor Botanic Garden (BBG) has an area of 87 ha with a height of 260 m above sea level. BBG is a living plant museum with the most complete collection of tropical plants in the world, built with a beautiful landscaping concept. BBG location is very strategic because it is easy to reach from anywhere. Its location close to the highway can be easily accessed by visitors from outside the city of Bogor. Bogor Botanic Garden is the oldest botanical garden in Southeast Asia.

2.2. Method of the study

This research was conducted using descriptive method consisting of historical reference search, and compare from initial map origin of Bogor Botanic Gardens in 1826, 1892, 1927 and current condition, and field observation to explore space, functions, landscape elements and layout. Map selection based on the occurrence of additional areas.

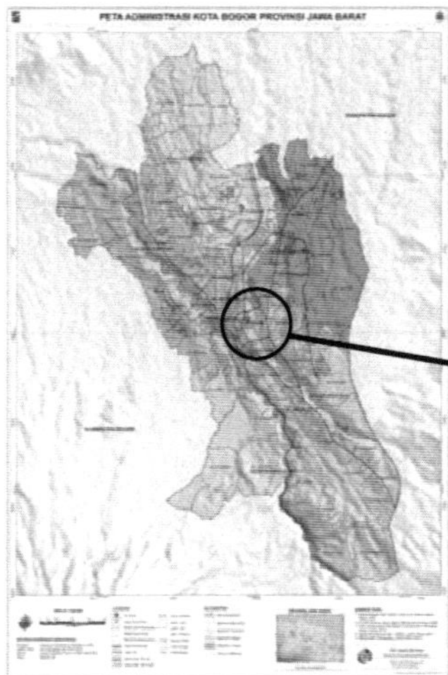

Map of Bogor City

Figure1. Map of detailed study area

3. Result and Discussion

3.1. History and Development of Bogor Botanic Garden

The History begun when the Kingdom of the Netherlands sent Cornelis Theodorus Elout, and G.A.G.P. Baron van der Capellen to Indonesia with Prof. Caspar George Carl Reinwardt was a Director of Agriculture, Arts, and Education for Java in 1816. On April 15, 1817 Reinwardt sparked the idea of establishing a botanical garden to Governor-General G.A.G.P. Baron van der Capellen, the idea was then approved. Finally, on May 18, 1817 formally established a Botanical Garden in Bogor City, under the name of 'Lands Plantentuin te Buitenzorg. After independence, 1949's Lands Plantentiun te Buitenzorg renamed Jawatan Penyelidikan Alam, then became the Lembaga Pusat Penyelidikan Alam (LLPA) for the first time managed and led by the Indonesian nation, Prof. Ir. Kusnoto Setyodiwiryo. At that time LPPA had 6 subsidiary institutions, namely Bibliotheca Bogoriensis, Hortus Botanicus Bogoriensis, Herbarium Bogoriensis, Treub Laboratory, Zoologicum Bogoriensis Museum and Marine Investigation Laboratory. In 1956 for the first time the leadership of the Botanical Garden was held by the Indonesian people, Sudjana Kassan [6].

Bogor Botanic Gardens which was originally only ±40 ha in the course of his life experienced several times the expansion, so now the extent has reached ±78 ha (Figure 2). The addition of the BBG area occurred in 1892 and 1927. This expansion was important because of the increased of research and expedition in botany, more area was needed. Then gradually established branches of botanical gardens in various regions, such as Cibodas Botanical Gardens in West Java (1862). Purwodadi Botanical Gardens in East Java (1941) and Eka Karya Botanical Gardens in Candi kuning/Bali (1959). The Bogor Botanic Gardens and their branches form a separate unit, also known as Kebun Raya Indonesia.

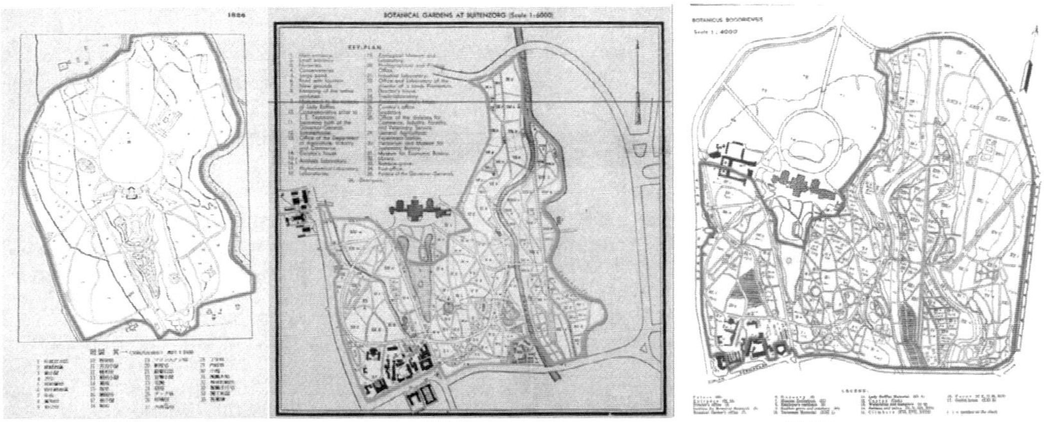

1826 1892 1927

Figure 2. Maps of development of 'Lands Plantentuin te Buitenzorg (Bogor Botanic Garden) (source: http://media-kitlv.nl)

3.2. Role and Function Bogor Botanic Garden

Establishment of the Bogor Botanic Garden was originally based on two main roles that are plan collections and research activities of cultivating various types of native plants of Indonesia and to help provide educational facilities. It turns out that in the further development of the function of the Botanical Gardens is increasingly widespread, among others by the introduction of economic crops (rubber, quinine, coffee, vanilla, palm oil etc.) from various other tropical countries, and the expansion of the scope of activities that ultimately cover the overall resources biology (microorganisms of plants and animals) and their environment [7].

There is a change / development of organizational structure. Various centers / research institutes were built, although then one by one eventually escape from its parent (Botanical Gardens). Therefore, it is not surprising that the Botanical Gardens are still recognized as the parent who gave birth to almost all research institutions in the field of biology and agriculture in Indonesia.

Currently Botanical Gardens have 5 main functions, that is:

1. Organize exploration / collection of living plants that have or are expected to have economic potential and / or are considered important in terms of the science of botany.
2. Conducting research in various fields of botany, such as morphology, genetics, physiology and ecology.
3. Plants introduction to support the development of agriculture in general.
4. Provide assistance in the form of guidance, facilities and plant material for educational purposes.
5. Provide a healthy recreation place for public, as well as stimulate the growing awareness and understanding of the meaning and role of natural resources, especially vegetation resources (flora).

BBG plays a role in the conservation of plants in Indonesia by working with the local government to build a regional botanic garden. Until 2015, there are 27 Botanic Gardens in Indonesia consisting of 5 Botanic Gardens managed by LIPI, and 22 Botanical Gardens managed by the Regional Government. In addition to conserving plants, the Botanical Gardens also play a role in reducing carbon emissions [8]. Existence of the Bogor Botanic Garden support the growth of Bogor City, because many institute were founded as development of BBG [9].

3.3. Plant Collections

KRB collection is the result of field expedition and donations from various countries. It increases the number of collections. Otherwise the deaths of various causes reduce the collection count. Based on the latest record number of Botanical Garden in 2016 reached 13.061 specimen representing 3.301 species, 1.227 genera or 218 tribes outside the orchid co-ordinates. The orchid collection itself recorded 6.169 specimen 572 species of 106 genera. In addition to orchids, a collection that also stands out is legumes (Leguminosae), Pinang-pinangan (Palmae), Talas-Talasan (Araceae) and Coffee-copies (Rubiaceae). In addition, the collection of various types of bamboo is also interesting to see considering its very important role, especially in the countryside. Changes in the number of collection plants in KRB can be seen in Figure 3. BBG has thematic gardens that focus on specific types of plant collections, such as the Mexican Garden, Teijsman Garden, Medical Plants Garden, Water Garden, Soedjana Kassan Garden, and Araceae Garden.

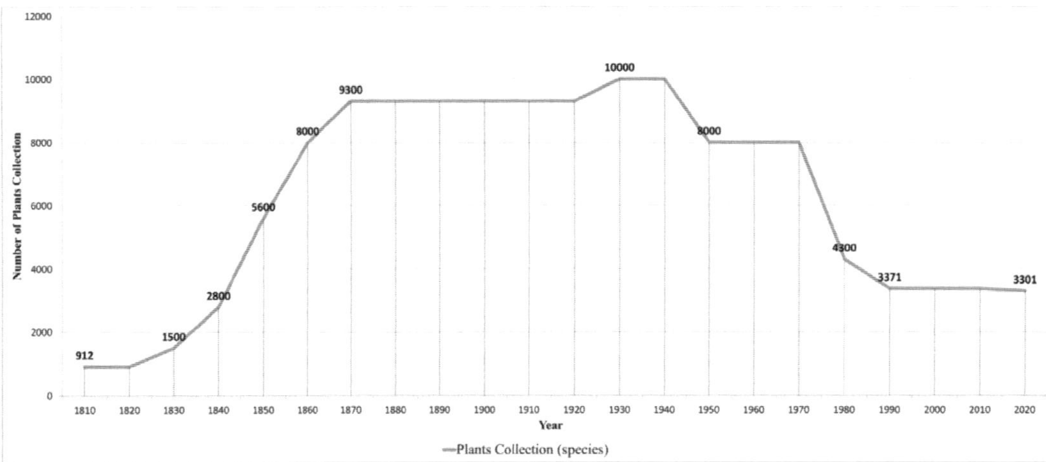

Figure 3. Plants collection of Bogor Botanic Garden
(source: http://www.krbogor.lipi.go.id/id/Jumlah-Koleksi-Kebun-Raya-Bogor.html)

3.4. Design of Bogor Botanic Garden

According to the history at the earlier 1816Caspar George Carl Reinwardt began research in the field of plant science and interested in investigating the various plants which used for treatment and considered the exploration of plants and agricultural problems as well as his duties in the Dutch East Indies. The collections of all those plants were located in a garden around the courtyard of the Bogor Palace (previously occupied by Lieutenant-Governor Thomas Stamford Raffles). Through the help of the Dutch gardener William Kent (1779-1827), the courtyard of the Bogor Palace developed into a beautiful garden with a classic English garden. This is then recognized as the beginning of the original design of The Bogor Botanic Garden. KRB is a colonial botanic gardens. Menurut Rutherford (2015), the main colonial botanic gardens were generally near towns and cities. They had complex ornamental landscapes like public parks, while containing botanical plant collections and systematic display beds, and became valued public gardens [10].

During the leadership of Teysmann (1830), the plant collections were established based on taxonomy patterns. The plants are arranged in large natural groups according of the same family (Figure 4). The most famous are the Canaries avenues (planted in 1832 and 1896), the trees are covered with climbing plants, orchid collection, the palm and rattans, pandanus, cactus collection and vines that are currently the emerging characters. The ponds with their water lilies, amongst which the splendid *Victoria amazonica*, belong to beautiful scenery. Identity of a botanical garden depends not only on its academic association, but also on its function, which is revealed among other things through its structure and the arrangement of its content.

Rare and unique plants are things that make visitors come to the botanical garden. Research on visitor preferences to features and points of interest in Bogor Botanical Gardens has been done by Hadi et all that there were 45.91% photos consist of plants as focused objects. The plants that mostly captured on those places are: 1) Orchid flowers in Orchid Interior garden, 2) Canna flowers as median of Astrid Avenue, and 3) Cactus, Agave, Yucca, Bromelia, Euphorbia and drought resistant plants in Mexican Garden and. These plants were arranged well with other materials and landscape elements in a good design [11].

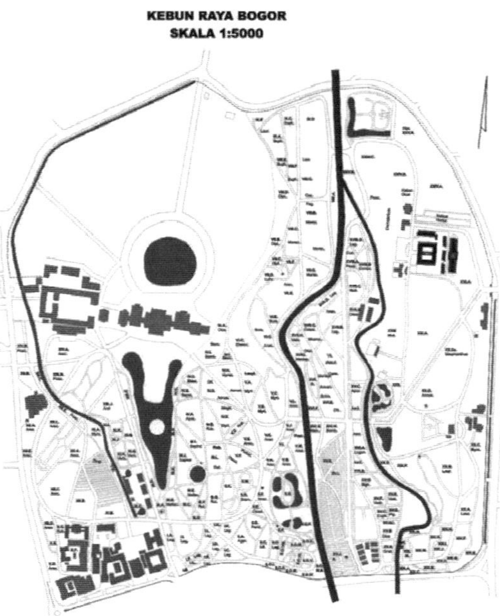

Figure 4. Layout of Bogor Botanic Garden based on plant taxonomy
(source: Subdivision Registration-BBG, 2010)

4. Conclusion

The history of the botanical garden is closely related to the history of botany itself. The botanical garden is an example of the influence of science on the garden art. The design changes of the botanical garden are in line with the change in function. At the beginning of the establishment, the aim of the Botanic gardens is to collect and make easily accessible within its ground plans of the Malay Archipelago and other tropical areas.The design of the garden is simple; there is no arrangement in planting the collection. With its development, the function of the garden increased as a place of research and education, so began to be applied systematic arrangement of plants based on taxonomy, this is followed by the increase of garden area to the north and south of the Bogor Palace. Supporting facilities such as laboratories and herbaria began to be built well. Currently the botanical garden also serves as a place of recreation, and then the space needs and design of the botanical garden changed as the lawn for picnics and garden site furniture.

Acknowledgments
The authors would like to thank the management of The Bogor Botanic Garden for the valuable input and granting us permission to conduct our research.

IOP Conf. Series: Earth and Environmental Science **179** (2018) 012026 doi:10.1088/1755-1315/179/1/012026

References

[1] Republik Indonesia 2011 Peraturan Presiden Republik Indonesia Nomor 93 Tahun 2011 Tentang kebun raya 1–8

[2] Holttum R . E . 2014 The Historical Significance of Botanic Gardens in S . E . Asia *Taxon Vol. 19, No. 5 (Oct., 1970), pp. 707-714*19 707–14

[3] Spencer R, Cross R, Botanic R, Victoria G and Avenue B 2016 Muelleria The origins of botanic gardens and their relation to plant science , with special reference to horticultural botany and cultivated plant taxonomy *Muelleria*35 43–93

[4] Mickulas P 2002 A scientific garden in a public space: the New York botanical garden, 1900–10 *Stud. Hist. Gard. Des. Landscapes*22 193–213

[5] Peter A. Kumble and Christopher C. Houston, Kumble P a and Houston C C 2009 Landscape Studies The elements of a conservation botanic garden for eco tourism : Belize Botanic Garden as a case study *J. Landsc. Stud.*2 1–15

[6] Rijnberg T F 1992 *'s Lands Plantentuin Buitenzorg 1817 - 1992 Kebun Raya Indonesia* (Jakarta Selatan: CV Wira Pratama)

[7] Dehgan B 2014 *Public Garden Management: a Global Perspective, Volume 1* (Florida: Xlibris Corporation LLC)

[8] Purnomo D W 2015 Peran Kebun Raya Indonesia dalam upaya konservasi tumbuhan dan penurunan emisi karbon 1 66–70

[9] Gunawan A 1995 *Preference and attitude for landscape and land use of Bogor City* (Kyoto University)

[10] Rutherford S 2015 *Botanic Gardens* vol 23(Great Britain: Shire Publications Ltd.)

[11] Hadi A A, Mizuuchi Y, Setyanti D, Honjo T and Furuya K 2017 Identifying Visitor Preferences For Locations And Features In Bogor Botanical Garden , Indonesia , Using Gps Tracking And Geotagged Photos *Archit. Environ.*16 47–60

Assessment of the quality of public Green Open Space (GOS) in the urban fringes in response to urban sprawl phenomenon (case study District of Tanah Sareal, Bogor City)

J T Hidayat[1*] and M Ridwan[2]

[1]Lecturer, Departement of Regional and City Planning, Pakuan University
[2]Alumny, Departement of Regional and City Planning, Pakuan University

[*]E-mail; janthyhidayat@unpak.ac.id

Abstract : The growth of urban areas in Bogor City physically marked by rapid growth in the urban fringe, where its growth tend to be widespread and dispersed randomly as well getting out of control (urban sprawl phenomenon). Urban sprawl resulted in the loss of green space and has caused many environmental impacts associated with the reduction of public green open space. The public green open space (GOS) in urban fringe area are essential for guiding sustainable urban development in response to urban sprawl phenomenon. The objective of this study is to assess the quality of public GOS in district of Tanah Sareal Bogor City. Analysis were done by using Worldview 2 imagery year 2015 through NDVI method. The study demonstrated that in the current area of GOS still meets the rules, but the quality level of public (GOS) are varies where the best quality is public GOS in Kelurahan Sukaresmi and the worst is public GOS in district of Kedungwaringin.

Keywords: NDVI, public green open space (GOS), urban sprawl, urban fringe

1. Introduction

Urban sprawl is an expansion of urban areas characterized by the increase of built up areas as the result of land conversion from non-urban use into urban use. Due to the population growth and development of urban activities, the land development has to be undertaken in the surrounding urban area called by urban fringe area [1].The growth of urban areas in Bogor City physically marked by rapid growth in the urban fringe, where it's growth tend to be widespread and dispersed randomly as well getting out of control (urban sprawl phenomenon). It's caused to inefficiency and misery in land use [2]. The urban sprawl phenomenon is the growth of new residential areas on the urban fringe of the city to accommodate population growth, industry, housing developments and commercial activities. Urban fringe areas will be under increasing urban activity pressure due to the urban sprawl phenomenon resulting in environmental degradation, congestion, infrastructure crisis and disaster risks where the process of urban fringe development leads to unsustainability [3]. The growth trend of urban fringe will continue in the future, as population growth remains high [4]. Currently, urban spatial layout is still potentially as a provider of urban space but has not functioned optimally with due regard to ecological balance and sustainability [5]. It takes an effort to response the negative impact of the urban sprawl phenomenon.

Urban areas require public Green Open Space (GOS) to meet the needs of the community in conducting activities while controlling the comfort of microclimate and the aesthetic harmony of the city. Urban public GOS can essentially preserve the quality of the environment through prevention efforts or anticipate environmental problems without reducing urban development activities to meet human needs so as to achieve sustainable urban development [6]. Based on Law No. 26 year 2007 in order to realize a quality environment is done through the management of urban green open space (GOS) in accordance with the needs of the city [7]. GOS consists of public GOS and private GOS. The proportion of GOS in urban areas shall be at least 30 (thirty) percent of the total urban area. The proportion of public GOS in urban areas shall be at least 20% (twenty) percent of the total urban area while private GOS is 10% (ten) percent.

Based on the background, this study aims to identify the availability of public GOS and analyze the quality of public GOS in the urban fringe areas of Bogor City in order to response of urban sprawl phenomenon. The public GOS are essential for guiding sustainable urban development in urban fringe areas.

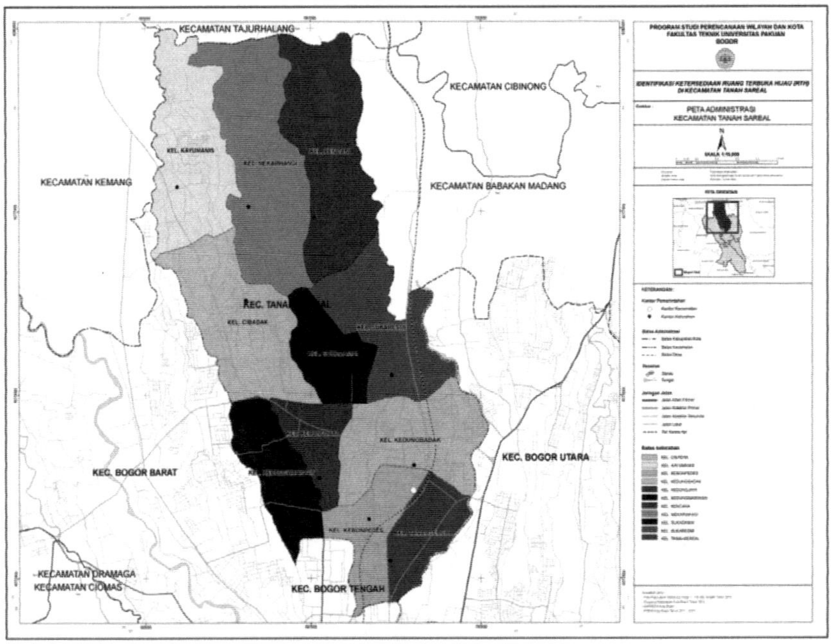

Figure 1. Research Location

2. Methodology

2.1. Research Location

The study was conducted in the administrative area of district of Tanah Sareal (figure 1). The location was selected with the most rapidlly occurrence of the urban sprawl phenomenon is along the main transport corridor and urban fringe areas in Bogor City. District ofTanah Sareal covers 11 (eleven) kelurahan of total area 2329,70 ha (Kelurahan Cibadak, Kayu Manis, Kebon Pedes, Kedung Badak, Kedung Jaya, Kedung Waringin, Kencana, Mekarwangi, Sukadamai, Sukaresmi and Tanah Sareal).

2.2. Methods

Methods use through analyzes of Worldview 2 Imagery coverage of year 2016 and Landsat 8 Imagery coverage of year 2016 which is based on Geographic Information System (GIS) software and Erdas Imagery software.

The NDVI Method (Normalize Deference Vegetation Index) is used to assess the quality of public GOS where it describes the greenish level and translates into a vegetation cover class showing the vegetation density [8]. This method is a mathematical combination between red band and NIR band that has long been used as an indicator of the existence and condition of vegetation [9]. Vegetation land cover is classified into 5 (five) classifications based on the greenish density level. Each classification is given weight value vegetation cover where weight value 1(one) for very thin greenish density, weight value 2 (two) for thin greenish density, weight value 3 (three) for medium greenish density, weight value 4 (four) for thick greenish density and weight value 5 (five) for very thick greenish density.

The quality of public GOS is grouped into 5 (five) classifications are as follow: very bad category for greenish density value with interval 0-12, bad category for greenish density value with interval 13-25, enough category for greenish density value with interval 26-37, good category for greenish density value with interval 38-50, and very good category for greenish density value with more than 50.

3. Result and Discussion

District of Tanah Sareal's GOS in year 2016 is 717,79 ha or 35,23% from total area, where the biggest GOS composition is field of 500,7 ha (21,6%) and the smallest is the office yard of 0.4 ha or (0.02%), see Table 1.

Table 1. Type and Area of Green Open Space District of Tanah Sareal Year 2016

Type GOS	Area (ha)	(%)
Office park	0,41	0,02
Fields	500,70	21,58
Sport fields	13,17	0,57
Road	8,86	0,38
Shrub	27,12	1,17
Road border	7,14	0,31
Railway border	21,61	0,93
River border	16,58	0,71
SUTET border	0,22	0,01
Urban parks	0,85	0,04
Vacant land	204,12	8,80
Public cemetery	17,01	0,73
Total	717,79	35,23

Source:Analyzes GIS 2016

The Public GOS in the study areas covered 10 (ten) type where dominated by the public GOS type of border ie Railway border, river border, road border and SUTET border. The total area of public GOS in the study area is only 85,44 ha or 3.68% of the total area of district of Tanah Sareal, presented in Table 2. Based on Undang-undang No. 26/2007 on spatial arrangement which requires every city area must have at least 30% of GOS from its total area, consisting of 20% public GOS and 10% private GOS. Currently, the availability of GOS in the research area is still 35.23% of the total research area. This condition still meets the applicable area requirements. Based on Perda Kota Bogor No. 8/2011 on Spatial Planning (RTRW) Bogor City Year 2011 - 2031, district of Tanah Sareal physical development directed to the use of residential land, trade and services, as well as municipal service facilities. Currently the growth of built area in the area of western Bogor City shows the existence of acceleration and changes are quite [high]. This condition shows the occurrence of urban sprawls phenomenon in the study area

To maintain the stability of the quality of the Environment and provide comfort of community activities one is through the availability of public GOS by 20% of the total area. Availability of GOS in the district of Tanah Sareal is conducted to know the percentage of area of eksisting public GOS. Looking at the above guidelines, it can be said that public GOS in District Tanah Sareal at year 2016 is still very less. which means that in the district of Tanah Sareal is needed public GOS of 464,14 ha to fulfill public GOS requirement 20%. While currently the provision of Public GOS in the district of Tanah Sareal is only 85.44 ha (3.68%), it is still needed for 378,7 ha to fulfill 20% public GOS. Represented in Table 2.

Table 2. The Type of Public GOS in the District of Tanah Sareal Year 2016

No	Clasification	Type	Area (ha)
1	GOS of Urban park and urban forest	Taman Kota	0,85
		Lapangan Olahraga	13,17
2	GOS of road border and median	Pulau Jalan	8,9
		Sempadan Jalan	7,1
3	GOS of Rail way border	Sempadan Kereta	21,61
4	GOS of river border	Sempadan Sungai	16,58
5	GOS of line high voltage electricity network	Sempadan SUTET	0,22
6	GOS of public cemetery	TPU	17,01
Total of Public GOS (Ha)			**85,44**
Total area of District of Tanah Sareal (Ha)			**2320,70**
Persentage of Public GOS (%)			**3,68**
Total needs of Public GOS (Ha)			**464,14**
Deficit of Public GOS(Ha)			**378,7**

Source : analysis results.

Efforts to fulfill the public GOS in the district of Tanah Sareal consider the availability of spaces where the pressure from urban sprawl phenomena is higher, therefore the assessment of the condition of vegetation/greenish quality of public GOS is very helpful to maintain and improve the environmental quality. The greenish quality of public GOS in every sub-district in the district of Tanah Sareal represented in Table 3.

Table 3. The area of greenish density level of public GOS

No	Kelurahan	The area of greenish density level (ha)					Total (ha)
		Very thin	Thin	Medium	Thick	Very thick	
1	Kel. Cibadak	1,83	0,88	0,61	0,27	0,42	**4,01**
2	Kel. Kayumanis	0,72	0,81	1,45	4,38	3,5	**10,86**
3	Kel. Kebonpedes	2,88	1,5	6,9	0,96	-	**12,24**
4	Kel. Kedungbadak	3,64	1,51	2,34	2,55	1,67	**11,71**
5	Kel. Kedungjaya	0,81	0,09	0,01	-	-	**0,91**
6	Kel. Kedungwaringin	0,28	0,07	0,01	-	-	**0,36**
7	Kel. Kencana	0,73	0,51	0,58	0,27		**2,09**

Source : Analysis Result

8	Kel. Mekarwangi	2,08	2,2	1,34	1,33	0,05	**7,00**
9	Kel. Sukadamai	0,3	0,5	0,19	0,13	0,01	**1,13**
10	Kel. Sukaresmi	1,2	1,39	2,52	3,24	7,37	**15,72**
11	Kel. Tanahsereal	4,16	3,95	6,81	3,37	1,12	**19,41**
Total		**18,63**	**13,41**	**22,76**	**16,5**	**14,14**	**85,44**
Proportion of total area Kec Tanah Sareal (%)		**0,80**	**0,58**	**0,98**	**0,71**	**0,61**	**3,68**

Assessment on the quality condition of public GOS in the District of Tanah Sareal based on table 4, that the sub district that has good quality condition of public GOS (very thick greenish) is located in the district of Sukaresmi with the value of 61,35 and the district of Tanah Sareal with the value of 51,57; while sub district that has worst quality condition of public GOS is located in the district of Kedungwaringin with the value of 0.45.

The distribution and score of greenish density level of public GOS in the district Tanah Sareal represented in Figure 2 and Table 4.

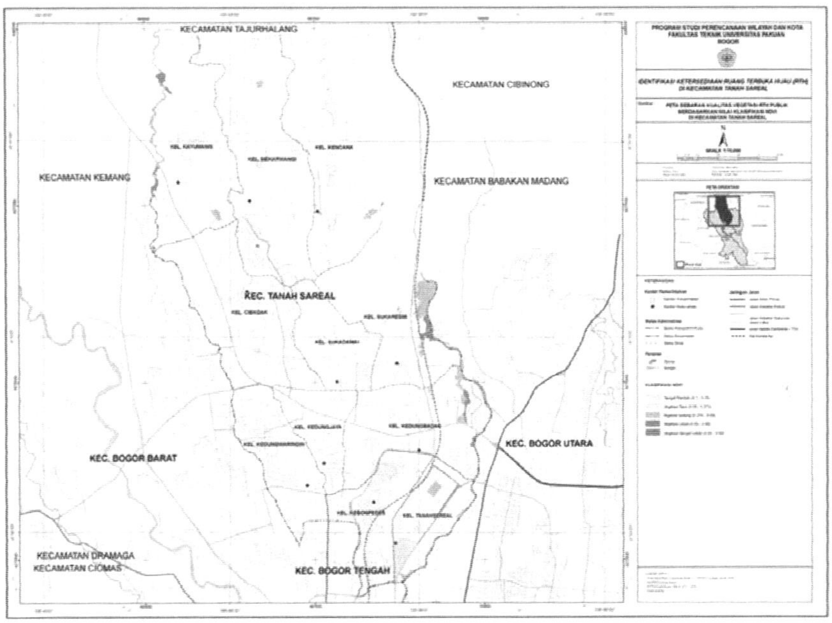

Figure 2. The distribution of public GOS greenish density level

Most of the urban fringe areas of Jabodetabek including the district of Tanah Sareal are experiencing rapid development and lead to unsustainability [5], especially for the environmental dimension. This is indicated by the decrease in GOS [11]. The public GOS can be a controlling environmental sustainability dimension because it can maintain the quality of the environment.

Curently district of Tanah Sareal's public GOS is still not sufficient, given the difficulties in meeting the needs,public GOS then can be done by improving the quality of public GOS through replanting so that the level of vegetation density becomes increased. In the district of Tanah Sareal there are 2 (two) districts that must be maintained the quality of vegetation density of public GOS because it has good quality and 5 (five) districts that must be improved the quality of vegetation density of public GOS because it has very bad condition. Considering that the public GOS in the district of Tanah Sareal is dominated by the type of GOS border (river border, road border, rail way border and Electricity

Transmission Tower border), then by maintaining and improving the quality of this public GOS the function and objective of public green space is achieved. Based on the results of this study is expected to improved the quality of public GOS in the urban fringe through the local government policy to controlling the land
use by detail planning and zoning regulation.

Table 4. The quality of public GOS greenish density level

District	Score for greenish density level					*Quality of public GOS*
	Very thin	Thin	Medium	Thick	Very thick	
	1	2	3	4	5	
Kel. Cibadak	1,83	1,76	1,83	1,08	2,1	8,6 (Very bad)
Kel. Kayumanis	0,72	1,62	4,35	17,52	17,5	41,71 (Good)
Kel. Kebonpedes	2,88	3	20,7	3,84	0	30,42 (Medium)
Kel. Kedungbadak	3,64	3,02	7,02	10,2	8,35	32,23 (Medium)
Kel. Kedungjaya	0,81	0,18	0,03	0	0	1,02 (Very bad)
Kel. Kedungwaringin	0,28	0,14	0,03	0	0	0,45 (Very bad)
Kel. Kencana	0,73	1,02	1,74	1,08	0	4,57 (Very bad)
Kel. Mekarwangi	2,08	4,4	4,02	5,32	0,25	16,07 (Bad)
Kel. Sukadamai	0,3	1	0,57	0,52	0,05	2,44 (Very bad)
Kel. Sukaresmi	1,2	2,78	7,56	12,96	36,85	61,35 (Very good)
Kel. Tanahsereal	4,16	7,9	20,43	13,48	5,6	51,57 (Very good)

Source : analysis results

4. Conclusion

Urban sprawl phenomenon that occurred at urban fringe areas of Bogor City especially in the district of Tanah Sareal has caused many environmental impacts associated with the reduction of green open space. In an effort to compensate for the pressure of urban sprawl phenomenon in the district of Tanah Sareal is through improving the quality of GOS. Currently, the need for public GOS in the district Tanah Sareal is still lacking and there are 5 (five) districts with bad public GOS condition, then by improving the quality of public GOS is expected to overcome the environmental problems. Local government policy (detailed plan and zoning regulation) is needed to control the availability and quality of green open spaces in response to the urban sprawl phenomenon. Further research is needed on the techniques and types of efforts that must be done in order to improve the quality of this public GOS.

References

[1] Yunus Hs 2006 Struktur Tata Ruang Kota (Jogyakarta : Pustaka Pelajar)

[2] Hudalah D, Winarso H and Woltjer J 2010 *Peri-urbanisation in East Asia; A new Challenge for Planning?* International Development Planning Review **29 (4)** 503-519

[3] Uguy M J H 2006 Pengembangan Lingkungan Peri Urban yang menuju Keberlanjutan : Suatu Anlisis tentang Urban Sprawl sebagai akibat Suburbanisasi (Jakarta : Universitas Indonesia)

[4] Firman T 2004 *New Town Development inJakarta Metropolitan Region : a Prespective of Spatial Segregation* Habitat International **28 (3)** 349-368

[5] Hidayat J T 2007 *Urban Sprawls Phenomenon and Sustaibality of Jakarta Metropolitan Area* Proc. Seminar Toward Sustainable Jakarta Metropolitan Area (Bogor: Indonesia)

[6] David N 2014 *Public Policies for managing Urban Growth and Protcting Open Space* Landscape and Urban Planning **65** 271-286

[7] Republik Indonesia 2007 Undang-undang No 26 Tahun 2007 Tentang Penataan Ruang

[8] Towson 2016 *NDVI (Normalized Difference Vegetation Index)* (London: Chesapeake)
[9] Dulbahri P Suharsono Hartono and Suharyadi 1997 *Penginderaan Jauh dan Interpretasi Citra* (Yogyakarta : Gajah Mada University Press)
[10] Fauzy R Hidayat JT Mulyadi E 2013 *Determination of Quality of Green Open Space Using Worldview 2 Satellite Imagery* Proc. 34th Asian Conference of Remote Sensing (Bali; Indonesia)
[11] Hidayat JT Sitorus SRP Rustiadi E and Machfud 2013 *Dinamika Pertumbuhan dan Status Keberlanjutan Kawasan Permukiman di Pingiran Kota Metropolitasn Jakarta* Majalah Ilmiah Globe **15 (1)** 93-100

Level of gas emissions of passenger car on climb road segment

N Sari[1*], S W Praja[1], and B Hidayat[1]

[1]Land Transport Academy, Ministry of Transportation, Jl. Raya Setu No.89 Cibitung Bekasi 1700, Indonesia.

[*]Email: miss_vita20@yahoo.com

Abstract. Each island in Indonesia has its own topographical characteristics, which generally consists of plains, plateaus, hills and mountains. This topography influences vertical alignment of the roads. Gradient of vertical alignment on road construction planning is maximum 10%, but in facts it is shown that in some areas they are reached more than 10%. While the vehicle gas emission test in Indonesia currently only apply for public transport and freight cars, where the test is only performed on a flat road and in idle conditions without considering geometric aspects. Passenger cars are not subject to this test contrary to the increase of their population of about 10-15% annually which is much higher than the public transport. This research aims to determine the level of gas emissions of passenger cars on climb road segment (vertical alignment). The methods of analysis are by using distribution frequency of the vehicle gas emissions data test of passenger cars in dynamic condition on climb road segment. This research conduct with Gas Analyzer Equipment which is measured CO, CO_2, HC, and Lambda (The composition of gasoline and Air). Sample of passenger cars used by two cars with the same specification and different age. The data used in this research are passenger cars speed with three variations are 30, 40 and 50 kph. The gas emissions test of passenger cars is performed on the road which have the gradient of vertical alignment is maximum 10%, and this test is regardless to the value of skid resistant, roughness and driver characteristics. The result shown that climb and slope road segment influence the level of gas emissions of passenger cars for CO and CO_2 has almost the same value. The condition climb and slope road segment for Lambda 0.99 % volume, the value of CO volume is 0.35% volume and CO_2 is 12.3% volume. While the content of HC has different characteristics that is when the vehicle climb, HC content lower than when the vehicle slope. When the mixed composition of fuel and air (Lambda 0.99% volume), the vehicle climbed about 11.5 ppm and when the vehicle sloped about 30.5 ppm. The ratio of HC between vehicles climbed and sloped is 1: 2.

Keywords: climb road segment, gas emissions, vertical alignment

1. Introduction

1.1. Background

Geographically Indonesia is an archipelago with have more than 17 thousand islands. Each island has its own topographical characteristics, which generally consists of plains, plateaus, hills and mountains. This topography influences the vertical alignment of the roads in Indonesia. On road construction planning in cities of Indonesia, the gradient of vertical alignment is maximum 10%, but in facts it is shown that in some areas they are reached more than 10%, for example crest and sag at Nagrek area.

The vertical alignment more than 10% impact to engine performance of the vehicle, this condition is increase the level of vehicle gas emissions.

Traffic represents one of the largest sources of primary air pollutants in urban areas. Large amount of vehicle exhaust emissions will have seriously adverse effects on human health. Assessing the effects of air quality management strategies in urban areas is a major concern worldwide. In addition, worldwide epidemiological studies show a consistent increase in cardiac and respiratory morbidity and mortality from exposure to vehicle exhaust pollution. Both in urban and residential regions, this has become the main areas of toxic compound emissions from the unrestrained use of vehicles burning fossil fuels. In these areas, the population is very sensitive to vehicular pollution [3].

In Indonesia vehicles increased from year to year, the exhaust gas produced from the vehicles causes air pollution about 70 to 80 percent, while industrial air pollution only 20-30 percent. Air pollution caused by many things, among others: vehicle fumes, factory smoke, garbage burning and etc. Vehicle exhaust is the biggest cause of air pollution because the development of technology in various fields, especially in the transportation field, resulting in the number of vehicles with a wide range type and brand increased quietly high [4].

1.2. Research problem

The vehicle gas emission test in Indonesia currently only apply for public transport and freight cars, where the test is only performed on a flat road and in idle conditions without considering geometric aspects. Passenger cars are not subject to this test contrary to the increase of their population of about 10-15% annually which is much higher than the public transport.

1.3. Research limitation

The limitation of this research is the specification vehicle that become sample test for this research are passenger cars with gasoline, capacity 1300 cc, two variant ages: vehicle with age 6 years and 8 years. The location test of this research is underpass at Tambun market, Bekasi. Equipment that used to calculated the CO_2, CO and HC value by using Gas Analyzer CAP 320.

2. Literature Review

2.1. Gas emission

Air pollution that cause by vehicle exhaust gas produced from incomplete fuel combustion. The gas emission cause by incomplete burning of fuel in engine vehicle. Elements contained in gas emission among others CO, NO_2, HC, C, H_2, CO_2, H2O and N_2, where it's become pollutant on air and can disturb human health to cause the dead at a specific level [5].

2.1.1. Carbon Monoxide (CO)

CO content from the vehicle exhaust depends on the comparison of fuel and air. But on perfect fuel combustion, the CO value can be zero. Carbon monoxide which content 0.03% can be dangerous toxin on air that is inhaled by human. This gas in human body will attack the erythrocyte, which bring the oxygen to whole human body. In the indoor room, percentage of CO volume 0.1% or more can be deadly [1].

On high level the concentration of CO inhaled humans could make fatal risks on human body, even can cause the death. The character of CO is a gas odorless and colorless and very toxic, then CO is often referred to as a silent killer. The effect of CO gas is a harmful gas for human body due to the binding capacity of CO gas to Hb is 240 times the bonding power of CO to O2. When the blood CO gas (HbCO) is enough high, it would start to occur symptoms include headache (HbCO 10 percent), nausea and shortness of breath (HbCO 20 percent), impaired vision and concentration decreased (HbCO 30 percent) unconscious, coma (HbCO 40-50 percent) and if would continue to cause death. On long term would showing symptoms of neurological disorders, cerebral infarction, cardiac infarction and infant death in the womb. High level of CO gas in the blood can come from cigarettes and smoke from motor vehicles. Air quality on indoor room, CO gas can also be a gas that causes building associated illnesses, with complaints of headache, nausea, and vomiting [4].

2.1.2. Carbon Dioxide (CO_2)
Carbon Dioxide (CO_2) in Indonesia increased, its share in total global CO_2 emissions from 0.6% in 1990 to 1.2% in 2005 and 1.4% in 2015. Since 1990, the only decline in CO_2 emissions has been recorded for 2008 (0.6%); in 2015, the increase was 4%, which was above the average over the last 10 years (3.4%). Coal consumption in Indonesia increased continuously, at an average rate of 15.5% per year, between 1990 and 2007, then decreased by 13% in 2008 and increased again by 14.4%, on average, between 2009 and 2015 [6].

2.1.3. Hydrocarbons (HC)
Hydrocarbons are gasoline compound element. Hydrocarbons (HC) are produced from incomplete fuel combustion. Hydrocarbons (HC) measured on ppm (part per million). It formed from incomplete fuel combustion on cylinders. Hydrocarbons (HC) may cause cancer, and some Hydrocarbons (HC) also can cause harmful effects on tissues and the human immune system [5].

2.2. Vertical alignment
Vertical alignment is elevation plan on road geometry which design on every axis. The vertical alignment of a road consists of gradients (straight lines in a vertical plane) and vertical curves. The vertical alignment is usually drawn as a profile, which is a graph with elevation as vertical axis and the horizontal distance along the center line of the road as the horizontal axis. Almost the passenger cars could past the vertical alignment with gradient 7-8% without changing speed comparison on flat area [2].

3. Research Methodology
The research method used is as follows:

3.1. Collecting data
This research collects data from primary data. The data are collected by testing passenger car's gas emissions in the field. The vehicle gas emissions testing which categories: the location has gradient more than 10%, emission gas test on three variation of car speed are 30 kph, 40 kph and 50 kph. The data analyzed are CO, CO_2, and HC value.

3.2. Method of analysis
The methods of analysis are by using distribution frequency of the vehicle gas emissions data test of passenger cars in dynamic condition on climb road segment (Figure 1). This research conduct with Gas Analyzer Equipment which is measured CO, CO_2, HC, and Lambda (The composition of gasoline and Air) (Figure 2). Sample of passenger cars used by two cars with the same specification and different age. The data used in this research are passenger cars speed with three variations are 30, 40 and 50 kph. The gas emissions test of passenger cars is performed on the road which have the gradient of vertical alignment is maximum 10%, and this test is regardless to the value of skid resistant, roughness and driver characteristics.

Figure 1. Location test at underpass Tambun market, Bekasi.

Figure 2. Setting up the gas analyzer CAP 320 on passenger cars with capacity 1300 cc.

4. Research Analysis

4.1. Result and discussion

Data from the measurement result in the field are presented in the table below:

Table 1. Level of gas emission car aged 6 in slope segment.

Speed	Level of Gas Emission						Rotation Per Minutes
	CO	CO_2	O_2	HC	Lambda	CO correction	
30	0.115	12.6	-0,09	20	0.99	0.14	2500
40	0.35	12.3	-0,08	30.5	0.98	0.37	3300
50	0.57	12.4	-0,09	43	0.97	0.66	3000

Table 1 shows from the results of the exhaust emission test experiment with car aged 6 on the slope segment, the CO and HC levels rise in proportion to the increase in vehicle speed while the CO_2 and O_2 levels do not increase and decrease significantly with increasing speed.

Table 2. Level of gas emission car aged 6 in climb segment.

Speed	Level of Gas Emission						Rotation Per Minutes
	CO	CO_2	O_2	HC	Lambda	CO correction	
30	0.325	12.4	-0,045	29	0.98	0.38	2500
40	0.15	12.6	0.095	11.5	0.99	0.18	3300
50	0.67	12.3	-0,02	23	0.97	0.77	3000

Table 2 shows the results of the exhaust emission test experiment with car aged 6 on climb segment, CO and HC levels increased and decreased with the lowest decrease at a speed of 40 km/h with CO and HC levels of 0.15% and 11.5 ppm and the highest increase at a speed of 60 km/h with CO and HC levels of 0.67% and 23 ppm.

Table 3. Level of gas emission car aged 8 in slope segment.

Speed	Level of Gas Emission						Rotation Per Minutes
	CO	CO_2	O_2	HC	Lambda	CO correction	
30	0.375	13.1	0.33	76	1	0.445	2500
40	0.56	13.3	0.06	57.5	0.985	0.61	3300
50	35.785	13.2	0.08	57	0.98	0.695	3000

Table 3 shows the results of the exhaust emission test experiment with car aged 8 on the slope segment of CO and HC levels increased and decreased with the highest decrease at a speed of 40 km/h with CO and HC levels of 0.56% and 57.5 ppm and the lowest increase at a speed of 60 km/h with CO and HC levels of 0.67% and 23 ppm.

Table 4. Level of gas emission car aged 8 in climb segment.

Speed	Level of Gas Emission						Rotation Per Minutes
	CO	CO2	O_2	HC	Lambda	CO correction	
30	0.48	13.2	0.195	75.5	0.99	0.52	2500
40	0.6	13.1	0.305	63.5	0.99	0.655	3300
50	0.99	10.75	3.68	66.5	1.22	1.365	3000

Table 4 shows from the results of the exhaust emission test experiment with car aged 8 on the climb segment, the CO level rises in proportion to the increase in vehicle speed while the level of HC levels decreases inversely with the vehicle speed. For CO_2 and O_2 levels there is no significant increase and decrease on increasing speed.

4.2. Correlation of emission gas with vehicle speed
Correlation of emission gas with vehicle speed with climb and slope segment conditions where the speed of the vehicle running at speeds of 30, 40 and 50 kph are shown in figures below.

4.2.1. Correlation between CO and vehicle speed in slope segment
The increase of vehicle speed will increase the CO level contents that produced from gasoline and air exhausted (Figure 3). Whereas the percentage of car aged 6 years increased average 62% and car aged 8 years increased 49%. The condition shown vehicle moved in slope segment.

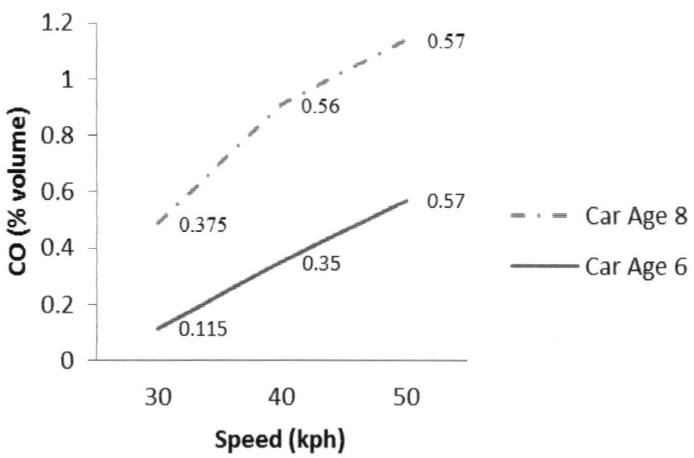

Figure 3. Correlation between CO and vehicle speed in slope segment.

As shown in the Figure 3, can be interpreted that the increase of vehicle speed will increase the CO level contents that produced from gasoline and air exhausted. Whereas the percentage of car aged 6 years increased average 62% and car aged 8 years increased 49%. The condition shown vehicle moved in slope segment.

4.2.2. Correlation between CO_2 and vehicle speed in slope segment
The increase of vehicle speed wouldn't effected on the CO_2 level contents that produced from gasoline and air exhausted (Figure 4). Regardless vehicle speed CO_2 level contents produced fluctuated at average 12-13% volume. The condition shown vehicle moved in slope segment.

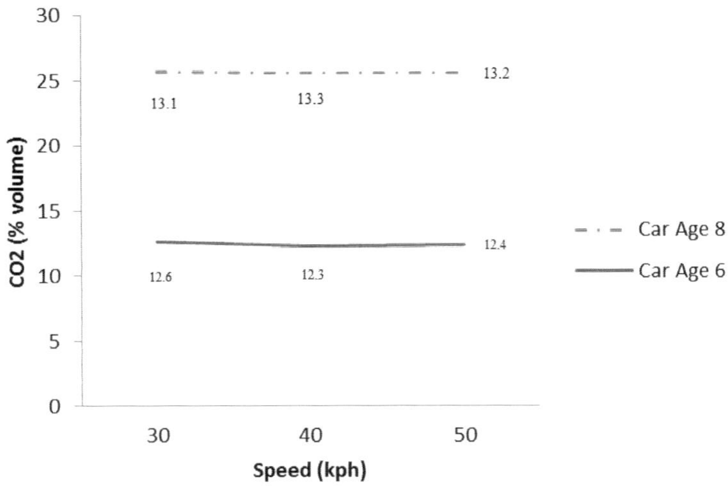

Figure 4. Correlation between CO_2 and vehicle speed in slope segment.

As shown in the Figure 4, can be interpreted that the increase of vehicle speed wouldn't affected on the CO2 level contents that produced from gasoline and air exhausted. Regardless vehicle speed CO2 level contents produced fluctuated at average 12-13% volume. The condition shown vehicle moved in slope segment.

4.2.3. Correlation between HC and vehicle speed in slope segment
The increase of vehicle speed will affect the HC level contents that produced from gasoline and air exhausted (Figure 5). The percentage of car aged 6 years increased average 41% and car aged 8 years decreased 15%. The condition shown vehicle moved in slope segment.

As shown in the Figure 5, can be interpreted that the increase of vehicle speed will affect the HC level contents that produced from gasoline and air exhausted. The percentage of car aged 6 years increased average 41% and car aged 8 years decreased 15%. The condition shown vehicle moved in slope segment.

4.2.4. Correlation between CO and vehicle speed in climb segment
The increase of vehicle speed will increase the CO level contents that produced from gasoline and air exhausted (Figure 6). Where is the percentage of car aged 6 years increased from 5% (30-40 kph) to 97% (40-50 kph) and car aged 8 years increased from 25% (30-40 kph) to 65% (40-50 kph). The condition shown vehicle moved in climb segment.

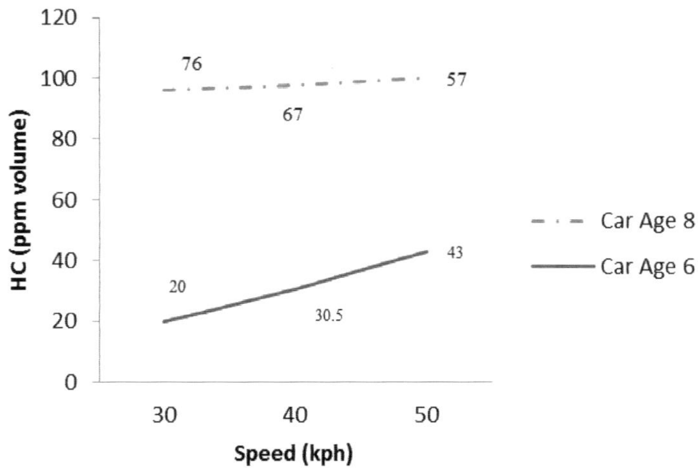

Figure 5. Correlation between HC and vehicle speed in slope segment.

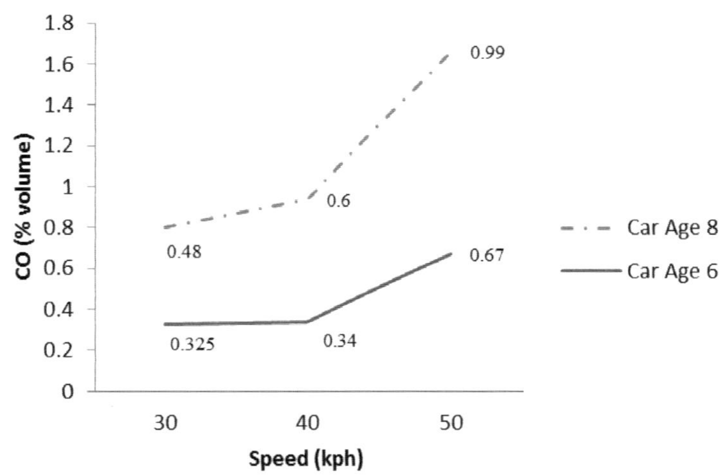

Figure 6. Correlation between CO and vehicle speed in climb segment.

As shown in the Figure 6, can be interpreted that the increase of vehicle speed will increase the CO level contents that produced from gasoline and air exhausted. Whereas the percentage of car aged 6 years increased from 5% (30-40 kph) to 97% (40-50 kph) and car aged 8 years increased from 25% (30-40 kph) to 65% (40-50 kph). The condition shown vehicle moved in climb segment.

4.2.5. Correlation between CO_2 and vehicle speed in climb segment

As shown in the Figure 7, can be interpreted that the increase of vehicle speed wouldn't effected on the CO_2 level contents that produced from gasoline and air exhausted. Regardless vehicle speed CO_2 level contents produced fluctuated at average 12-13% volume. The condition shown vehicle moved in climb segment.

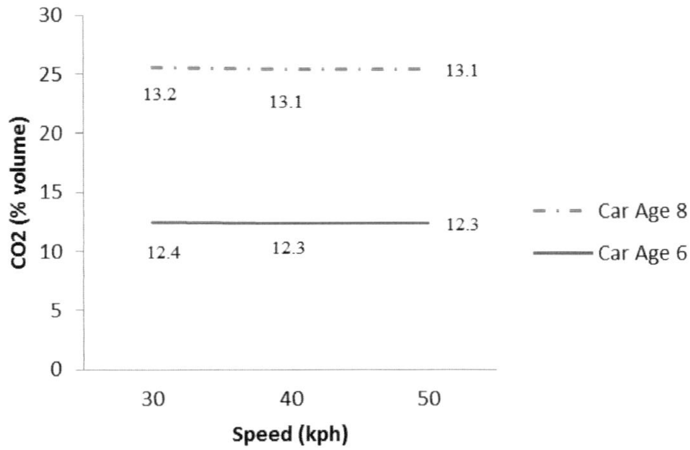

Figure 7. Correlation between CO2 and vehicle speed in climb segment.

4.2.6. Correlation between HC and vehicle speed in climb segment

The increase of vehicle speed will effected on the HC level contents that produced from gasoline and air exhausted (Figure 8). Where is the percentage of car aged 6 years changed with slope curve shape average from decrease 60% (30-40 kph) to increase 100% (40-50 kph) and car aged 8 years increased from decrease 16% (30-40 kph) to increase 5% (40-50 kph). The condition shown vehicle moved in climb segment.

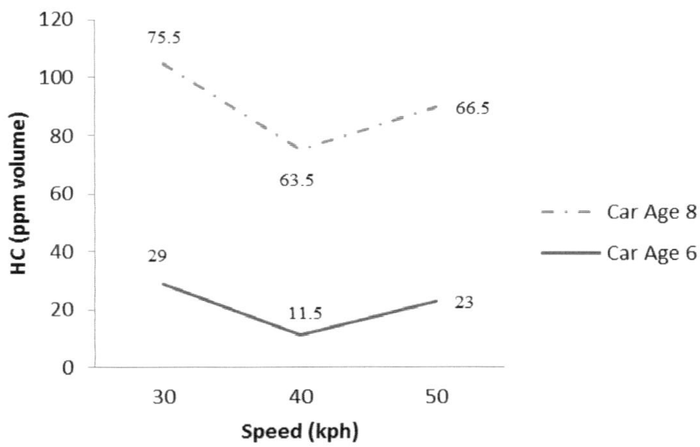

Figure 8. Correlation between HC and vehicle speed in climb segment.

As shown in the Figure 8, can be interpreted that the increase of vehicle speed will affected on the HC level contents that produced from gasoline and air exhausted. Whereas the percentage of car aged 6 years changed with slope curve shape average from decrease 60% (30-40 kph) to increase 100% (40-50 kph) and car aged 8 years increased from decrease 16% (30-40 kph) to increase 5% (40-50 kph). The condition shown vehicle moved in climb segment.

4.2.7. The influent of gasoline and air mixture in engine combution

The amount of exhaust emissions is determined from the combustion process in the Vehicle combustion chamber. Advance technology and vehicle maintenance affect the exhaust emissions produced. Another factor that also affects the level of exhaust emissions is the quality of fuel used (octane levels).

The results of the exhaust emission test apparatus for the composition of the fuel and air mixture are expressed in lambda. The lambda unit is % volume. If the lambda value close to 1 indicates that the composition of the fuel and air mixture better.

An emission test for vehicles aged 6 years and 8 years with the composition of fuel and air mixture of 0.99% volume, evidently 8 years old vehicle has CO content in the exhaust gas higher 0.26 to 0.30% volume than the vehicle age 6 years.

For the composition of the fuel mixture and the air of 0.99%, the CO_2 content in the emission gas for both test vehicles has approximately the same value in the range of 12-13% volume. This condition is different from the HC content of the combustion engine results indicate that the average age of 8 years vehicle has 60 ppm HC and for the age of 6 years vehicle on average has 30 ppm HC. HC levels are also affected by engine rotation speed (RPM) which has a relationship inversely proportional, if high RPM then low HC content and instead low RPM then high HC content.

4.2.8. The influent of climb or slope segment in level of gas emission

The influence of climb and slope on exhaust emissions (CO and CO_2) has almost the same value. Road condition climb and slope for Lambda 0.99% CO content volume is 0.35% volume and CO_2 is 12.3% volume. While the content of HC has different characteristics that is when the vehicle climbed HC content lower than when the vehicle sloped. When the mixed composition of fuel and air (Lambda 0.99% volume) the vehicle climbed about 11.5 ppm and when the vehicle sloped about 30.5 ppm. The ratio of HC between vehicles climbed and sloped is 1: 2.

5. Conclusion

The result shown that climb and slope road segment influence the level of gas emissions of passenger cars for CO and CO_2 has almost the same value. The condition climb and slope road segment for Lambda 0.99 % volume, the value of CO volume is 0.35% volume and CO_2 is 12.3% volume. While the content of HC has different characteristics that is when the vehicle climb, HC content lower than when the vehicle slope. When the mixed composition of fuel and air (Lambda 0.99% volume), the vehicle climbed about 11.5 ppm and when the vehicle sloped about 30.5 ppm. The ratio of HC between vehicles climbed and sloped is 1:2. The government regulation on the vehicle gas emissions test should not be on static condition, but it's should be considering real conditions, which is the vertical alignment impact to the level vehicle gas emissions.

References

[1] Arends BPM and Berenschot H 1980 Motor bensin translate of benzinemotoren Erlangga Jakarta
[2] Hendarsin S 2000 Perencanaan teknik jalan raya Politeknik Negeri Bandung: Bandung
[3] Jie Lu 2011 Environmental effect of vehivle exhausts Global and local effects – A comparation between gasoline and diesel, Thesis of applied environmental science school oof business and engineering Halmstad University Halmstad
[4] Maryanto D, Mulasari S A, and Suryani D 2009 Penurunan kadar emisi gas buang karbon monoksida (CO) dengan penambahan arang aktif pada kendaraan bermotor di Yogyakarta KES MAS Journal Vol 3 ISSN: 1978-0575 Yogyakarta
[5] Novita EK, Hakam M, and Santiasih I 2006 Emisi gas carbon monoksida (CO) dan hidrocarbon (HC) pada rekayasa jumlah blade turbo ventilator sepeda motor "supra X125 Rotasi Jurnal teknik mesin http://ejournal.undip.ac.id/index.php/rotasi
[6] Olivier GJ et al 2016 Trends in global CO_2 emissions 2016 report PBL Netherlands environmental agency Netherlands http://edgar.jrc.ec.europa.eu/news_docs/jrc-2016-trends-in-global-CO2-emissions-2016-report-103425.pdf

Monitoring land use change and urban sprawl based on spatial structure to prioritize specific regulations in Semarang, Indonesia

S Subiyanto[1*] and L Fadilla[1]

[1] Department of Geodesy Engineering, Faculty of Engineering, Diponegoro University, Jl. Prof. Sudarto SH, Tembalang, Semarang – Indonesia.

*Email: sawitrisubiyanto66@gmail.com

Abstract. Indonesia is undergoing rapid pace of urbanization and inadequacy of reliable data for environmental and urban planning, especially in the developing countries. Therefore, this paper aims to evaluate land use/cover change (LUCC) and urban spatial expansion, from 2005 to 2017, in Semarang City, Indonesia, using satellite images, field observations, and socio-economic data. The urban sprawl was also measured using Shannon's entropy based on its primary spatial form. As known, urban sprawl measurements based on its spatial forms would affect the policies and specific regulations in dealing with its dominant form. LUCC change and urban growth were simulated for 2029, using the CA-Markov model. The results show rapid growth of built-up areas that led to a significant decrease in the agriculture areas, gardens and wasteland, from 2005 to 2017. The obtained relative entropy values indicate Semarang city has experienced an increasing urban sprawl over the last few years. The CA-Markov model predicts that this unsustainable trend will continue in the future and built-up areas. The results in this study determine appropriate policies and regulations, especially to prevent linear glaciers along the main road.

Keywords: CA-Markov model, land use/cover change, Semarang City, urban sprawl

1. Introduction

Urban growth prediction is often based on the dynamic land use/cover pattern and its relationship with selected socio-economic factors. Land use change in metropolitan areas typically reflects economic development and population growth. Thus, the analysis of spatial-temporal patterns for land use/cover provides an objective basis for understanding the relationships between urban growth and related economic, population and environmental factors (Irwin & Geoghegan, 2001). The growth of the city was caused by many factors, such as technology, environment and several spatial components, socio-economic and politic. However, the mapping and monitoring of urban sprawl and LULC changes using GIS and remote sensing techniques has attracted more interests and has largely proved to be effective and valuable tools for monitoring and estimating urban sprawl over a time period [3]. Urban growth in Indonesia is caused by social activities e.g. port, trade, manufacturing industries and services. When increased populations in built-up areas such as settlement, industrial, trade and service are expanded wider, agriculture lands decrease in mostly intensive ways, the city becomes dense and allows population distributions and settlement spread to suburb, it is also known as urban sprawl.

Semarang is developing with over-bounded administration caused by increased population. It raised the settlement distribution and urban sprawl in Semarang. Semarang is the capital of Central Java Province and has an area of 373.70 km^2. In recent years, the development of Semarang is also marked by the appearance of several skyscrapers in corners of the city. With populations of about 1.5 million of people and 76.06% working in the service sectors, resulting Semarang to grow rapidly, causing the high growth of residential areas and industrial areas of trade. This study aims to determine the direction of physical distribution of Semarang City from 2005-2017 and predict physical distribution of Semarang City in 2029 with Cellular Automata (CA) Markov Model towards the Spatial Plans of Semarang.

2. Study Area and Data Sources

2.1. Study area

This study area is located in Semarang City, Central Java, Indonesia. Semarang is the capital of Central Java Province, the only city in Central Java, which can be classified as a metropolitan city. As the provincial capital, Semarang is looked up to by other cities in Central Java Province. The largest proportion of land use in Semarang's is settlements (33.12%). This shows that the land still has a dominant function as a domestic service. Based on the land use map, it shows that the distribution of land use settlements were spur-lane main roads located mainly in the central of city. The large proportions of residential land area indicated the magnitude of the demands of public service and this proves that the city of Semarang totally urbanized. Geographically Semarang is located 7.0051° S, 110.4381° E with an area of 373.70 km^2 with boundaries.

2.2. Data sources

2.2.1. Landsat Thematic Mapper (ETM+)

Landsat Thematic Mapper (ETM+) images were acquired for 2005, 2011and 2017. In addition, a land use map, compiled in 1997 at a field survey scale of 1:25,000, was used for the accuracy assessment. This study uses Landsat 7 ETM + image which has acquisition of recording data on June 10, 2005 and June 11, 2011, where in that year there was damage to the section of Scan Line Correction (SLC) which stopped functioning normally (OFF). Geometric correction in this research is done using image to image method by giving GCP (Ground Control Point) on 2015's Quickbird image that has been geometry corrected. The master scene (Quickbird image of 2015) was geometrically corrected and registered to the land use map, using 33 Ground Control Points (GCP) and second order polynomial transformation with the nearest neighbour resampling method.

3. Methods

3.1. Multi-temporal analysis method for change detection

Multi-temporal analysis was carried out with the purpose of identifying the changes of land use that occurred between the periods 2005–2011 and 2011–2017 in the study area. This analysis was done by overlaying Landsat ETM and OLI images previously classified by the supervised method. Processing software, ENVI was used and the maps obtained multi-temporal analysis result, shown in a cross-tabulation array also known as transition matrix [8]. This matrix identified the most important transitions that occurred in a given period. From this matrix, a percentage of the areas for each category (land cover and land use) was identified. With this information, the covers that suffered significant transitions could be distinguished, and land use changes were identified. Changes in land use between successive dates were detected by post-classification comparison. The post-classification comparison leads to a categorical map that indicates the land use classes at the two successive observation years for every pixel. The traditional post-classification cross-tabulation [6] was employed to establish ''from'' and ''to'' categories for the two dates of the images, which was essential for the definition of land use change trajectories.

3.2. Urban sprawl measurement

Urban sprawl over the period of 12 years (2005-2017) was determined by computing the area of all the settlements from the digitized topography calculations of 2005-2017 and comparing it with the area obtained from the classified satellite imagery for the built-up theme. The original classification of land-use of 12 categories was aggregated to vegetation, built-up (residential & commercial), agricultural lands & open, and water bodies. Area under built-up theme was recognized and the whole built-up theme from that imagery was digitized; this vector layer gave the urban area of 2017. Further, by applying vector analyses, the built-up area under each village was calculated.

The percentage of an area covered by impervious surfaces such as asphalt and concrete is a straightforward measure of development. It can be safely considered that developed areas have greater proportions of impervious surfaces, i.e. the built-up areas as compared to the lesser-developed areas. Considering the built-up area as a potential and fairly accurate parameter of urban sprawl has resulted in making considerable hypothesis on this phenomenon. Since the sprawl is characterized by an increase in the built-up area along the urban and rural fringe, this attribute gives considerable information for understanding the behavior of such sprawls. This is also influenced by parameters such as, population density, population growth rate, etc.

3.3. CA Markov model

Cellular Automata (CA) Markov is a dynamic model used for spatial simulation (geo-simulation) at a given time. The superiority of the CA Markov model is that it can be used to study a simple pattern to a complex pattern with a simple principle. The CA Markov model is widely adopted and applied in the field of earth science, one of which is to study land use change. Markov Chain in Markov concept is, in essence, a mathematical model designed to describe a process in progress in sequence. The modeling process is done on Idrisi software by running Cellular Automata-Markov (CA-Markov) module. This module is processed by combining the Markov Chain module which produces Transitional Probability and Multi-Objective Land Allocation (MOLA) which performs iteration process to obtain the final composition. Predicted land use change is processed based on early year land use, land suitability and neighbors.

3.4. Geometric corrections

In this research geometric correction is done by using *image to image* method with Quickbird image as the basic image and completed by giving 33 points of GCP which spread evenly at research location. Accuracy result of geometric correction (RMSE) must be less than 1 pixel in order to be classified as corrected image.

3.5. Analysis of field data validation

Validation of field data was done by random sampling method by taking 120 point of land use samples. Field validation is needed to determine the accuracy of the results of the image interpretation and the result of digitization with the actual situation in the field. Field validation was conducted on 13-14 July 2017.

4. Results and Discussions

4.1. Gap and fill processing

This study uses Landsat 7 ETM + image which has acquisition of recording data on June 10, 2005 and June 11, 2011, where in that year there was damage to the section of Scan Line Correction (SLC) which stopped functioning normally (OFF) since May 31, 2003. The result shown in Figure 1, the SLC-off is showing the occurrence of black lines due to the satellite sensor (stripping) causing large gap, so the image has no spectral value on the areas. This will certainly be a problem in classifying an object that is in the image. One way to reduce the gap is to fill in the main Landsat image gap with another Landsat image, which has different gap sections.

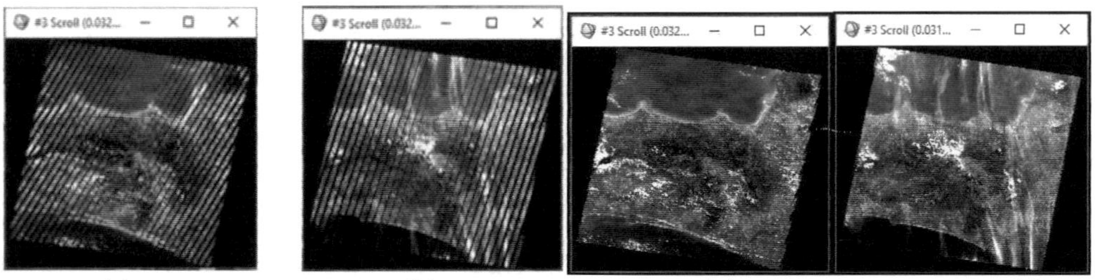

(a) (b) (c) (d)

Figure 1. (a) imagery of 2005, (b) 2011 experiencing slc-off and (c) imagery of 2005, (d) 2011 after processed using frame and fill.

4.2. Geometric correction results

The image generated from the satellite recording process is not free from errors. The errors can be satellite movement, earth rotation, sensor mirror movement and earth topography. Geometric correction is intended to minimize the errors. The RMSE results are shown in Table 1. The error in 2011 satellite image was slightly higher than the other years.

Table 1. Results of RMSE GCP.

No.	Year	RMSE GCP (Pixel)
1	2005	0.429
2	2011	0.521
3	2017	0.401

4.3. Land usage change year 2005-2017

The analysis of land use change is done quantitatively by calculating the amount of land area change each time the research is based on the results of the analysis of overlapping or Overlay Identity done on the GIS software to know the magnitude of changes in the area of change. Analysis overlapping or Overlay Identity performed on the ArcGIS software to determine the magnitude of changes in the area of change.

Table 2 shows that Semarang City within twelve years i.e. from 2005-2017 experienced significant land use change for 6067.674 hectares. The largest land use changes are mostly settlement land which is up to 54% of the total area. This additional settlement area is the result of conversion of paddy fields, ponds, services, offices and commercial and heat land. Industrial estates have increased by 3.2% of the conversion of paddy fields, heat land and ponds. There are lots of abundant lands that turn into industrial lands, services, offices and commerce, fields to settlements by 55%. This indicates the growth of non-built land into wake land. Based on the result of the research, it is found that the width of land change is not built into landholdings of 59% (3573.213 Ha). The growth of this settlement land is due to the rapid growth of the population so that the need for shelter is higher. Given the need for shelter is a primary need that is owned by every human being. Another consequence of the spike in population is the demand for goods and services so that the growth of office space and commerce and services both in the field of education, worship, sports and recreation is increasing.

Table 2. Changes in area of land use year 2005-2017.

Change classification of land use in the year		Change Area	
2005	2017	Ha	%
Rice fields	Fishpond	12.115	0.199
Rice fields	Industry	8.088	0.133
Rice fields	Airport	44.029	0.725
Rice fields	Offices and commerce	4.410	0.072
Rice fields	Heat Land	85.419	1.407
Rice fields	Settlement	121.579	2.003
Services	Offices and commerce	113.040	1.862
Services	Heat Land	109.869	1.810
Services	Settlement	0.518	0.008
Forest	Heat Land	1113.309	18.348
Offices and commerce	Services	4.632	0.0763
Offices and commerce	Heat Land	3.501	0.057
Offices and commerce	Settlement	0.865	0.014
Settlement	Services	87.112	1.435
Settlement	Offices and commerce	28.943	0.477
Settlement	Heat Land	862.819	14.219
Heat Land	Industry	155.136	2.556
Heat Land	Services	20.879	0.344
Heat Land	Offices and commerce	18.258	0.300
Heat Land	Rice fields	43.573	0.718
Heat Land	Settlement	3091.635	50.952
Fishpond	Industry	27.221	0.448
Fishpond	Airport	38.843	0.640
Fishpond	Rice fields	28.746	0.473
Fishpond	Settlement	43.135	0.710
Amount		6067.674	100

4.4. Physical distribution direction year 2011-2017

Semarang City with dense residential conditions made it very clear that the growing of developed sites were not too significant anymore, so it is possible to experience the growing of developed sites in the western and southern regions of Semarang City. Lands that are classified as agricultural land and forests threats the growing of the developed sites. Meanwhile in the coastal areas of the city that has a built land, they tend to dense so that its development could not be significant due to limited available lands.

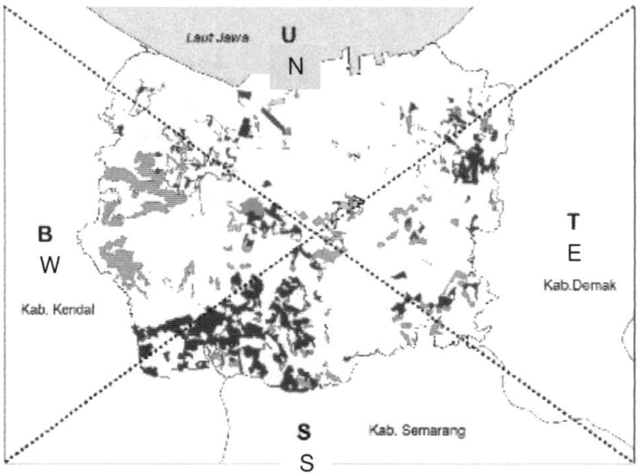

Figure 2. Results of direction of physical distribution of territories year 2011-2017.

From Figure 2 it is known that the change in settlements marked by the purple color is seen to spread. However, the most heavily settled settlement to the south is approaching Semarang regency and border between Semarang regency and Kendal regency. In addition, the change of settlements is quite widespread leads to the East approached the industrial area Terboyo and Demak District. Settlement changes are also encountered in the western part of Semarang City, where its direction approaches industrial area. This is due to the expansion of urban development into the periphery, and the increase of residential areas due to the high urbanization. Pursuant to result of analysis of physical distribution of area of Semarang city of year 2005 until 2017 can be seen that the sprawl of Semarang leads to suburb.

4.5. Analysis of conformity of CA Markov model with spatial plans of Semarang City in 2011-2031
The result of CA Markov Model processing is the prediction of physical distribution of the region in 2029. The output file is a raster file divided into 12 classes of land use. Analyzing the suitability of modeling results of CA Markov Model in the form of map of prediction of physical spreading of Semarang City area in 2029 with map of spatial plants Semarang city in 2011-2031 is done quantitatively by doing analysis of Overlap or Overlay Identity done on GIS software. Both are said to be appropriate if there is a similarity of land use in one area.

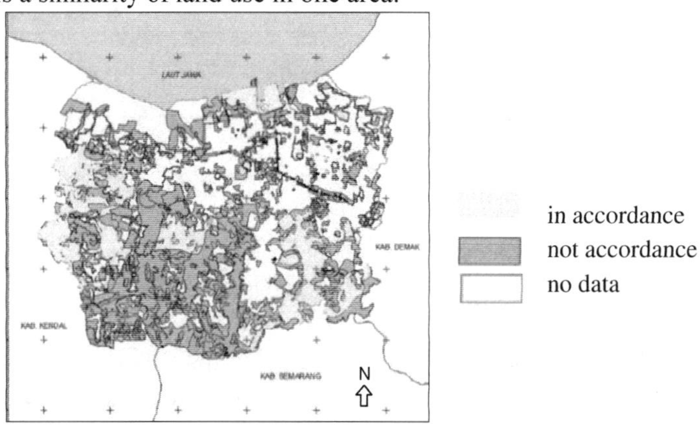

Figure 3. Conformity of CA Markov model and spatial plants.

Based on the result of identity overlap analysis done on Markov Model prediction map with spatial plant of Semarang City obtained 108 land use declared unsuitable to spatial plants Semarang City in 2011-2031. While 28 land use is declared as appropriate. On a large settlement land the changes occurred by 18% of the total area with the designation of 16157.615 Ha of settlement land. While the designation of Heat Land area of 470.588 Ha, and industrial estate area of 1048.776 Ha. Seen in Figure 3 above, the pink color represents the unsuitable land of yellow color represents the area corresponding to the spatial plants.

4.6. The prioritize specific policies and regulations of Semarang

The growth of Semarang's developed site will reach an unbalanced condition (limited) starting in 2025, where non-developed site is only 27% left. As for achieving ideal conditions have been missed, developed site at the present time (2010) is already under 50%. In the year 2020 where the limit of the year 2010-2030 spatial plans on built space stayed 27%. Under the minimum standards referring to the the provisions of Regulations No. 27 Year 2006, it requires a building compaction policy and encourages the construction of a two-story multi-storey house should be done.

5. Conclusions

The change of Semarang City from 2005 to 2011 is quite dynamic with value of 3254.416 Ha. Since 2011 until 2017 the land use change was significant, i.e. 6067.674 Ha. The result of physical analysis distribution of area of Semarang City of year 2005 until 2017 can be seen that sprawl of Semarang leads to suburb. This is in accordance with the Multiple Nucleus Theory, which is proposed by Harris and Ullman. Based on the analysis of conformity of model results of CA Markov and Spatial Plans of Semarang City 2011-2031 on class of settlement area, industrial estate and Heat Land with value of 56.931% are declared unsuitable and the other 43.07% are suitable. With a confidence level >40% (0.40) is stated quite well.

References

[1] Arifin S and Ita C 2016 Implementasi penginderaan jauh dan SIG untuk inventarisasi daerah rawan bencana longsor Jurnal penginderaan jauh LAPAN Vol 3
[2] Banowati E 2006 Geografi permukiman Semarang: Universitas Negeri Semarang
[3] Dadras HZ, Shafri N, Ahmad B, Pradhan S, Safarpour 2015 Spatio-temporal analysis of urban growth from remote sensing data in Bandar Abbas city, Iran, Egypt J Remote sens space sci., 18 1 pp. 35-52
[4] Irwin EG and Geoghegan J 2001 Theory, data, methods: developing spatially explicit economic models of land use change Agriculture ecosystems & environment 85: 7–23
[5] Ilyas A 2006 Pengaruh perkembangan Kota banjarmasin terhadap penggunaan lahan di kota kertak Tesis (S2) Pasca sarjana jurusan perencanaan wilayah dan kota UNDIP Semarang.
[6] Lunetta RS and Elvidge CD 1999 Remote sensing change detection London Taylor & Francis
[7] Malingreau JP 1997 Penggunaan lahan pedesaaan penafsiran citra untuk inventarisasi dan analisisnya Yogyakarta: puspics UGM
[8] Pontius JRG, Shusas E, and McEachern M 2004. Detecting important categorical land changes while accounting for persistence Agric Ecosyst Environ 101, 251–268 http://dx.doi.org/10.1016/j.agee.2003.09.008
[9] Peraturan Pemerintah RI No. 24 Tahun 2009 Tentang Kawasan Industri

Identification of landscape disturbance in the Parangtritis sand dune area for sustainable environment

Sunarto[1*], M N Malawani[1], H Fatchurohman[2], and A Cahyadi[1]

[1] Department of Environmental Geography, Faculty University of Gadjah Mada, Yogyakarta, Indonesia.
[2] MPPDAS Faculty of Geography, University of Gadjah Mada, Yogyakarta, Indonesia.

*Email: sunartogeo@gmail.com

Abstract. In 2014, the Parangtritis sand dunes area has been established as a geological natural preserve. Due to its distinctive characteristics and uniqueness, the Parangtritis sand dunes area is included to the privilege values of Yogyakarta Province. This research aims to 1) identify various anthropogenic activities that possibly affecting the natural process of the Parangtritis sand dunes, 2) formulate appropriate management actions to ensure the continuity of aeolian process. Several anthropogenic activities were identified in the research area such as settlements, agriculture, fishponds, and coastal forest cultivation. Among those anthropogenic activities, coastal forest cultivation is the most threatening activity to the aeolian process of the Parangtritis sand dunes. Beside the density of the vegetation, much also inhibit the sand transport as the main controlling factor for the formation of sand dunes. For precise management actions, sand dunes zonation is urgently required. The whole area in the supply zone must be cleared from any anthropogenic activities to keep the sand sufficient. The transport zone, where the sand is actively transported also should be cleared from anthropogenic activities, especially from coastal forest cultivation. In this zone, indigenous vegetation are allowed to grow as biodiversity icon. The accumulation zone is suitable for coastal forest cultivation as a barrier and protection for the settlement area and agriculture area outside the Parangtritis sand dunes.

Keywords: landscape disturbance, sustainable environment, Parangtritis sand dune

1. Introduction

Coastal sand dunes are typical landform that occur by the aeolian processes. Aeolian process is a geomorphological process that have two aspects which are erosion and deposition. The sand accumulation work is very dependent on the grain size of the sands [1]. Wind is the main transport agent in the aeolian process. The wind blown out the sand particle and naturally sort the sands based on its grain size [2]. The sediment transport sediment in three different ways i.e. rolling, saltation, and suspension [3]. One of the results of sand accumulation process is the formation of sand dunes. Barchans dunes or crescent moon dunes is very distinctive sand dunes type. Parangtritis coastal area is one of the locations in the world where the barchans dunes are formed.

The uniqueness of Parangtritis coastal area is the existence of barchans sand dunes, which commonly formed in a region with dry climate. However, the coastal area of Parangtritis is considered as wet tropical climate region. Parangtritis coastal area is very suitable for the formation of sand dunes because of several controlling factors such as 1) the wind blowing from the sea to the hinterland; 2) the existence of wind corridor or natural wind tunnel; 3) sufficient sands supply; 4) unconsolidated materials; 5) beach morphology; 6) beach sloping factors; 7) beach width; and 9) the tidal range [4]. The composing material of Parangtritis sand dunes were originated from Merapi Volcano, and mainly composed by magnetite, volcanic glass, andesitic rock fragment, plagioclas, augit, hyperstin, and some ilmenite [5].

The existence of sand dunes are becoming one interesting factors for tourism activities in coastal area. Sand dunes phenomenon in the coastal area is a very interesting and attractive factors in terms of tourism development [6]. The development of tourism in Parangtritis coastal area lead into rapid anthropogenic development within the coastal area. This research aims to identify several anthropogenic activities that possibly disturb the process of sand dunes formation and propose a suitable micro zonation to protect the sand dunes sustainability.

An example of serious protection effort by the government, formal regulation has been released (governmental legislation). The formal legislation expressed in Government Regulation number 26 year 2008 about Spatial Planning of National Region and Regional Regulation of Yogyakarta Province number 2 year 2010 about Province of Yogyakarta Spatial Planning for 2009-2029, as provincial strategic region of protection and cultivation. Therefore, several aspects related to detailed landscape planning in Parangtritis Coastal Area is necessarily important and must consider the natural process to keep the sustainability of the sand dunes.

2. Method

The main data for this research was collected by field survey and measurement of several natural processes such as wind and sand movement in sand dunes area. The observation began with aerial photograph and sattellite images interpretation to identify anthropogenic activities in the reseach area. The research area map can be seen in Figure 1. The identification results then validated on the field to identify the real condition in the research location. Further observation carried out to determine the impacts of each anthropogenic activities to the sand dunes formation.

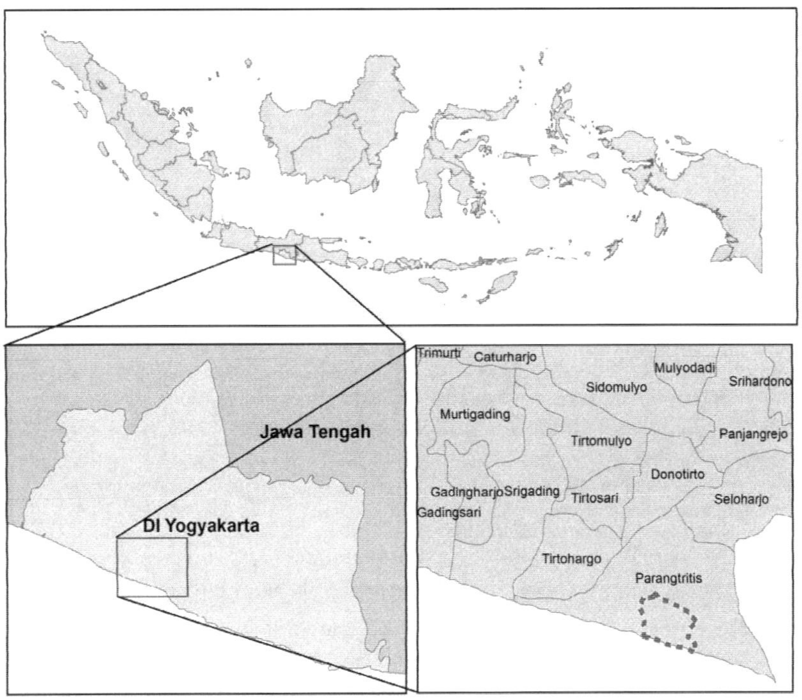

Figure 1. The research location map at the primary zone of Parangrtitis sand dunes.

In addition wind velocity measured by hand held anemometer and wind direction measured by compass, while the sand movement measured using vertical sand trap or sand sampler [7]. Result from wind measurement presented as m/s for wind speed, while sand movement (transport) presented as gr during one hours measurement. Others weather condition measured by automatic data recorder installed in sand dune area near research location. The results from the measurements then used as the main data that provide scientific consideration for proposing microzonation in the primary zone of Parangtritis sand dunes. The microzonation could be very useful for more detailed landscape management.

3. Result and Discussion

According to the interpretation of 2014 aerial photograph, it can be recognized that the sand dunes area occupied by several land use. Land use is considered as anthropogenic activity within the sand dunes area, especially in the primary zone. Several land use type that occupy the primary area are coastal forest, paddy field, husbandry, settlement, fishponds, bare land, pasture, and dry land agriculture. Coastal forest is the most dominating land use in the primary zone. Coastal forest occupy 86 hectares of the primary zone. The most dominating vegetation in the coastal forest is *Causarina equisetifolia* and *Acacia mangium*. The delineation of land use by 2014 aerial photograph shown in Figure 2.

Figure 2. Primary zone land use in 2014 from the sand dune area of Parangtritis.

To determine which land use with the most considerably impact on sand transport process, therefore the land use change pattern over years should be analyzed. From 1980 to 2014, the land use change was very rapid, especially the vegetation cover. Forest plantation program initiated by certain department causing the increasing of vegetation cover. Uncontrolled vegetation cover might interfere the sand transport as the main process of sand dunes formation. The measurement results from the vertical sand trap also show the different between open space area and area that covered by vegetation cover. Result from sand transport measurement shows in Table 1. Based on the measurement there was different sand mass transporting by wind between fore dune and main dune. Sand mass transported on the fore dune are greater than main dune. Heaviest sand mass recorded on the fore dune was 40 gr transported by wind with speed 5 m/s. Smallest recorded sand mass transported by wind was 2 gr with wind speed 1.1 m/s on the main dune. Based on Figure 3, also inferred that sand transport reduced by anthropogenic disturbance between fore dune and main dune. Besides that, wind speed also reduced. Average wind speed reduced by anthropogenic disturbance between fore dune and main dune is 3 m/s. However, sand transport has greater reduction with average sand mass is 10 gr. This data provide some facts that anthropogenic disturbance significantly affected to the natural Aeolian process, especially sand movement by wind.

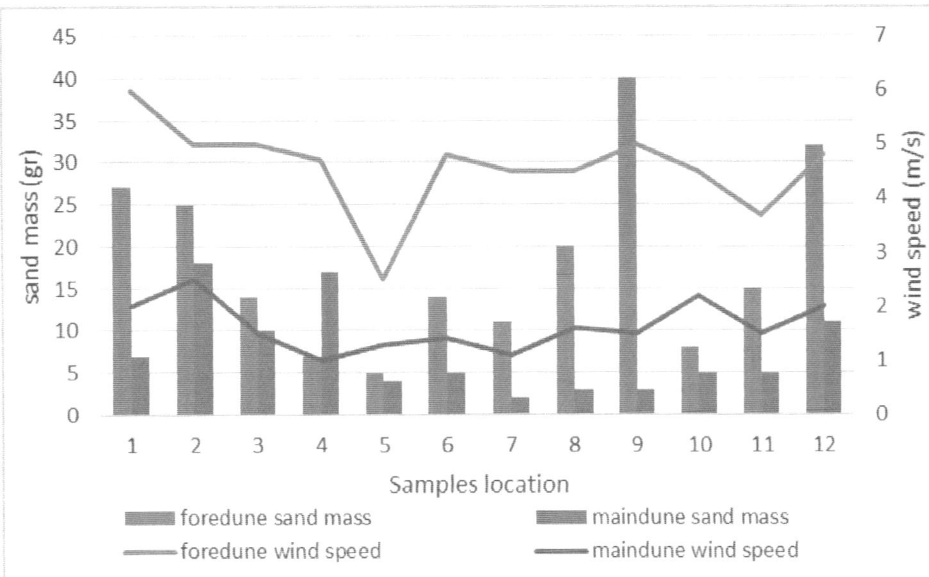

Figure 3. Comparison sand transport field measurements on the fore dune and main dune area of Parangtritis.

Weather condition on sand dune are of Parangtritis has unique condition especially some parameters related to sand transport process. The most influenced parameter is wind speed. From weather data recorded from automatic weather station (AWS) installed on sand dune area managed by collaboration between Parangtritis Geo-maritime Science Park (PGSP) and Faculty of Geography of Gadjah Mada University, showed that maximum wind gust recorded on December 2017 to March 2018 is 44 km/h and maximum wind speed is 9 km/h (Fig. 3). Based on Beaufort's scale wind gust classified as strong breeze and wind speed mostly classified as light breeze. Those wind are dominance blown to northern direction. Other parameter that influenced to sand transport process is precipitation. When sand became wet due to heavy rainfall, sand cannot move because their density is increase. The maximum rainfall recorded is 61.5 mm per day on February. However, beside all recorded data is dominance by rainy day, a couples days on March recorded as no rainfall events. This events might be the best events for sand transport process on sand dune area. Therefore, sand transport will effectively occurs on dry season (Figure 4).

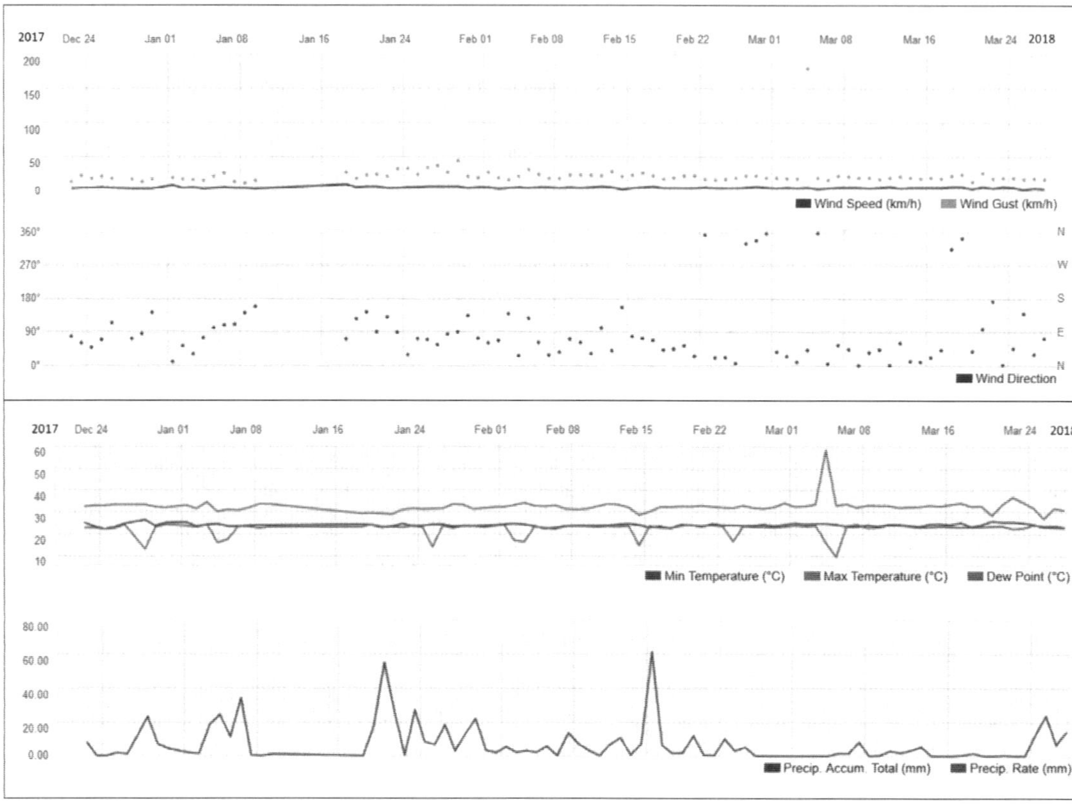

Figure 4. Some weather conditions recorded form AWS installed on sand dune are of Parangtritis. The AWS recorded data can be easily downloaded on: https://goo.gl/pdn69W

To manage the sustainability of Parangtritis sand dunes ecosystems, microzonation based on the effect of anthropogenic activity and intensity is required. The main consideration for this microzonation is sand movement process. Based on the identification, it is recognized that the sand transport zonation divided into three zone namely supply zone, transport zone, and deposition zone. The supply zone is where the sand material source located and being transported by the wind. The sand material originated from marine deposition that located in fore dune area. The transport zone is where the sands actively transported by the wind. The dynamic wind movements transport the sand with different movement types. According to the measurement, 60% of the sands transported by creeping, 35 % by jumping, and the rest transported by floating in the air (Table 1). The deposition zone is where the sands accumulated and become the last location for the sand movement. This zone is adjacent with the other zone outside the sand dunes primary zone. The sand dunes zonation shown in Figure 5.

Not only the spatial distribution, but activities management inside the primary zone also necessarily important especially related to land use management. In the supply zone, the anthropogenic activities that should be restricted are fishponds, settlement, and agriculture. The vegetation existence are needed in several locations to reduce the impacts of high tide and marine erosion. In the transport zone, all the anthropogenic activities that already mentioned before should be restricted. Only indigenous vegetation that allowed to grow, which is *Siwalan*. In the deposition zone, the anthropogenic activities that must be restricted are agriculture, fishponds, and settlement. The indigenous vegetation are allowed as a barrier to prevent the sand movement to the settlement area outside the primary zone.

Figure 5. Microzonation of primary sand dune area of Parangtritis.

4. Conclusion

Sand transport measured on filed provide that anthropogenic disturbance significantly affected to the Aeolian processes, this is indicated that landscape disturbance has occurred on the study area. There are four type of anthropogenic intervention that caused landscape disturbance in Parangtritis sand dune area, i.e. settlement, fishpond, coastal forest, and agriculture. Coastal forest cultivation is the most influence due to environmetal change. Primary zone need to divided into three micro zonation based on geomorphological process. Each microzonation created for proper landscape and environmental sustainability, especially for sand dune natural processes.

References

[1] Joyosuharto S 1985 Fundamentals of Land Clasification Yogyakarta Fakultas Geografi Universitas Gadjah Mada
[2] Summerfield MA 1991 Global Geomorphology, An Introduction to the Study of Landform London Longman Scientific&technical
[3] Reineck HE and Singh IB 1973 Depositional Sedimentary Environments Berlin Springer
[4] Sunarto 2014 Geomorphology and its contribution in conservation of Bergumuk coastal aeolian from Agrogenic and Urbanogenic Threat Yogyakarta, University of Gadjah Mada
[5] Verstappen H 1983 Applied Geomorphology Amsterdam: Elsevier
[6] King CAM 1966 Techniques in Geomorphology London Edward Ardnold
[7] Goudie A, Anderson M, Burt T, Lewin J, Richards K, Whalley B, and Worsley P 2005 Geomorfological Techniques Second Edition London and New York Routledge Taylor and Francis Group

Improving quality of Bermuda Grass (*Cynodondactylon* var. *Evergreen*) by adding of bentonite to growing media and proper irrigation frequency

A Permatasari[1] and N Nasrullah[*,2]

[1]Bachelor Student of Landscape Architecture Department, Bogor Agricultural University (IPB), Indonesia
[2]Landscape Architecture Department, Faculty of Agriculture, Bogor Agricultural University (IPB), Indonesia

*Email : nizarnasrullah@yahoo.com

Abstract: Generally sand is used as root zone of turf grass in golf course. High permeability of sand caused irrigation inefficient. This research was conducted to find the effect of addition of bentonite and frequency of irrigation to improve qualities of Bermuda grass (*Cynodondactylon* var *evergreen*). Experiments using factorial completely randomized design with two treatment factors with 3 replications. The first factor was the addition of bentonite 25 mesh which consisted of three levels i.e. 100% sand, 87.5% sand + 12.5% bentonite, and 75% sand + 25% bentonite. The second factor was the frequency of irrigation which consisted of three levels i.e. every day, every two days, every three days. Results of research showed interaction of media mix and the irrigation frequency was significantly affected all variable of grass qualities. Mix of sand 87.5% + bentonite 12.5% with watering of once every two days produced the best visual quality, and functional quality at the height of grass, clipping dry weight, shoot dry weight, as well as on the efficiency of irrigation water use (EIWU). It is recommended to use mix of sand 87.5% + bentonite 12.5 % with every two days watering that generally produced the best qualities and spends water consumption.

Keywords: bentonite, Bermuda Grass, functional quality, irrigation, visual quality,

1. Introduction

The game of golf will be interesting and fun if it is backed by a good quality of turf grass. The quality of grass is divided into two aspects, namely visual quality and functional quality [1]. Visual quality consists of density, texture, color uniformity, presence of particles on the surface, and grass type purity. While the functional quality consists of grass height, shoot dry weight, root dry weight, root length, and grass elasticity.

One important factor in the growth of grass is the availability of water and nutrients that must be available in sufficient quantities. This is because the golf course root zone is based on sand, so it has a low water and nutritional retention capacity. One thing that can be done to maintain the availability of water and nutrient is to modify the root zone. The advantage of modification of sand media has reported. Pre plant incorporation of granular subtances aided root development [2], while addition of biochar improved water retention and increased overall plant growth [3].Bentonite is a rock with very fine grains and contains a lot of silicate clay minerals. The silicate clay mineral in bentonite is dominated by montmorillonite, a clay mineral formed from the basic components of silicon-oxygen tetrahedron and aluminum octahedron [4]. High clay mineral content in bentonite is expected to reduce the level of sand media permeability so that the water content and nutrients needed by the grass will be fulfil.

Irrigation arrangement is one of the important management of grass. Turf typically requires 1 to 1.5 inches of water per week for normal treatment conditions [5]. To maintain the quality of crops managed in large areas such as golf courses, irrigation can be a factor that can increase production costs. Improper irrigation arrangements may affect the visual and functional quality of turf grass. In addition, by determining the exact frequency of irrigation, it can reduce the cost of management without reducing the growth and quality of turf grass. Besides proper quantity of watering will ensure good water conservation which important for sustainable golf management. This study aims to find out the best combination of bentonite and sand mixture media and frequency of irrigation to the visual and functional quality of the Bermuda grass (*Cynodondactylon* var evergreen).

2. Research Method

Experiment was conducted in green house of experimental farm of Bogor Agricultural University Darmaga Campus from April to July 2015. Experiments using factorial completely randomized design with two treatment factors with 3 replications. The first factor was the addition of bentonite to sand which consisted of three levels i.e. B0 (100% sand), B1 (87.5% sand + 12.5% bentonite), and B2 (75% sand + 25% bentonite). Size of bentonite particle used in this experiment was 25 mesh). The second factor was the frequency of irrigation which consisted of three levels i.e. A1 (once every day), A2 (once every two days), and A3 (once every three days) with a volume of 400 ml/basin. Hence there were 27 experiment units such as B0A1 (treatment of 100% sand + once every day irrigation) and B1A3 (treatment of mix 87.5% sand and 12.5 % bentonite) + once every 3 days irrigation). Data were analyzed by Anovatest, if the results is significant than analyzed by *Duncan Multiple Range Test* (DMRT).

Bermuda grass (*Cynodondactylon* var evergreen) was propagated by stolon and cultivated in the basin with size 40 cm diameter, and 25 cm height. Growing media consisted of 5 cm gravel in the base of basin, and treatment media above the gravel. After planting the stolon, irrigation was done every day with 400 ml/basin until 2 weeks after planting (WAP), after that irrigation was done according to treatments of irrigation frequency.

Initial fertilizer 10 gram Nitrogen/m^2 or 3.73 gram/basin of NPK (15-15-15) was added in 3 WAP based on previous experiment .After that, fertilizer 3.73 gram NPK/basin were added every two weeks until 11 WAP. That fertilizer was dissolved in water than pour to growing media. Insect (*Spodopteraexigua*) and weeds (*Cyperusrotundus*) which appeared were manually removed.

Respond of turf to treatment was evaluated by observation of visual and functional qualities of turf. Besides, physical character of growing media were also analyzed including bulk density, porosity and permeability. Visual qualities were observed including grass coverage, height, shoots density and color. Color was measured using Munsell Color Chart for Plant Tissue. Functional qualities were observed including clipping dry weight in 7-12 WAP, than canopy dry weight, root length, root dry weight in the end of experiment. Shoot density, clipping dry weight were observed based on 10cm x 10cm square sampling area. Root, shoot and leaves dry weight were measured after 48 hours draying in 80°C oven.

3. Research Results and Discussion

3.1. Physical Characteristic of Media

Result of experiment showed treatment was significantly influenced bulk density, porosity and permeability of media. The highest (1.45 gram/cm^3) bulk density was found in B2A1 treatment and the lowest one (1.22 gram/m^3) was in B0A3 treatment. Generally media of 100% sand has lower bulk density. Than the higher bentonite portion, the higher value of bulk density. Based on USGA [6], good bulk density for root zone is 1.4 cm/m^3, but 1.20-1.6 cm/m^3 is suitable range for root zone.

The highest value of media porosity (53.9%) was found in treatment of B0A3 and the lowest value (46.6%) is obtained in B1A1 treatment. Generally whatever of mixture media with irrigation frequency once every day showed the lowest porosity, and frequency once every 3 days showed the highest porosity.

As the same as porosity trend, the fastest value of permeability was obtained in treatment of B0A3, and the slowest value was obtained in B2A1. Generally value of permeability tend to decrease when portion of sand decrease and frequency of watering decreased. Based on US Soil Conservation Services, treatments in this research indicate permeability values were 9.05-12.18 cm/jam which categorized as moderately fast permeability.

Table 1. Interaction of media mixture and irrigation frequency treatment to physical Characteristic of Media

Treatment	Bulk density gram/cm^3	Porosity (%)	Permeability (cm/jam)
B0A1	1.42de	46.7b	9.12ab
B0A2	1.33bc	49.7f	10.72e
B0A3	1.22a	53.9i	12.18g
B1A1	1.4de	46.5a	9.17b
B1A2	1.37cd	49.0e	10.12c
B1A3	1.29b	52.1g	12.07g
B2A1	1.45e	47.3c	9.05a
B2A2	1.39de	48.7d	10.33d
B2A3	1.3b	52.4h	11.85f

3.2. Visual qualities of grass

3.2.1. Grass coverage

The result of the research shows that there is a significant interaction between planting media and irrigation frequency at grass coverage from 3-10 weeks after planting (WAP). Treatment of 75% sand and 25% bentonite with frequency once every 3 days showed the best results from 9-10 WAP compared to other irrigation frequencies. However on 7-10 WAP, treatment of 100% sand with watering frequency once every day was not significantly different from the treatment of sand 75% + 25% bentonite. In other words watering more efficiently done every two days with a mixture of media 75% sand + 25% bentonite.

Table 2. Interaction of media mixture and irrigation frequency treatment to grass coverage

Treatment		Week rd							
Media	Watering Frequency	3	4	5	6	7	8	9	10
B0	A1	58.87bc	76.13h	78.94f	80.88e	82.98e	87.17e	94.43e	100d
	A2	57.80ab	67.77d	75.84e	78.62d	80.90d	86.02d	92.30d	97.63c
	A3	60.93d	66.17c	68.41b	70.88a	73.87a	77.07a	81.13a	89.53a
B1	A1	59.10c	72.16e	81.75g	84.85g	87.36g	91.20g	96.90f	100d
	A2	62.77e	72.17e	82.75h	83.84f	86.07f	89.07f	93.50de	99.33d
	A3	59.10c	64.97b	70.15c	75.80c	78.17c	81.00c	84.83b	92.80b
B2	A1	64.83f	75.00g	85.68i	86.75h	88.53h	93.90h	98.53g	100d
	A2	63.53e	73.77f	72.71d	78.41d	82.47e	87.47e	94.53e	99.47d
	A3	58.17a	62.43a	67.14a	72.66b	76.03b	79.03b	86.13c	93.40b

Notes: Values which followed by the same character in the same column indicate no significant difference by Duncan Multiple Range Test in alfa 5%.

3.2.2. Grass Height

At the end of study, the lowest grass height was obtained on 100% sand media treatment (B0). At 7 - 12 WAP, daily irrigation frequency treatment (W1) indicated the best results on grass height. At 11-12 WAP, treatments W1 and W2 was not significantly different. This shows that daily irrigation becomes inefficient because it produces the same grass height as on W2. The significant interaction occurred between plant media and irrigation frequency treatment at 7-12 WAP. At 8 - 12 WAP, the highest grass height was obtained on the mixture of sand 87.5% + bentonite 12.5% with irrigation frequency every two days (B1A2). At 12 WAP, the lowest result was obtained at B0A3 treatment. This indicates that the grass height is not maximal if done under less water conditions with sand medium.

Table 3. Effect of media and irrigation frequency treatment to grass height

Treatment	Week after planting					
	7	8	9	10	11	12
Media						
B0	4.14a	3.83a	3.06a	2.88a	2.73a	2.27a
B1	3.92a	3.91a	3.50b	3.27b	3.00a	2.66b
B2	3.80a	3.99a	3.39b	3.14b	2.91b	2.59b
Irrigation frequency						
A1	4.24b	4.10a	3.81b	3.61c	3.33b	2.92b
A2	4.12b	3.86a	3.64b	3.36b	3.24b	2.84b
A3	3.50a	3.78a	2.50a	2.32a	2.07a	1.74a

Table 4. Effct of interaction of media mixture and irrigation frequency treatment to grass height

Treatment		Week after planting							
Media	Irrigation frequency	3	4	5	6	7	8	9	10
B0	A1	58.87bc	76.13h	78.94f	80.88e	82.98e	87.17e	94.43e	100d
	A2	57.80ab	67.77d	75.84e	78.62d	80.90d	86.02d	92.30d	97.63c
	A3	60.93d	66.17c	68.41b	70.88a	73.87a	77.07a	81.13a	89.53a
B1	A1	59.10c	72.16e	81.75g	84.85g	87.36g	91.20g	96.90f	100d
	A2	62.77e	72.17e	82.75h	83.84f	86.07f	89.07f	93.50d e	99.33d
	A3	59.10c	64.97b	70.15c	75.80c	78.17c	81.00c	84.83b	92.80b
B2	A1	64.83f	75.00g	85.68i	86.75h	88.53h	93.90h	98.53g	100d
A2		63.53e	73.77f	72.71d	78.41d	82.47e	87.47e	94.53e	99.47d
A3		58.17a	62.43a	67.14a	72.66b	76.03b	79.03b	86.13c	93.40b

3.2.3. Shoot density

The significant interaction occurred between the combination media and the frequency of irrigation at 7 - 12 WAP on shoot density (table 5). All treatment combinations increased shoot density up to 12 WAP. At 7 WAP, the lowest shoot density was obtained by B0A3 treatment, while the highest result was obtained by B2A1 treatment at 7 - 12 WAP. However on 11-12 WAP, treatment of B2A1 was not significantly different from B2A2. This shows that irrigation every two days is more efficient to produce high shoot density.

Based on shoot grade density criteria [6], at 10 WAP the treatment of sand 87.5% + bentonite 12.5% 25 mesh and treatment of sand 75% + bentonite 25% with daily and every two days irrigation frequency produced the shoots which including in the medium density class. While at 12 MST, 100% sand with three-day irrigation frequency (B0A3) and mixed sand media treatment 87.5% + 25% mesh bentonite with three-day irrigation frequency (B1A3) produced shoos which including in low density class.

Table 5. Effect of interaction of media mixture and irrigation frequency treatment to shoot density

Treatment		Week-					
Media	Irrigation	7	8	9	10	11	12
B0	A1	74.00cd	80.00b	87.67c	92.33c	114.00cd	121.67c
	A2	71.67bc	79.67b	85.00bc	94.33c	110.67c	120.33c
	A3		70.67a	79.33a	81.00a	85.00a	91.33a
B1	A1	78.00de	85.33c	92.67d	100.67d	120.00e	125.33cd
	A2	77.67d	84.67c	87.67c	101.67d	117.00de	124.67cd
	A3	70.03bc	77.67b	82.67ab	85.00ab	88.00ab	94.67a
B2	A1	82.33e	91.33d	98.00e	112.67e	120.67e	129.33d
	A2	74.33d	77.67b	86.33bc	104.67d	122.00e	128.33d
	A3	67.67ab	73.33a	79.33a	86.33b	90.33b	103.67b

3.2.4. Color

At 7-12 WAP, in general the lowest color score was obtained at the treatment of planting medium with daily irrigation frequency (B0A1, B1A1, B2A1 (table 6) .The highest color score was obtained at the treatment of planting media with irrigation frequency every two days (B0A2, B1A2, B2A2), whereas treatment with three-day irrigation frequencies produced a higher color score than the frequency of each day. This indicates that irrigation frequency greatly affects the quality of the grass color. The more frequent the frequency of irrigation, the decreasing the quality of the grass color. In this study, intermediate frequency of irrigation (every two days), could improve the quality of grass color.

Table 6. Interaction of plant media and irrigation frequency to leaf color (Munsell Color Chart)

Treatment		**Week rd-**					
Media	Irrigation Frequency	7	8	9	10	11	12
	A1	3.33a	3.33a	3.67a	4.00a	4.00a	3.33a
B0	A2	4.00bc	4.00ab	4.33ab	4.67ab	4.33ab	5.00cd
	A3	4.00bc	4.00ab	4.00ab	4.00a	4.00a	5.00cd
	A1	3.67ab	3.33a	3.67a	4.33ab	4.00a	4.00ab
B1	A2	4.00bc	4.67b	4.67b	5.00b	5.00bc	5.33cde
	A3	4.00bc	4.33b	4.33ab	4.33ab	4.33ab	5.33cde
	A1	4.33b	4.00ab	4.00ab	4.33ab	4.67abc	4.67bc
B2	A2	4.67c	4.67b	4.67b	5.00b	5.33c	6.00e
	A3	4.00bc	4.33b	4.33ab	4.67ab	4.67abc	5.67de

Notes: The numbers followed by the same letter in the same column show results that are not significantly different according to the DMRT test at the 5% level.
B0: sand 100%
B1: mixture of sand 87,5% + bentonite 12,5% 25 mesh
B2: mixture of sand 75% + bentonite 25% 25 mesh
A1: Irrigation frequency everyday
A2: Irrigation frequency every two days
A3: Irrigation frequency every three days

At 7-12 WAP, the highest color score was obtained at combination of sand 75% + bentonite 25% 25 mesh with every 2 days of irrigation frequency (B2A2), while the lowest score was obtain at medium of sand 100%. This suggests that a mixture of sand with bentonite can improve grass color quality. The higher the bentonite mix, the higher the quality of the grass color.

3.3. Functional Qualities

3.3.1. Clipping dry weight
The best combination of mixed plant media treatment with irrigation frequency to dry weight of clipping occurred at 9 WAP, i.e. combination of B1A2 (mixture of sand 75% + 25% 25 mesh bentonite with irrigation frequency every other day). At the end of the study, the lowest weight was obtained by B0A3 (100% sand with every three-day irrigation frequency), while the highest weight was obtained by B1A2 (mixture of sand 87.5% + bentonite 12.5% 25 mesh with irrigation frequency every 3 days).
Combination of treatment of bentonite media with irrigation frequency greatly influence dry weight of clipping. The weight results show that watering is not necessary every day, because watering every two days is more efficient and effective in producing the best dry clipping weights.

3.3.2.. Crown dry weight
The real interaction takes place between the combination of mixed media treatment and the irrigation frequency to the crown dry weight. The lowest weight was obtained by a combination of 100% sand with irrigation frequency every three days (B0A3). While the highest result was obtained in combination of mixed of sand 87,5% + bentonite 12,5% 25 mesh with irrigation frequency every two days (B1A2).

In general, the combination treatment of irrigation frequency with 100% sand media produced the lowest crown dry weight. This suggests that the bentonite mixture is preferable in producing canopy dry weights. In addition, the combination of mixed plant media with irrigation frequency every three days, produced the lowest weight. The density of shoots and grass height is thought to affect the dry weight of the crown. It shows that water shortages can inhibit metabolism, so that less nutrients can be absorbed and transported in grass canopy.

3.3.3. Root Dry Weight
At the end of the study, there was a significant interaction between the combinations of mixed plant media with irrigation frequency to root dry weight.

Combination of mixed plant media treatment with daily irrigation frequency resulted in lowest root dry weight. However, the same result is obtained by a combination of irrigation frequencies once every 3 days with 100% sand and a mixture of sand of 87.5% + bentonite 12.5% (B0A2 and B1A2).

The highest root dry weight result was obtained by B2A3 treatment (mixture of sand 75% + 25% bentonite). In general, the combination of mixed plant media treatment with irrigation frequency every three days resulted in the highest root dry weight. This shows that in the condition of water shortage, the grassroots grew longer to find the water needs in the planting medium.

The canopy dry weight ratio with roots shows the proportion of transported nutrient distribution. The weight ratio will be greater if the root dry weight is smaller, and vice versa. In this study, more nutrients are transported to canopy than root. The highest weighted ratio was produced by combination of B0A1 treatment (100% sand with daily irrigation frequency), while the lowest weight ratio was produced by B2A3 (75% sand + 25% bentonite). In general, the combination of media treatment with irrigation frequency each day produced the highest weight ratio, while the combination of media with every three days irrigation frequency resulted in the lowest weight ratio.

3.3.4. Root Length

The significant interaction occurs between the combination treatment of planting medium and the irrigation frequency to the root length. The shortest root was obtained at B0A1 treatment (100% sand medium with irrigation frequency every day). In general, the combination of irrigation frequency treatment with 75% sand + 25% bentonite mixture resulted in longer roots than 100% sand treatment and 87.5% sand + 12.5% bentonite. This suggests that more bentonite blends can increase the length of the roots.

The longest root is produced at B2A3 treatment (75% sand + 25% bentonite mixture). It is not advisable to administer too frequent water, as this may cause grass to form roots just near the surface of the soil [7]. If the soil is made slightly dry, it forces the grass to grow deeper roots. The deeper roots are stronger in grasping the growing media, so the grass is not easily uprooted when a golf ball be beat.

3.3.5. Efficiency of Irrigation Water Use (EIWU)

Significant interaction occurs between mixture of planting media and irrigation frequency on efficiency of irrigation water use (Table 2). Observation of efficiency of irrigation water usage based on dry weight of plant can be seen in EIWU column. EIWU based on plant dry weight is also called the total dry weight ratio (DW) / irrigation volume. Total DW Ratio / Irrigation Vol, shows the amount of irrigation water used to produce the dry weight of the plant. A good ratio is the ratio the greater it is. A large ratio shows more efficient use of water in the formation of dry weights.

Table 7. Interaction beetween mixture planting media with irrigation frequency to Total Dry Weight

Treatment		Total dry weight (gram/sample)	Total irrigation volume (ml/sample)	**EIWU** (gram/ml)
Media	Irrigation frequency			
B0	A1	1.99	552.72	0.0036a
	A2	2.03	276.36	0.0073d
	A3	1.35	184.24	0.0073d
B1	A1	2.13	552.72	0.0038b
	A2	2.29	276.36	0.0083g
	A3	1.38	184.24	0.0075e
B2	A1	2.28	552.72	0.0041c
	A2	2.17	276.36	0.0078f

Description: Notes: The numbers followed by the same letter in the same column show results that are not significantly different according to the DMRT test at the 5% level.

B0: sand medium 100%
B1: mixed sand media 87.5% + bentonite 12.5%
B2: mixed sand media 75% + 25% bentonite
A1: irrigation frequency every day
A2: irrigation frequency every two days
A3: irrigation frequency every three days

The lowest water efficiency was obtained in B0A1 treatment (100% sand with daily irrigation), while the highest value was obtained in B1A2 treatment (87.5% sand + 12.5% bentonite with irrigation frequency once every two day). In general, the combination of media with irrigation every two days can produce the highest dry weight ratio every 1 ml of water. This shows that water usage for irrigation every two days is more efficient and effective than every day or every three days.

4. Conclusion

Based on the result, the treatment of planting medium with irrigation frequency gives significant effect to all observation variables. In general, mixed of sand with bentonite media produce better visual and functional quality of grass than without bentonite media. Every two days of irrigation frequency treatment, resulting in better grass quality compared to every days irrigation frequencies and three day irrigation frequencies. In general, the frequency of irrigation every two days applied to the mixture of sand media of 87.5% + 12.5% bentonite give the best visual and functional quality of grass.

References

[1] Turgeon A J 2005. *Turfgrass Management* **7**th ed. New Jersey: Prentice – Hall Press.

[2] Keminski J E, Dernoeden PH., Bigelew C.A. 2004. Soil amendments and fertilizer source effects on creeping bentgrass establishment, soil microbial activity, thatch, and disease(Article). HortScience **39**: 620-626

[3] Vaughn SF, D Dinelli, B T, N Joshee, MM Vaughan and SC Peterson 2015.Creeping bentgrass growth in sand-based root zones with or without biochar. Scientia Horticultura **197:**592-596.

[4] Foth HD 1994. *Dasar-dasar Ilmu Tanah*. Purbayanti ED, Lukitawati DR, Trimulatsih R, penerjemah; Hudoyo SAB (Editor). Yogyakarta (ID): Gadjah Mada University Press. Terjemahan dari: *Fundamentals of Soil Science*. Ed

[5] Christians N 2004. *Fundamental of Turfgrass Management*. New Jersey (US): John Wiley & Sons Inc.

[6] Beard JB. 1982. *Turfgrass Management for Golf Courses*. Minneapolis: (US). Burgress Publ. Co.

[7] Emmons R. 2000. *Turfgrass Science and Management* 3rd ed. New York (US) : Delmar Thompson Learning Inc.

IOP Conf. Series: Earth and Environmental Science **179** (2018) 012032 doi:10.1088/1755-1315/179/1/012032

Community participation in upgrading slum area: study of the implementation of "Kampung Deret" in Petogogan, Jakarta

D E P Lestari[*] and T Kurniawan[2]

[1]Undergraduate Alumni of Public Administration Study Program, Universitas Indonesia, Depok 16424, Indonesia
[2]Department of Public Administration, Universitas Indonesia, Depok 16424, Indonesia

[*]Email: dianpriharsetyo@gmail.com

Abstract. The implementation of development that places the community as the main actor is an approach that is in line with the concept of urban governance in which the community must participate in planning and managing urban affairs. One of the problems faced by many big cities in the world including Jakarta is slum areas. This paper attempts to discuss how community participation in upgrading slum areas that are implemented by the Provincial Government of Jakarta through a "KampungDeret" program, especially in Petogogan. By using literature studies and in-depth interviews, this paper tries to explain how community participation based on five key principles of urban governance, namely participation, accountability, transparency, responsiveness, and efficiency. Based on data and information collected can be seen that the community was involved in the implementation of the program from planning to implementation. However, the community is not given periodic information regarding the use of funds, thus raising accountability issues. In addition to problems in terms of efficiency of timeliness in the implementation of the program.

Keywords: Pentogogan, slum area, urban governance

1. Introduction

Jakarta is one of the cities in Java Island that has the largest population and density in Indonesia. According to the National Population and Family Planning Board report, Jakarta as the nation's capital has the highest urbanization rate of 100 percent in 2000 and 2010, while the population density is 12,592 inhabitants per square km in 2000 and increased to 14,469 inhabitants per square km in 2010 [1]. The concentration of urban population in Indonesia is in metropolitan Jakarta as much as 20% [2]. Population density in Jakarta Province needs attention because it can cause problems in various fields. One of the problems that arise and need to be alerted by the Provincial Government of Jakarta is the emergence of slum areas in some areas of Jakarta. The increasing number of slums is still an obstacle that is still difficult to handle for the Provincial Government of Jakarta in managing the city area.

Slum dwelling locations in Jakarta continue to grow to 6,000 hectares with an annual growth rate of about 1 percent. The number of slums in Jakarta not only creates problems for the disorder of the city's order but also can have an impact on the environment. One of the efforts that can be done by the city government to reduce the urban problem is through urban management. According to Leman [3], urban management includes environmental management, transportation management, land management, the role of the private sector in urban development, financial management, and housing development management. In the procurement of urban housing especially for low-income people, there are two types of government roles, namely as providers and enablers. As a provider or service provider, the government acts as a housing producer of the city. Whereas as an enabler, the government acts as a party providing assistance and encouragement for low-income communities to participate in the procurement of housing [4]. Government as enablers is appropriate with the current development paradigm, i.e., shifting from the dominant role of government to development that encourages community participation [5]. Implementation of shifting development by placing society as the main actors of development also influence development in urban scope. In the implementation of good governance, there is the concept of urban governance, i.e., various ways from individuals and institutions, public and private, in planning and managing the general affairs of the city [6]. Furthermore, urban governance later developed into good urban governance by the United Nations Center for Human Settlements. The program aims to explore the attributes of government systems that will help manage cities better, and better in the face of rapid urban developments. According to the United Nations Center for Human Settlements, urban governance has principles such as good governance, namely participation, accountability, transparency, responsiveness, and efficiency. According to Dwiyanto [7], the principles of good governance in the implementation of public services namely transparency, participation, nonpartisan, accountability, responsiveness and efficiency. Nevertheless, in this study, the researcher uses five principles, namely participation, accountability, transparency, responsiveness, and efficiency according to the United Nations Center for Human Settlements and see the indicators of these principles based on the proposed Dwiyanto. With good urban governance mechanism expected all development stakeholders can be involved from the beginning of planning, implementation, utilization of the maintenance of development results.

In the arrangement of slums in urban areas requires the participation of the community. Community participation in urban slum settlement arrangement existed in KampungDeret Program. One of the goals of the KampungDeret Program is to increase community participation and responsibility to manage slum settlements. The program of KampungDeret is the workplan of Governor JokoWidodo which derived from the vision of Governor and Vice Governor mission period 2012-2017 based on Jakarta Governor Regulation No. 64 of 2013 on Home Improvement Assistance in Slums through Sub-District Planning. KampungDeret Program in Petogogan is the earliest established and a pilot in the implementation of the program. The implementation of the KampungDeret Program in Petogogan took the local community as a development committee of KampungDeret. Based on this, it can be seen that in organizing slum settlements in Petogogan there is community participation in the process as in the perspective of urban governance. However, further research is still needed on the KampungDeret Program in Petogogan according to the perspective of urban governance. Therefore, the researcher is interested in studying the implementation of Kampung Derek Program according to the perspective of urban governance.

2. Methods of research

This research used post-positivist paradigm to see the natural facts that happened in KampungDeret Petogogan and analyze the implementation of KampungDeret Program using the perspective of urban governance theory. The instrument of data collection used is by using in-depth interview and literature study. Data and information obtained from in-depth interview and literature study were then analyzed qualitatively. The research instrument in this research is to compile the research question after establishing the main research variables.To ensure the reliability and validity of the findings, then the triangulation process is used in this research.

This research used questionnaire tool and interviews were done directly (direct interview). In-depth interviews were conducted on some parties including the Housing and Building Agency of the Provincial Government of Jakarta which handled the KampungDeret Program, sub-district officials, community leaders such as Petogogan Urban Village Secretary, former Chairman of RW 05, Head of RT (neighbourhood) and Chairman of Pokmas RW 05 Petogogan Urban Village, non-governmental organizations (Pokmas Petogogan), and academics who understand the topic of this research and concerned with urban management arrangement, Yayat Supriyatna, and RW 05 Petogogan Urban Village.

3. Results and discussion

In looking at the implementation of the KampungDeret Program in Petogogan according to the perspective of urban governance, this study uses five principles, namely participation, accountability, participation, responsiveness, and efficiency. The first principle is the principle of participation seen based on eight indicators. The first indicator is the role of key groups in planning, decision making, implementation, monitoring, and evaluation. The key group in the implementation of the KampungDeret in Petogogan is the community group. The community group is responsible for the implementation of the KampungDeret until making the report on the implementation of the program and the accountability of the development implementation. The second indicator is media freedom and the presence of local media. In the implementation of the KampungDeret Program in Petogogan, there have been many prints and local electronic media that publish the program. The Provincial Government of Jakarta does not specifically publish but cooperates with local media "Berita Jakarta" in publicizing the implementation of the KampungDeret Program. The third indicator is the percentage of people involved in voting (decision-making). In determining the decision to implement the KampungDeret Program in Petogogan, the whole community was invited to the sub-district office twice for socializing the program and bringing together the aspirations of the KampungDeret Program. Based on this, it is known that the community is involved in determining the decision to participate in KampungDeret Program and related to its implementation. The fourth indicator is the process of public discussion on public issues. The process of public discussion begins with socialization. The Provincial Government of Jakarta previously conducted a socialization at the municipal level to introduce the KampungDeret Program which was then forwarded to the districts up to the sub-district. The socialization can be seen as an effort to equalize the vision and ideas at the provincial level to the sub-district and community. The realization of the KampungDeret Program will be undertaken if the community agrees to participate in the program. To unite the aspirations of the Petogogan community related to the approval of the program and its implementation, the community is invited to the sub-district office. The community is given a choice to follow the program or not, and the choice if not willing to follow the program.

The fifth indicator is the availability of legal umbrella for community participation. Governor Regulation No. 64 of 2013 provides a strong legal umbrella for the Petogogan community to be involved in the implementation of the KampungDeret Program in their neighborhood. In the governor's regulation, the participation of the community becomes one of the goals of the home improvement assistance program through the KampungDeret Program, namely in article 3 point b. The sixth indicator is community involvement in policy formulation. The policy used as the basis for the implementation of KampungDeret Program is the Governor Regulation No. 64 of 2013 on Home Improvement Assistance in Slums through Sub-District Planning. The policy is top-down so that the public is not involved in formulating policies. The seventh indicator is based on the availability of media/forums for community participation. In the implementation of KampungDeret Program, there are no media/forums that accommodate the involvement of the community, but if people want a forum, then the community can coordinate themselves. In Petogogan there is a forum that embraces community involvement,i.e., community groups. The last indicator is the involvement of the community in monitoring the implementation of local government. In the implementation of KampungDeret Program in Petogogan, the community participates in supervising the development process and always asking about the progress of the development of the house series.

The second principle is the principle of accountability. The first indicator is the reference service used by the bureaucratic apparatus in the process of program implementation. The duties and authority of Housing and Building Agency in the implementation of KampungDeret are regulated in Governor Regulation No. 64 The year 2013 Article 25 paragraph 2, which is to conduct monitoring together with Regional Financial Management Agency and a supervisory unit. Also, the implementation of

KampungDeret Program also uses the existing standard operating procedure in each institution concerned without any special operating procedure standard to implement the program. In addition to the Housing and Building Agency, the implementation of the KampungDeret Program in Petogogan also involves the South Jakarta Mayor and Head of Petogogan Sub-District. Mayor of South Jakarta performs its duties and authority as evaluation team including development committee referring to Governor Regulation No. 64 The year 2013 Article 22 paragraph 2. The second indicator is action seen by bureaucratic apparatus if there is a community that does not meet the specified requirements. The Housing and Building Agency does not impose sanctions on the community that does not meet the requirements. The action taken by the Housing and Building Agency against the unqualified public is by not providing financial assistance to the next term to the community by the provisions of Governor Regulation No. 64 of 2013. On the other hand, for the requirements of land tenure or land ownership, the Housing and Building Agency continues development if there are people who do not meet the requirements because the arrangement is made based on the concept of the region. Communities in Petogogan mostly have no evidence of land ownership. The local community has traditionally occupied the land which is owned by the government.

The third indicator that the interests of the community get priority from the bureaucratic apparatus. This indicator can be seen from how the government accommodates the interests of the community. The settlement environment in Petogogan until September 2013 is shabby and unfit for habitation. Based on these conditions, then the arrangement of sub-districts through KampungDeret Program according to the direction of Governor Joko Widodo. Considering the environmental conditions, the KampungDeret Program in Petogogan Sub-District is done using the concept of rejuvenation. Seeing the plan of rejuvenation, it can be a solution in improving the Petogogan environment that was initially shabby and unfit to be habitable. Rejuvenation shows the efforts of the Provincial Government of Jakarta in accommodating the needs of the Petogogan community for the needs of residential neighborhoods that are habitable. The fourth indicator is the accountability media carried out periodically. The accountability of financial issues and the progress of the physical implementation of the development of KampungDeret is entirely handed over to consultants who are reported periodically to the Housing and Building Agency so that the public claims to be unaware of the question of accountability. Based on this, it can be seen that there are accountability reports carried out periodically by the consultant, but no media informs the accountability report to the public.

On the principle of transparency, the first indicator to be seen is access to information on the amount of budget allocated for a slum settlement activity. Information on the budget allocation that has been spent for the implementation of the KampungDeret Program is not published in detail by the Housing and Building Agency as well as the allocation for the implementation of the KampungDeret Program in Petogogan. This is because there is no interest to publish. However, the public can access information on the amount of budget allocation by asking directly to the Housing and Building Agency. The second indicator is the openness in procedures and requirements of program participation. In the Governor Regulation No. 64 of 2013 mentioned criteria and requirements of objects that can get home improvement assistance. In determining the location of the implementation of KampungDeret, the Housing and Building Agency disseminates information to the mayor to propose the location to be built. The Housing and Building Agency also provides guidance on the location criteria that may be proposed by the mayor which can then be disseminated to the community to know the information if they want to follow the program. The third indicator is the openness in the cost and time required to follow the program. In the implementation of KampungDeret in Petogogan, through the Chair of the Community Group, the Petogogan community was informed that the cost of the development of KampungDeret Petogogan came from the assistance of the Provincial Government of Jakarta sourced from the Regional Revenue and Expenditure Budget. Also, the timing of the development of KampungDeret in Petogoganis also informed to the community, which is planned for 3-4 months so that the community is asked to rent other settlement temporarily during the development process. Users and other stakeholders can understand the fourth indicator of procedure and regulatory requirements. Based on the interviews conducted with one of the residents, it was found that the people were quite familiar with the procedures and regulations related to the implementation of the KampungDeret Program. Residents explained about the KampungDeret Program, criteria, and indicators that can be said to be slim and can follow the program. The familiarity of the community to the procedure shows that the Housing and Building Agency openly provides clear information to stakeholders.

The fifth indicator is the ease of obtaining information about program procedures and requirements. Procedures and requirements regarding the KampungDeret Program are described in detail in Governor Regulation No. 64 of 2013. The public can access information about the KampungDeret Program through the Provincial Government of Jakarta website,i.e., jakarta.go.id. Also, the Provincial Government of Jakarta also conducted a socialization to introduce the program and the presence of agents in the community, namely the public actor as a liaison agent between the government and the community. The sixth indicator seen is the ease of obtaining information about the financing and time required for the implementation of the program. Information on the financing of the KampungDeret Program or home improvement assistance is described in Governor Regulation No. 64 of 2013 which can be accessed through the Provincial Government of Jakarta portal. In the Governor Regulation, it is explained that the source of funding for the KampungDeret Program comes from the Regional Revenue and Expenditure Budget which is an indirect spending group and is included in the type of social assistance expenditure. Disbursement of home improvement assistance is done through the beneficiary account. The time of fund disbursement and implementation is done as many as three stages. The seventh indicator is the openness of the work unit and officials authorized to provide an understanding of the program. Regarding providing an understanding of the KampungDeret Program, the Housing and Building Agency is assisted by its staffs to socialize first with the community. The Housing and Building Agency together with the mayor, head of sub-district, and community actors provided information on what and how the KampungDeret Program will be conducted in Petogogan. Meanwhile, according to the urban planning expert who became the informant in this research, in the implementation of KampungDeret Program, there is a lack of transparency about the people who are entitled to get assistance from the land ownership status is not clear. Second, the transparency in the use of the budget. Also, transparency can also be seen in the absence of evaluation or monitoring of programs that have been implemented.

The fourth principle is the principle of responsiveness. The first indicator to be seen is the presence or absence of complaints from the community regarding the program. The Petogogan community complained about the narrowness of built houses, the distance between adjacent houses, the absence of adequate drying places, the substantial expense of completing unfinished house construction, poor quality of buildings and environmental conditions, poor water quality, and land tenure issues that remain unclear. The second indicator is the attitude of the bureaucratic apparatus in responding to complaints from the public. In response to public complaints, the government is not responsive. In the effort to obtain the status of land owned, it is seen that inter-government agencies throw each other responsibility. On the other hand, related to clean water problem also no improvement made by the Provincial Government of Jakarta. The third indicator is the use of complaints from service users as a reference to improve the implementation of the program. The number of complaints from the community regarding the implementation of the program to be one of the reasons for the evaluation of the policies underlying the KampungDeret Program as an effort to improve the implementation of the program. The fourth indicator is the various actions of the bureaucratic apparatus to provide service satisfaction in the future. The action of the Housing and Building Agency to provide the satisfaction of service in the future is to conduct policy evaluation. On the other hand, in the problems Petogogan society complained about the results of the program implementation, the Housing and Building Agency did not respond as an attempt to address the problem. Unresponsive of the agency shows that there is no action of the Housing and Building Agency in giving the satisfaction of service to Petogogan community who follow the program.

On the principle of efficiency, the first indicator is about the certainty of implementation costs to be provided to the community and which must be issued by the community. In the implementation of KampungDeret Program, it is explained in detail the calculation of monetary aid that will be obtained by the community that is by the area of the beneficiary's house. The Petogogan community is not charged for the implementation of the KampungDeret Program. The second indicator is the certainty of program execution time. In the Governor Regulation No. 64 of 2013, there is no time for implementation of the program. On the other hand, the Housing and Building Agency determines the execution time determined for 3-4 months. The implementation of the KampungDeret Program in Petogogan was delayed by up to six months due to technical obstacles in the field. The next indicator is that it does not require much effort in following the program. Petogogan community does not need much energy in following the program. The community only participates in the program socialization, while the consultant does the development. The fourth indicator is the cost of implementation that lightens the community. The cost of implementing the KampungDeret Program using the grants derived from the Regional Revenue and Expenditure Budget. The Petogogan community gets funding to repair their own houses without having to pay their expenses so that it can alleviate the community. The fifth indicator is the execution time that lightens the community. The implementation time of the KampungDeret Program in Petogogan which retreats up to 6-8 months is quite burdensome to the citizens because the residents have to rent another house longer with the cost of rent is quite expensive. The sixth indicator is the provision of aid or program by the bureaucracy without any coercive action to the public to spend extra cost of program implementation. The KampungDeret program is a program of assisting in home improvement from the Regional Revenue and Expenditure Budget. The community is not required to pay for the expenses, so it is known that there is no coercion for the community to incur additional costs.

4. Conclusions

In the implementation of KampungDeret Program in Petogogan, there has been community participation. The participation of the community one of them is shown from the participation of community leaders as a development committee of Kampung Deret Petogogan. On the other hand, seen from the implementation of the principle of accountability, this study shows that the community does not receive periodic reporting related to the use of funds received in the implementation of the program. Also, the implementation of the Kampung Deret Program in Petogogan is inefficient when viewed from the timeliness. Inefficiency happens because of constrained problems of providing difficult construction materials provided quickly. Despite the establishment of a habitable environment, there are unresolved issues and anxiety that is the status of land ownership.

Based on research conducted, the researcher suggested before the development program of Kampung Deret should be done carefully so that the results obtained can be maximized. In the planning should be done land consolidation in advance so that the issue of land ownership and thefuture building is not a problem. The second thing to note, namely the need for clear publication of the program budget allocation so that people know the allocation of aid received. In addition, researchers also see the need for collaboration with various other agencies in the Provincial Government of Jakarta in socializing and implementing the program. The researcher also suggested to the Provincial Government of Jakarta to pay attention to the legality aspect in undertaking the rejuvenation of slums, especially in the case of land and building ownership.

Acknowledgements

This paper is a concise version of the results of research conducted in the completion of the undergraduate thesis on the Public Administration Study Program, University of Indonesia.

References

[1] Badan Kependudukan dan Keluarga Berencana Nasional 2013 *Profil Kependudukan dan Pembangunan di Indonesia* (Jakarta: Badan Kependudukan dan Keluarga Berencana Nasional)

[2] Firman T 2004 Demographic and Spatial Patterns of Indonesia's Recent Urbanization *Popul Space Place* **10**421-434

[3] Leman E 1994*Urban Management: A Primer* (Bangkok: Urban Management Programme Regional Office for Asia-Pacific, United Nations Human Settlements Programme)

[4] Panudju B 1999 *Pengadaan Perumahan Kota dengan Peran Serta Masyarakat Berpenghasilan Rendah* (Bandung: Penerbit Alumni)
[5] Rustiadi E, Saefulhakim S and Panuju D R 2009 *Perencanaan dan Pengembangan Wilayah* (Jakarta: Yayasan Pustaka Obor Indonesia)
[6] United Nations Human Settlements Programme 2002 *The Global Campaign on Urban Governance* (Nairobi: United Nations Human Settlements Programme)
[7] Dwiyanto A 2005 *Mewujudkan Good Governance Melalui Pelayanan Publik* (Yogyakarta: Gadjah Mada University Press)

Combating climate change through community participation in preserving the environment: a case from Hutan Organik (Organic Forest) in Megamendung, Bogor Regency, Indonesia

T Kurniawan

Department of Public Administration, Universitas Indonesia, Depok 16424, Indonesia

E-mail: teguh.kurniawan@ui.ac.id

Abstract: According to Hansen and Biringer, environmental threats like climate change require that we extend conservation planning beyond the boundaries of protected areas, and into a future in which ecosystems and biomes may be quite different than they are today. One of the important ecosystems to be conserved in efforts to overcome climate change is forests. Forests can be managed to mitigate climate change because forests fulfill many important environmental functions and services. According to CIFOR, forests are important in determining the accumulation of greenhouse gases in the atmosphere. Although it has a very important role, forests have big problems as a result of deforestation, including in Indonesia. This paper tries to discuss how community participation has an important role in improving environmental conditions, especially in maintaining forest sustainability. The case discussed is "Hutan Organik" or organic forest that is an effort from the community in making forests in a critical area in Megamendung, Bogor. Hutan Organik is an artificial forest made and managed personally by Bambang Istiawan and family organically and currently has an area of approximately 27 hectares. Through this paper is expected to provide an overview of the importance of conservation efforts by the community as one of the strategies that must be optimized by the government in building environmental resilience in combating climate change.

Keywords: climate change, Hutan Organik, Megamendung

1. Introduction

Climate change occurs as a result of increased greenhouse gasses in the atmosphere such as carbon dioxide, methane, and several other artificial chemicals. Regarding carbon dioxide, according to the United States Environmental Protection Agency, carbon dioxide accounts for 82.2% of the greenhouse gas emissions in the United States by 2015 [1]. The composition of carbon dioxide according to Pittock increases as a result of human activity primarily derived from the burning of fossil fuels such as coal, oil and natural gas, forest destruction and carbon-rich soil and the manufacture of cement from limestone [2].

Associated with forest destruction as one of the causes of increasing carbon dioxide content in the atmosphere, according to Hansen and Biringer, environmental threats such as climate change require that we extend conservation planning beyond ecosystems and biomes may be quite different than they are today [3]. One of the important ecosystems to be conserved in efforts to overcome climate change is forests. Forests can be managed to mitigate climate change because forests fulfill many important environmental functions and services. According to CIFOR, forests are important in determining the accumulation of greenhouse gases in the atmosphere [4].

Although it has a very important role, forests have big problems as a result of deforestation, including in Indonesia. The average deforestation is occurring in Indonesia from 2000 to 2012 ranges from 0.8 million hectares/year [5]. This figure is of course very worrying to see the role and function of forests that are very important in storing carbon. Forest ecosystems around the world include its soil storing approximately 1,200 giga tons of carbon in which this amount is estimated to exceed the amount of carbon currently in the atmosphere [6].

Increased carbon dioxide composition that causes increased greenhouse gases is a result of human activity. Therefore, efforts to reduce the carbon dioxide composition associated with forest conditions will not be released from efforts related to human activities as well, one of them through reforestation. In other words, it is necessary to encourage community participation in preserving nature, especially forests. Related to community participation in reforestation, then in Megamendung Bogor, there is a very extraordinary effort made Bambang Istiawan and his wife Rosita Istiawan in converting the critical land into an organic forest. The forest that was built in August 2001 by Istiawan and his wife is then known as "Hutan Organik" or organic forest [7].

What is done by Istiawan and his wife is a best practice undertaken by individuals in making forests from a critical land. This effort, of course, needs to get attention and can be an inspiration for other people who want to contribute real in coping with climate change. Therefore, this paper tries to discuss how environmental participation is important in improving environmental sustainability so that it can contribute to tackling climate change. Through this paper is expected to provide an overview of the importance of conservation efforts by the community as one of the strategies that must be optimized by the government in building environmental resilience in combating climate change by learning from the making of artificial forests organically as done by Istiawan and his wife.

2. Methods of research

To achieve the intended objectives, this paper uses a qualitative approach through literature studies on some publications, reports, and news on Hutan Organik. Also, data and information are also obtained from the in-depth interviews of Bambang Istiawan and his wife Rosita Istiawan and by visiting and seeing the conditions in the Hutan Organik area.

The data and information obtained are analyzed qualitatively and then used to describe how community participation in maintaining the environment especially in reforestation can be done as well as the challenges and obstacles it faces. The analysis will also be linked to efforts that can be made by central and local governments in encouraging the emergence of community initiatives to voluntarily conserve the environment through tree planting activities to become forests.

3. Results and discussion

Results will be discussed in two parts. The first section will describe "Hutan Organik" and its current condition as a best practice of individual contribution in creating a forest that can make a significant contribution to climate change. While the second part will analyze the efforts that can be done by both central and local governments to optimize community participation in protecting the environment so that it can contribute in efforts to tackle climate change, especially by endorsing the making of forests by private individuals.

3.1. Hutan Organik: a best practice of nature conservation by private individual

Hutan Organik is the official name of a human made forest which is a voluntary initiative from Bambang Istiawan and his wife Rosita Istiawan located in Megamendung Sub-district, Bogor Regency. Currently, the area of Hutan Organik is estimated to reach 27 hectares consisting of 2 parts of 12 hectares in the area that goes into the area of Megamendung Village and 15 hectares in the area of Bojong Koneng Village.

IOP Conf. Series: Earth and Environmental Science **179** (2018) 012033 doi:10.1088/1755-1315/179/1/012033

The making of Hutan Organik begins with the Istiawan's dream to spend his old age on the edge of the forest, where his dreamland no longer exists forests, which eventually builds his forest in Megamendung. The activity of the "Hutan Organik" started from August 2001 until today in making "Ecosystem and Critical Rehabilitation Model" and implemented in the "Voluntary Initiative" pattern independently. In general, the beginning of this forest is built on instinct and spirit of desire to find a way to live on the side of the forest as soon as possible, without following the formal rules and aspects of science related to the field of forestry.

At the commencement of the initiative, the area for which the forest is made is an area with damaged environmental conditions and can be categorized as critical land. This region has an extreme contour from 15 degrees to 80 degrees, so it is shaped like a bowl. In 2002/2003, there was one dead spring in this area, 70% of the area was reeds with very little vegetation, soil acidity condition had pH 2 ½ - 4, and no worms were found on the soil in that location. A description of the condition of the area at the commencement of the initiative can be seen in Figure 1 below.

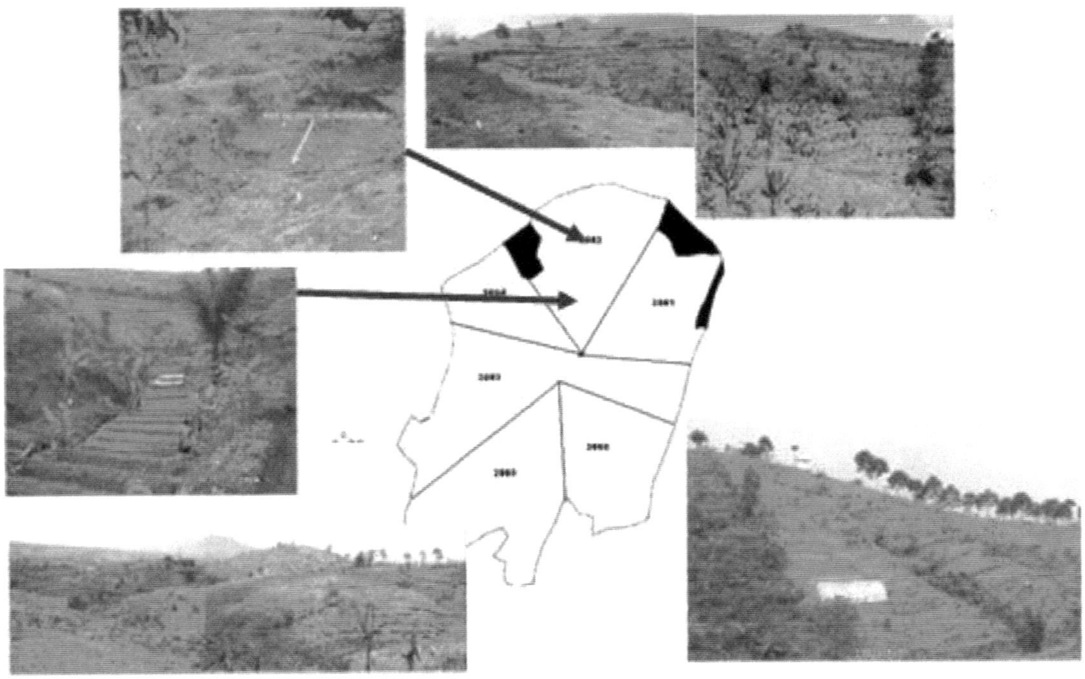

Figure 1. Initial condition at the location of Hutan Organik (source: presentation of Istiawan)

In developing its artificial forest, the method employed by Istiawan is a combination of frontier trees with endemic trees where the composition is more frontier trees than endemic trees. The chosen frontier tree is African wood (*Maesopsis eminii*) while the endemic tree is Rasamala (*Altingia excelsa Noronha*) or Puspa (*Schima wallichii*). African wood is chosen as a frontier with consideration of the growing speed as well as the absorption capacity of the leaves as well as the strength of its roots in the soil. Also, African wood is also selected because the seeds are easy to obtain, the nursery is very easy, the speed grows very well, and the leaves can be used for goat feed. In tree planting, fertilizer method used is to use organic fertilizer and then allowed to grow naturally. The use of this method which then makes this artificial forest initiative is named as Hutan Organik or organic forest. In the period of tree growth, agroforestry activities also take place in the Hutan Organik.

The initiative undertaken by Istiawan in their journey is a very significant one in the effort to tackle climate change. The forests created by them have contributed to absorbing carbon dioxide as well as improving critical land conditions into an ecosystem environment that has an impact on the lives of living things. Regarding carbon dioxide absorption, a study conducted by a student from the Bogor Agricultural University in 2009 for example, provides information on the contribution of Hutan Organik where within 4.2 hectares of land can absorb an average carbon dioxide of 84.9 tons [8].

In addition to contributing to efforts to tackle climate change, the presence of Hutan Organik is also able to contribute to the improvement of environmental conditions. During the first four years (2001 - 2005) the speed of the reforestation process looks quite good, but there has been no significant change in soil quality. Nevertheless, the ecosystem improvements that occur are significant, among others, in the functioning of re-spring of dead springs and the emergence of new springs, whereas this area has no vein/water flow in the soil. Therefore, the return of springs is the impact of vegetation. A picture of significant vegetation development of Hutan Organik can be seen in Figure 2 below.

Photo satellite area of Hutan Organik in 2003, taken from the files of Faculty of Forestry, Bogor Agriculture Institute

Photo satellite area of Hutan Organik in 2007, taken from google earth 2007

Photo satellite of Hutan Organik in 2013 and 2017

Figure 2. Vegetation development of HutanOrganik (source: presentation of Istiawan)

The existence of Hutan Organik has also led to the emergence of various types of fauna found in the location of various insects, reptiles, birds, rabbits, mouse deer and others. Also, according to Mr. Istiawan, Hutan Organik is also visited by wild animals that prey on goats that he raised as many as two times of arrival, last on October 3, 2014. It is estimated that goats are eaten by "Meong Congkok" / Felisbengalensis type of cats may even be a root tiger.

The participation of Istiawan in preserving the environment is certainly an example and imitated by other components of society so that it can participate significantly in overcoming climate change. Related to this, then based on the observation and analysis of the author, some things became the key to the success of Istiawan in making and developing Hutan Organik, as follows.

First, because Istiawan has a passion so willing to voluntarily use their own money to acquire land and then plant trees on the land that has been mastered. Acquiring the land and then planting it with trees is not a low-cost thing to do. Therefore financial ability is of course very decisive. Because of the passion they have, they have a high commitment and consistency to carry out the agenda that they have designed to create forests and save critical lands.

Second, the existence of good networking with various circles. Tenure of land in the Hutan Organik location at the commencement of the initiative is also not an easy thing to do. Given the large number

IOP Conf. Series: Earth and Environmental Science **179** (2018) 012033 doi:10.1088/1755-1315/179/1/012033

of land brokers who offer land to other parties. However, by building good networking with various circles, Istiawan able to overcome various obstacles faced both related to the control of the land and various other obstacles.

Third, in establishing Hutan Organik, Istiawan also use the advancement of science and technology and by building an etwork with some academics from various universities. These academics then helped them to analyze some of the physical problems they faced Istiawan also have a willingness always to learn including about crops, forestry, agroforestry and other knowledge that is very supportive in the development of Hutan Organik.

Fourthly, Istiawan also learn from nature and then choose the method of planting using frontier trees, a combination of planting, organic methods and planting endemic trees. They also have the willingness to share experiences so that then start many parties who visit and learn the activities that exist in Hutan Organik. By sharing this experience, they are also able to add networking and support from various parties who visit the Hutan Organik.

3.2. Optimizing community participation in preserving the environment to combat climate change by endorsing private individual contribution in forest making

Based on the experience of Mr. and Mrs. Istiawan which is the best practice of community participation in preserving the environment, there are a number of things that can be done by the Government both in the Central and Regional so that it can grow the community participation as done by Mr. and Mrs. Istiawan in making Hutan Organik.Related to that, the author sees the Government can do some thingsin the following description.

First, the government must build a common awareness in the community that carbon dioxide is a common enemy and planting trees is the main weapon to be used against the enemy. Community groups that should be the main targets are middle-class groups with knowledge, capital, and passion. This middle-class group is usually a group that has sufficient financial ability, so that if awakened awareness will be easier to move and uses their funds. Other groups of people should also be concerned, so that awareness of climate change becomes the main consciousness of the community.

Second, to plant trees requires land, so the government should grant access to land and land use for environmental conservation purposes and not just for the sake of the economy. The government should map the landowners who let their land not be used to force it than to plant trees. The approach is a land function instead of land ownership. Anyone who owns the land is obliged to plant trees as long as the land has not been used for other designations.

Third, the government also needs to provide support and guarantees through various programs including through community forestry programs that are currently being promoted. Another program that can be done is a critical land improvement program.

Fourth, campaigning through social media to the success of the program is also very necessary to do as well as to build community integrity that the main purpose is to save the environment and not for economic interests. Even if there are economic benefits such as from agroforestry and livestock, then that is not the main factor.

4. Conclusions

In conclusion, it can be concluded that community participation in preserving the environment can have a significant impact on climate change. The case of Hutan Organik can provide confidence to us that efforts to encourage community participation to preserve the environment among others through the creation of artificial forests by individuals are an important strategy that must be done by the government. The existence of Hutan Organik has provided important information on the contribution of artificial forests in tackling climate change.

To encourage massive participation of the community in the effort to address climate change, some advocacy from the Government is required. One of the advocates that can be done is to campaign carbon dioxide as a common enemy and the effort to plant trees as the main weapon to overcome the enemy. The government should also provide support through its various policy instruments such as regulating land tenure and use as well as incentives and disincentives for the community. Campaigns on social media also need to be strengthened to reach many potential communities to participate.

Acknowledgements
The author would like to thank Bambang Istiawan and his wife Rosita Istiawan who is willing to be key informants in writing this paper and sharetheir experience in making "Hutan Organik." You are both a great source of inspiration.

References

[1] US EPA 2017 *Inventory of U.S. Greenhouse Gas Emissions and Sinks 1990-2015* (Washington DC: The United States Environmental Protection Agency)
[2] Pittock A B 2005 *Climate Change: Turning Up The Heat* (Collingwood: CSIRO Publishing)
[3] Hansen L and Birringer J 2003 *Buying Time: A User's Manual for Building Resistance and Resilience to Climate Change in Natural Systems* ed L J Hansen, J L Biringer and J R Hoffman (Washington DC: World Wildlife Fund)
[4] CIFOR *Forests and Climate Change* http://www.cifor.org/forests-and-climate-change
[5] Margono BA, Potavop P V, Turubanova S, Stolle F and Hansen M C 2014 *Nat. Clim. Change* **4** 730-735
[6] Freer-Smith P H, Broadmeadow M S J and Lynch J M 2007 *Forestry and Climate Change* (Oxfordshire: CAB International)
[7] http://www.hutanorganik.com/en/
[8] Rused E S 2009 *Nilai Ekonomi Kegiatan Rehabilitasi dalam Menghasilkan Air dan Menyerap Karbon di Blok S Cipendawa Megamendung, Bogor* (Bogor: Department of Forestry Management, Faculty of Forestry, Bogor Agricultural Institute)

Jumping to the top: catalysts for leapfrogging to a water sensitive city

C Brodnik[1*], J Holden[2], R Marino[3], A Wright[4], V Copa[1], B Rogers[1], H S Arifin[5], R Brown[2], K Djaja[11], M Farrelly[1], R L Kaswanto[5], D Marsudiantoro[6], D Marthanty[6], L Maryonoputri[12], E Payne[4], M Purwanto[7], D R Lovering[3], Y Suharnoto[7], J Sumabrata[6], R Suwarso[8], Y Syaukat[9], C Urich[4], and D Yuliantoro[10]

[1] School of Social Science, Monash University, Australia.
[2] Monash Sustainable Development Institute, Monash University, Australia.
[3] Monash Art Design and Architecture, Monash University, Australia.
[4] Department of Civil Engineering, Monash University, Australia.
[5] Department of Landscape Architecture, Bogor Agricultural University, Indonesia.
[6] Department of Civil Engineering, University of Indonesia, Indonesia.
[7] Department of Civil and Environmental Engineering, Bogor Agricultural University, Indonesia.
[8] Centre for Election and Political Party, University of Indonesia, Indonesia.
[9] Department of Resource and Environmental Economics, Bogor Agricultural University, Indonesia.
[10] Directorate of Research and Innovation, Bogor Agricultural University, Indonesia.
[11] Graduate School for Strategies and Global Studies, University of Indonesia, Indonesia.
[12] Directorate of Research and Community Engagement, University of Indonesia, Indonesia.

*E-mail: Christoph.Brodnik@monash.edu

Abstract: Climate change, population growth and rapid urbanisation have severe implications for cities and the way in which they interact with water. As a response to these challenges the water sensitive cities concept emerged, which supports cities to become more resilient to these challenges while making them more prosperous, sustainable and liveable. A water sensitive city harnesses the whole water cycle through integrated water management solutions, designs beautiful blue and green urban spaces and comprises healthy communities who are strongly connected with each other and with their local environment. Indonesian cities have an opportunity to 'leapfrog' towards a water sensitive city and to bypass the negative consequences that have resulted from urbanisation and growth that developed countries have gone through. Contributing to this growing field of research, this paper synthesizes key insights from the transformative change, sustainable urban water management and leapfrogging literature. The paper defines what leapfrogging to a water sensitive city means and describes three catalysts that facilitate this transition: trans-disciplinary science, cross sectoral collaboration and innovation experiments. The paper also introduces a joint Australian-Indonesian research program that develops water sensitive city leapfrogging strategies by translating these catalysts into practice.

Keywords: leapfrogging, sustainable urban water management, water sensitive city

1. Introduction

Cities are the arena in which the battle for sustainability and resilience will be won or lost [1]. This holds especially true in Asia where 65% of urban development in the 21st century occurred [2]. Asian cities in particular are faced with severe challenges such as climate change, environmental degradation and social inequalities. In many ways cities in Asia have exceeded the carrying capacity of the natural environment to provide the necessary life supporting resources for their inhabitants in a sustainable way [3]. Researchers, politicians, decision makers, citizens and others are therefore urgently looking towards new development frameworks that help increase the resilience of their cities to these threats.

The Water Sensitive City (WSC) concept [4,5] emerged as a response to these dangers to environmental health and social stability. The framework has been used by governments (e.g. Australia, Singapore and China) and international organizations (e.g. Asian Development Bank), as a vision and strategic guidance for investment and planning. The WSC concept describes an urban development paradigm where water is managed in an integrated manner to increase the productivity, resilience, sustainability and liveability of urban areas. Cities in developed countries have usually gone through different development stages on their way towards achieving greater water sensitivity, but these stages often resulted in negative social and environmental consequences [5]. Rather than having to go through these less than desirable stages, cities in developing countries can leapfrog stages to avoid making the same mistakes developed countries made in their past [5-7]. In general terms, the concept of leapfrogging suggests that undesirable conditions that increase levels of pollution and vulnerability can be bypassed to reach the more desirable sustainable and resilient conditions. Cities in developing countries are particularly well positioned to takes these leaps forward because less resources have been invested in traditional urban water management infrastructure and institutions which makes them more receptive to water sensitive practices [6,7].

To date, most leapfrogging studies focus on the adoption and use of particular technologies, such as communication technology [8] and describe how more advanced technologies (e.g. mobile phones) can lead to the bypassing of less advanced levels of technological development (e.g. landlines). While these studies highlight the importance of innovation for leapfrogging specific technology stages, what remains less clear is: how leapfrogging at an entire sector (e.g. water, energy) or city-scale unfolds, what enabling conditions initiate and sustain leapfrogging, and how it could be best steered into a desired direction [7,9].

This paper addresses this gap by putting forward three catalysts of leapfrogging that support cities to bypass traditional development stages associated with negative social and environmental consequences and to move towards a more productive, sustainable, resilient and liveable water sensitive city. The paper begins by describing the WSC concept and how it relates to urban resilience. This is followed by an outline of the leapfrogging literature with a focus on cities and their interactions with water. On this basis, the paper defines leapfrogging to a WSC. The paper then describes two important prerequisites for leapfrogging: contextualized strategies and system capacities. To this end, the paper introduces three leapfrogging catalysts by synthesizing key insights from the literature on transformative change and sustainable urban water management (SUWM). Lastly, a collaborative research project between Australia and Indonesia is introduced. Its objective is to develop water sensitive city leapfrogging strategies by translating these catalysts into practice. Thus, this paper builds upon and contributes to leapfrogging research in developing countries and provides practical insights regarding how leapfrogging towards water sensitive cities can be catalyzed.

2. Water Sensitive Cities

Many cities throughout the world have to adapt to rapid urbanization, population growth and climate change by increasing their resilience to these challenges. These cities are increasingly recognizing that traditional urban water management approaches are ill-equipped to deal with these challenges due to deep uncertainties, changing societal values and negative environmental consequences [5,10].

In looking for a suitable way to address these challenges the water sensitive city concept emerged as a unifying vision and strategic guidance towards an urban water management approach that meets a city's water needs, but also delivers on a range of associated liveability, resilience and sustainability benefits[1]. The water sensitive cities framework has been increasingly adopted by governments (e.g. Australia, Singapore and China) and international organizations (e.g. Asian Development Bank) to guide urban development to wards this vision. At its core, a WSC is based on managing the water cycle in a holistic and integrated way rather than focusing on the different elements of the water cycle separately. This ensures that basic human needs are met while protecting and enhancing receiving waterways, reducing flood risk and creating beautiful green and blue urban spaces for healthy and happy communities. As such, the approach delivers on a range of objectives critical for the prosperity, liveability and resilience of a city such as: capacity to buffer extreme events, public green and blue space, clean and healthy waterways, pleasant and cool urban areas, connected and healthy communities, cultural significance and many others [4,5,11].

A WSC is built on three principles of practice that weave through the social and technical fabric of an urban area [4]. The principles (or 'pillars') provide the essential foundation for infrastructure development, urban design and social capital seeking to optimize the use of water resources, increase resilience to extreme events and protect ecosystem services in the urban landscape [6]. According to Wong and Brown [4] the three principles are:

a) Cities as water supply catchments: This principle emphasizes the concept of not relying on a sole source of water but instead develop more resilient, diversified system that uses multiple sources at different scales while matching the source of the water to its intended use.
b) Cities providing ecosystem services: This principle emphasizes the idea that urban landscapes actively supplement and support the natural environment and the services it provides.
c) Cities comprising water sensitive communities: This principle emphasizes the importance of institutional and social capacity, support and participation in sustainable urban water management.

A cities path towards greater water sensitivity has traditionally followed a sequential way in which each 'state' is building on the development of the previous stage. This is captured in the Urban Transitions Framework of Brown et al. [5] (Figure 1). Based on a historic analysis of the technical and institutional arrangement in urban water management over time the framework identifies six distinct city developments stages that cities go through when they progress towards greater water sensitivity.

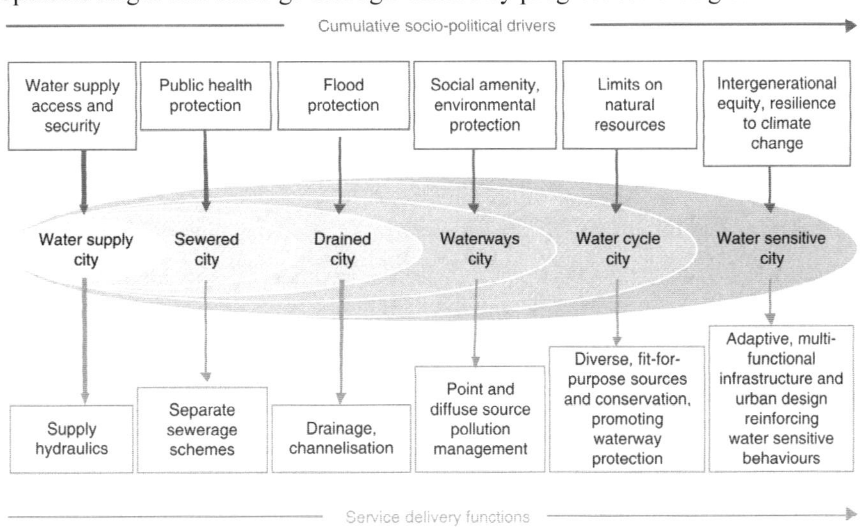

Figure 1: Urban water transition framework by Brown et al. [5]

[1]https://watersensitivecities.org.au/

The six states are mapped against two dimensions:

i. Cumulative Socio-Political Drivers: "the socio-political drivers (demands and expectations) that emerge from society's growing environmental awareness, amenity expectations and evolving attitudes toward water management" [6].
ii. Service Delivery Functions**:** "the increasingly diverse services required to address those drivers as cities transition to greater sustainability" [6].

The different stages of development. Each definition is taken directly from Brown et al. [6]:

a) Water Supply City
"The most basic state of modern water management, whereby a centralized system provides water to a growing urban population that expects cheap and equitable water for all. Large quantities of water are extracted from the environment using infrastructure such as pipes and dams. The public expects that water is cheap, harmless to the environment and limitlessly available."

b) Sewered City
"Building on the previous state, the Sewered City is drive by a desire for better public health and hygiene. Diseases caused by domestic and industrial waste effluent leads to the development of sewerage systems that divert effluent away from housing and into waterways outside of cities. As in the earlier state, it is assumed that the discarding of effluent does not harm the environment."

c) Drained City
"A need to protect homes and infrastructure from flooding is the driver behind the Drained City. The channeling of rivers enables the development of floodplains for housing and rapid urban growth. Like effluent, stormwater is directed away from urban areas and into waterways, generally thought of as dumping grounds for waste. The community expects water supply, sewerage and drainage services to be provided cheaply."

d) Waterways City
"The environmental impacts of both water extraction and waste processing are taken into account for the first time. As the social and aesthetic values of clean waterways are extolled, urban planning begins to integrate water as an important consideration. The unfettered extraction of freshwater is now being curbed, and receiving waterways are protected by filtering stormwater through bio-filtration systems such as rain gardens and artificial wetlands distributed throughout the city."

e) Water Cycle City
"In this state, water is actively conserved and supplies from diverse sources such as stormwater, greywater and recycled wastewater are used in a fit-for-purpose manner. Sustainability is now widely embraced, and the former hydro-social contract, in which government was expected to deliver risk-free water supply services, has been replaced with co-management arrangements between government, business and community."

f) Water Sensitive City
"Based on holistic and integrated water cycle management that meets the city's water needs while also delivering a range of associated liveability benefits. A Water Sensitive City manages water in a way that protects the health of receiving waters, mitigates flood risk and creates green public spaces that also harvest and recycle water. Infrastructure, technology and urban design will be flexible, recognizing the link between society and technology. The community is actively engaged with water, through recreational enjoyment of irrigated green spaces throughout the city, and have opportunities for more active involvement in the water system."

3. Leapfrogging

The concept of leapfrogging describes the potential to jump over or skip undesirable technologies or development stages such as those that are pollution intensive and to directly establish a more sustainable one [9,12]. Literature on leapfrogging has been particularly focused on how industries of developing countries can skip particular development stages that developed countries have gone through, consequently avoiding the social and environmental consequences that came with these pathways. This is reflected in the definition of Binz et al. [9] who define leapfrogging as "a situation in which a newly industrialised country learns from the mistakes of developed countries and directly implements more sustainable systems of production and consumption, based on innovative and ecologically more efficient technology". Other definitions however highlight that the concept is not only relevant for developing countries. For example, leapfrogging as described in Jefferies and Duffy [13] is conceptualised as "the idea that there are new paths to higher standards of living which bypass the mistakes that other communities made." Seen this way, the concept pertains to any community who wants to learn from another community in order to avoid making the same mistakes, consequently improving their lives. More directly related to urban water management Wong [14] states that leapfrogging is about "capturing and building on advancements and innovations in policies and technologies achieved in other places and avoiding the traditional evolutionary approach to infrastructure development and management."

For cities and their interactions with water, this means to skip over development stages which have been shown to create environmental, social and economic vulnerabilities by managing water in a fragmented, technocratic way and to advance to a stage that is characterized by greater productivity, resilience, sustainability and liveability [14] (See Figure 1). The difficulty to advance to a WSC in developed countries is that technological and institutional path dependencies lead to changes being incremental often resulting mere optimizations of unsustainable practices with limited potential for systematic change. Sunk costs and vested interests are very high through decades of investment which have aligned organizations, legislations and infrastructure with a particular set of practices and 'locking-in' the status quo [15]. The opportunity to leapfrog in developing countries comes from their relatively low levels of investment in traditional infrastructure and institutions which makes existing practices less entrenched. These conditions constitute a major opportunity for latecomer countries as technological and institutional lock-ins and path dependencies are not as strong, making them more receptive to adopt water sensitive practices [6,16]. This makes these cities well-placed to leapfrog directly to an urban water management system that enhances their productivity, resilience, sustainability and liveability rather than following the incremental evolution of urban water infrastructure and institution many cities in developed countries have gone through.

Based on the understanding outlined above we define leapfrogging for the purpose of this paper as systemic, vision led and irreversible change that bypasses traditional practices, infrastructure and institutions associated with less desirable city states to achieve a more productive, resilient, sustainable and livable city state.

3.1. Prerequisites for leapfrogging

The presence of contextualised strategies and absorptive capacity are considered to be imperative for leapfrogging to unfold [12,16]. To develop these prerequisites three particularly relevant leapfrogging catalysts are put forward in this paper: a) transdisciplinary science b) cross-sector partnerships and c) innovation experiments.

3.1.1. Contextualized strategies

The leapfrogging literature highlights a range of barriers that leapfrogging strategies face in developing countries. These barriers are predominantly found in the institutional, cultural or organisational domains within a sector [17, 18]. A core element of these barriers is the misalignment of problems and solutions between developed and developing countries. Hence, a barrier to these leapfrogging strategies is that of "contextual fit" where learned experiences and recommendations from elsewhere may prove dysfunction or even counter-productive [18]. Echoing this insight Sauter and Watson [12] highlight that there is no standard model of leapfrogging. The authors caution against a 'one size fits all' approach and emphasize that leapfrogging pathways are always unique and that case specific resources, strengths and weaknesses need to be considered when leapfrogging strategies are developed. In addition to the importance of spatial context, Shah et al. [18] recommend that leapfrogging strategies should consider the temporal context and in particular the necessary short (e.g. resource allocations), mid (e.g. legislative system) and long term (e.g. cultural norms) changes as well as interactions between these layers.

3.1.2 Absorptive capacity

In addition to contextualised strategies, successful leapfrogging also requires conducive conditions that enable these strategies to be adopted, to take hold and to grow. Underlying this notion is the conceptualisation of cities as a complex adaptive system [19]. In such systems, leapfrogging can, per definition, not be the outcome of a specific action or top-down implementation program as the complexity of the system blurs direct cause and effect relationships between what individuals and organisations do and system level outcomes [20]. This is compounded by the fact that leapfrogging requires efforts by many stakeholder groups which contribute in different ways and at different times to the leapfrogging process. Based on these insights, scholars propose to focus on developing the potential of a system that allows change to unfold into a desired direction [16, 21]. In the context of leapfrogging this potential has also been described as absorptive capacity or a systems' "ability to learn and implement the technologies and associated practices of already developed countries" [22].

3.2. Leapfrogging catalysts

3.2.1. Transdisciplinary science

Historically, mono-disciplinary and engineering dominated urban water management sectors worldwide. This led to a command and control approach to water management which in turn led to the development of large scale and centralized infrastructure. Over time engineering became supplemented by economic rationalities which supported narrowly focused economic efficiency agendas and short-term cost optimizations [23]. A pattern that has also been observed in Indonesian cities [24]. Whilst such predominately mono-disciplinary approaches have managed to improve water related health and safety to a certain extend they fall short in delivering on basic and higher order societal needs without severely increasing vulnerabilities to future uncertainties or leading to environmental degradation [25].

In response to these developments SUWM scholars point to the importance of transdisciplinary approaches to advance towards a water sensitive city [26-28]. Advancing towards a WSC requires solutions in different domains (e.g. infrastructure, community, governance) which no single discipline can develop by itself. To deliver social, environmental and economic outcomes the cross fertilization of ideas from multiple disciplines is better suited to develop solutions that fit the desired outcomes [27,28]. Ruiz et al. [28] point to the importance of harnessing and strategically combining formal knowledge (that is expert knowledge validated by scientific methods) as well as informal knowledge (that is tacit, practical or experiential knowledge from specific local contexts) which transcends sectors and particular stakeholder groups. The authors argue that transdisciplinary approaches are the key avenue for building knowledge and envisioning solutions from a holistic point of view instead of an isolated technocratic point of view and are consequently essential for the realization of a WSC. As such, transdisciplinary approaches are required for developing a shared vision of water sensitivity that is grounded in local aspirations and for devising solutions that take these into account. Furthermore, the advantage of transdisciplinary research that includes political scientists or community experts for example, is the translation and operationalization of research findings for policy makers and community representatives, thereby facilitating the uptake of new insights in policy and practice. Context specific leapfrogging strategies therefore require transdisciplinary approaches as the connection of WSC principles to local policy and practices requires knowledge from different fields and backgrounds.

3.2.2 Cross sector collaboration

Partnerships between stakeholders from different sectors (e.g. academia, business, government or civil society) usually involve a type of structured form of working together towards converging interests and the development of aligned objectives [29]. Such partnerships have a long history in public sectors and have also been identified as critical for advancing towards a water sensitive city [30]. However, they are difficult to set up and sustain because of diverging aims, power imbalances, different modes of operating and different ways of communicating. Scholars therefore propose a process model for such partnership which suggests that they change over time with changing requirements [31]. Research from the urban water management sector in Australia illustrates this. Brown et al. [30] found that partnerships between academia and industry were particularly important when new water sensitive technologies were introduced to the sector. At this early stage, these partnerships allowed the testing of innovative solutions, their evaluation and adaption in close collaboration between universities and industry to ensure that solutions are high performing while being suitable for real world application. At later stages, these partnerships became wider and increasingly involved policy makers to evaluate and support water sensitive practices at a more strategic level across the entire planning, implementation and management chain [30]. This highlights that partnerships are critical for developing solutions and implementation strategies that take local biophysical and institutional particularities into account.

As previously described, leapfrogging involves many stakeholders with different value frames and is based on learning from the mistakes and successes of others. As such, it places a premium on cross sectoral partnership that enable social learning which is described as a collective process between different stakeholders in which a number of actor develop shared meanings, values and understandings through interactions [32]. While social learning can take on different forms (e.g. informal, formal) bridging organization or intermediaries are often highlighted in the literature as key conduits for creating such learning situations and for long lasting partnerships [33]. They broker knowledge between otherwise separate stakeholders, facilitate technical or managerial skill development and nurture trust and relationships between partners which are all critical processes for creating the capacity of a system to move towards water sensitivity [6,21].

3.2.3 Innovation experiments

Innovation experiments are considered a critical means for developing contextualized solutions for a specific problem while supporting the development of a professional and organizational culture that embraces reflection and learning [34]. The key to achieve this is their setup. While they can either focus on the technical (e.g. decentralized flood mitigation) or social domain (e.g. new governance arrangements for watershed management) their most critical design parameter is that they are implemented and monitored outside the traditional isolated environments of research laboratories. Instead, such experiments involve the application of innovative ideas in a real-world context specifically designed for experimentation, reflection and learning based on participation and end-user involvement [35]. They represent a learning by doing and doing by learning approach that can significantly enable searching, learning and knowledge sharing opportunities and thereby provide an important means for finding new or adapting existing solutions to the developing country context. Additional benefits are that they build stakeholders confidence for different solutions, develop trust amongst the different parties involved and ultimately influence capacity building and policy formulation [36].

Because such experiments implement often new and radically different urban water management practices outside of traditionally save university settings they usually face a range of different barriers. These were found to be predominantly situated in the socio-institutional setting rather than in the technical realm and relate to an inherent conservatism that underpins a 'fear of failure' for such projects [37]. More specifically, cost burden for correcting a 'failed' experiment and potential unintended consequences, personal and organizational reputation as well as political and legal liabilities are major hurdles. On the other hand, a number of success factor for such water sensitive experiments have been identified. For example, Farrelly and Brown [34] highlight that an early engagement of key stakeholders, the sustained involvement of champions and a continuity of engagement from different stakeholder groups as well as mechanisms for defining roles and responsibilities and risk sharing are critical success factors for innovation experiments to be carried out.

4. Australia – Indonesia Research Collaboration: The Urban Water Cluster

A collaborative research program, the "Urban Water Cluster", has been established between universities in Australia and Indonesia, to support the transition or leapfrogging of Indonesian and Australian cities towards more sustainable, productive, resilient and liveable conditions through mutual learning and rapid uptake of context specific water practices. Funded by the Australian Department of Foreign Affairs and Trade through the Australia Indonesia Centre, this 2 million Australian Dollar investment builds on established research in Australia and Indonesia through a multi-institute research collaboration.

While traditional project delivery involves planning and scheduling activities to achieve outcomes on-time, on budget and with clearly defined performance indicators, the Urban Water Cluster has adopted a framework that better harnesses the momentum that each leapfrogging catalyst creates and delivers focus projects and networks of multi-organizational peer groups to enhance innovative environmental practice [36]. The multi-institutional partnership arrangement shown in Figure 2, and integrated cross-sectoral engagement framework developed by the partners shown in Figure 3, represent a translation of the leapfrogging catalyst described in this paper to an Indonesian context.

Through the involvement of social and political scientists, civil engineers, hydrologists, architects, urban planners and agriculture scientists across six institutes (Figure 2) the Cluster is well presented by transdisciplinary science. The cluster brings together leading researchers in Australia and Indonesia to work together on the issue of SUWM and practice in both countries. The first year of the program has been dedicated to building collaboration agreements to operationalize the research, establish and connect research teams, select case study sites and co-develop research work plans. Time and resources are key enablers of social learning [36] and ensuring a strong foundation for innovation.

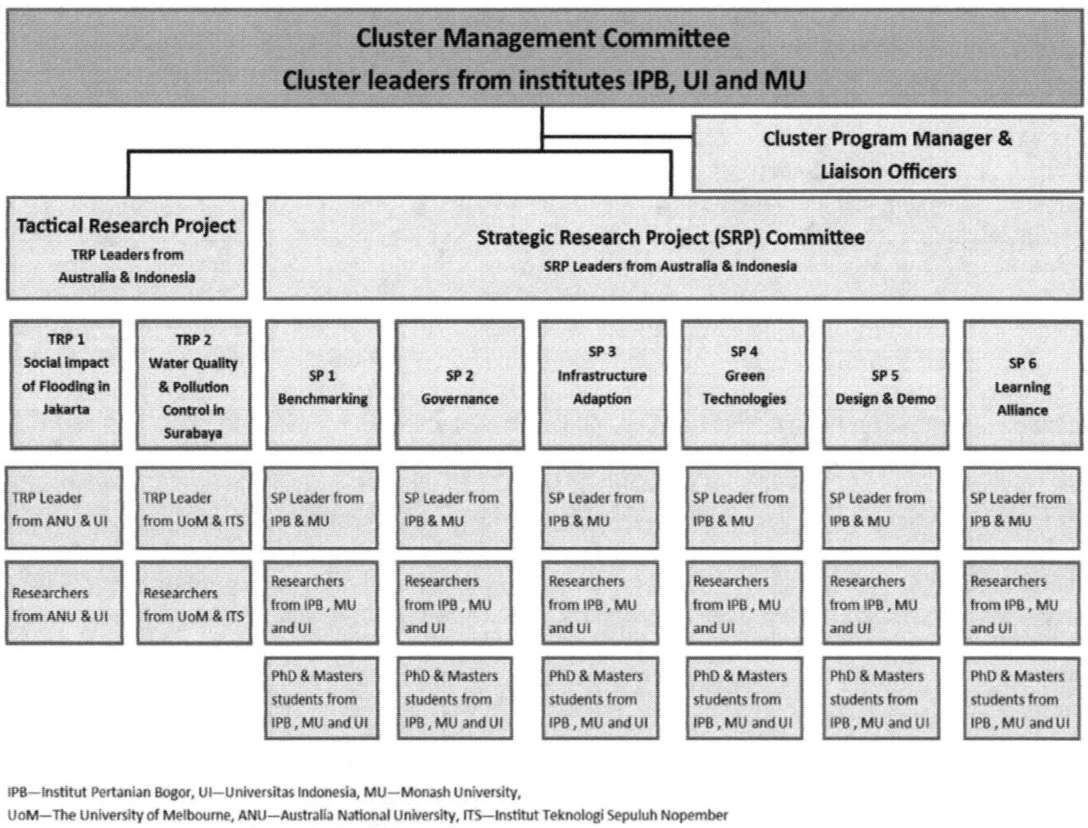

Figure 2. Urban water cluster research collaboration and partnerships.

The urban water innovation experiment includes both strategic and tactical elements. The Strategic Research Program (SRP) aims to deliver a suite of interrelated socio-technical tools to support the leapfrogging of Indonesian cities to a WSC state, including (SP1) city benchmarking, (SP2) transition governance, (SP3) infrastructure adaptation, (SP4) technological innovation, (SP5) urban design, and (SP6) creation of learning alliances (Figure 2). These tools, applied to four case study sites around Bogor and in the Ciliwung and Bekasi river catchments, will inform the leapfrogging strategies for each site. The Tactical Research Projects (TRPs), also have socio and technical elements that broaden the research to other Indonesian cities, including Jakarta and Surabaya. In its entirety, the interdisciplinary collaboration of researchers include 3 Australian Universities and 3 Indonesian Universities, involving 39 academics and 30 higher degree students (PhD and Masters).

Cross-sector collaboration and social learning has been intentionally integrated into the cluster framework throughout the design of the research cluster (Figure 2 & Figure 3) and in particular the creation of a Learning Alliance (SP 6) which will deliver the learning agenda. The learning alliance aims are to ensure the innovative experiments are guided by the experience, expertise and organizational settings relevant to sustainable urban water management practice, and emphasizes sharing socio-technical, organizational and cultural knowledge, understanding the local context at individual and organization levels and across multi-disciplines. Specifically, the alliance brings researchers together to co-design innovation experiments and engage with local, regional and national governments, developers, donors/banks, utilities, communities and Non-Governmental Organizations (NGOs). This engagement of expertise from Indonesia and Australia in water and sanitation, water policy, urban planning, green infrastructure build and management, is achieved through governance arrangements, masterclasses, focus group discussions and workshops.

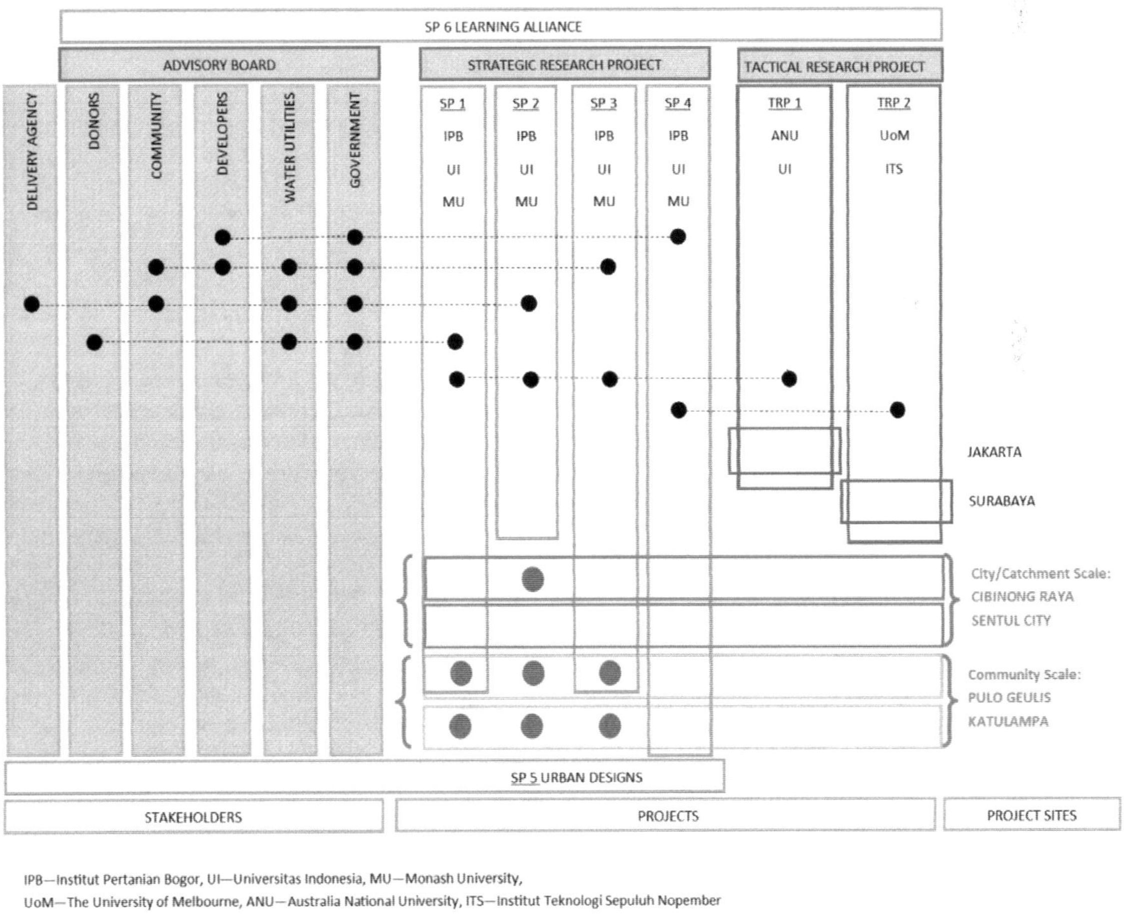

Figure 3. Integrated research framework for the urban water cluster.

Governance of the overall program is provided by an advisory board comprised of champions from each country, discipline and sector. Endorsement of the board and research program by Indonesian Bogor city and regency mayors and the Australian Department of Foreign Affairs and Trade (donor), develops "political capital and commitment" for change and provides legitimacy for the framework [36]. Program governance by the cluster management committee and SP Leaders from Indonesia and Australian leaders, ensures each institute have balanced input into the research program (Figure 2). Cluster leadership and institutional partnership is viewed as crucial to leveraging and strengthening relationships and provide the impetus to lead conceptual thinking in all areas of urban water management and city development. Coordination of the program by a project manager and two liaison officers, strengthens cross-institute relationships and collaboration efforts and ensures governance and administration of the program is well resourced.

Outside of the governance arrangements, cross-sector engagement is achieved through meetings, masterclasses, symposiums, workshops and focus group discussions, that aim to create, share, synthesize and disseminate knowledge in water sensitive design and practices. Consistent with the leapfrogging catalysts, the cluster engagement strategy has been carefully designed to actively create collaboration opportunities between Indonesian and Australian academic, government and industries, through the learning alliance network. The integrated framework (Figure 3) provides a holistic platform to understand urban water and WSC from the perspectives of different actors and deliver practice-based urban design and demonstration forums. These two activities will assure that the cluster achieves on ground practice change.

5. Concluding Remarks

This paper set out to describe catalysts for leapfrogging and to explain their potential role for achieving a water sensitive city in a developing country context. It drew on leapfrogging literature to define leapfrogging towards a water sensitive city and to highlight two key prerequisites. Building on this understanding the paper synthesized key insights from the transformative change and sustainable urban water management literature to describe three catalysts for leapfrogging (transdisciplinary science, cross sector partnerships and innovation experiments) which support the development of both prerequisites simultaneously. The paper also introduced a joint research project between Australian and Indonesian universities that represents a strategic translation of these catalysts into practice.

How the project, its processes and outcomes will facilitate leapfrogging of Indonesian cities towards a Water Sensitivity City remains subject to ongoing research. Nevertheless, two important points for leapfrogging towards water sensitivity in a developing country context can already be made based on this paper and the projects current implementation phase. First, the transformative change and SUWM literature provides a promising scholarly basis for future research on leapfrogging towards water sensitivity. Both bodies of literature with their shared normative basis for sustainable development, their focus on systemic change processes as well as their view on achieving real-world impact make them particularly well suited from a research and practice change perspective. Second, the role of international collaboration between developed and developing country research institutions provides a strong basis for robust leapfrogging research, the development of leapfrogging strategies and their implementation. None of these challenging tasks could be achieved by any side on its own and it is through this type of work that cultural exchange and mutual learning emerges from which both countries benefit.

References

[1] Frantzeskaki N, Broto V, Coenen L, and Loorbach D (Eds) 2017 Urban sustainability transitions New York USA: Routledge
[2] UN - Habitat 2012Prosperity of cities: state of the world's cities 2012/2013
[3] Singh RB (Ed) 2015 Urban development challenges, risks and resilience in asian mega cities Tokyo Japan: Springer
[4] Wong T and Brown R 2009 The water sensitive city: principles for practice Water Sci Technol

60(3) 673–82

[5] Brown R, Keath N and Wong T 2009 Urban water management in cities: historical, current and future regimes Water Sci Technol 59(5) 847–55

[6] Brown R, Rogers B and Werbeloff L 2016 Moving toward Water Sensitive Cities: A guidance manual for strategists and policy makers

[7] Poustie M, Frantzeskaki N, and Brown R 2016 A transition scenario for leapfrogging to a sustainable urban water future in Port Vila Vanuatu Technol Forecast Soc 105 129–139

[8] Steinmueller E 2001 ICTs and the possibilities for leapfrogging by developing countries Int Labour Rev 140(2) 8–23

[9] Binz C, Truffer B, Li L, Shi Y, and Lu Y 2012 Conceptualizing leapfrogging with spatially coupled innovation systems: The case of onsite wastewater treatment in China Technol Forecast Soc 79(1) 155–171

[10] Pahl-Wostl C, Vörösmarty C, Bhaduri A, Bogardi J, Rockström J, and Alcamo J 2013Towards a sustainable water future: shaping the next decade of global water research Curr Opin Env Sust 5(6) 708–714

[11] Wong T, Allen R, Brown R, Deletic A, Gangadharan L, GernjakW, ... and Walsh C 2013Blueprint 2013 – stormwater management in a water sensitive city Melbourne, Australia

[12] Sauter R and Watson J 2008 Technology leapfrogging: a review of the evidence A report for DFID TyndalCenter for Climate Change Report

[13] Jefferies C and Duffy A 2011The SWITCH transition manual - managing water for the city of the future

[14] Wong T 2016 How developing cities can meet the challenges of the 21st century Retrieved from https://blogsadborg/blog/how-developing-cities-can-meet-challenges-21st-century

[15] Unruh G 2002 Escaping carbon lock-in Energ Policy 30(4)317–325

[16] Poustie M 2014 Enabling socio-technical transitions to sustainable urban water management in the south west pacific region Monash University

[17] Gallagher K 2006 Limits to leapfrogging in energy technologies Evidence from the Chinese automobile industry Energ Policy 34(4)383–394

[18] Shah T, Makin I and Sakthivadivel R 2000 Three limits to leapfrogging: Issues in transposing successful river basin management institutions in the developing World Irrigation and river basin management: options for governance and institutions 89–114

[19] Dunn G, Brown R, Bos J and Bakker K 2017 Standing on the shoulders of giants: understanding changes in urban water practice through the lens of complexity science Urban Water J14(7) 758 – 767

[20] Westley F, Olsson P, Folke C, Homer-Dixon T, Vredenburg H, Loorbach D ... and Leeuw S 2011 Tipping toward sustainability: emerging pathways of transformation Ambio 40(7) 762–780

[21] Wolfram M, Frantzeskaki N and Maschmeyer S 2016 Cities, systems and sustainability: status and perspectives of research on urban transformations CurrOpinEnv Sust 22 18–25

[22] Dahlman C and Nelson R 1995 Social absorption capability, national innovation systems and economic development In D Perkins and B Koo (Eds) Social capability and long-term growth Basingstoke: Macmillan Press

[23] Brodnik C, Brown R and Cocklin C 2017 The institutional dynamics of stability and practice change: the urban water management sector of Australia (1970–2015) Water Resour Manage 31(7)

[24] Kooy M and Bakker K 2008 Splintered networks: the colonial and contemporary waters of Jakarta Geoforum 39(6)1843–1858

[25] de Haan FJ, Ferguson B, Adamowicz R, Johnstone P, Brown R and Wong T 2014 The needs of society: a new understanding of transitions, sustainability and liveability Technol Forecast Soc 85 121–132

[26] Barron N, Kuller M, Yasmin T, Castonguay A, Copa V, Duncan-Horner E, ... and Deletic A 2017 Towards water sensitive cities in Asia: an interdisciplinary journey Water Sci Technol

76(5)1150 – 1157
[27] Brown R, Deletic A and Wong T 2015 How to catalyse collaboration Nature 525 7–9
[28] Ruiz A, Dobbie M and Brown R 2017 Insights and future directions of transdisciplinary practice in the urban water sector Journal of Environmental Studies and Sciences 7(2) 251–263
[29] Hamann R and April K 2013 On the role and capabilities of collaborative intermediary organisations in urban sustainability transitions J Clean Prod 50 1–10
[30] Brown R, Farrelly M and Loorbach D 2013 Actors working the institutions in sustainability transitions: The case of Melbourne's stormwater management Global Environmental Change, 23 701–718
[31] Clarke A and Fuller M 2010 Collaborative strategy management: strategy formulation and implementation by multi-organizational cross-sector social partnerships J Bus Ethics 94 85–101
[32] Pahl-Wostl C, Sendzimir J, Jeffrey P, Aerts J, Berkamp G and Cross K 2007 Managing change toward adaptive water management through social learning Ecol Soc 12(2)30
[33] Brown R and Clarke J 2007 Transition to water sensitive design: the story of Melbourne, Australia Facility for Advancing Water Biofiltration, Monash University
[34] Farrelly M and Brown R 2011Rethinking urban water management: experimentation as a way forward Global Environmental Change 21 721–732
[35] Luederitz C, Schäpke N, Wiek A, Lang D, Bergmann M, Bos J, … and Westley F 2016 Learning through evaluation - A tentative evaluative scheme for sustainability transition experiments J Clean Prod 169 61-76
[36] Bos J, Brown R and Farrelly M 2013 A design framework for creating social learning situations Global Environmental Change, 23(2) 398–412
[37] Brown R and Farrelly M 2009 Delivering sustainable urban water management: a review of the hurdles we face Water Sci Technol 59(5) 839–46

Study about factors influencing transition of green open spaces based on analysis of land use in Depok City, West Java, Indonesia

M Matsuda[1*], N Nasrullah[2], and B Sulistyantara[2]

[1] Graduate School of Horticulture, Chiba University
[2] Department of Landscape Architecture Faculty of Agriculture, Bogor Agricultural University

[*]E-mail: acha1305@chiba-u.jp

Abstract. Most green open spaces in Jakarta, the capital city of Indonesia, are susceptible to conversion into urban use. In Jakarta, land development is progressing to the point where there is no room for additional land development, resulting in a shift to land development in other cities, especially the urban areas surrounding Jakarta, Jabodetabek. This study focused on land development in Depok City located south of Jakarta. Depok is influenced strongly by land development in Jakarta, and further economic growth is expected. The purpose of this study is to reveal the changes in the area of green open spaces to investigate the transition of green open spaces in Depok and to identify the factors influencing the transition of green open spaces in Depok. ERDAS 9.1 Software was used to analyze the changes in the area of green open spaces in 1995–2015. To identify the factors influencing the transition of green open spaces, binomial logistic regression analysis was used. It was revealed that the area of green open spaces has rapidly decreased from 2005 (9789.28Ha) to 2015(5157.80Ha) in Depok. The results of logistic regression analysis showed Logit $Y = 2.575 - 0.118\ X1 - 0.012\ X2$. Two factors particularly influenced the transition of green open spaces, the proximity to educational areas or main roads was associated with reductions in green open spaces.

Keywords: green open spaces, land use, Depok City

1. Introduction

Indonesia has the fourth largest population in the world with a fast-growing economy, and further economic growth is expected in the near future. With continued growth, urban land development is progressing rapidly and green open spaces are changing to urban land use areas at an increasing rate. Most green open spaces in Jakarta, the capital city of Indonesia, are susceptible to conversion into urban use. Therefore, it is necessary to increase the protected areas in the future. According to Law No. 26-the spatial planning law, established in 2007, each city must have green open spaces with an area of at least 30% of the total area of the city. The government of Jakarta has been promoting measures for the expansion of green open spaces, such as planting roadside trees. However, a large part of Jakarta is covered by impervious surfaces such as pavement and buildings and the city has only a few green open spaces left; thus, the achievement of the Law No. 26 target does not seem possible [1]. In Jakarta, land development is progressing to the point where there is no room for additional land development, resulting in a shift to land development in other cities, especially the urban areas surrounding Jakarta, Jabodetabek.

The environment, land use management strategies, and socio-economic factors are the major causes of land use changes [2]. Some factors are caused by local and social changes and land conditions [3]. When new jobs are created by establishing factories and commercial facilities, housing demand in that location may increase. There is a possibility that new houses are built around existing transportation routes, stations, schools, hospitals, and commercial buildings. The environmental factor, such as being located in the outside of flood zone, is also considered when constructing new houses.New housing construction reduces the area of available green open spaces [4]. The population growth, the emergence of new town projects and large manufacturing centers, greater job opportunities, and the availability of easy transportation all contribute to urbanization and suburbanization [5].

In the Jabodetabek metropolitan area, the distribution of land use within the city should be studied to solve land use problems by proposing new plans for keeping green open spaces while maintaining the economic growth. In Indonesia, urbanization may progress in cities other than the Jabodetabek metropolitan area in the near future. By identifying the factors influencing the transition of green open spaces to other uses by land development in Jakarta, it is possible to create timely and effective plans for the other cities.

2. Study Purpose

The purpose of this study is to reveal the changes in the area of green open spaces to investigate the transition of green open spaces in Depok and to identify the factors influencing the transition of green open spaces in Depok.

3. Study Method

3.1. Study Area

This study focused on land development in Depok Citylocated south of Jakarta, geographically located at coordinates 6°19′00″–6°28′00″S and 106°43′00″–106°55′30″E. Depok is a part of the Jabodetabek metropolitan area, which is the largest and the most strategic metropolitan area in Indonesia. Jabodetabek consists of Jakarta, Bogor City, Bogor Regency, Depok City, Tangerang City, South Tangerang City, Tangerang Regency, Bekasi City, and Bekasi Regency [6]. This area will be further developed as a strategic region to support national economic growth [7].

Depok is one of the newly developed areas in West Java, formed by Law No. 15 in 1999. Depok has a tropical climate and the difference in seasonal rainfall amounts is small. The tropical climate supports agriculture by providing rainfall all year round but it is necessary to consider the distribution of land use within the city, especially, the availability of green open spaces. Depok is a commercial and a public utilities center, where water is stored to supply groundwater to the city of Jakarta. Moreover, it is a hotspot for educational and residential development. Depok is strongly influenced by land development in Jakarta, andfurther economic growth is expected. Further economic growth in Jakarta will also put extra pressure on Depok.

Based on the data from the Central Statistics Agency of Depok [8], the total population of Depok was approximately 2,179,810 in 2016 and the annual population growth rate of 2010–2016 was 3.53%. The total land area of Depok is 200.29 km^2, which includes 11 districts, Beji, Bojongsari, Sawangan, Pancoran Mas, Cipayung, Sukma Jaya, Cimanggis, Tapos, Limo, Cilodong, Cinere.

3.2. Materials and Methods

In this study, ERDAS IMAGINE 9.1, ArcGIS 10.2.2, and IBM SPSS Statistics19 tools were used for analysis. ERDAS IMAGINE 9.1 was used to analyze the changes in the area of green open spaces in 1995–2015, ArcGIS 10.2.2 was used to analyze the land use in Depok, and IBM SPSS Statistics19 was used for statistical analysis. The Landsat satellite images were processed using the steps of layer stack, geometric correction, subset image (cutting the image of the study area), land cover classification, and accuracy assessment.

To identify the factors influencing the transition of green open spaces, binomial logistic regression analysis was performed:

Logit (Y) = a+ $b_1X_1 + b_2X_2 + b_3X_3 + ... + b_iX_n$
Description: Y = Dependent Variable, X = Independent Variable, a = Constant, b_i = Coefficients of Independent Variables, i for 1,2,3, ... n

The logistic regression formula was developed using data from 2005 to 2015. In this study, the dependent variable is defined as the area of green open spaces converted into non-green open spaces. The independent variables included the distance from commercial areas, industrial areas, educational areas, hospitals, rivers, lakes, main roads, a highway, a highway entrance, railroad stations, and inundation-prone areas.

In order to separate the areas where the green open space is decreasing and the areas where the green open space is maintained, the analysis was conducted at 1-km resolution. The number of cells was 239; 78 cells were extracted as the area where the green open space was greatly reduced from 2005 to 2015 and 14 cells were extracted as the area where the green open space was maintained. Land use data for 2015 are provided by the Depok Municipal Spatial Planning Bureau.

4. Results and Discussion

4.1. Changes in the area of green open spaces

The processing of Landsat imagery revealed the percentage of the land cover in Depok with 80% accuracy. Landsat images for the years 1995 (TM), 2005 (TM), and 2015 (ETM+) were used; the image for 1995 revealed the extent of Depok with 87.02% accuracy, the image for 2005 with 88.24% accuracy, and the image for 2015 with 85.16% accuracy. The results of land cover classification (green open space and non-green open space) are shown in Table 1 and the green open space maps of Depok are shown in Figure 1. The results showed that the area of green open spaces has rapidly decreased from 2005to 2015in Depok. The area of green open spaces was 9789.28 Ha in 2005, 48.88% of the total area of the city.The area of green open spaces was 5157.80 Hain 2015, 25.75% of the total area of the city.

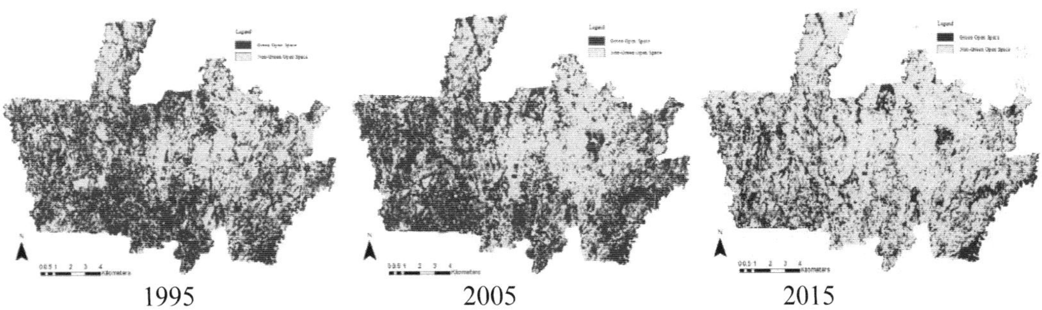

Figure 1. Green Open Space Maps of Depok

Table 1. Land Cover in Depok

Land Cover	1995		2005		2015	
	Area (Ha)	Percentage (%)	Area (Ha)	Percentage (%)	Area (Ha)	Percentage (%)
Green Open Space	10002.49	49.94	9789.28	48.88	5157.80	25.75
Non-Green Open Space	10026.51	50.06	10239.72	51.12	14871.20	74.25
Total	20029	100	20029	100	20029	100

4.2. Factors influencing the transition of green open spaces

The result of the binomial logistic regression analysis can be expressed by the following equation:

Logit (area of green open spaces converted into non-green open spaces) = 2.575 − 0.118 × Distance from educational areas − 0.012 × Distance from main roads

Using the p-value of each independent variable, the distance from educational areas and distance from main roads were found to be statistically significant at the 5% level, but the distance from commercial areas, industrial areas, hospitals, rivers, lakes, a highway, a highway entrance, railroad stations, and inundation-prone areas were not statistically significant at the 5% level. The coefficients, odds ratio, odds ratio atthe 95% confidence interval, and p-value of each variable are shown in Table 2.

Table 2. Results of logistic regression analysis

Variables	Coefficient	Odds Ratio	Odds Ratio at the 95% Confidence Interval	P Value
Distance from educational areas	−0.118	0.889	0.807 − 0.979	0.017
Distance from main roads	−0.012	0.988	0.978 − 0.999	0.031
Intercept	2.575			

The results suggest that the distance from educational areas and main roads werethe main factors influencing the transition of green open spaces in Depok. By using each main road, access from Depok to Jakarta, Bogor, and Tangerang is possible. The number of motor vehicles, especially the number of motorcycles, is rapidly increasing in Depokeach year. The spatial structure of the city along with the transportation network that depends mostly on private vehicles leads to a decrease in green spaces and in urban environmental quality [9]. No relation between the reduction in the area of green open spaces and the distance from railroad stations was detected by the results. There is also no relation between the reduction in the area of green open spaces and the distance from the entrance of a highway. It is possible to easilytravel to other cities by using motor vehicles on main roads in Depok. The number of schools, as well as the number of students in elementary schools, junior high schools, and high schools has increased in recent years.

The large number of people who work in Jakarta but live in Depok has tremendously increased the property development in this city. Most of the residential area in Depok has high building density and widespread asphalt pavement cover [10]. The reduction of the area of available green open spaces has actually been caused by new housing construction in Depok. The number of residential buildings has increased in areas where the living conditions have improved, and distance from educational areas and distance from main roads are among the major factors reducing the area of green open spaces in Depok.

5. Conclusion

The area of green open spaces has rapidly decreased in Depok from 2005 to 2015. The proximity to educational areas or main roads was associated with reductions in green open spaces because the number of residential buildings has increased in areas where living conditions have improved. It is important to develop green open spaces (e.g., city parks) that children and families can easily access to. Currently, there are only a few city parks in Depok that cover 294.38 Ha, only 1.47% of the total area of the city, according to the Depok Municipal Spatial Planning Bureau [11]. It is also important to increase the area of green open spaces by planting street trees. The street trees in Depok cover an area of 56.39 Ha, 0.28% of the total area of the city.

There is still vacant land available with an area of 692.48 Ha in eight districts of Depok, 3.46% of the total area of the city. Although new buildings may occupy that land in the future,it is necessary to develop green open spaces in that land. The regulations on green open spaces in the spatial plan of Depok for the provision and utilization of public green open spaces include the development of roof gardens and vertical gardens in densely populated areas. It is necessary to put effort into the development of roof gardens and vertical gardens to improve the environmental quality in Depok.

By developing green open spaces within a short distance of educational areas, the function of the space for environmental education could be emphasized. This study reveals that the distance from the educational areas influences the reduction of green open spaces as new knowledge; thus, this factor should be considered in future development plans as a survey component.

References

[1] Sulistyantara B 2009 Some problems faced in the developing green open spaces at Jakarta Indonesia. Organization for Landscape and Urban green Infrastructure (72) 18-21
[2] Yu W, Shuyi Z, Changsan W, Wen L and Xiaodong N 2014 Analyzing and modelling Land Use Land Cover Change (LUCC) in the Daqing City, China. Applied Geography (31) 600-608
[3] Ikebe K 2006 Research on factors influencing changes in land utilization for the purpose of extracting indicators contributing to the formulation of land utilization plans. Rural Planning (25) 449-454
[4] Rini, E F, Sulistyarso H and Pamungkas A 2014 Factors influencing the availability of green open space in East Surabaya. Architecture & Eevironment **13** (1) 75-92
[5] Pravitasari, A E, Saizen I, Tsutsumida N, Rustiadi E and Pribadi D O 2015 Local spatially dependent driving forces of urban expansion in an emerging Asian Megacity: The case of Greater Jakarta (Jabodetabek). Journal of Sustainable Development. **8** (1) 108-119
[6] Hasibuan H S, Soemardi T P, Koestoer R and Moersidik S 2014 The role of transit oriented development in constructing urban environment sustainability, the case of Jabodetabek, Indonesia. Procedia Environmental Sciences (20) 622-631
[7] Pribadi D O and Pauleit S 2015 The dynamics of peri-urban agriculture during rapid urbanization of Jabodetabek Metropolitan Area. Land Use Policy (48) 13–24
[8] Central Statistics Agency of Depok City 2017 Depok Municipality in Figures 2017 Depok (Indonesia): Government of Depok City
[9] Asri D U and Hidayat B 2005 Current transportation issues in Jakarta and its impacts on environment. Eastern Asia Society for Transportation Studies (5) 1792-1798
[10] Zaina A FM, Permatasarib P A, Ainya C N, Destrianaa N, Mulyatia D F and Edia S 2015 The detection of urban open space at Jakarta, Bogor, Depok, and Tangerang – Indonesia by using remote sensing technique for urban ecology analysis. Procedia Environmental Sciences (24) 87-94
[11] Depok Municipal Spatial Planning Bureau. 2016. Spatial plan of Depok city 2012-2032. Depok (Indonesia): Government of Depok City

The actual use of urban forest for Jakarta's dwellers

N D Permata[1*], Syartinilia[2], and A Munandar[2]

[1]Graduate Student of Department of Landscape Architecture, Faculty of Agriculture, Bogor Agricultural University (IPB);
[2]Department of Landscape Architecture, Faculty of Agriculture, Bogor Agricultural University (IPB).

*E-mail: nike.dyah@ymail.com

Abstract: Urban forest is urban landscape feature serves as recreation area for urban dwellers. Jakarta has 15 urban forests spread all over Jakarta. Urban forest is urban landscape feature serves as a recreation area for urban dweller. How the actual use of urban forest by Jakarta city's dwellers is important to evaluate. The objective of this study were to analyse the extent of each urban forest coverage can reach by walk and how many dwellers's villages were visiting the urban forest from inside and outside Jakarta. This study combined field based interviews of urban forest's visitors with GIS analysis. As a result, University of Indonesia (UI) Forest (55 ha) and Dukuh Forest (0,58 ha) were identify as the largest (36 villages, 6.696,38 ha) and smallest urban forest (2 villages, 38,20 ha), respectively. Overall 73,33% visitors came from within the urban forest coverage area, whereas the rest (26,66%) came from outside. Therefore the actual use of urban forest has proven that urban forest as one of recreation area for Jakarta city's dwellers.

Keywords: dwellers, urban forest, urban landscape

1. Introduction

Urban forest is the expanse of land with cohesive and dense trees in urban areas, both on state land and authority land that have been set by the goverment. Urban forest can serve as a dweller recreation area [3]. As a recreational area, urban forests were used as a gathering place and doing their daily activities [1]. People who visits urban forests will feels happier, relaxed, calm, and feel closer to nature. Urban forests also have a positive influence on visitors for the activities they do [9].

Jakarta is one of the largest cities in Indonesia (66,150 ha). Jakarta has 15 urban forests with the total area 9.97% spread all over Jakarta [2, 6]. Urban forest is one of the green open space for dweller as a recreation area. Urban forest should be easily accessible and close to dwelling area, especially for the dweller who lives around the urban forest in radius 5-10 minutes by walk. The distance between urban forest and dwelling area is between 300 - 500 m [5].

Urban forest should be located in the settlement area and place for dweller activities [7]. Dwellers will use urban forests based on the facilities and design of the urban forests [11]. Urban forest has important thing to increase the urban social side. For example, if urban forest has good accessibility, crop structure is preferred by the dweller, good facilities, and complete infrastructure, it will increase visitors attract. In this study, we tried to analyze the extent of each urban forest coverage can reach by walk and how many dwellers were visiting the urban forest from inside and outside Jakarta.

2. Methodology

This study was conducted in 15 urban forest located in Jakarta (fig 1), at coordinates 607'0 " latitude - 6024'00 " latitude and 106040'30 " longitude - 106058'30 " longitude. This research consisted of data collection, field survey, interview, and data analysis. Visitors sampling method was done by non-probability sampling approach through convenience sampling method, thats mean respondent was in the research location and willing to be interviewed. Total visitors sampling for this research were 120 visitors.

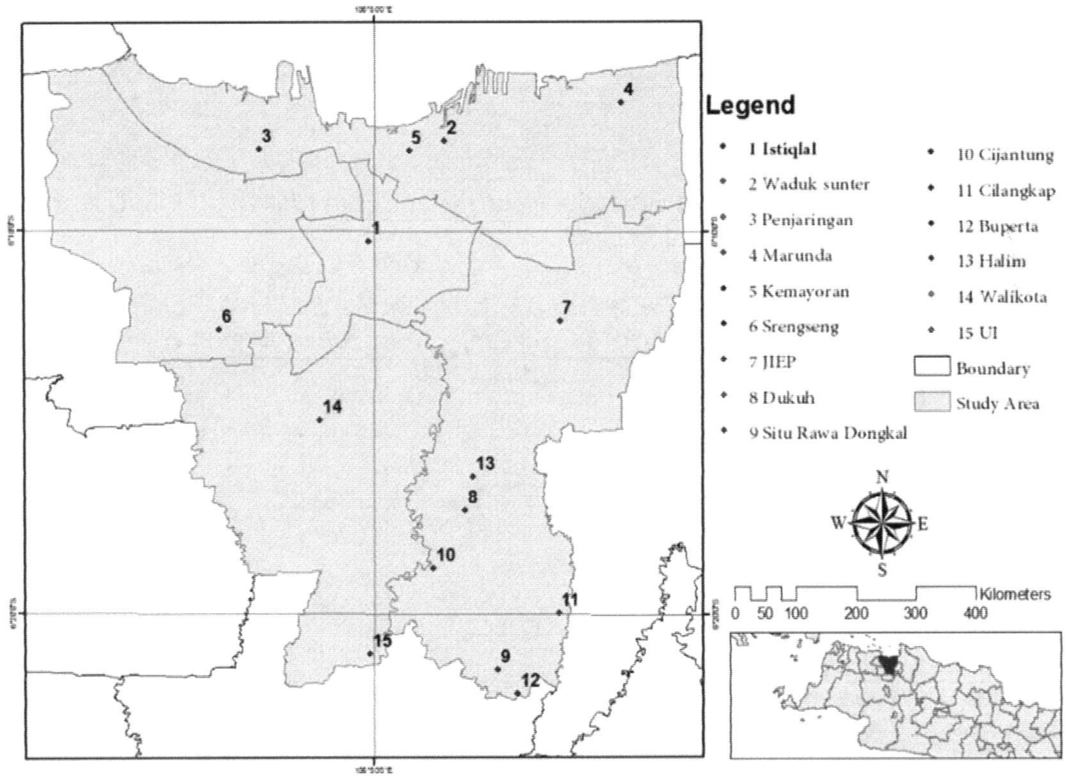

Figure 1. Study Area

The analysis was done using ArcGIS tools with buffer analysis. A buffer analysis is a method of analysis in the form of a region or zone of a certain distance around a physical entity, such as a point, line or polygon that has been defined [8]. The required information in this stage of analysis was a map of urban forest distribution in Jakarta and the area of urban forest. Each map was done by overlay analysis. Then buffer analysis was used to find out the extent to which urban forests can be driven by urban communities by walk. Buffer analysis for each urban forest are different depend on the extent of urban forest area, modified (Table 1).

Table 1. Area and range of urban forest by walk

No	Urban Forest Area (ha)	Radius Buffer (km)
1	>20	>4
2	20	4,0
3	18	3,5
4	16	3,0
5	14	2,5
6	12	2,0
7	10	1,5
8	8	1,0
9	2	0,5
10	<2	0,3

Source: [12]

3. Results

The results shown that urban forest can coverage many villages in Jakarta (fig 2). The largest urban forest in Jakarta is Univesity of Indonesia forest (55,46 ha). University of Indonesia forest can coverage 36 villages with total area coverage is 6.696,38 ha (table 2). Furthermore, the second largest urban forest is Bumi Perkemahan Cibubur forest (27,32 ha). Bumi Perkemahan Cibubur forest can coverage 25 villages around urban forest with the total coverage area is 6.312,09 ha.

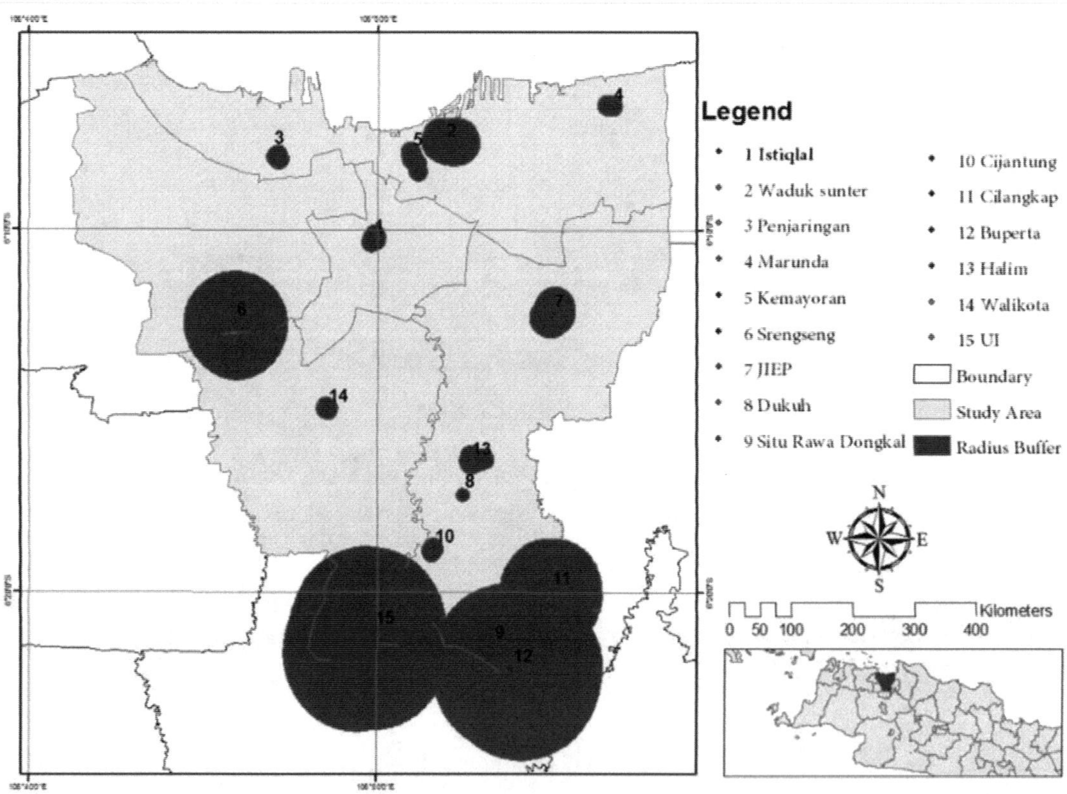

Figure 2. Buffering Area of Urban Forest

The smallest urban forest in Jakarta is Dukuh forest (0,58 ha). Dukuh forest can coverage only 2 villages around urban forest. The total area that Dukuh forest can coverage is 30,20 ha. The second smallest urban forest is Blok P forest (1,64 ha). Blok P forest can coverage 3 villages with the total coverage area is 106,94 ha.

In the Fig 5 shown that 73,33% of urban forests in Jakarta can coverage less than 10 villages. They are Istiqlal Mosque forest (4 villages), Waduk/Danau Sunter forest (5 villages), Kanal Banjir Barat forest (4 villages), Barikat Nusantara forest (3 villages), Kemayoran forest (4 villages), PT JIEP forest (6 villages), Dukuh forest (2 villages), Situ Rawa Dongkal forest (2 villages), Komplek Kopasus Cijantung forest (3 villages), Lanud Halim forest (5 villages), and Blok P forest (3 villages). While only 26,66% of urban forest that can covered more than 10 villages, they are University of Indonesia forest (36 villages), Bumi Perkemahan Cibubur forest (25 villages), Srengseng forest (17 villages), and Mabes TNI Cilangkap forest (18 villages).

Table 2. Urban Forests Coverage

Name of Urban Forest	Urban Forest Area (Ha)	Radius Buffer (Km)	Total Coverage Villages Area (Ha)
Istiqlal Mosque	1,08	0,5	140,70
Waduk/Danau Sunter	8,20	1	620,33
Kanal Banjir Barat	2,49	0,5	114,17
Barikat Nusantara	1,59	0,5	116,85
Kemayoran	4,60	0,5	200,73
Srengseng	15,00	2,5	2.366,26
PT JIEP	8,90	1	475,05
Dukuh	0,58	0,3	38,20
Bumi Perkemahan Cibubur	27,32	4	6.312,09
Situ Rawa Dongkal	3,28	0,5	157,64
Komplek Kopasus Cijantung	1,75	0,5	111,35
Mabes TNI Cilangkap	14,43	2,5	2.372,59
Lanud Halim	3,50	0,5	200,05
Blok P	1,64	0,5	106,94
University of Indonesia	55,40	4	6.696,38

Table 3 shown that 73,33% visitors came from coverage villages area (Fig 3). There are Waduk/Danau Sunter forest, Kanal Banjir Barat forest, Srengseng forest, PT JIEP forest, Dukuh forest, Bumi Perkemahan Cibubur forest, Situ Rawa Dongkal forest, Komplek Kopasus Cijantung forest, Lanud Halim forest, Blok P forest, and University of Indonesia forest. Furthermore, 26,66% visitors came from outside coverage villages area. There are Istiqlal Mosque forest, Kanal Banjir Barat forest, Srengseng forest, and Bumi Perkemahan Cibubur forest.

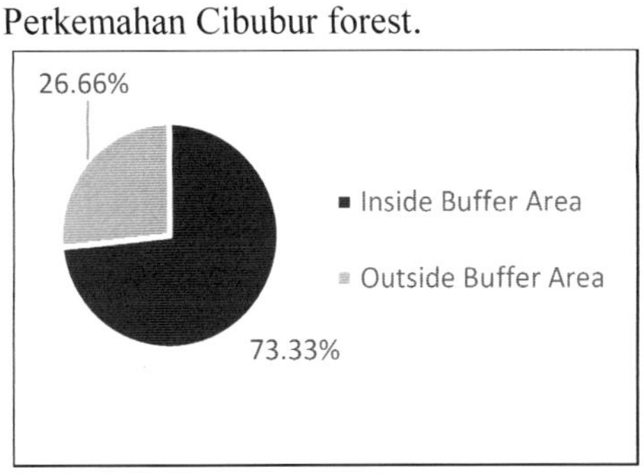

Figure 3. Visitor

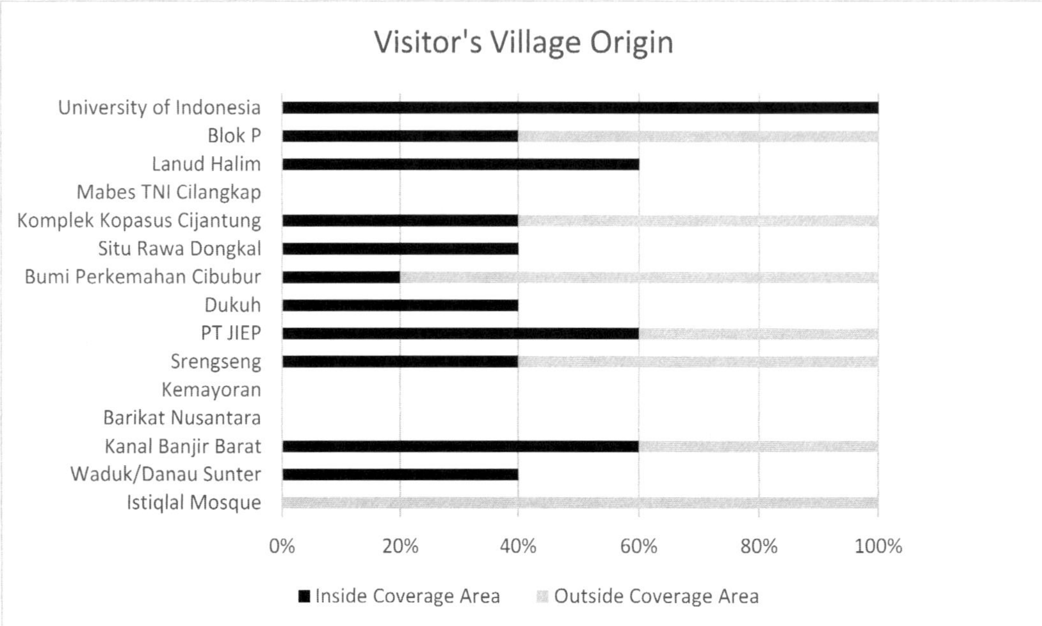

Figure 4. Visitor's Village Origin of Urban Forest

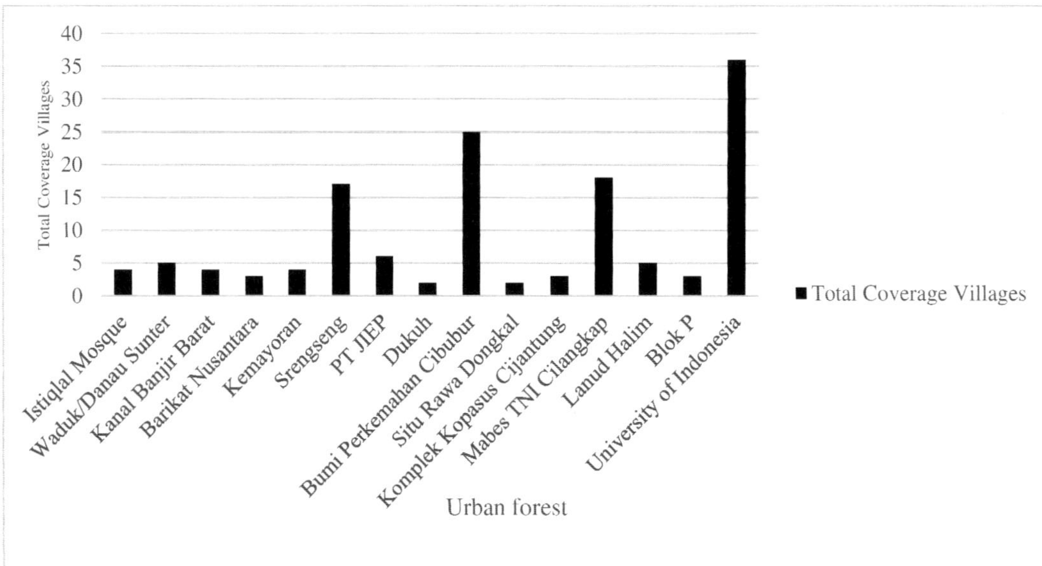

Figure 5. Total Coverage Villages

4. Discussion

Based on table 2 and figure 2, we can see there are 4 urban forests coverage more than 10 villages, while there are 11 urban forests coverage less than 10 villages. There are only 28,47% areas in Jakarta have been coverage by urban forest. It means most of urban forests in Jakarta has not been coverage many villages around urban forest. Urban forests coverage villages based on their areas. In Jakarta there are only 4 urban forests has large area, there are University of Indonesia forest, Bumi Perkemahan Cibubur forest, Mabes TNI Cilangkap forest, and Srengseng forest. Each urban forest has more than 10 ha areas, so they can coverage many villages around them.

Based on table 3 it can be seen that 73,33% of urban forest visitors are dweller who live in the coverage villages. This shown that many urban forest visitors are from the urban forest coverage villages. Visitors prefer to visit the urban forest that is close to their location. This is because the distance is close so it is easy to access and does not take a long time to get to the urban forest [10] [12].

A total of 26,66% of urban forest is not visited by the dweller in the coverage area. Among them are Istiqlal Mosque forest, Barikat Nusantara forest, Kemayoran forest, and Mabes TN Cilangkap forest.

Visitors of Istiqlal Mosque forest are generally came from outside the coverage villages, which came from the Cempaka Putih Barat Villages, Kelapa Gading Villages, Harapan Jaya Villages, Bekasi, and Depok. Visitors came to the Istiqlal Mosque forest aims to pray in the Istiqlal Mosque. This is why many of Istiqlal Mosque forest visitors are came from outside the coverage villages.

Barikat Nusantara forest is one of the urban forests located in an industrial area. When researchers visited the urban forest, there was no one visited the urban forest. Based on the results of observations of this research is due to inadequate facilities and infrastructure, not accessible, and included in the area that less secure because it was to quiet. This is why people rarely to visits the Barikat Nusantara forest.

Kemayoran forest and Mabes TNI Cilangkap forest are the urban forest that is closed for public. Visitors are prohibited to enter the urban forest, unless they have permission to visit the urban forest. Kemayoran forest is devoted to research activity for researcher, while for Mabes TNI Cilangkap forest located inside Indonesian National Armed Forces area.

Urban forest is not only visited by the dweller from the coverage villages, there are also some urban forests visited by visitors who are not from the coverage area. A total of 26,66% of urban forest is visited by people who are not from coverage area. They are Istiqlal Mosque forest, Banjir Kanal Barat forest, Srengseng forest, and Bumi Perkemahan Cibubur forest.

Banjir Kanal Barat forest, Srengseng forest, and Bumi Perkemahan Cibubur forest are urban forest that functioned as conservation area. Although the urban forest functioned as a conservation area, many of visitors also came from outsite coverage area not olny from coverage area. This is becouse the urban forest offers some attraction and complete facility. One important factor to attract visitors is attraction [4]. Banjir Kanal Barat forest, Srengseng forest, and Bumi Perkemahan Cibubur forest offers a wide range of attractions and facilities that visitors can enjoy. Urban forest should be offers many attractions and facilities [5]. Such as jogging track, lake, parking lot, mosque, gazebo, toilet, park bench, sports field, hall, children's playground, and camping ground. The number of facilities provided by urban forest manager so that many visitors coming from outside the coverage villages.

5. Conclusion

As a result, it is found that University of Indonesia forest is the urban forest with the most coverage villages of villages, that are 36 villages. Also, as many as 73.3% of urban forest visitors are the community who are included in the coverage villages Visitors choose the urban forest based on the closest to their location. This is because it is more efficient by visitors when compared to having to visit places far from their location. In addition, not only dweller who are in the coverage area who visited the urban forest, there are also some urban forest visitors came from outside the coverage area. At least 26.6% of urban forest are visited by visitors from outside the coverage villages. The urban forest is designated as urban forest with a function as a place of recreation, so that existing facilities are adequate for the visitors.

References

[1] Cranz G 1978 Changing Roles Of Urban Parks-From Pleasure Ground To Open Space *Landscape* 22(3) pp 9-18.

[2] Farisanto D 2012 *Evaluasi Program Konservasi Guna Melestarikan Kelangsungan Ekologi Di Taman Tegallega* Universitas Pendidikan Indonesia.

[3] Government Regulation of Republic Of Indonesia 2002 *Urban Forest* Number 6 Year 2002.
[4] Gunn C A 1993 *Tourism Planning Basics, Concepts, Cases third Edition* Taylor and Francis USA.
[5] Hörnsten L 2000 Outdoor recreation in Swedish forests – Implications for society and forestry *Silvestria* p 196 Acta Universitatis Agriculturae Suecia [Swedish]
[6] Imansari N and Khadiyanta P 2015 Penyediaan Hutan Kota dan Taman Kota sebagai Ruang Terbuka Hijau (RTH) Publik Menurut Preferensi Masyarakat di Kawasan Pusat Kota Tangerang *Jurnal Ruang* (1)3 pp 101—110.
[7] Kanijnendijk CC 2008 *The Forest and The City The Cultural Landscape of Urban Woodland* Springer [Denmark].
[8] Pramudia E 2008 *Evaluasi Potensi Obyek Wisata Aktual Di Kabupaten Agam Sumatera Barat Untuk Perencanaan Program Pengembangan* Bogor Agriculture University Bogor [ID]
[9] Solecki W D and Welch J M 1995 Urban parks: green spaces or green walls? *Landscape and Urban Planning* (32) pp 93-106.
[10] Tyrväinen L, Mäkinen K, Schipperijn J 2007 Tools For Mapping Social Values Of Urban Woodlands And Other Green Areas *Landscape and Urban Planning* (79)1 pp 5–19.
[11] Velasco E, Roth M, Norford L, Molina L T 2016 Does urban vegetation enhance carbon sequestration? *Landscape and urban planning* (148) pp 99-107.
[12] Van Herzele A and Wiedemann T 2003 A Monitoring Tool For The Provision Of Accessible And Attractive Urban Green Spaces *Landscape and Urban Planning* 63 pp 109–126.

Study of blue open space perception in riparian landscape of Ciliwung river

F Ulfah[1], Nurhayati[2], and H S Arifin[2]

[1] Graduate Student of Department of Landscape Architecture, Faculty of Agriculture, Bogor Agricultural University (IPB), Indonesia;

[2] Department of Landscape Architecture, Faculty of Agriculture, Bogor Agricultural University (IPB), Indonesia

*E-mail: fittriaulfah@gmail.com

Abstract: Bogor is a city with a high population density where located along of the middle part of the Ciliwung river watershed. The need sattlement of low income community lead to the occupation and unplanned settlement on the riparian. The using of the riparian river as undeveloped area has impact to safety community and ecological system at the riparian and the blue open space (BOS). The objectives of this study were, to analysis the situation of the activities and perceptions community in two different locations. This research was conducted in Bogor City area, in Pulo Geulis and Griya Katulampa. The data perception were colected by purposive sampling method and analyzed by using Chi-Square test analysis. Based on landscape character and socio cultural of the community has different condition and situation between in Pulo Geulis ang Griya Katulampa. Based on the the result of perception in Pulo Geulis's, the background of age has the most influence while at Griya Katulampa were the educational and the education background the most influential aspect to the material aspects being tested. Based on its preference, in Pulo Geulis, the background of respondents' most influential character in responding it was the background of kind of work while in Griya Katulampa the background of origin, length of stay and education. Based on its preference, in Pulo Geulis, background of character respondents' that most influencial it was type of work while in Griya katulampa were origin, length of stay, and education.

Keyword: blue open spaces, riparian, Ciliwung River

1. Introduction

Bogor City potential to become a residential area and housing development supported by the regulation of Bogor City No 1 year 2000 about Spatial Planning, where Bogor City has function as residential city, commerce, industrial, tourism and education. The high demand of urban housing causes the need for land also increases exploited for development. Limited conditions of urban land lead to irregular population distribution. This has an impact on the conversion of protected lands into constructed land as in the Ciliwung River riparian section. This is due to the extent of land change patterns over a period of 30 years in the Ciliwung watershed, changes from water bodies to settlements of 76.39% of total extent of water bodies changed [1] as well as the economic condition of low-income communities which resulted in the inability of the community to build or rent a decent dwelling [2].

Communities utilize protected areas as a living activity space to perform various of social interaction in it and can change the face of a landscape. The spatial use pattern is the spread of cultivation and protection activities and their linkages to achieve social, economic and cultural development targets

IOP Conf. Series: Earth and Environmental Science **179** (2018) 012037 doi:10.1088/1755-1315/179/1/012037

based on the potential of natural, human and artificial resources [3]. People with different educational backgrounds and economic conditions will influence people's perception of the existence of blue open space (BOS) in their daily life. The formation of a landscape is the result of a combination of interactions between various aspects of human resources, natural resources, artificial resources, social, cultural, economic, technological, information, administration, security defense, aquaculture protection function and environmental aesthetics, space and time dimensions which is in complete unity as well as qualified to form spatial planning [4]. The purpose of this study is to analyze the public perception of RTB.

2. Methods

2.1. Study Area and Sampling Technique
This research was conducted in Bogor City, in Pulo Geulis and Griya Katulampa (Figure 1). The study was conducted in four months, starting from January to April 2017.

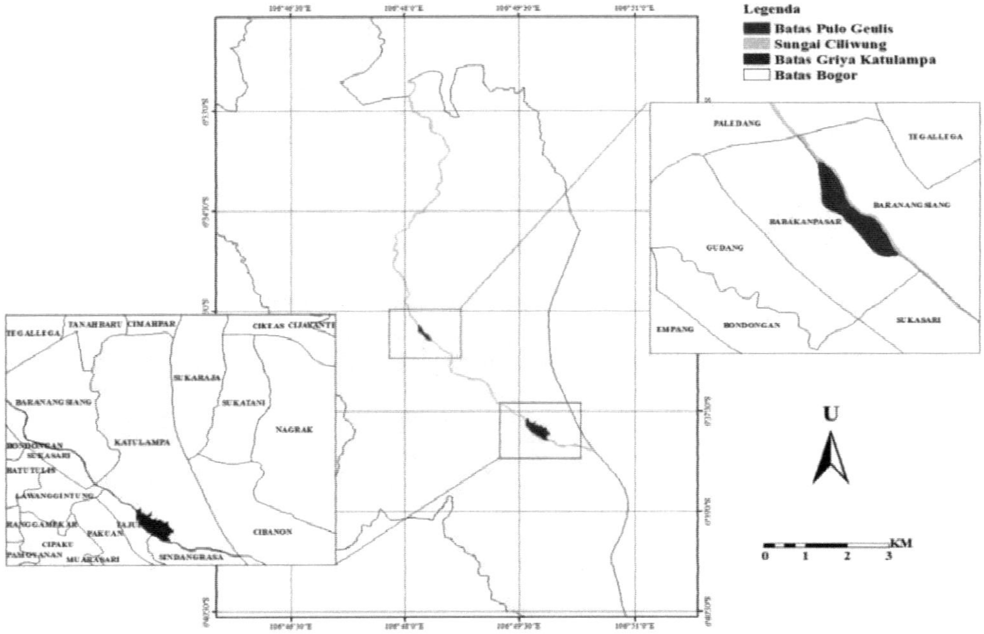

Figure 1. Research location

This research used descriptive qualitative method to describe systematically and analyze public perception. The data were collected by literature study, documentation, interviews with resource persons (community and managers), and field survey. Data collected in aspects of physical condition, demography (population, population density, age, gender), social, culture, maps (location) and people's perceptions.

Method of collecting data perception by spreading the questioner to respondent by using purposive sampling method. The criteria of respondents that must be a local citizen, willing to be a respondent, is quite close to riparian river with a distance of about 10 m-30 m from the river, and limit of age respondents at least 17 years. The number of questionnaires distributed at the research sites was 60 copies each location getting 30 copies.

2.1. Data analyses
Perception data was analyzed using descriptive statistic method, Chi-Square test as a tool. This test was done with Statistical Product and Service Solution (SPSS) with $\alpha = 0.05$. The variables tested were seven respondents' background (gender, age, education, type of work, origin, tribe and length of stay) and questionnaire material consisting of five aspects (level of community knowledge, utilization, comfort, visual, and policy). The percentage of repondent characteristics backgrounds in Pulo Geulis and Griya katulampa can be seen in Table 1. The result of the respondent's perception is seen from the influence of the respondent's character background on the questionnaire material tested in Pulo Geulis and Griya Katulampa.

Table 1. Percentage of respondent characteristics

Respondent's character	Category	Percentage (%)	
		Pulo Geulis	Griya Katulampa
Gender	Male	43.33	53.33
	Female	56.67	46.67
Age (year)	17-25	6.67	6.67
	26-65	90	86.67
	66-75	3.33	6.67
Education	No school	0	0
	Elementary school	40	6.67
	Junior high school	36.67	6.67
	Senior high school	16.67	50
	Diploma	6.67	10
	College	0	26.67
Type of work	Does not work	3.33	0
	PNS/ TNI/ POLRI	0	13.33
	Entrepreneur	23.33	20
	Student / Student	333	10
	The Informal Sector	30	16.67
	Housewife	40	40
Origin	Native	73.33	16,67
	Comer	26.67	83,33
Tribe	Sundanese	86.67	53.33
	Chinese	6.67	10
	Javanese	3.33	23.33
	Others	3.33	13.33
Length of Stay (year)	01-10	6.67	40
	10-20	10	13.33
	20-30	13.33	43.33
	30-40	20	3.33
	40-50	23.33	0
	50-60	16.67	0
	60-70	10	0

3. Results and Disscusion
The result of the respondent's perception is seen from the influence of the background of the respondent's character toward the perception of the questionnaires tested on Pulo Geulis and Griya Katulampa. The seven backgrounds of the respondents are gender, age, education, type of work, origin, tribe and length of stay. The questionnaires tested consisted of five aspects: level of community knowledge, utilization, comfort, visual, and policy (regulation). The background of respondent's gender affects the material aspects of the questionnaires tested at the Pulo Geulis site in the aspects of utilization, comfort and policy while in Griya Kayulampa affect the visual and policy aspects. Background age of respondents influence on the questionnaire material tested in Pulo Geulis is on the material level of knowledge, utilization, comfort, and policy while in Griya Katulampa is on the material level of knowledge of

respondents. Respondent's educational background influences the questionnaires that are tested in Pulo Geulis that is on the material level of knowledge, utilization, convenience, and policy while in Griya Katulampa that is influential on material utilization, visual and policy. The background of the respondent's work affects the questionnaires tested in pulo geulis on the material of utilization, comfort, and policy while in Griya Katulampa that is on the comfort and visual material. Background of the respondent's influence on the questionnaires tested in Pulo Geulis is on the material level of knowledge, comfort, and policy while in Griya Katulampa that is only on the material utilization and visual. The background of the respondent's tribe has no effect on the material being tested either on location at Pulo Geulis and Griya Katulampa. Background of respondents' long life affect the questionnaires material tested in Pulo Geulis is on the material level of utilization and comfort while at the Griya Katulampa that is on the material utilization, comfort, visual and policy.

In knowledge aspects, the character of respondents background that influential in Pulo Geulis were education, age and origin while in Griya Katulampa the influence background was age (Table 2). Related aspects of knowledge in Pulo Geulis, on the material the function of blue open space is influenced by the educational background of junior high school graduates provide the highest response 26.67% do not understand. This showed that the educational background of respondents in the age of productive age was still relatively low affects the weak knowledge of responders related of BOS either from definition, type, or function. Material of river as one of blue open space influenced by background respondent's of age and origin, age of 26-65 years give highest response 60% do not understand and indigenous people give highest response 40% strongly agree. One's perception of the quality of a landscape was determined by the strong interaction between one's variables and one's knowledge of the landscape [5]. Related to the level of knowledge in Griya Katulampa, the material of water as the source of life was influenced by the background of age 26-65 years gives the highest response 66.67% strongly agree. Water is not only to demand of the needs of human life but also for all living things. The presence of blue open space has a positive impact on improving people's mental health [6]. River utilization material has been used in accordance with the function influenced by the age background, age 26-65 years gave the highest response 50% strongly agree. Based on the results shown by productive age (26-25 years) in Griya Katulampa was very responsive to knowledge about water which become an important part in life must be utilized in accordance with its function in daily life. The productive age has the highest level of knowledge in a community that was influenced by the number of interaction activities with its environment so that it can see, hear and obtain more the information [7].

In utilization aspects, the character of respondents background that influential in Pulo Geulis were gender, age, education, occupation and length of stay while in Griya Katulampa were education and origin (Table 3). Related to the aspect of utilization in Pulo Geulis, on the material of water source used in daily life age of 26-65 years 63.33% choosed to use river, and elementary and junior high school background as much as 36.67% have used PDAM. Based on the results, productive age was an active user of the river in daily activities other than the river as a source of water utilized, educational background it affects respondents to utilize PDAM water sources mainly for consumption in daily life. Materials of river as household waste and garbage was influenced by background of age 26-65 years give the highest response 87.67% disagree. Public perceptions were reflected of behavior that affects to the environment, due to the cleanliness caused by many residents who throw garbage in the river causing the impression of a slum and smell [8]. Material of utilization of water in the right way affects the quality of the environment, age group 26-65 years gives the highest response 86.67% agreed. However, the condition of the river in Pulo Geulis was very in bad conditions so its was not to be used as a source of water in daily life caused by the amount of waste in the river. Related to the material of river used in daily life, 16.67% of men use river as gathering place, 26.67% of men and 6.67% do not work as fishing grounds, and 6.67% entrerpreneur use river for fishing and keramba, 30% housewives and 16.67% length of stay 50-60 years using the river as a bath, and 10% length of stay 60-70 years using the river as a toilet. Activities carried out by the community in the river, especially men with background of freelance, visited the river either just to sit or spending time to fishing, while housewives and also people who have lived over 50 years come to the river to using river to daily activities (washing, bath and toilet).

Table 2. Result of Chi-Square test of respondent's perception on knowledge aspects

No	Aspects of Knowledge	Gender		Age		Education		Type of work		Origin		Tribe		Length of Stay	
		Pulo	Griya	Pulo	Griya	Pulo	Griya	Pulo	Griya	Pulo	Griya	Pulo	Griya	Pulo	Griya
1	Water as a source of life	0.10	0.61	0.36	**0.04***	0.54	0.54	0.46	0.32	0.83	0.49	0.26	0.93	0.66	0.43
2	The function of the blue open space	0.06	0.17	0.94	0.52	**0.00***	0.50	0.23	0.16	0.14	0.29	0.87	0.92	0.48	0.30
3	River as one of the blue open spaces	0.11	0.13	**0.02***	0.82	0.17	0.89	0.15	0.18	**0.03***	0.46	0.93	0.16	0.55	0.25
4	Utilization of the r iver has been in accordance with the function	0.64	0.35	0.55	0.06	0.06	0.80	0.59	0.34	0.12	0.82	0.88	0.46	0.17	0.53

Description Value α = 0.05 if test value> 0.05, then the tested background is not mutually tied (free) with questionnaire material in question. when the test value <0.05, then the tested background is tied to the questionnaire material in question. bold print (*), showing test values α <0.05.

Table 3. Chi-Square test results of perceptions of respondents on utilization aspects

No	Utilization Aspects	Gender		Age		Education		Type of work		Origin		Tribe		Length of Stay	
		Pulo	Griya	Pulo	Griya	Pulo	Griya	Pulo	Griya	Pulo	Griya	Pulo	Griya	Pulo	Griya
1	Benefits of water in daily life	0.37	0.92	0.16	0.70	0.86	0.10	0.07	0.28	0.27	0.51	0.98	0.14	0.66	0.93
2	Water resources in daily life:														
a	River	0.25	0.28	**0.00***	0.79	0.67	0.91	0.49	0.82	0.54	**0.02***	0.98	0.82	0.16	0.72
b	wells	0.87	0.74	0.81	0.19	0.22	0.59	0.86	0.38	0.14	0.13	0.82	0.53	0.44	0.33
c	PDAM	0.35	0.92	0.31	0.69	0.05	0.11	0.11	0.52	0.42	**0.00***	0.45	0.39	0.18	**0.02***
d	Water springs	0.43	0.70	0.36	0.92	0.98	0.72	0.21	0.72	0.61	0.74	0.57	0.68	0.16	0.58
3	River as a place of disposal of household waste and garbage	0.11	constant	**0.00***	constant	0.67	constant	**0.00***	constant	0.43	constant	0.99	constant	0.69	constant
4	Utilization of water in the right way affects the quality of the	0.85	0.24	**0.00***	0.90	0.73	**0.02***	0.08	0.41	0.70	**0.03***	0.10	0.36	0.12	0.83

environm ent

Description Value $\alpha = 0.05$ if test value> 0.05, then the tested background is not mutually tied (free) with questionnaire material in question. when the test value <0.05, then the tested background is tied to the questionnaire material in question. bold print (*), showing test values $\alpha < 0.0$

In comfort aspects, the character of respondents background that influential in Pulo Geulis were gender, age, origin, and length of stay while in Griya Katulampa length of stay and type work (Table 4). In Pulo Geulis related to the material of the comfort level around the river was influenced by the background of male which gives the highest response 26.64% less comfortable. Related to the material of comfort level when interacting with the river was directly affected by the ages of 26-65 year age provides the highest response 43.33% comfortable. The perception of the individual depends on his psychological state affecting his vision, taste, smell, hearing and touch [5] . Related to the comfort level material aspects at Griya Katulampa, the material level of comfort when interacting with the river was directly influenced by the background of long stay, the length of stay 20-30 years provide the highest response 26.67% comfortable. The Griya Katulampa community in general were comers with a length of stay more than 20 years still quite comfortable with the river existing condition, it was because the spatial condition of the housing located not directly on the edge of the river then no decreased direct interaction with the river. The comfort level of residential homes, the length of stay 1-10 years provides the highest response 30% comfortable. For the condition of the ressidential, people felt still quite comfortable and worthy to be occupied because of the environment that was still in good condition, especially protected area like the green line on riparian rivers and springs. The reason for choosing the location of the current residence of 36.67% housewife felt quite comfortable with the current occupancy because there has high society tolerance, due to high diversity and heterogeneity of society around there.

In visual aspects, the character of respondents background that influential in Griya katulampa were gender, education, occupation and origin (Table 5). Related to the visual material aspects in Griya Katulampa, the material of the river was influenced by educational background, high school graduates provide the highest response 26.67% less comfortable. Related to the level of comfort visual spasial arrangement, educational background of high school graduates provides the highest response 30% less comfortable. The level education of community was very influential on the assessment of the visual quality of the river, it can be concluded that the difference background status of educational that was part of the internal factors affected perception of the visual quality of the landscape. The perception of a person was not only influenced by internal factors but also external factors that was in the form of the stimulus itself and the circumstances in which the perception takes place [9]. Housewives were as a lower-frequency group doing activities outside the home compared to workers who work outdoors daily so that it affects their perception with what was seen and felt when looking at the environment. In addition, good visual conditions greatly affect the quality of life of the community. People living with BOS view shown that they have a more comfortable life with a more positive emotional level [10].

In policy aspects, the background of influential character of respondents in Pulo Geulis were gender, age, occupation, and origin while in Griya Katulampa were gender, education, and length of stay (Table 6). Related materials of the have been aware of policies and regulations regarding the use and protection of river areas, borders and the surrounding environment, the native people providing the highest response 40% already know. Based on this it can be seen that most people in Pulo Geulis already know and realize that the policy that protects the river area legally. The materials of the application of policies and regulations in the right way were influenced by the background of age and occupation, age of 26-65 years gave the highest response 36.67% and 13.33% of housewives rate was not appropriate. It also proves that the level of public awareness in running the rules and policies that have been set still very far from the word accordingly so that there are still many violations that cause environmental contamination. In Griya katulampa related to the material of the extent to which policies and regulations have been affected were influenced by gender and type of work backgrounds, 26.67% of men, 20% of housewives, and the informal sector rate the policies that were only partially implemented.

Table 4. Chi-Square test results of respondents' perceptions of comfort aspects

No	Comfort aspects	Gender		Age		Education		Type of work		Origin		Tribe		Length of Stay	
		Pulo	Griya	Pulo	Griya	Pulo	Griya	Pulo	Griya	Pulo	Griya	Pulo	Griya	Pulo	Griya
1	Comfort levels around the river	**0.02***	0.26	0.53	0.14	0.78	0.08	0.59	0.37	0.17	0.15	0.65	0.07	0.68	0.40
2	Comfort level when interacting with the stream directly	0.50	0.72	**0.02***	0.12	0.97	0.14	0.77	0.90	0.55	0.58	0.34	0.73	0.87	**0.02***
3	The level of comfort of residential homes today	0.64	0.72	0.14	0.06	0.53	0.09	0.76	0.85	0.64	0.21	0.71	0.69	0.88	**0.01***

Description Value α = 0.05 if test value> 0.05, then the tested background is not mutually tied (free) with questionnaire material in question. when the test value <0.05, then the tested background is tied to the questionnaire material in question. bold print (*), showing test values α <0.05.

Table 5. Chi-Square test results of perceptions of respondents to the visual aspects

No	visual aspects	Gender		Age		Education		Type of work		Origin		Tribe		Length of Stay	
		Pulo	Griya	Pulo	Griya	Pulo	Griya	Pulo	Griya	Pulo	Griya	Pulo	Griya	Pulo	Griya
1	Visuals of the river affect the quality of the environment	0.25	0.75	0.08	0.26	0.13	0.21	0.39	0.77	0.08	0.53	0.50	0.44	0.14	0.93
2	The visual level of the river	0.64	0.57	0.62	0.99	0.59	**0.00***	0.67	0.28	0.29	0.60	0.41	0.10	0.79	0.91
3	Level of comfort of visual arrangement	0.69	0.33	0.22	0.86	0.97	**0.00***	0.12	0.53	0.62	0.74	0.23	0.27	0.43	0.40

Description Value α = 0.05 if test value> 0.05, then the tested background is not mutually tied (free) with questionnaire material in question. when the test value <0.05, then the tested background is tied to the questionnaire material in question. bold print (*), showing test values α <0.05.

Table 6. Chi-Square test results of perceptions of respondents on the policy aspects

No	Policy Aspects	Gender		Age		Education		Type of work		Origin		Tribe		Length of Stay	
		Pulo	Griya	Pulo	Griya	Pulo	Griya	Pulo	Griya	Pulo	Griya	Pulo	Griya	Pulo	Griya
1	Knowing the policies and regulations regarding t he use and protection of river areas, borders and the surrounding environment	0.09	0.26	0.53	0.23	0.62	0.35	0.45	0.28	**0.03***	0.14	0.62	0.89	0.14	0.25

2	Conformity in the application of policies and regulations	0.27	**0.03***	**0.00***	0.60	0.92	0.43	**0.00***	0.21	0.54	0.74	0.30	0.42	0.45	0.27
3	The extent to which policies and regulations are in place	**0.03***	0.75	0.96	0.75	0.56	0.08	**0.01***	0.37	0.44	0.82	0.09	0.22	0.58	0.54

Description Value α = 0.05 if test value> 0.05, then the tested background is not mutually tied (free) with questionnaire material in question. when the test value <0.05, then the tested background is tied to the questionnaire material in question. bold print (*), showing test values α <0.05.

Conclusion

Based on the physical condition of the landscape and socio-cultural background of the community, there are differences in landscape character and community behavior between Pulo Geulis and Griya Katulampa on the utilization of water in daily activities and their interaction with RTB. Based on the perception, in Pulo Geulis, the background of age is the background that has the most relevance to the four aspects of the five aspects that are tested on the aspects of knowledge, utilization, comfort, and policy while in Griya Katulampa educational background that has the highest linkage to the three aspects of the five aspects tested. Educational backgrounds have relevance to aspects of utilization, visual, and policy.

References

[1] Arifin HS, Kaswanto, Nurhayati HSA. 2013. Analisis Lanskap Agroforestri: Analisis Ketersediaan *Green Water* dan *Blue Water* dalam Manajemen Lanskap yang Berkelanjutan di Daerah Aliran Sungai Ciliwung [Penelitian Strategis]. IPB Press. Bogor (ID): Institut Pertanian Bogor.

[2] Cahya DL, Nana J. 2012. Penataan Kawasan Kumuh (Pulo Geulis) Kelurahan Babakan Pasar Kecamatan Bogor Tengah Kota Bogor. *Planesa*. 3(1): 1-10.

[3] Haryanti DT. 2008. Kajian Pola Pemanfaatan Ruang Terbuka Publik Kawasan Bundaran Simpang Lima Semarang. [Tesis]. Program Pasca Sarjana. Universitas Diponegoro. Semarang (ID): Universitas Diponegoro.

[4] Mokodongan BK, Rieneke LE, Sela, Hendriek H, Karongkong . 2014. Indentifikasi Pemanfaatan Kawasan Bantaran Sungai Dayanan di Kotamobagu. *J Sabua*. 6(3): 273-283.

[5] Nassar JL. 1988. *Environmental Aesthetics Theoory, Research and Applications*. Cambridge (UK): Cambridge University Press. 259p

[6] Nutsford D, Amber LP, Simon K, Femke R. 2016. Residential Exposure to Visible Blue Space (But Not Green Space) Associated with Lower Psychological Distress in a Capital city. *Health and Place*. 3(2): 70-78.

[7] Hutabarat BTF, Ronald IO, Iyone H. 2015. Gambaran Perilaku Masyarakat Terhadap Pengelolaan Sampah Padat di Kelurahan Malayang II Kecamatan Kota Manado. *Jurnal kedokteran Komunitas dan Tropik*. 3(1): 41-47.

[8] Cesarin BT, Chorina G. 2015. Persepsi Masyarakat Terhadap Permukiman Bantaran Sungai. *Prosiding Temu Ilmiah IPLBI*. 117-122.

[9] Mayasari AD, Aji RK, Syahrani. 2014. Persepsi Masyarakat Terhadap Kebijakan Relokasi Penduduk Bantaran Sungai Karangmumus Samarinda Kalimantan. *eJournal Administrative Reform*. 2(4): 2422-2434.

[10] White M, Smithh A, Humphryes K, Pahl S, Snelling D, Depledge. 2010. Blue Open Space: The Importance of Water for preference, Affect, and Restorativeness Ratings of Natural and built Scene. *J. Enviroment Psychology*. 30: 48

IOP Conf. Series: Earth and Environmental Science **179** (2018) 012038 doi:10.1088/1755-1315/179/1/012038

Study about forest management planning by CSR activities: a case study of the Kishiwada Hilly Development District in Japan

A Yamamichi[1*], T Nakahara[2], K Furuya[3], and Y Shimomura[1]

[1] Osaka Prefecture University, 599-8531 Osaka-fu Sakai-city Naka-ku Gakuen-cho 1-1, Japan
[2] Issei Co., Ltd., 675-1217 Hyogo-ken Kakogawa-city Kamisou-cho Kusukuri 416, Japan
[3] Chiba University, Graduate school of Horticulture 271-8510 Chiba-ken, Matsudo-city Matsudo 649, Japan

[*]Email: cay24706@osaka-fu.ac.jp

Abstract. The objective of this study was to clarify the method of forest activities of CSR for collaboration between companies and local communities, focusing on the Kishiwada Hilly Development district in Japan. The research method is mesh analysis by GIS, a questionnaire, interviews and a field survey. First, this area was subdivided as divided into 2625 meshes, and each of the meshes was evaluated via the natural conservation importance index and recreational usability index. Second, we examined the zoning estimation method based on spatial characteristics and the difference in human activities. Third, we identified the beliefs and challenges for CSR forest activities by the four companies participating in forest activities in the study area. The result was, first, bamboo forest accounts for 26.2%, such as forests without administrators. Second, suitable forests for recreation totalled 52.3%, occupies a majority of the land. Third, the purpose of participation in corporate forest-making activities varied depending on the departments. A common problem was a lack of knowledge regarding forest management. In conclusion, these findings suggest that the development of forest management plans considering the needs of CSR forest activities and the conditions throughout the region would improve the resilience and the benefits of Satochi-*Satoyama*.

Keywords: CSR, GIS, Kishiwada Hilly Development District

1. Introduction

Recently, regarding resilience, socio-ecological production landscapes (SEPLs) have attracted attention. SEPLs are shaped through long-term harmonious interactions between humans and nature in a manner that fosters well-being while maintaining biodiversity and ecosystem services [1] [2]. Moreover, after the Natural Capital Declaration by the UNEP FI, natural environment conservation activities and community contribution activities are becoming more important for companies.

In Japan, since the 1960s, most community woodlands, namely "*Satochi-Satoyama*", have faced conservational problems, such as a decrease in area and an increase in abandonment. The social background for why forests are no longer managed includes the fuel revolution, changes in industrial structure, increasing depopulated areas, and a lack of personnel that support local industries. In addition, as public eco-awareness increases, people reaffirm the importance of the environmental conservation benefits and amenity benefits that *Satochi-Satoyama* provides. Some companies in Japan are trying to regenerate the environment of *Satochi-Satoyama* as part of their corporate social responsibility (CSR) activities. Although they are aiming to restore and use *Satochi-Satoyama* as natural, cultural, and recreational resources, many challenges remain for companies due to often lack the viewpoint of adaptive management of the natural environment. The objective of this study was to clarify the method of forest activities of CSR and construct a land use model for collaboration between companies and local communities, focusing on the Kishiwada Hilly Development district in Japan.

2. Material and Method

The study area, the Kishiwada Hilly Development district, is included in the hill region of the city of Kishiwada, Osaka Prefecture in Japan. The study district is about 159.0 ha and contains a housing development project district with the aim of restoring *Satochi-Satoyama*. In the period of high economic growth, some regional development was planned, but this business collapsed following the burst of the economic bubble. For that reason, forests and land without administrators remained.

In Osaka Prefecture, the administrative and municipalities introduce groups and forest owners who want forestry activities, as the system of adopt forest.

The research method was as follows. First, the study area was divided into 25 m × 25 m mesh, with 2625 meshes extracted via a geographical information system (ArcGIS ver. 10.3) [3, 4, 5]. Each of the meshes was evaluated via the natural conservation importance index and recreational usability index. Both these indices include vegetation naturalness and type, micro topographic features, slope, distance from facility, water area, mountain ridge, and view area. The collected data were reclassified through each indicator referring to previous research.

Second, we examined the zoning estimation method based on components of spatial characteristics [6, 7]. Applying the future land use plans to forest management based on the difference in human activities, four thematic maps were generated. They were based on a score from 1 to 4 points by weighting how much each indicator was overlaid. The four types of multiple forest management were recreation, environmental learning, ecological, and landscape formation. In addition, two types of land use model were developed by weighting the results of the integrated maps. Maps of the land use models were created to identify area preferences and values, from the perspectives of the conservation of the landscape and the promotion of recreational use.

Third, to identify the beliefs for CSR forest activities by the four companies participating in forest activities in the study area, a questionnaire survey was conducted. The main items in each survey were the factors for participating in CSR forest activities, form of participation, actual situation of activities, and future problems and issues. Interviews were also conducted with the company and relevant organizations, NPOs, and the local government (from 9th to 16th, November 2015). They were specifically about the needs and seeds of each stakeholder. In addition, a field survey was conducted based on the map created in the second step.

3. Result and Discussion

3.1. Evaluation of meshes via GIS (Table 1, Table 2)

The study district is located at the foot of the mountain north of Kono-yama in the city of Kishiwada, the ridges of which are seen throughout the district. The altitude, including the area around the district, is 42.5 to 295.8 m, with the high section on the south side. Regarding the area, 63.9% is sloping more than 5° (Table 1). After dividing the 29 ground vegetation zones into 11 types, bamboo forest accounts for 26.2% of this area (Table 2). Deciduous forests comprise 16.8% and evergreen forests 1.2%. The district has two main roads: one east to west and one north to south. In the entire district, a narrow path has supported the survival of *Satochi-Satoyama*. Ponds and waterways for agriculture comprise 17.1%.

Table 1. The Slope Level of Study area

degree	level (*e.g.*)	mesh	%
0~3	*easy to walk, cycling*	340	13.0%
3~5		606	23.1%
5~7	*experience and activity in nature*	592	22.6%
7~9		480	18.3%
9~11		336	12.8%
11~13	*hiking, sleigh ride*	104	4.0%
13~15		76	2.9%
15~17		52	2.0%
17~19	*mountain climbing*	19	0.7%
19~27		20	0.8%
	total	2625	100.0%

Table 2. The Zone types of Ground Vegetation

zone types	mesh	%
deciduous forest	442	16.8%
evergreen forest	31	1.2%
plantation	17	0.6%
wet grassland	4	0.2%
drying grassland	619	23.6%
bamboo forest	687	26.2%
cultivated land	419	16.0%
artificial grassland	43	1.6%
artificial structures	91	3.5%
bare land	270	10.3%
waters	2	0.1%
total	2625	100.0%

3.2. Land use modeling

3.2.1. Evaluation of future forest type (Table 3)

A land use model was constructed by considering the current situation and future land use in adaptive management of forests. Forest for recreation was evaluated by *utility* and vegetation type. The *utility* is combining the slope and distance from facilities. A suitable recreation rank I comprised 447 meshes (17.0% of the area), which were distributed in the north and west. Rank II, which is easy to utilize for recreation, comprised 927 meshes (35.3%) and was distributed throughout the district.

Forest for environmental learning was evaluated by *biodiversity* and distance from facilities. The *biodiversity* is combining the vegetation naturalness and water area. A suitable environmental learning rank I, with high biodiversity, comprised 322 meshes (12.3%). The meshes are distributed around the valley. Rank II for environmental learning, which is easy to access and expected to improve biodiversity, comprised 583 meshes (22.2%). These meshes are distributed along the agricultural waterways and in the western part of the district.

Ecological forest was evaluated by combining the vegetation naturalness and micro topographic features. Rank I of this type, which has high vegetation naturalness and soil that is rich in nutrients, comprised 141 meshes (5.4%) and was distributed in the north. Rank II, with the prospect of recovery of the ecosystem, comprised 950 meshes (36.2%), distributed along the valleys.

Forest for landscape formation was evaluated by combining the view area, mountain ridge, and vegetation type. Rank I, such as deciduous forests that enjoy high seasonality and high visibility, comprised 426 meshes (16.2%). Rank II, of which the visibility is low but located near the ridge, comprised 1018 meshes (38.8%). Both ranks of landscape formation forest are distributed from the southwest to the north.

3.2.2. Construction of land use models (Table 4, Table 5, and Figure 1)

These results (*3.1.1.*) were integrated from the perspective of the conservation of landscapes and the promotion of recreational use [8]; [9]. From the viewpoint of conservation, we applied the biosphere reserve (BR) of UNESCO's Man and the Biosphere Program. The results were divided into three zones: the core zone, which is important for conserving biodiversity; the buffer zone, which surrounds the core zone; and the transition zone, which require policies that are in harmony with environmental conservation. The core zone was composed of 380 meshes (14.5%) and located high visibility northern and valley lines, near the main road. The buffer zone included 1017 meshes (38.7%) (Table 4).

However, from the viewpoint of recreational use, emphasis was placed on the suitability for active use of the *Satochi-Satoyama*. The area targeting the recreation forest type totaled 1374 meshes (52.3%), occupies a majority of the land, and is distributed in a mosaic pattern throughout the entire area, including the northern part, which has high visibility (Table 5).

IOP Conf. Series: Earth and Environmental Science **179** (2018) 012038 doi:10.1088/1755-1315/179/1/012038

Table 3. The Evaluation of Future Forest type

suitable forest types		rank Ⅰ / point 4	Ⅱ / 3	Ⅲ / 2	Ⅳ / 1
recreation	mesh	**447**	**927**	675	576
	%	**17.0%**	**35.3%**	25.7%	21.9%
environmental learning	mesh	**322**	**583**	1403	317
	%	**12.3%**	**22.2%**	53.4%	12.1%
ecological	mesh	**141**	**950**	1130	404
	%	**5.4%**	**36.2%**	43.0%	15.4%
landscape formation	mesh	**426**	**1018**	244	937
	%	**16.2%**	**38.8%**	9.3%	35.7%

n=2625

Table 4. The Evaluation of Land Use for Conservation

for conservation n=2625

	core		buffer				transiton			others
	Ⅰ	Ⅱ	Ⅰ	Ⅱ	Ⅲ	Ⅵ	Ⅰ	Ⅱ	Ⅲ	
mesh	83	297	536	187	195	99	386	254	370	218
%	3.2%	11.3%	20.4%	7.1%	7.4%	3.8%	14.7%	9.7%	14.1%	8.3%
sum	*380(14.5%)*		*1017(38.7%)*				*1010(38.5%)*			

Table 5. The Evaluation of Land Use for Active Recreation

for active recreation n=2625

	Ⅰ	Ⅱ	Ⅲ	Ⅳ	Ⅴ	Ⅵ	Ⅶ	others
mesh	447	927	43	140	186	459	106	317
%	17.0%	35.3%	1.6%	5.3%	7.1%	17.5%	4.0%	12.1%
sum	*1374(52.3%)*		*183(7.0%)*		*186(7.1%)*	*565(21.5%)*		

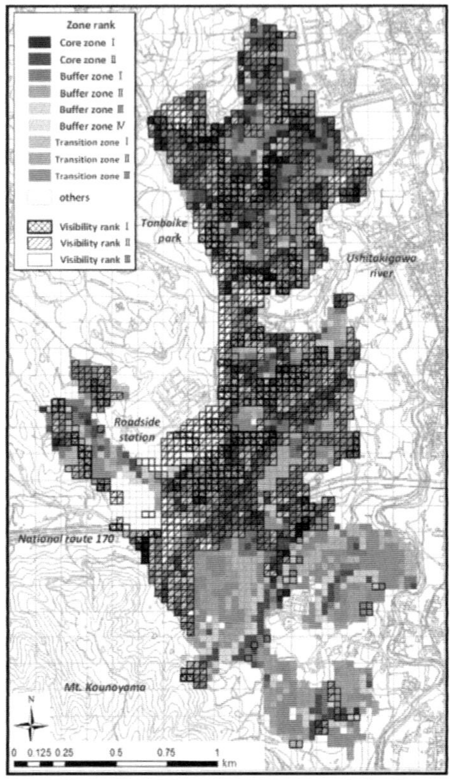

Figure 1. The Land Use model of Evaluation for Conservation

3.3. CSR Evaluation

3.3.1 Evaluation of CSR activity area (Table 6)

The *K company* activity area was 20.2% for slopes below 6° and 42.3% for slope of 6° to 18°. Approximately 60% are these slopes afford easy operation, so recreation is expected to be utilized. Currently, forest maintenance, such as mowing and bamboo cutting, is proceeding in 46.6% of these areas. However, bamboo forest accounts for nearly half, at 47.8%. Part of the area will be utilized as managed bamboo forest; otherwise, it would be necessary to switch to deciduous forest.

The *O bank*'s activity area was 17.8% for slopes less than 6° and 51.1% for slopes of 6° to 18°. It accounted for nearly 70% of active slopes. Currently, forest maintenance is carried out in 42.5% of these areas. As 39.6% are *vacant* open spaces like those of the *K Company*, recreation is expected to be utilized in a wide range, mainly in the open space.

The *H company* activity area was 27.7% with a loose slope less than 6°. Most of it is distributed to the east side of the route and is expected to be utilized as an activity base, such as like a plaza. On the west side of the street, there were almost steep slopes above 18°. However, owing to the distribution of deciduous forests, these areas are expected to be utilized as environmental learning forest via forest maintenance and management activities.

Regarding the *S company* activity area, bamboo forest comprised 51.5%. To consider recreational use, it is necessary to switch to deciduous forest in a wide range. In addition, 61.2% accounted for steep slopes of 18° or more. However, it is also expected to be used as a place for environmental learning, such as by using ponds in the active area, and by improving the roadside in the forest.

Table 6. The Evaluation of CSR Activity area

name (sector)	area (ha)	slope		vegetation (e.g.)		recent activity (work area)	forest rank (e.g.)*	
K Co., Ltd. (Other Products)	0.89	6°>	20.2%	bamboo	47.8%	planting, mowing,	conservation	b.
		6-18°	42.3%	*vacant*	25.0%	bamboo cutting	recreation	*(others)*
		18°<	37.5%	evergreen	15.2%	(46.6%)	visibility	Ⅲ, Ⅱ
O Bank (Banks)	0.8	6°>	17.8%	*vacant*	39.6%	bamboo cutting,	conservation	b., c.
		6-18°	51.1%	bamboo	31.1%	recreation activity	recreation	Ⅰ, Ⅱ, Ⅶ
		18°<	31.1%	*mixed*	20.2%	(42.5%)	visibility	Ⅱ, Ⅰ, Ⅲ
H Co., Ltd. (Other Products)	0.94	6°>	27.7%	deciduous	44.0%	planting, mowing	conservation	t., b., c.
		6-18°	19.7%	evergreen	17.6%	bamboo cutting	recreation	Ⅳ, Ⅱ, Ⅰ
		18°<	52.6%	*vacant*	14.4%	(23.2%)	visibility	*(others)*
S Co., Ltd. (Pharmaceutical)	0.45	6°>	11.6%	bamboo	51.5%	mowing,	conservation	b., t., c.
		6-18°	18.9%	deciduous	13.4%	bamboo cutting	recreation	Ⅰ, Ⅱ, Ⅳ
		18°<	61.2%	*vacant*	12.8%	(17.9%)	visibility	Ⅱ, Ⅲ

* c.: core zone, b.: buffer zone, t.: transition zone

3.3.2. Companies recognition for CSR forest activities

In corporate forest-making activities, the purpose of participation varied depending on the departments in charge of the CSR activities. For the *O bank*, which had a special CSR department in charge of the activities, the purpose of participation was to raise interest in environmental problems. Regarding the *H Company*, whose general affairs department was responsible for their CSR activities, the purpose of participation was to improve employees' ability and welfare benefits. The General Affairs Department/Environmental Affairs Department of the *K Company* was in charge of the CSR activities, with the purpose of participation to improve the environmental awareness of the company's employees and to improve the ability. Three companies other than *H Company* were the in-house commemoration of the founding ceremony etc. as a trigger to participate in the activity.

The reason for choosing the location of a CSR activity is that the *K Company* is closer to its headquarters, and the *O bank* is located closer to its branch offices. Besides, a roadside station (*Aisai-land* by *Japan Agricultural cooperatives*) was the base of activities. In the form of participation, the *K Company* established a new payment system for activities and was supporting the CSR activities. Meanwhile, the other three companies participated in in-house events and training. The *O bank* also saw activities by circles by employees in the company.

The current activity content was common to the four companies. It was a recreational activity with bamboo cutting and grass cutting, with assistance from NPOs. For the degree of satisfaction of the activities, the *K Company* and the *O bank* obtained the expected effect. However, the *H Company* had a short activity period, so the current satisfaction level was low. Future activities and future images are expected to be diversify the recreation through the collection of fruit trees and the observation of nature. A common problem of the present situation was a lack of knowledge regarding *Satochi-Satoyama* and forest management. Two companies, the *K* and *H* companies, replied that they cannot image the future in five years or the image of ideal *Satochi-Satoyama*. The challenge is to share knowledge by collaborating with other companies, unifying the management goals of the district, and sharing the roles of each activity area. On the other hand, the *S company* is hoping for cooperation from local governments and NPOs, it was reluctant about direct interaction among companies.

Evaluation by local governments and NPOs was mentioned as a merit that local governments does not cost the administrative expense of the land, and NPOs can obtain the reward by guidance.
It can be said that forestation is an attractive activity for companies as an approach to CSR, as it can meet social needs and nurture employees through such activities. On the other hand, it is also difficult to participate without a chance. It is expected that the number of participating companies will increase by local governments positively attracting companies and creating opportunities for forest-making activities. In the future, cooperation of local governments and NPOs will be necessary to implement satisfactory forest making for both companies and regions. Furthermore, it is also vital that firms that are the subject actors focus on the future image of the forests and clarify management policy. In addition,

local governments have grasped the current state of the entire district and found that it is important to systematically implement forest making that responds to corporate intentions. In the future, as the value of CSR activities within each company is recognized, it should be said that companies actively engage in CSR activities and revitalize forests.

4. Conclusion

In this study, we tried to search for possibilities of collaboration between companies and local communities in adaptive management of forests, from the perspective of CSR activities. From the result and discussion, however there are many bamboo forests left unmanaged now, it would be possible to switch to deciduous forest. Second, from the viewpoint of conservation, the land use models was distributed combining three zoning pattern throughout the entire area. Then suitable forests for recreation totaled 52.3%, occupies a majority of the land. Third, the purpose of participation in corporate forest-making activities varied depending on the departments. Even though there are differences included the reasons for their participation and the forest types of activity areas, a common problem of forest activities was a lack of knowledge regarding forest management.

These findings suggest that the development of forest management plans and processes considering the needs of CSR forest activities and the conditions throughout the region would improve the resilience and the benefits of *Satochi-Satoyama* [10, 11]. Furthermore, by cooperating with various management entities, both adaptive and long-term strategies should be taken in CSR forest management for the recovery of forest ecosystem services.

References

[1] Hashimoto S, Koriki C, Nakamura S, Hoshino S and Shimizu N 2015 *Hotspots Mapping of the Social-ecological Production Landscapes of the Noto Peninsula*. Landscape Research Japan Online. 8(0); 31-36

[2] Gu H and S M Subramanian 2014 *Drivers of change in socio-ecological production landscapes: implications for better management*. Ecology and Society. 19(1); 41

[3] Aikoh T. and Tomidokoro Y 2012 *Examination of Zoning Methodology Considering the Natural Resources and the Recreational Resources*. Landscape Research Japan Online. 5(0); 96-103

[4] Katagiri Y, Yamashita H and Ishikawa M 2004 *A Study on the Database of the Watershed Unit Based on the Common Data, and the Analysis and Evaluation Method for the Openspace Planning*. journal of the Japanese Institute of Landscape Architecture 67(5); 793-798

[5] Ostrom E 2009 *A General Framework for Analyzing Sustainability of Social-Ecological Systems*. Science. 24(325); 419-422

[6] Ministry of Land, Infrastructure, Transport and Tourism. (accessed 15 July 2015). *National Land Numerical Information*. available at: http://nlftp.mlit.go.jp/ksj-e/index.html

[7] Masuda N, Shimomura Y and Abe T 1990 *Study on the Environmental Planning of Reservoir based on the Conservative Utilized Potential*. journal of the Japanese Institute of Landscape Architecture 53(5); 257-262

[8] Kohsaka R, Togoe S, Tomiyoshi M, Iwata M, Fujihara Y and Matsuoka H 2015 *Quantitative Analysis on the Forestry Activities by Small and Medium-Sized Companies: Comparison between Kaga and Noto Areas in Ishikawa Prefecture Based on Postal Survey*. Journal of Forest Economics. 61(2); 13-22

[9] Tanaka N 2000 E*stimation of Tourism/Recreational Potential for Forest Management on a Watershed Scale*. journal of the Japanese Institute of Landscape Architecture. 63(5); 607-612

[10] Takeuchi K, Washitani I and Tsunekawa A 2001 *Satoyama : the traditional rural landscape of Japan*. University of Tokyo Press. Japan. 257pp

[11] Matsui T, Takeuchi K and Tamura T 1990 *Natural Environment of the hills: the characteristics and conservation*. Japan. 202pp

[12] Land Readjustment Union of Kishiwada city hills. 2015. *Southern Osaka city planning project: Environmental impact assessment document of Land Readjustment Project*. Japan.

Spatial and temporal distribution of nitrate concentration in Ciliwung River, Bogor City

A J Sutrisno[1*], R L Kaswanto[2], and H S Arifin[2]

[1] Master Student of Landscape Architecture, Graduate School of Bogor Agricultural University, Jl. Meranti Dramaga IPB Campus, Bogor 16680, Indonesia.
[2] Landscape Architecture Department, Faculty of Agriculture, Bogor Agricultural University, Jl. Meranti Dramaga IPB Campus, Bogor 16680, Indonesia.

*Email: alfredjsmanurung@gmail.com

Abstract. Water is very important as a source of life for human and other living organism. Bogor City has two main rivers, i.e. Cisadane River and Ciliwung River. Those rivers, particularly Ciliwung River has low in water quality, that has been affected by land use changes and human activities. Nitrate concentration is one of hazardous chemical observed in this research, due to Nitrate contamination is very dangerous for human health. Focus of this paper is how the spatial distribution of Nitrate concentration in Ciliwung River. There were 13 sites of water sampling in Ciliwung River. Water samples were collected in December 2016, January 2017, and February 2017. Inverted Distance Weighted method was used to identify spatial distribution of Nitrate concentration. According to the analysis results, Nitrate had different concentrations in December 2016, January 2017, and February 2017. The highest concentration of Nitrate in December 2016 has occurred in three sites, i.e. Kedung Halang, Cibuluh and Sukaresmi. Nitrate concentration had increased from 4.30 to 5.70 mg/L in January 2017, and the highest had occurred in Sukaresmi. Further more, Nitrate concentration had decreased become 4.80 mg/L in February 2017, where the highest concentration of Nitrate had occurred in Cibuluh, Sukaresmi, Bantar Jati, and Kebun Raya Bogor.

Keywords: Ciliwung River, nitrate concentration, spatial distribution, water quality

1. Introduction

Water pollution is big issue worldwide, especially in urban landscape. Hazardous chemicals are detected in rivers due to global rapid population growth and intensive domestic activities in urban landscape [1, 2, 3]. One of hazardous chemical in river is Nitrate concentration. Nitrate concentration is very soluble in water [4]. This condition affected contamination in river, because river flow lead Nitrate concentration spread along the river. Nitrate contamination is an environmental issue of considerable concern, because the potential health hazards to both human and animal populations [5, 6, 7, 8]. Limit concentration of Nitrate for drinking water is 10 mg/L [9].

Ciliwung River is one of aquatic environment in Bogor City. Population growth resulted in land use changes. Land use in riparian Ciliwung River dominated by settlement (38.69 ha) and housing (20.67 ha). In addition, human activity in riparian Ciliwung River affected water quality in the river. Nitrate concentration coming from human waste and agricultural waste. Composition of water pollution on

Ciliwung River is industrial 7.3%, detergent 40%, human waste 44%, and household waste 6.7% [10]. The purpose of this paper is determined spatial and temporal variation of Nitrate concentration in Ciliwung River.

2. Site Description and Methodology

2.1. Site description

This study was conducted at Ciliwung River, which is located in Bogor City with elevation 157 – 367 m above sea level (asl). There were 13 sites in Ciliwung River, all site in this river divided based on villages around the Ciliwung River (Table 1). Length of Ciliwung River, which flows in the part of Bogor City is 14.5 km (Figure 1).

Table 1. Sampling plot location and coordinate position.

	Site	Longitude	Latitude
S1	Katulampa	106^0 50'16.014"	$6^0$38'1.2210"
S2	Tajur	106^0 49'22.824"	$6^0$37'27.856"
S3	Sukasari	106^0 48'49.458"	$6^0$36'59.814"
S4	Baranangsiang	106^0 48'36.461"	$6^0$36'38.510"
S5	Babakan Pasar	106^0 48'21.552"	$6^0$36'26.205"
S6	Kebun Raya Bogor	106^0 47'57.837"	$6^0$35'27.875"
S7	Sempur	106^0 47'51.792"	$6^0$35'4.1190"
S8	Tanah Sereal	106^0 48'17.830"	$6^0$34'25.884"
S9	Bantar Jati	106^0 48'28.089"	$6^0$34'13.179"
S10	Kedung Badak	106^0 48'31.575"	$6^0$33'42.262"
S11	Kedung Halang	106^0 48'13.020"	$6^0$33'3.2380"
S12	Cibuluh	106^0 48'19.756"	$6^0$33'7.4350"
S13	Sukaresmi	106^0 48'5.0890"	$6^0$32'51.727"

2.2. Sampling and analysis

Water samples collected 50 ml from Ciliwung River in December 2016, January 2017, and February 2017. Water samples from each site analyzing with Ion Chromatography (LC-10A, Shimadzu, Japan). Ion chromatography used to determine the concentrations of Nitrate (NO_3^-). Before determine Nitrate concentration, water samples stored in refrigerator with temperature under 4°C.

2.3. Data analysis

This study used spatial interpolation. Spatial interpolation is used determined the distribution of Nitrate concentration in Ciliwung River furthermore, the basic principles of spatial interpolation is to create a surface that models the sampled phenomenon in the best possible way. There were 13 sampling point in this study, the point is determined by a random distance where each point represents a village based on the administrative boundaries of the village. This study uses one of spatial interpolation method that is inversed distance weighted.

Inverted distance weighted interpolation (IDW) is an interpolation method commonly used in spatial interpolation analysis [11]. In IDW method, it is assumed substantially that the rate of correlations and similarities between neighbours is proportional to the distance between them that can be defined as a distance reverse function of every point from neighbouring points [12]. Inverted distance weighted method is used to predict the value of a point to be known by using values from known points [13]. Where, $Z(x)$ is the value of the point to be known, n is the number of sample points, Z_i is the value of the known sample point. The distance between the known sample point and the sample point you want to know is d_i, and u is the weight value based on the distance.

$$Z(x) = \sum_{i=1}^{n} \frac{Z_i}{d_i^u} \bigg/ \sum_{i=1}^{n} \frac{1}{d_i^u} \tag{1}$$

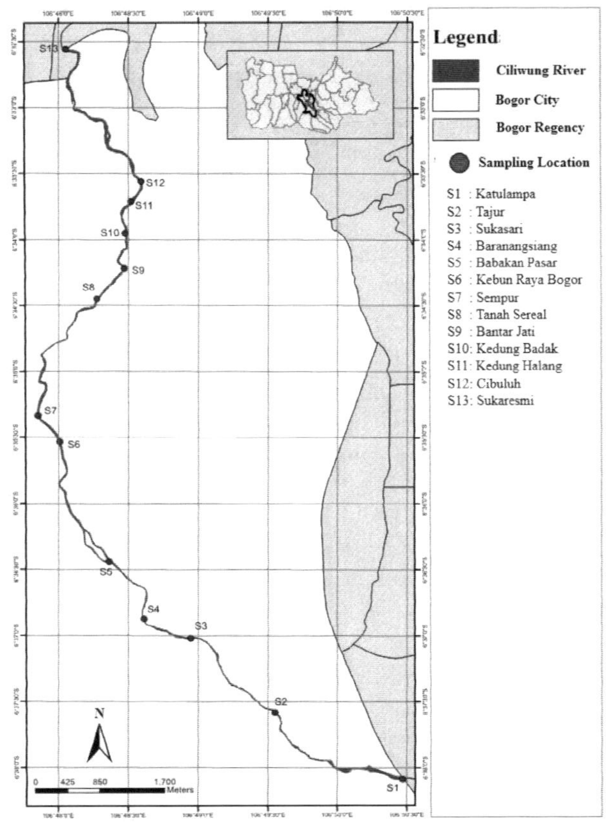

Figure 1. Study area and sampling plot in Ciliwung River.

3. Result

3.1. Spatial variation of Nitrate

River water flows from Katulampa (S1) to Sukaresmi (S13), this means the concentration of Nitrate flows from Katulampa (S1) to Sukaresmi (S13). Spatial variation of Nitrate influenced by land use. Land use of Ciliwung river riparian in the Bogor city is dominated by settlements 38.69 ha and housing 20.67 ha (Figure 2). Kedung Halang (S11), Cibuluh (S12), and Sukaresmi (S13) are villages dominated by settlements and housing. The highest Nitrate concentration was in all three villages and the lowest concentration was in Katulampa (S1). So, the concentration of Nitrate increases from Katulampa (S1) to Sukaresmi (S13).

IOP Conf. Series: Earth and Environmental Science **179** (2018) 012039 doi:10.1088/1755-1315/179/1/012039

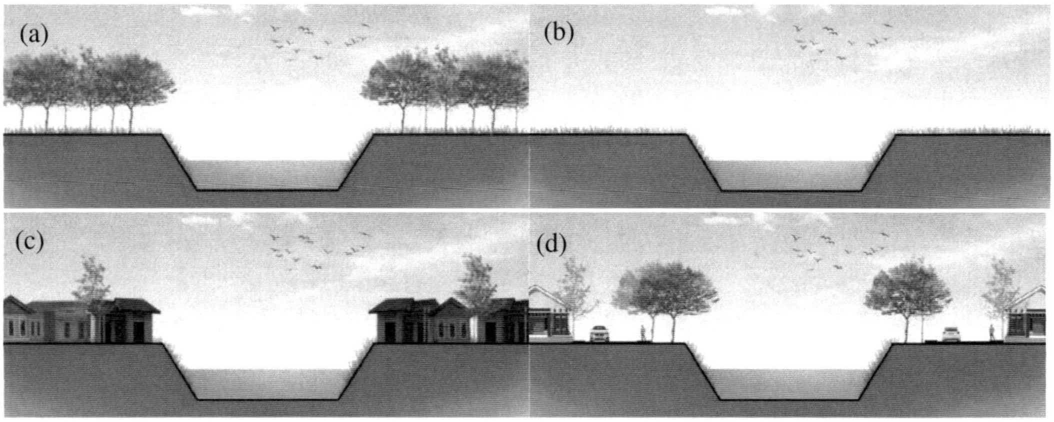

Figure 2. Land use in riparian Ciliwung River:(a) dominated by vegetation, (b) dominated by non-built area, (c) dominated by settlements and housing, (d) dominated by road and vegetation.

3.2. Temporal variation of Nitrate

The average of Nitrate concentrations in December 2016 was 4.30 mg/L with a maximum concentration was 4.90 mg/L in Kedung Halang, Cibuluh and Sukaresmi. The minimum concentration was 2.79 mg/L in Katulampa (Figure 3). In January 2017, the avarage of Nitrate concentration had increased from 4.30 to 5.70 mg/L with maximum concentration was 6.80 mg/L in Sukaresmi and minimum concentration was 4.09 mg/L in Katulampa (Figure 4). But, the average of Nitrate concentrations in February 2017 had decreased from 5.70 mg/L to 4.80 mg/L where, the maximum concentration was 5.39 mg/L in Cibuluh, Sukaresmi, Bantar Jati, dan Kebun Raya Bogor, and the minimum concentration was 4.00 mg/L in Katulampa (Figure 5).

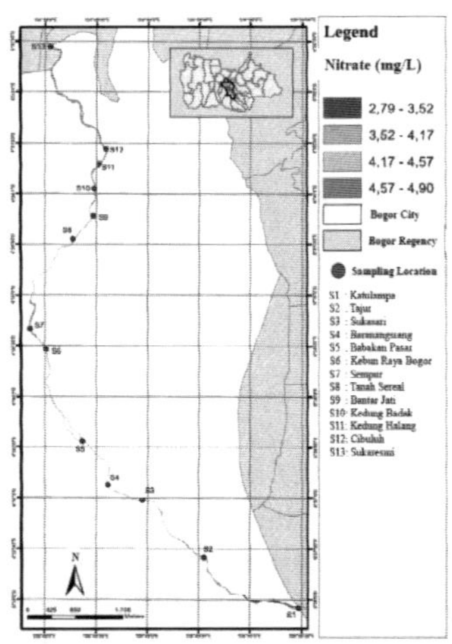

Figure 3. Distribution of nitrate in December 2016.

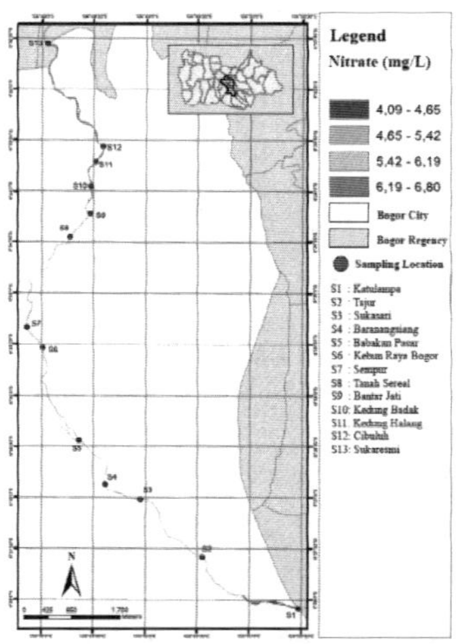

Figure 4. Distribution of nitrate in January 2017.

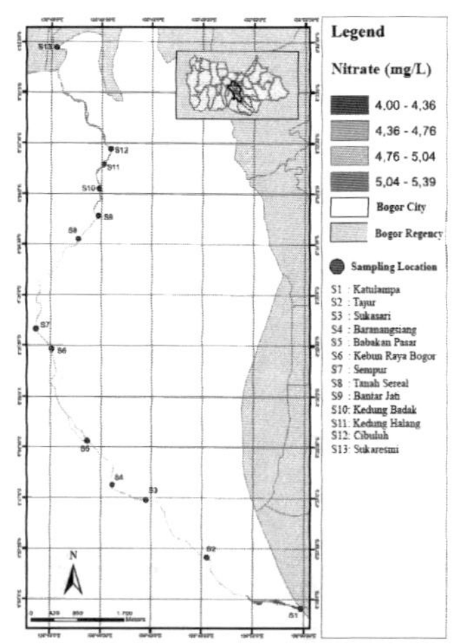

Figure 5. Distribution of nitrate in February 2017.

3.3. Variation of water temperature

Water temperature is a physical parameter that can encourage the occurrence changes in concentration on chemical parameters, this depends on height and low water temperature [14]; [15]; [16]. In December 2016, the average of water temperature in the Ciliwung River was 26.9 ^{0}C. In January 2017, the average of water temperature had decreased became 24.9^{0}C. It's identified that in February 2017 had decreased become the lowest, i.e. 22.3^{0}C (Figure 6).

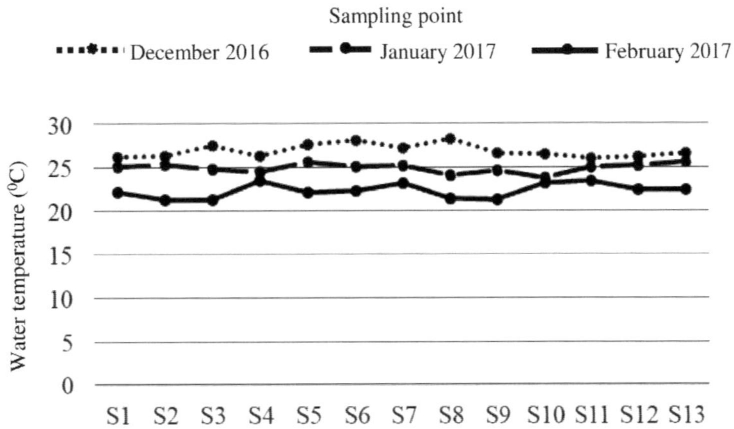

Figure 6. Water temperature.

4. Discussion

The decreasing and increasing of Nitrate concentration in Ciliwung River is influenced by several factors. In the term of temporal variation, seasonally water temperature has affected to decreasing and increasing of Nitrate concentration. When the water temperature decreases, it can decrease the concentration of Nitrate [17]. In this research, the average of water temperature was different for each month. It's 26.9^{0}C, 24.9^{0}C, and 22.3^{0}C in December 2016, January 2017 and February 2017, respectively.

Other factors that affect the decrease and increase of nitrate concentration are water discharge and rainfall. When, water discharge and rainfall were high, this condition can affect the decrease of Nitrate concentration, because the occurrence of dilution [18, 19].

In the term of spatial variation, land use may affect the decrease and increase of Nitrate concentration [20, 21]. Nitrate concentration was very high in Kedung Halang (S11), Cibuluh (S12), and Sukaresmi (S13), since the land use of these villages are settlements and housing predominantly. Household waste from settlements or housing can degrade water quality [22, 23, 24, 25].

The Nitrate concentration in Kedung Halang (S11), Cibuluh (S12), and Sukaresmi (S13) was different with Katulampa (S1). Nitrate concentration in Katulampa is the lowest one, due to the land use in this village is not dominated by settlements and housing.

5. Conclusion

It's well known that spatial and temporal distribution of Nitrate concentration was different. Water temperature, rainfall, and water discharge were affected distribution of Nitrate concentration through temporal variation. This condition influenced nitrate concentration in December 2016, January 2017, and February 2017. Where, Nitrate concentration in January 2017 was higher than Nitrate concentration in December 2016 and February 2017. On the other hand, the type of land use affects the distribution of

Nitrate concentrations in spatial variation. Where, the highest concentration in Kedung Halang (S11), Cibuluh (S12), and Sukaresmi (S13) with the use of land dominated by settlements and housing.

Reference

[1] Srebotnjak T, Carr G, De Sherbinin A, and Rickwood C 2012 A global water quality index and hot-deck imputation of missing data Ecol Indic 17 108–119
[2] Su S, Xiao R, Mi X, Xu X, Zhang Z, and Wu J 2013 Spatial determinants of hazardous chemicals in surface water of Qiantang River, China Ecol Indic 24 375–381
[3] Islam M, Han S, Ahmed M, and Masunaga S 2014 Assessment of trace metal contamination in water and sediment of some rivers in Bangladesh J Water Environ Technol 12 109–121
[4] Sutrisno A J, Han Z, Satake S, and Fukumoto K 2017 The role of wetland ecosystems to reduce the concentration of nitrate in groundwater IOP Conference Series: Earth and Environmental Sci (Bogor:2nd Int Sym Sustainable Landscape Development) 91 012004
[5] Canter L W 1997 Nitrates in Groundwater (New York: CRC Press Inc)
[6] Carlson J R and Breeze R G 1984 Ruminal metabolism of plant toxins with emphasis on indolic compounds J Anim Sci 58 1040–1049
[7] Davidson K L, Hansel W, Krook L, McEntee K, and Wright M J 1964 Nitrate toxicity in dairy heifers I Effects on reproduction, growth, lactation and vitamin A nutrition J Dairy Sci 47 1065–1073
[8] Sindelar J J and Milkowski A L 2012 Human safety controversies surrounding nitrate and nitrite in the diet Nitric Oxide 26 259–266
[9] World Health Organisation (WHO) 2006 Guidelines for Drinking Water Quality vol 2, Recommendations -3rd ed (Geneva:World Health Organization) p 417
[10] Satmoko Y 2010 Kondisi kualitas air Sungai Ciliwung di Wilayah DKI Jakarta ditinjau dari parameter organic, amoniak, fosfat, deterjen, dan bakteri coli J Air Indonesia 6 34–42
[11] Yaxin C, Ruimin L, Chengchun S, Peipei Z, Chenghong F, and Zhenyao S 2012 Spatial and temporal variations in nitrogen and phosphorous nutrients in the Yangtze River Estuary Marine Pollution Bulletin 64 2083–2089
[12] Agung Setianto and Tamia Triandini 2013 Comparison of kriging and inverse distance weighted (IDW) interpolation Methods in lineament extraction and analysis J SE Asian Appl Geol 5 21–29
[13] Xie Y F, Chen T B, and Lei M 2011 Spatial distribution of soil heavy metal pollution estimated by different interpolation methods: accuracy and uncertainty analysis Chemosphere 82 468–476
[14] Shiller A M 1997 Dissolved trace elements in the Mississippi river: seasonal, interannual and decadal variability Geochim Cosmochim Acta 61 21–30
[15] Tripti M, Gurumurthy G P, Balakrishna K, and Chadaga M D 2013 Dissolved trace element biogeochemistry of a tropical river, Southwestern India Environ Sci Pollut Res 20 4067–4077
[16] Bu H, Wang W, Song X, and Zhang Q 2015 Characteristics and source apportionment of dissolved trace elements in the Jinshui river of the South Qinling Mts , China Environ Sci Pollut Res 22 14248–14257
[17] Thomas J, Joseph S, and Thrivikramji K P 2017 Hydrochemical variations of a tropical mountain river system in a rain shadow region of the Southern Western Ghats, Kerala, India Applied Geochemistry J 1 1–15
[18] Kirchner J W, Feng X H, Neal C, and Robson A J 2004 The fine structure of water – quality dynamics: The (high – frequency) wave of the future Hydrol Process J 18 1353–1359
[19] Llyod C E M, Freer J E, Collins A L, Johnes P J, and Jones J I 2014 Methods for detecting change in hydrochemical time series in response to targeted pollutant mitigation in river catchments Hydrology J 514 297–312
[20] Junaidi E and Tarigan S D 2011 Pengaruh hutan dalam pengaturan tata air dan proses sedimentasi

daerah aliran sungai (DAS): studi kasus di DAS Cisadane J Penelitian Hutan dan Konservasi Alam 8 155–176
[21] Wang Y, Li Y, Liu, X, Liu F, Li Y, Song L, Li H, Ma Q, and Wu J 2014 Relating land use patterns to stream nutrient levels in red soil agricultural catchments in subtropical central China Environ Sci Pollut Res 21 10481–10492
[22] Lei P, Zhang H, and Shan B Q 2012 Dynamic characteristics of nitrogen and phosphorus in the representative input tributaries of Danjiangkou Reservoir Environ Sci J 33 3038–3045
[23] Karima A, Kaswanto RL 2017 Land use cover changes and water quality of Cipunten Agung Watershed Banten. IOP Conf Series: Earth and Environmental Science 54(1):012025
[24] Wilson C O 2015 Land use/land cover water quality nexus: quantifying anthropogenic influences on surface water quality Environ Monit Assess 187 424
[25] Noviandi TUZ, Kaswanto RL, Arifin HS 2017 Riparian landscape management in the midstream of Ciliwung River as supporting Water Sensitive Cities program with priority of productive landscape. IOP Conf Series: Earth and Environmental Science 91(1):012033

IOP Conf. Series: Earth and Environmental Science **179** (2018) 012040 doi:10.1088/1755-1315/179/1/012040

Bird community in urban residential area: Which species sustained after five elapse years?

A Mardiastuti[1], Y A Mulyani[1*], A T Asmoro[1], and M S K Putra[1]

[1] Department of Forest Resources Conservation and Ecotourism, Faculty of Forestry, Bogor Agricultural University, Darmaga Campus, Bogor, West Java, Indonesia

[*]E-mail address: ani_mardiastuti@ipb.ac.id, aniipb@indo.net.id

Abstract. Birds are able to sustain almost everywhere, including in urban residential areas where habitat components were limited. Some birds may survive, while others might not. The objective of the research was to identify bird species which still sustained in an urban area, after 5 years elapsed time. The study site was Sentul City (West Java), a large residential area that arranged into clusters of different housing and tree planting arrangement. Birds were sampled in 2012 and 2017 within 5 housing clusters and boulevard, using a point count method. There were 42 bird species found in 2012 and 35 species in 2017, and thus there were a decrease of 7 species (16%). Eurasian Eurasian Tree sparrow, Sooty-headed Bulbul, and Scarlet-headed Flower pecker were the most abundant birds. Two new species were recorder in 2017. About 60% of bird species were sustained as exploiters and adapters in the study sites, including tree sparrows, bulbuls (sooty-headed), flowerpeckers and sunbirds, and munias. The semi-natural cluster within the residential area harbour avoiders that preferred shrubby old-growth habitat such as babblers, cuckoos, and nightjars.

Keywords: Bogor, housing complex, housing clusters, Sentul City, urban birds

1. Introduction

Cities worldwide are the most heterogeneous landscapes, resulting from highly fragmented environments, and create a mosaic of patches of land use types of various sizes [1, 2]. Cities in Indonesia is no exception. Unfortunately, scientific research related to the impact of urban fragmentation on urban biodiversity in Indonesia is lacking.

In Indonesian cities, most cities were dominated by residential areas, although the average figure is unavailable. With the current movement toward green development, residential areas in cities have been designed and managed toward more environmental-friendly areas. The typical important activity by the developer of residential areas is providing environmental-friendly habitat by planting trees and creating/maintaining green open spaces.

Trees along the road and among housing, planted in the backyards, as well as parks and other green open spaces in the residential areas provide good habitat for urban wildlife, including birds. Some research on urban bird diversity have been conducted in big Indonesian cities in Java, e.g. Jakarta [3], Bandung [4], and Semarang [5]. However, there has been no research on temporal basis, and thus there has been no information about the species that are able to sustain in residential areas after some period of time.

In order to understand how bird community responds to urban development and fragmentation in residential areas after a period of time, a field research in Sentul City residential area was conducted. During the last five years, the residential area in Sentul City has been rapidly developed. Several housing clusters, apartment complex, and malls were and have been built, eliminating areas that might have been served as bird habitat in previous years. The objective of this research was to reveal the bird community in Sentul City in 2012 and again in 2017, in particular to find out which species were able to sustain in the residential areas and which ones were lost or extirpated.

2. Methods

2.1. Description of the study site

The study area was Sentul City residential area, located in the periphery of the City of Bogor, West Java. The whole complex occupied an area of around 2,400 ha, consists of 13 housing complexes (covers most area), malls, hospital, amusement park, mosques and churches, hotels, offices, apartment, market, house-shop complex, convention centre, schooling, bus terminal, and 18-hole golf course. The whole area basically is a landscape of mosaic small patches for birds.

Six sites were selected as study area, consisted of five housing clusters (representing different housing age and house size) and the main boulevard. The housing clusters were Northridge, Bukit Golf Hijau-Mediterania I, Venesia-Pasadena-Sakura, Bali Cluster, and Cluster Argenia Cluster. Trees, palm, and shrubs (mostly are indigenous species, some are non-local) have been planted mainly along the main road (boulevard) and between houses within clusters, except for Northridge Cluster.

The Northridge is very spacious, has big houses on large-sized land. There are many unbuilt areas in between houses, and this cluster overall resembles semi-natural area. There is also a small lake in this cluster. Bukit Golf Hijau-Mediterania I was built about 20 years ago, consists of large to medium sized houses. Mature raintree *Samanea saman* characterized this cluster and creates heavy shading along the major roads. Venesia-Pasadena-Sakura have medium to small houses, and located in the valley, surrounded by some open spaces. Bali Cluster (Tapaks iring, Legian, Besakih, Udayana) has small houses, mostly town-house style. Argenia Cluster is relatively new cluster of medium-sized houses, built approximately 9 years ago. Trees planted in this cluster are almost reach their maturity.

2.2. Field method

Birds within each cluster and along the boulevard were observed by using the standard point count method [6] in 2012, and again in 2017. In each cluster, 10 random points (radius 25 m) were established to sample the bird communities. Duration of field observations varied, and the observations were ended when there was no additional new species, which took between 4 to 10 days for each cluster. Nomenclature and species identification follows [7].

Duration for bird observation in each point was 10 mins, after a 2-min. wait. Data recorded were species name and its number. Each point was replicated at least three times in different days, to ensure a representative result. Observation was conducted twice a day, in the morning (05:30-9:00 am) and in the afternoon 3:30-6:00 pm). Diurnal bird survey was not conducted in heavy claudy days, rainy days or when thick mist occurred. Nocturnal birds were intermittently surveyed and identified mostly from their calls.

Based on the occurrence of the birds in the clusters, birds were categorization as ‘exploiters’ and ‘adapters’, and ‘avoiders’, followed the definitions by [8] and [9] for the birds’ response to urbanization. Modified to the study area, urban exploiters were defined as species that are most successful (i.e. widespread, can be found in 5 to 6 sites) in the residential area being studied. Urban adapters were species that commonly found (i.e. found in 2 to 4 sites), while urban avoiders were those that can be found in a semi-wild habitat (i.e. tend to avoid urban environment), in this case was the Northridge Cluster. Exploiters and adapters were considered as sustained in the residential areas.

Bird species found in 2017 for each category were compared with those of 2012 to find out about the species that still sustained between the two periods. Newly recorded bird(s) in 2017 was identified, and so do the lost or extirpated species.

3. Results

A total of 43 bird species was recorded in all sites (data of 2012 and 2017 combined) (Table 1). There were 35 bird species in 2017, slightly lower than bird species found in 2012, which was 42 species. In 2017, two new species were recorded in the Northridge Cluster (semi-natural area), namely Black-crowned Night Heron and White-headed Munia.

Omitting the aerial foragers that might not be resident to the study area, the seven most common species, in decreasing order of abundance in 2012 and 2017 observations (combined), were: Eurasian Tree sparrow (*Passer montanus*), Sooty-headed Bulbul (*Pycnonotus aurigaster*), Scarlet-headed Flowerpecker (*Dicaeum trochileum*), Javan Munia (*Lonchura leucogastroides*), Olive-backed Sunbird (*Nectarinia jugularis*), Fluvous-breasted Woodpeker (*Dendrocopos macei*), and Common Tailorbird (*Orthotomus sutorius).* These species can be categorized as exploiters of residential areas in the study sites.

About 60.4% of the bird species were able to sustain in the residential areas, mostly insectivores, and nectarivores, and granivores (seed-eaters). Surprisingly, some piscivores (fish eaters) and bark borers were also part of the sustained species. In contrast, avoider species that might not sustain in the residential areas were generally raptor and other forest-dependent species (e.g. babbler, green pigeon).

Comparing between clusters, in both years Northridge Cluster had the highest number of bird species, followed by Bukit Golf Hijau, Argenia, Bali and Venesia. Along the main boulevard were also provide a good habitat for birds, comparable with other clusters. Except for the Venesia-Pasadena-Sakura, the species number in all study sites was decreasing. Even with the existence of the semi-natural habitat, Sentul City has been losing 7 bird species in the past five years, although one new species was recorded in of the Northridge Cluster.

Table 1. Presence (1) and absence (0) of birds in 2012 and 2017, organized based on their response to urbanization; bold – not found in 2017, underlined – new record in 2017; sequence and nomenclature follows [7]

Response to Urbanization	Common Name	Latin Name	2012	2017
Exploiters	Spotted Dove	*Streptopelia chinensis*	1	1
	Cave Swiftlet	*Collocalia linchi*	1	1
	Fulvous-breasted Woodpecker	*Dendrocopos macei*	1	1
	Pacific Swallow	*Hirundo tahitica*	1	1
	Striated Swallow	*Hirundo striolata*	1	1
	Sooty-headed Bulbul	*Pycnonotus aurigaster*	1	1
	Common Tailorbird	*Orthotomus sutorius*	1	1
	Olive-backed Sunbird	*Nectarinia jugularis*	1	1
	Scarlet-headed Flowerpecker	*Dicaeum trochileum*	1	1
	Eurasian Tree Sparrow	*Passer montanus*	1	1
	Javan Munia	*Lonchura leucogastroides*	1	1
	Scaly-breasted Munia	*Lonchura punctulata*	1	1
Adapters	Rusty-breasted Cuckoo	*Cacomantis merulinus*	1	1
	Plaintive Cuckoo	*Centropus bengalensis*	1	1
	Lesser Coucal	*Alcedo meninting*	1	1

	Black-nest Swiftlet	*Collocalia maximus*	1	1
	Blue-eared Kingfisher	*Halcyon cyanoventris*	1	1
	Javan Kingfisher	*Dendrocopos macei*	1	1
	Little Swift	*Apus affinis*	0	1
	Small Minivet	*Pericrocotus cinnamomeus*	1	1
	Common Iora	*Aegithina tiphia*	1	1
	Yellow-vented Bulbul	*Pycnonotus goiavier*	0	1
	Great Tit	*Parus major*	1	0
	Black-capped Babbler	*Pellorneum capistratum*	1	0
	Golden-bellied Gerygone	*Gerygone sulphurea*	1	1
	Olive-backed Tailorbird	*Orthotomus sepium*	1	1
	Bar-winged Prinia	*Prinia familiaris*	0	1
	Long-tailed Shrike	*Lanius schach*	1	1
	Oriental White-eye	*Zosterops palpebrosus*	1	1
Avoiders	Black-crowned Night-heron	*Nycticorax nycticorax*	0	1
	Spotted Kestrel	*Falco moluccensis*	1	0
	White-breasted Waterhen	*Amaurornis phoenicurus*	1	1
	Pink-necked Green Pigeon	*Treron vernans*	1	0
	Rock Pigeon[f]	*Columba livia*	1	0
	Zebra Dove	*Geopelia striata*	1	1
	Banded Bay Cuckoo	*Cacomantis sonneratii*	1	0
	Asian Drongo-Cuckoo	*Surniculus lugubris*	1	0
	Collared Scopsowl	*Otus lempiji*	1	1
	Large-tailed Nightjar	*Caprimulgus macrurus*	1	1
	Collared Kingfisher	*Todiramphus chloris*	1	0
	Asian House-martin	*Delichon dasypus*	1	0
	Pied Triller	*Lalage nigra*	1	0
	Horsfield's Babbler	*Malacocincla sepiarium*	1	0
	Chestnut-capped Babbler	*Timalia pileata*	1	1
	Common Pipit	*Anthus novaeseelandiae*	1	0
	White-breasted Wood-swallow	*Artamus leucorhynhus*	0	1
	Little Spiderhunter	*Arachnothera longirostra*	1	0
	White-headed Munia	*Lonchura maja*	0	1
	Total number of species		42	35

Table 2. Number of bird species in each cluster in Sentul City residential areas in 2012 and 2017

	North-ridge	BGH[1]	Venesia[2]	Bali[3]	Argenia	Boule-vard	Sentul City
Number of species; 2012	33	23	14	16	17	21	42
Number of species; 2017	27	15	17	13	12	16	35

Number of new species	3	0	6	1	3	4	2
Decreased in species number [%]	11	8	3	7	5	11	7 [16.7%]

[1]Bukit Golf Hijau-Mediterania I; [2]Venesia-Pasadena-Sakura; [3]Bali Cluster consists of Tapaksiring, Legian, Besakih, Udayana; f feral species

4. Discussion

Comparing with other research in other urban areas (e.g. [3, 4, 5]), bird community found in Sentul City residential areas were resembled to other cities in Java. Bird species in the studied clusters and boulevard showed dynamic changes, in which some species were lost and others were newly established. As predicted by the theory of island biogeography [10], over a long time period, there would be a turnover in bird community, including in urban bird community as also reported in Bogor Botanical Garden [11], Singapore [12], and in the isolated island of Barro Colorado in Panama [13].

Despite the fact that the bird community was dynamic over the five-year period, some bird species were able to sustain in the residential areas. The mosaic heterogenous patches in the residential areas still support birds that do not required specific habitat components. While some bird species require large habitat patches and forest-resembled habitat (e.g., thick canopy, dense shrub), small open patches may be sufficient for others.

Previous research [14] shown that urbanization typically leads to decrease in species number, an increase of biomass of abundant species, and a shift toward more generalist species. The habitat loss due to the rapid and vast development of Sentul City contributed to the decrease of the bird species and the loss of some non-adaptive species.

Remnants of semi-natural green open space in urban landscape, including in a residential area, have larger core areas that are not influenced by edge effect, such as higher anthropogenic disturbance and predation [15, 16, 17]. Therefore, in semi-natural area of Northridge will provide species with a more specific habitat requirement, as also reported by [16] in Madrid, Spain. Raptors, water-associated bird (waterhens, night-herons), birds that have brood parasites breeding strategy (cuckoos), birds that preferred bush-shrub habitat (babblers), and top canopy fruit eaters (green pigeons) were urban avoiders that can only be found in the semi-natural habitat of Northridge Cluster. These species may not sustain in the residential area if the semi-natural habitat is no longer available.

The loss of habitat due to the development of the residential areas of Sentul City can be compensated mainly by and fulfilling birds' habitat components. Some suggested activities are maintaining the semi-natural habitat as refuge area, intensifying planting of birds' food sources, and providing additional shrub layers in the existing parks/boulevard as new additional micro-habitat.

Acknowledgments. The Authors would like to thank the management of Sentul City (PT Sukaputra Graha Cemerlang) for their permission to conduct research in Sentul City.

References

[1] Faeth SH, Bang C, Saari S 2011 Ann. N.Y. Acad. Sci. 1223: 69–81
[2] Wu J 2010 Landscape Ecol. 25:1-4
[3] Mardiastuti A 2016 Considerations for Biodiversity Conservation in Modern Megacities: Lessons from Jakarta, Indonesia Pp. 60-62 in Tropical Conservation: Perspectives on Local and Global Priorities Aguirre AA and Sukumar R (Eds.) Oxford: Oxford University Press
[4] Fardila D and Sjarmidi A 2012 Res. J. Recent. Sci. 1:23-32
[5] Rahayuningsih M and Priyono B 2016 Int. J. Ecol. Dev. 31(1): 63-72
[6] Bibby C, Jones M, Marsden S 2000 Expedition Field Techniques Birds Surveys London: Expedition Advisor Centre
[7] MacKinnon J, Phillipps K, van Balen B 1998 Seri Panduan Lapangan Burung-Burung di Sumatera, Jawa, Bali dan Kalimantan Bogor: Birdlife International-Indonesia Program–Pusat

Penelitian dan Pengembangan Biologi LIPI
[8] Blair RB 1996 Ecol. Appl. 6:506-519
[9] Luniak M 2004 Synurbization: Adaptation of animal wildlife to urban development Proc. 4th Int. Urban Wildl. Symp., Warsaw, Poland
[10] MacArthur RH and Wilson EO 1967 The Theory of Island Biogeography Princeton, New Jersey: Princeton University Press
[11] Diamond JM, Bishop KD, van Balen S 1987 Conserv. Biol. 1(2): 132–142
[12] Chong KY 2013 The Effects of Urban Greenery on Biodiversity (Doctoral Dissertation) Singapore:
[13] Department of Biological Sciences - National University of Singapore
[14] Robinson WD 1999 Conserv. Biol. 139(1): 85-97
[15] Chace JF and Walsh JJ 2006 Landsc. Urban Plan. 74:46–69
[16] Saunders DA, Hobbs RJ, Margules CR. 1991 Conserv. Biol. 5(1):18–32
[17] Fernandez-Juricic E 2001 Biodivers. Conserv. 10:1303–1316
[18] Schneider NA, Low M, Arlt D, Prart T 2012 PLoS One 7:e31517

Urban cemeteries as habitat for birds

A Meltriana[1*], A Mardiastuti[1], and Y A Mulyani[1]

[1] Department of Forest Resources Conservation and Ecotourism, Faculty of Forestry, Bogor Agricultural University, Darmaga Campus, Bogor 16680, West Java, Indonesia

[*]E-mail: ipehaha610@gmail.com

Abstract. Cemeteries in an urban area may occupy a large area and harbor many bird species. The aim of this paper was to identify bird species found in cemeteries of urban area in Bogor, West Java. Three largest cemeteries were selected as study sites: Gunung gadung (36.0 ha), Dreded (6.5 ha), and Blender (6.7 ha). Point count method was used, followed by calculation of Shannon diversity indices, Jaccard similarity index, and species dominance. A total of 33 bird species were found in all sites, of which 21 species (H' 1.86) were found in Gununggadung, 20 species (H' 1.74) in Dreded, and 14 species (H'1.77) in Blender, all considered to have low species diversity. Similarity among sites were quite high (0.64-0.72), all showed typical urban bird species although the three sites are far apart. Dominant bird species were aerial insectivores (swiftlet) and granivores (especially munias). The low bird diversity in cemeteries was due to the poor tree species and coverage. Planting with more diverse tree plants might increase bird species in cemeteries, as well as increase their other ecological function in urban area.

Keywords: Birds, Habitat, Urban Area

1. Introduction

The growth of a city has resulted in the declining of available habitats for urban birds. The requirement to set aside an area for green open spaces in the city development has provided an opportunity for urban bird conservation; however, in general the size of urban green open spaces are relatively small. The model of island biogeography developed by MacArthur and Wilson [1] has shown that there is a positive correlation between size and bird species richness [1]. This relationship has also been shown in many studies, e.g [2, 3, 4]; therefore large green open spaces are needed to conserve birds in urban areas. One type of urban green spaces that may occupy a large area is cemetery. It is expected that urban cemeteries would harbor many bird species. Lussenhop [5] found that there was a positive correlation between urban birds with area size in cemetery and surrounding areas.

Most cemeteries in urban areas are also planted with vegetation, especially lush canopy trees such as rain tree (*Samanea saman*), and frangipani (*Plumeria accuminata*). The Regulation of Ministry of Public Works of Indonesia No. 5 Year 2008 gives guideline on the selection of plant species to be planted in cemeteries. In addition providing shades, the species should also capable in improving the ecological roles of cemeteries. Identification of bird species inhabiting cemeteries and knowledge on the use of cemeteries by birds will be useful in encouraging bird conservation outside conservation areas. Therefore, the objective of this study was to identify bird diversity in three different cemeteries in Bogor City.

2. Study site and method

2.1. Description of the study site

There were eight public cemeteries managed by the City of Bogor, but only three were chosen as the study sites based on their size (> 50,000m^2): Gunung Gadung Public Cemetery, Dreded Public Cemetery, and Blender Public Cemetery (Figure 1). Gunung Gadung Public Cemetery is the largest public cemetery in Bogor (360,000m^2), and located in the Sub District of South Bogor (06°38'65''S 106°48'53"E). The tree species planted include frangipani, pine (*Pinus merkusii*), umbrella-tree (*Maesopsis eminii*), leucaena (*Leucaena leucocephala*), candle nut (*Aleurites moluccana*) and mahogany (*Swietenia macrophylla)*, with open canopy. Unlike the other two cemeteries, this cemetery is not surrounded by housings, but there is a quiet public road pass through this cemetery. Gununggadung public cemetery is also dominated by Chinese style tombs, which consists of buildings resembling small houses surrounded by miniature garden. Dreded Public Cemetery (64,8m^2) is located in Sub District of South Bogor (06°37'05''S 106°48'16"E) and planted with raintree (*Samanea saman*), frangipani (*Plumeria acuminata*) and figs (*Ficus* sp) with rather close canopy, giving an impression of shady cemetery. This cemetery is surrounded by human settlement, bordered by road, close to railway (120m) and Cisadane River (30m). Blender Public Cemetery (66,7m^2) is located in Sub District of Tanah Sareal (06°34'02''S 106°47'49"E) and planted with frangipani, Spanish cherry (*Mimusops elengi*), loquat (*Eriobotrya japonica*), figs and blackboard tree (*Alstonia scholaris*) with the overall heights varies between 8 and 20 m. The vegetation was dominated by palm lily (*Cordyline fructricosa*). This quite open cemetery is surrounded by housings, close to Pakancilan River (50m), has a busy road nearby, and about 800m from railway. Local community also uses the area in the cemetery for other activities, such as jogging and hanging out during weekend mornings.

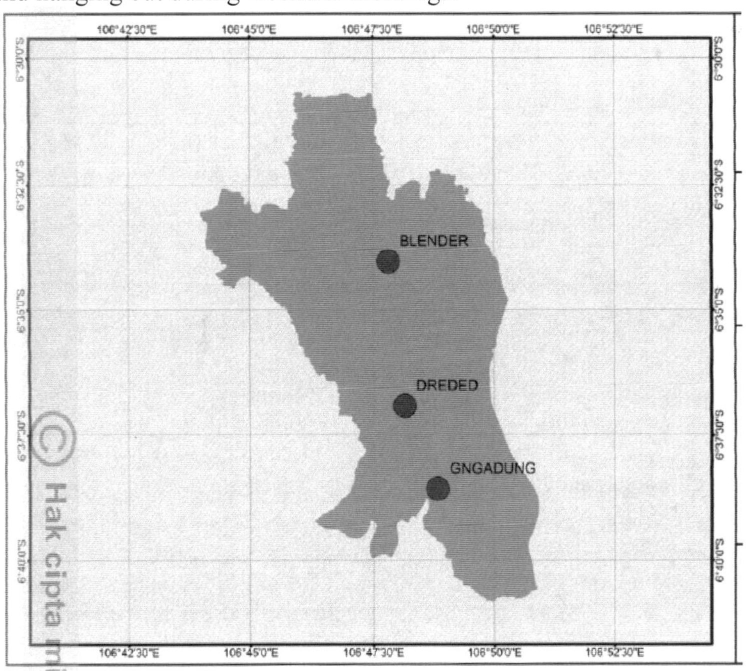

Figure 1. Location of the study sites in Bogor City

IOP Conf. Series: Earth and Environmental Science **179** (2018) 012041 doi:10.1088/1755-1315/179/1/012041

2.2. Method

2.2.1. Data collection

Bird observations were done in March and May 2015 during morning (06.00-09.00) and afternoon (16.00-18.00). Bird observations were conducted for six days in each site. Observation in Dreded Public Cemetery was conducted in March 2015, while observations in Blender and Gunung Gadung Cemeteries were conducted in May 2015. Point counts with a radius of 50m [6] were conducted to calculate diversity indices of the birds in all sites. The number of points varied due to different size of the cemeteries; there were 6 points in Dreded, 4 in Blender and 10 in Gunung Gadung. All birds seen or heard during 20 minutes observation period at each point were recorded. A binocular (8X42), a camera and a field guide were used to help in bird identification.

2.2.2. Data analysis

Total species diversity was expressed by the number of species (*S*). Shannon indices of diversity following Magurran [7] was calculated based on data obtained from point count method. Birds observed outside the radius of 50m were also recorded but were not taken into calculation for abundance and indices. Jaccard similarity index was also calculated to describe the difference in bird communities among sites [8]. Species dominance in each site was expressed as the proportion of the number of a certain species to the total bird number of all species. A species is categorized as dominant in one location if the dominance value is more than 5% [6].

3. Results

A total of 33 species from 19 families were observed during the study, while point count method recorded only 30 species of 17 families. The three species that were not found during point count were Purple Heron *(Ardea purpurea),* Rusty-breasted Cuckoo *(Cacomantis sepulcralis)*, and White-vented Myna *(Acridotheres javanicus)*. The highest species richness was found in Gunung Gadung Public Cemetery, although the number of families was highest in Dreded Public Cemetery (Table 1). Shannon diversity indices showed that the diversity in the study sites are generally low (ranged between 1.74 – 1.86), with moderate evenness indices (ranged between 0.59 – 0.69). The number of species in each family varied between 1 to 3 species. Families with highest number of species were Cuculidae (cuckoos), Columbidae (pigeon and doves), Sylviidae (prinia and tailorbirds), and Estrildidae (munias).

Table 1. Bird species richness and diversity indices in the study sites

Public Cemetery	Number of Families	Number of Species	H'	E'
Gunung gadung	10	21	1.86	0.62
Dreded	14	20	1.74	0.59
Blender	9	14	1.77	0.69

Jaccard similarity indices showed that all sites had high similarities (Table 2) with the highest similarity found between Blender and Dreded Cemeteries.

Table 2. Bird similarity indices among sites

	Gununggadung	Dreded	Blender
Gununggadung	1	0.64	0.71
Dreded	-	1	0.72
Blender	-	-	1

Based on the dominance criteria, there were six species that were dominant in Gunung Gadung Public Cemetery, whilst in each of the the other sites only four species were dominant. Dominant bird species in all sites were granivores (especially munias) and aerial insectivores (insectivores), but bulbuls were also dominant in Gununggadung and Blender (Table 3). Eurasian tree sparrow was dominant only in Dreded Public Cemetery.

Table 3. Dominance value (%) of bird species in the study sites

No	Bird Species	Common Name	Gununggadung Public Cemetery	Dreded Public Cemetery	Blender Public Cemetery
1	*Streptopelia chinensis*	Spotted Dove	1.97	0.25	2.25
2	*Streptopeliabitorquata*	Sunda Collared Dove		0.06	
3	*Geopeliastriata*	Zebra Dove	0.20		
4	*Cacomantis merulinus*	Plaintive Cuckoo	1.38	1.00	
5	*Centropus bengalensis*	Lesser Coucal	0.30		
6	*Collocalialinchi*	Linchi Swiftlet	11.32	19.60	24.08
7	*Apus affinis*	Little Swift	5.61		2.52
8	*Halcyon cyanoventris*	Javan Kingfisher	0.79		0.36
9	*Halcyon chloris*	Collared Kingfisher	2.66	0.69	
10	*Megalaima haemacephala*	Coppersmith Barbet		0.62	
11	*Dendrocoposmacei*	Fulvous-breasted Woodpecker	0.20		
12	*Dendrocoposmoluccensis*	Sunda Woodpecker	0.10	0.06	0.09
13	*Hirundotahitica*	Pacific Swallow			0.72
14	*Hirundostriolata*	Striated Swallow	3.05		
15	*Pericrocotuscinnamomeus*	Small Minivet		0.12	
16	*Pycnonotusaurigaster*	Sooty-headed Bulbul	8.96	4.24	8.84
17	*Pycnonotusgoiavier*	Yellow-vented Bulbul	0.89	4.68	7.30
18	*Pellorneumcapistratum*	Black-capped Babbler	0.20		
19	*Orthotomussutorius*	Common Tailorbird		0.06	0.36
20	*Orthotomussepium*	Olive-backed Tailorbird	0.59	0.19	
21	*Priniafamiliaris*	Bar-winged Prinia	0.10		
22	*Muscicapadauurica*	Asian Brown Flycatcher		0.06	
23	*Anthreptesmalacensis*	Plain-thoroated Sunbird		0.06	
24	*Cinnyrisjugularis*	Olive-backed Sunbird	0.49	0.81	
25	*Dicaeumtrochileum*	Scarlet-headed Flowerpecker	0.20	1.19	3.16

No	Bird Species	Common Name	Gununggadung Public Cemetery	Dreded Public Cemetery	Blender Public Cemetery
26	*Zosteropspalpebrosus*	Oriental White-eye		0.19	
27	*Passer montanus*	Eurasian Tree Sparrow	4.33	10.61	4.15
28	*Lonchuraleucogastroides*	Javan Munia	48.92	13.17	2.80
29	*Lonchurapunctulata*	Scaly-breasted Munia	7.78	42.32	42.20
30	*Lonchuramaja*	White-headed Munia			1.17

There were four species that are protected by the Indonesian Government Regulation recorded during the study; those are Javan Kingfisher (*Halcyon cyanoventris)*, Collared Kingfisher (*Halcyon chloris*), Brown-throated Sunbird (*Anthreptes malacensis*), and Olive-backed Sunbird (*Cinnyris jugularis*).

4. Discussion

Gunung Gadung Public Cemetery had the highest diversity index although the number of species was almost similar to those in Dreded Public Cemetery. Larger area in Gunung Gadung provides more varied resources for different bird species that occupy the habitat. Furthermore, open canopy in Gunung Gadung area allowed the observer to detect birds more easily than in habitat with close canopy. Similar result in the tropical rain forest were reported by Wisubudi [9] and Ahmadi [10].

Cemeteries as public open green spaces have been reported to harbor a high diversity of birds [11]. However, this study showed that bird diversity in three cemeteries was relatively low. These sites also had lower bird diversity compared to other studies on birds in green open spaces in Bogor and the surroundings. At least 63 species were reported in various urban habitats of Bogor by Ontario *et al.*[12], 42 species in housing complex by Asmoro [13], and 41 species in orchard by Nurfaizah [14]. However, lower result was reported from Cinangsi Public Cemetery in Bandung where only 14 species of birds were recorded [15]. The vegetation in the cemeteries in the study sites is more homogeneous than those in the other studies. Bird diversity has been reported to be higher in more heterogeneous habitats [16], [17]. Tews [18] reviewed studies on the relationships between animal diversity and habitat heterogeneity and reported that habitat heterogeneity may have different impact on different groups.

5. Conclusion

The low bird diversity in cemeteries was due to the poor tree species and coverage. Planting with more diverse tree plants might increase bird species in cemeteries, as well as increase their other ecological function in urban area.

Acknowledgments

The authors wish to thank the Agency for Sanitazion and Parks of Bogor City for permit to use the public cemeteries as study sites. We also would like to extend our thanks to Karim, Ilham, Sukri, Imam and Richsy who had helped in data collection.

References

[1] MacArthur RH and Wilson EO 1967 *The Theory of Island Biogeography*. (NewJersey: Princeton University Press)
[2] Kim J, Chae J, Koo T 2007 *Acta Ornithol*. 42:39-42
[3] Rahayuningsih M 2009*Komunitas Burung di Kepulauan Karimunjawa Jawa Tengah: Aplikasi Teori Biogeografi Pulau*PhD Dissertation (Bogor: Institut Pertanian Bogor)
[4] Schütz C, Schulze CH 2015*Ecology and Evolution* 5(22):5230–39 doi: 10.1002/ece3.1778

[5] Lussenhop J 1977 Condor 79: 456-61
[6] van Helvoort B 1981 *Bird Population in the Rural Ecosystem of West Java* (The Netherlands: Nature Conservation Department)
[7] Magurran AE 2004 *Measuring Biological Diversity* (USA: Blackwell Publishing Company)
[8] Krebs CJ 1978 *Ecology: The Experimental Analysis of Distribution and Abundance.* 2nd Ed (New York: Harper and Row Pub)
[9] Wisnubudi G 2009 *Vis Vitalis* 2(2): 41-49
[10] Ahmadi R. 2014.*Komunitas Burung pada Beberapa Habitat dengan Gangguan Berbeda di Hutan Lambusango, Pulau Buton, Sulawesi Tenggara.* BS Thesis (Bogor: Institut Pertanian Bogor)
[11] Chamberlain DE, Gough S, Vaughan H, Vickery JA and Appleton GF 2007 *Bird Study* 54(1): 87-97 DOI:10.1080/00063650709461460
[12] Ontario J, Hernowo JB, Haryanto and Ekarelawan 1990 *Media Konservasi*3(1) 15-28
[13] Asmoro AWT 2012 *Keanekaragaman Jenis Burung pada Beberapa Cluster Perumahan di SentulCity, Bogor Jawa Barat* BS Thesis (Bogor: Departemen Konservasi Sumber Daya Hutan Fakultas Kehutanan Institut Pertanian Bogor)
[14] Nurfaizah I 2017 *Keanekaragaman Burung di Area Kebun Buah, Taman Buah Mekarsari* BS Thesis (Bogor: Departemen Biologi, Fakultas Matematika dan Ilmu Pengetahuan Alam Institut Pertanian Bogor)
[15] Nur'aini 2012 *Keanekaragaman Jenis Burung dalam Pemanfaatan Areal Pemakaman sebagai Habitat Hidupnya* BS Thesis (Bandung: Biologi Fakultas Sains dan Teknologi) geri. Bandung.
[16] Katayama N, Amano T, Naoe S, Yamakita T, Komatsu I et al 2014 *PLoSONE* 9(3):e93359 doi:10.1371/journal.pone.0093359
[17] Mohd-Azlan J, Noske RA and Lewis MJ 2015 *Diversity* 7:118-36 doi:10.3390/d7020118
[18] Tews J, Brose U, Grimm V, Tielbo K, Wichmann MC, Schwageri M and Jeltsch 2004 *J. Biogeogr.* 31:79-92

Birds activities at urban greenways in Bogor

T Y Syaqina*[1], Y A Mulyani[1], and R Hermawan[1]

[1] Department of Forest Resources Conservation and Ecotourism, Faculty of Forestry, Bogor Agricultural University, 16680 ID

*Email:thrys23@gmail.com

Abstract. Greenway is one of available habitats for birds in Bogor. Birds that occupied greenways are more susceptible to traffic noise disturbance because birds use acoustic signals to communicate. This research aims to identify noise levels and bird activities in four Bogor greenways with different traffic frequencies. Four greenways observed in this research were Jalan Ahmad Yani, Jalan Paledang, Jalan Semeru and Jalan Tentara Pelajar. Methods used in this research were measuring noise with sound level meter, and O'Donnell and Dilk's five-one minute observation on birds. The result showed that Jalan Semeru had the highest noise value. Foraging activities, singing, and perching had lower frequencies in the noisiest greenway, while breeding activities was never recorded.

Keywords: birds, greenways, noise levels

1. Introduction

Cities have potentials as habitat for birds in urban areas. Vegetation in urban green open spaces provides food, resting and nesting sites for birds. Greenways are among the types of urban green spaces suitable for bird habitats [5], however this type of habitat is not free from disturbance. Urban greenways are located adjacent to road; therefore noise from traffic is a major disturbance for birds. Noise in urban areas can be generated by industrial activities, traffic, markets, schools, and others. Some studies suggest that noise generated by traffic is higher than those generated by other factors such as schools [20] and settlement [12].

Birds that utilize greenway habitats might be disturbed by noise that comes from the traffic because birds use acoustic signals to communicate. Lower number of bird species in greenways compared to those in city parks due to higher level of noise had been reported, and it was suggested that birds may avoid the dangers that exist on the habitat adjacent to the road [5]]. Bogor has a variation of greenways that are occupied by birds. Some of the greenways were located along arterial roads that are passed by vehicles more frequently than other road types; therefore the levels of noise pollution in those greenways might be higher. This study examined noise levels at different types of greenways and their effect on the activities of birds. This study is important because research related to noise and its impact to bird, especially in Indonesia, is still lacking.

2. Methods

2.1. Study area

A preliminary survey at 8 greenways of Bogor City was done to select study sites (Table 1). The criteria used to determine study site include traffic condition (heavy and light traffic), vegetation (composition, structure and density), and road characteristics. Heavy and light traffic conditions were determined by the numbers of motor bikes in each lane per 30 minutes. The motor bikes were chosen as reference because they pass more continuously compared to cars. Information on the number of passing vehicles at each location was obtained by counting vehicles manually. The calculation was carried out to obtain the average number of vehicles on the congested lanes traversed by >1000 units per 30 minutes, while the deserted lanes traversed by <1000 units per 30 minutes.

Greenway vegetation characteristics chosen were those with relatively similar structure and density, and the species composition of vegetation were diverse. Road characteristics include length (>1 km), no median, and roads whose noise comes mainly from traffic. Based on the survey greenways selected were Jalan Semeru, Jalan Paledang, Jalan Ahmad Yani and Jalan Tentara Pelajar. Jalan Semeru and Jalan Paledang were traversed by >1000 units of motor bike per 30 minutes while Jalan Ahmad Yani and Jalan Tentara Pelajar were the opposite.

Tabel 1. Road length and number of passing vehicles at study sites

Location/Street	Length (m)*	Vegetation	Number of Vehicles/30 minutes	
			2-wheel (unit)	≥4-wheel (unit)
Semeru	2075	Dense	1808	764
Paledang	1041	Dense	1452	540
Ciremai Ujung	1650	Dense	98	54
Lawang Gintung	1004	Dense	880	596
BatuTulis	1761	Sparse	1272	672
RE Martadinata	1060	Sparse	1560	548
Tentara Pelajar	1392	Dense	780	332
Ahmad Yani	2111	Dense	794	426

*Source: Bogor Public Work Office 2017

2.2. Equipment

Equipment used in this study was sound level meter LT Lutron SL-4012 for noise measuring, binoculars Nikon Aculon A211 16x50, camera Nikon CoolPix L840, and bird observation field guidebook [10].

2.3. Sites and observation times

This research was conducted at Jalan Ahmad Yani, Jalan Tentara Pelajar, Jalan Semeru and Jalan Paledang, Kota Bogor. The data collected in April 2017. The observations were conducted in the morning at 6:00 to 09:00 am and afternoon from 3:00 to 18:00 pm, based on the time of peak activities. Observations were made on weekdays and only during clear weather.

2.4. Noise measurements

Noise measurements were carried out in accordance to the Decree of the State Minister for The Environment No.Kep-48/MENLH/11/1996. The measurement was done for 10 minutes in each measurement, taken in the morning and afternoon. The noise level was measured at three points in the observation path that was at the point 0 km, 0.5 km and 1 km.

Measurement of noise is done at two altitudes at each point i,e at an altitude of 0.5 m (below) and 3-5 m (above). The height of 0.5 m is chosen because it is adjusted to the height of the vehicle, while the height of 3-5 m represents the mean of free height of tree branches in all locations. The resulting noise value was the result of morning and afternoon measurements at 3 points on each road, where each point has 2 altitude locations, so that 12 noise measurements per day is done. Two Sound Level Meters were used in order to measure the noise value at two different heights at the same time.

2.5. Bird activities

Bird observation was conducted using the five-one minute observation method [13]. The observer walked slowly along the 1 km transects while recording bird activity seen, and follow it for one minute. If the bird could be followed, four further observations were made for one minute, giving a maximum of five observations per bird. The parameters recorded were bird species, number of individuals, activities (Table 2), time used to perform the activity, and substrate. Individual bird was recorded when birds use the trees within the greenway. Any bird that just flew over the greenway was not recorded. Species belonging to Apodidae family was not considered in the analysis because they always fly above the trees and never perched.

Table 2. Description of bird activities recorded in greenways

Activities	Description
Calling and Singing	Chirp a note or a full song
Breeding	Calls to attract a partner, mating, courtship; bring materials to the nest or when inside the nest
Interaction	Interspecific and intraspecific interaction
Perching	Steady, looking around
Locomotion	Move from one substrate to another substrate
Resting	Bath, drinking, excretion, browse
Feeding (on vegetation)	Searching and taking food from the substrate of vegetation
Feeding (on the ground)	Searching for food on the ground

2.6. Data analysis

2.6.1. *Bird activities*

Data on observation of birds in the morning and afternoon were analyzed separately to examine whether there is any effect of time. Frequency of activities is described as a histogram. Analysis was only conducted for individuals that were recorded more than three times during the overall observations. The frequency of bird activity was calculated using the following formula:

$$\text{Frequency of activity (\%)} = \frac{\text{Number of each activity at one period in each location}}{\text{Sum of all activities at all locations}}$$

2.6.2. *Statistical analysis*

Mann-Whitney test was used to examine the difference between the average noise value of each location with heavy and light traffic. To examine whether there is any difference in bird activities among different noise level the Chi-Square test was used.

3. Results and Discussion

3.1. Noise level

Noise value at Jalan Semeru was the highest of all locations, while the lowest was at Jalan Paledang (Figure 1). ; but the value of heavy and light traffic noise has no significant difference (U test = 0.88; $p > 0.05$). This can be caused by the small number of samples.

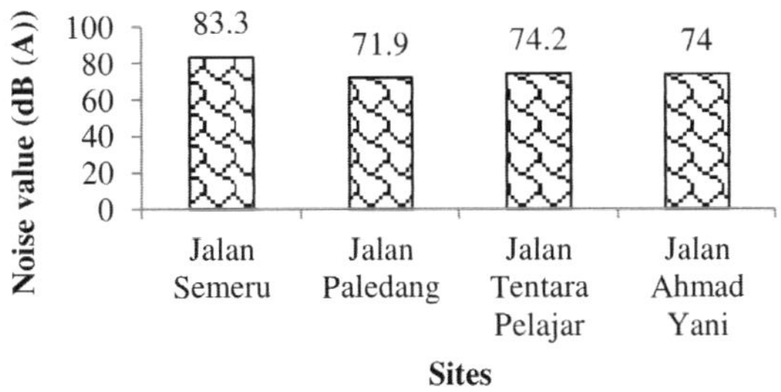

Figure 1. Noise level at each sites

3.2. Bird activities

Species that were recorded more than three times during the observation, thus were analyzed for its activities, were Spotted-dove (*Streptopelia chinensis*), Sooty-headed Bulbul (*Pycnonotus aurigaster*), Yellow-vented Bulbul (*Pycnonotus goiavier*), Olive-backed Tailorbird (*Orthotomus sepium*), Scarlet-headed Flowerpecker (*Dicaeum trochileum*), Oriental White-eye (*Zosterops palpebrosus*), Javan Munia (*Lonchuraleu gastroides*), and Eurasian Tree Sparrow (*Passer montanus*).

3.2.1. *Spotted-dove (Streptopelia chinensis)*

Spotted-dove often used trees within the greenway for locomotion (figure 2). Its activities were more performed in the morning. Spotted-dove was most prevalent in Jalan Ahmad Yani and Jalan Semeru. The species was usually seen inside a lush canopy, and never seen on the ground, while usually the Spotted-dove often seen on the ground, looking for food [10]. Spotted-dove in this study seemed to avoid roadways due to traffic or human disturbances. Spotted dove using higher strata in urban area to avoid human disturbances was also reported in Lampung [15].

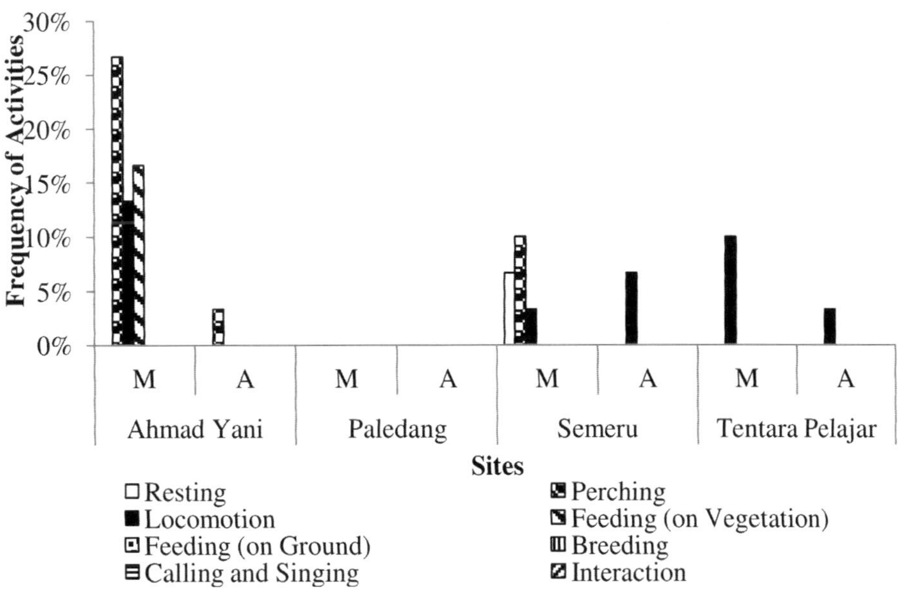

Figure 2. Frequency of activities of Spotted-dove; (M: morning; A: afternoon)

3.2.2. *Sooty-headed Bulbul (Pycnonotus aurigaster)*

Sooty-headed Bulbul was a dominant species at Jalan Ahmad Yani, Jalan Paledang and Jalan Semeru. Its activities were more performed in the morning. This species was observed doing all daily activities except for feeding on the ground. Feeding activities were mostly performed at Jalan Paledangwhich the noise level was the lowest (figure 3). Perching was the most common activities performed by this species, with the least were recorded at Jalan Semeru. It can be caused by this species' intolerance to exposing to noise for a long period. However, the presence of Sooty-headed Bulbul in Jalan Semeru was less frequent compared to those in the three other locations.

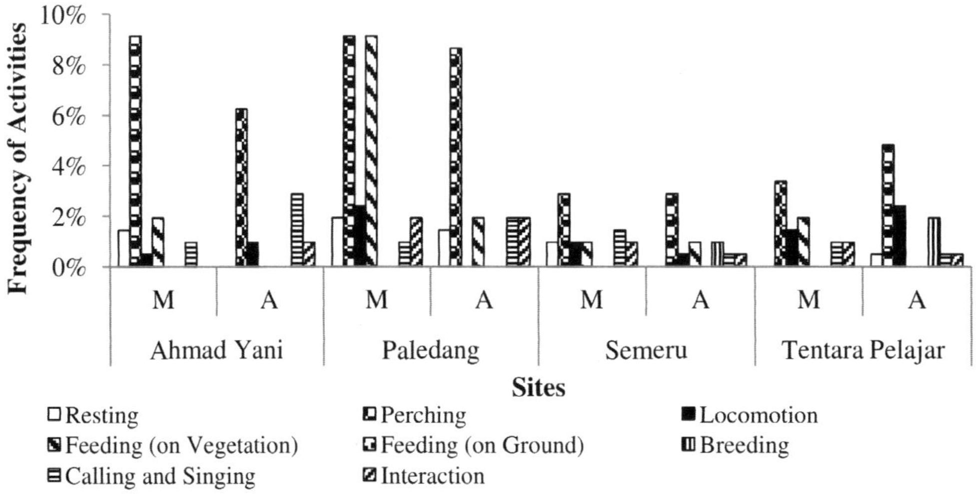

Figure 3. Frequency of activities of Sooty-headed Bulbul; (M: morning; A: afternoon)

3.2.3. Yellow-vented Bulbul (*Pycnonotus goiavier*)

The most frequent activities of Yellow-vented Bulbul was feeding on vegetation and perching (Figure 4). Its activities were more performed in the morning. Variety of its activities was higher at Jalan Paledang. The frequency of activities was lower in Jalan Semeru which had the highest noise level. In addition, factors such as illegal hunting suspected to be the cause of the low occurrence of Yellow-vented Bulbul in Jalan Semeru. Yellow-vented Bulbul was one of the most widely sold birds in Jakarta after Oriental White-eye [2].

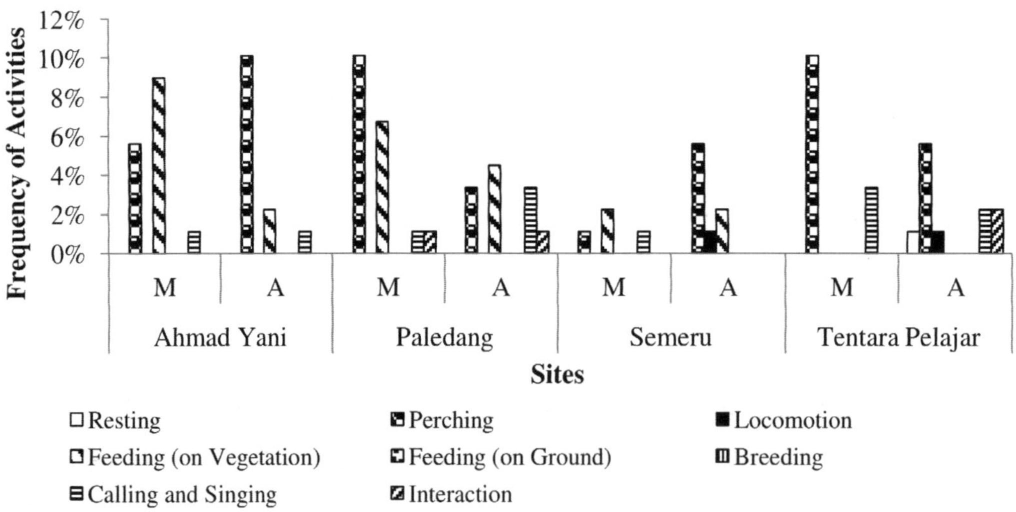

Figure 4. Frequency of activities of Yellow-vented Bulbul; (M: morning; A: afternoon)

3.2.4. *Olive-backed Tailorbird (Orthotomus sepium)*

This species rarely recorded at Jalan Semeru, but it was quite often observed with various activities in the three other sites. Its activities were more performed in the morning. This species was recorded foraging only at Jalan Ahmad Yani and Jalan Tentara Pelajar (Figure 5). Frequent activities were calling and singing. These birds were more often found in pairs, although no breeding activity was ever recorded. Some species of birds were found in the greenway only to perform foraging activity but do not breed [5]. It can be influenced by the needs of the area to breed which should be wider than the greenway. Singing was less frequent at Jalan Semeru. Birds used song to attract a partner, limiting the territory from competitors, communication between individuals in a group, calls from chicks to the parents, and warning of the dangers of predators that lurk [3].

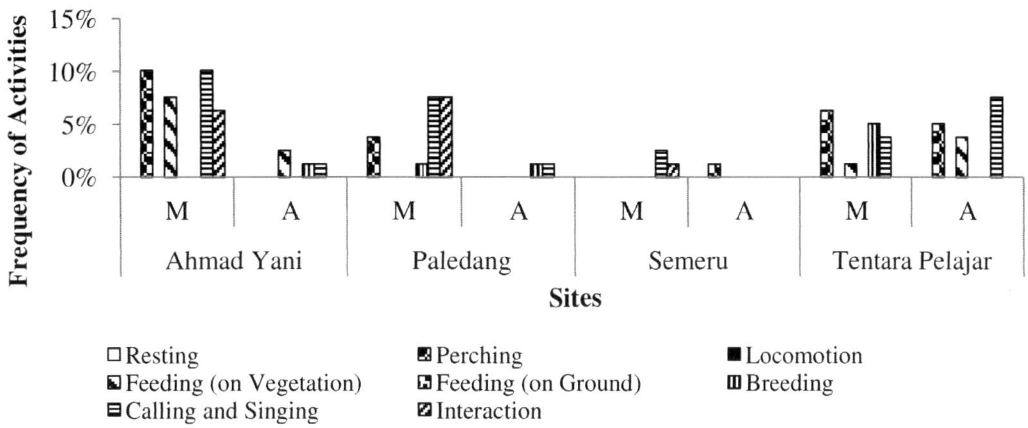

Figure 5. Frequency of activities of Olive-backed Tailorbird; (M: morning; A: afternoon)

3.2.5. *Scarlet-headed Flowerpecker (Dicaeum trochileum)*

The presence of this species was on Jalan Ahmad Yani and Jalan Paledang at most (figure 6). Its activities were more performed in the morning. Scarlet-headed Flowerpecker was frequently feeding at Jalan Ahmad Yani and Jalan Paledang. This bird was never seen feeding and breeding in Jalan Semeru. Noisy condition at Jalan Semeru allegedly impacted on the feeding activity of Scarlet-headed Flowerpecker. It also happened to *Turdus merula* that was seeking for food in a shorter time to raise awareness of human-derived disturbance (noise and pedestrians) [6]. It was also observed in *Charadius melodus* which hunt for a shorter meal because of the noise coming from ships [1]. Birds increased vigilance due to noise in order to avoid predators whose arrival may be difficult to detect due to masking effects [5].

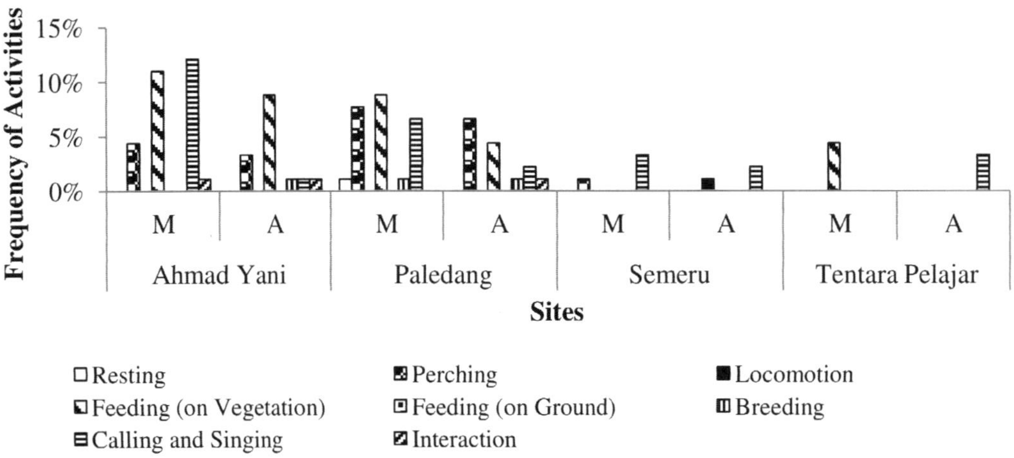

Figure 6. Frequency of activities of Scarlet-headed Flowerpecker; (M: morning; A: afternoon)

3.2.6. *Oriental White-eye (Zosterops palpebrosus)*

Smaller species with louder and higher song frequency were not susceptible to noise [17,[18]. In this study, this small sized bird was never seen in Jalan Semeru. Oriental White-eye was most visible at Jalan Paledang and had diverse activities (Figure 7). Its activities were more performed in the morning. Other factors than noise that affect the species at the study site were food availability. Oriental White-eye was often found in Jalan Paledang and Jalan Ahmad Yani because there were plants *Tabebuia crysantha* and *Bauhinia purpurea* which was a source of food such as nectar [4]. The presence of Oriental White-eye at the study site was thought to be influenced by noise levels and food availability. The presence of Oriental White-eye was not distributed evenly along the road at the study site. Each encounter with Oriental White-eye when observation was no more than 3 individuals, whereas Oriental White-eye supposed to form large groups [10], and have high relative abundance in Bogor [14]. Another factor that was suspected to affect the least of Oriental White-eye encounter was the illegal hunting due to high market demand for the species. Oriental White-eye were one of the most vulnerable bird species which hunted in their natural habitat [11] and were the most sold birds in Jakarta [2].

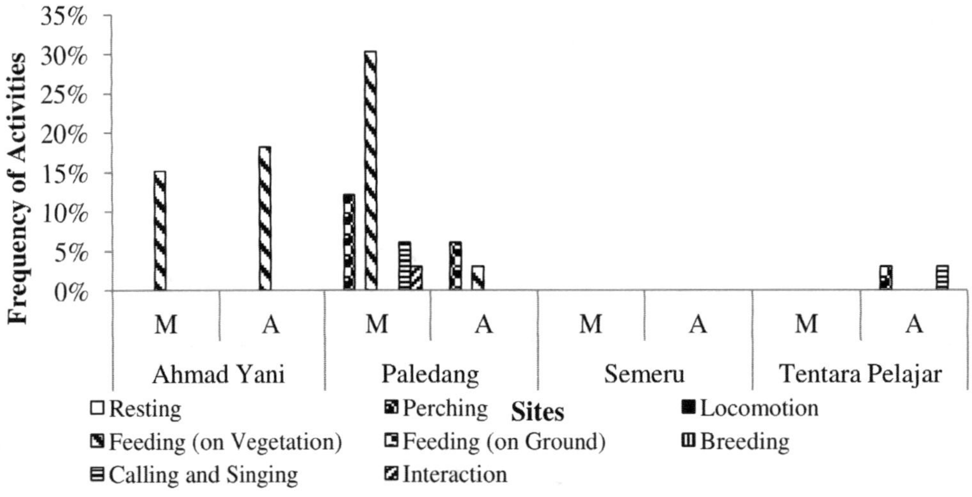

Figure 7. Frequency of activities of Oriental white-eye; (M: morning; A: afternoon)

3.2.7. *Javan Munia (Lonchuraleug astroides)*

The most varied activity of Javan Munia was performed at Jalan Tentara Pelajar. Its activities were more performed in the morning. Perching was the most frequent activity performed by Javan Munia. Making nest as breeding activity was only recorded at Jalan Tentara Pelajar (figure 8). Javan Munia prefers habitat such agricultural land and open grassland due to its primarily feeding on rice and grass seeds [19]. This was related with the conditions around Jalan Tentara Pelajar which was a moor-shaped area and open land belongs to the Department of Agriculture. The area around the Jalan Tentara Pelajar which formed as agricultural land was thought to be the reason why Javan Munia presented more, although the species was capable to present in the area with a high noise level like Jalan Semeru. This was supported by the presence of Javan Munia on the pavement within Jalan Semeru at the height of 1 meter.

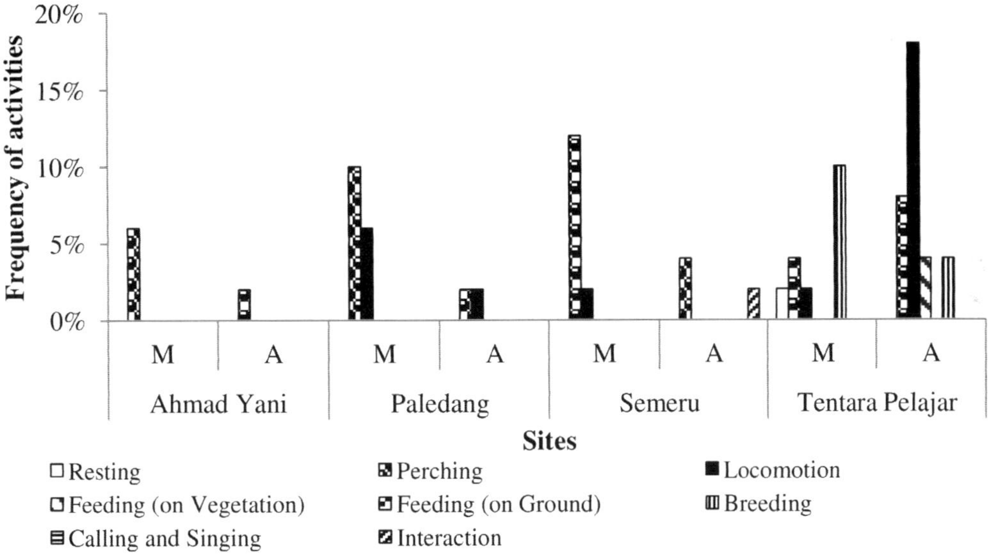

Figure 8. Frequency of activities of Javan Munia; (M: morning; A: afternoon)

3.2.8. *Eurasian Tree Sparrow (Passer montanus)*

Most recorded activities by Eurasian Tree Sparrow were perching and feeding on the ground. Its activities were more performed in the morning (figure 9). This species was often seen as group to search for food. Perching was more often performed on a building. This species were rarely seen at Jalan Ahmad Yani. The noise level was not expected to affect the Eurasian Tree Sparrow because this species was often observed on the pavement along the roadway. Foraging behavior of Eurasian Tree Sparrow causing these birds was often at the pavement. Some species were not affected by the presence of the highway because of its differences in sensitivity to noise [17]. This bird was rarely seen taking advantage of the trees on the greenway and more frequent to use the building wall. Eurasian Tree Sparrow frequently used roofs for nesting, seeds on the roadside and gardens as food [11]; therefore, this bird was never recorded feeding at Jalan Ahmad Yani because the site was dominated by trees with large seeds like canary.

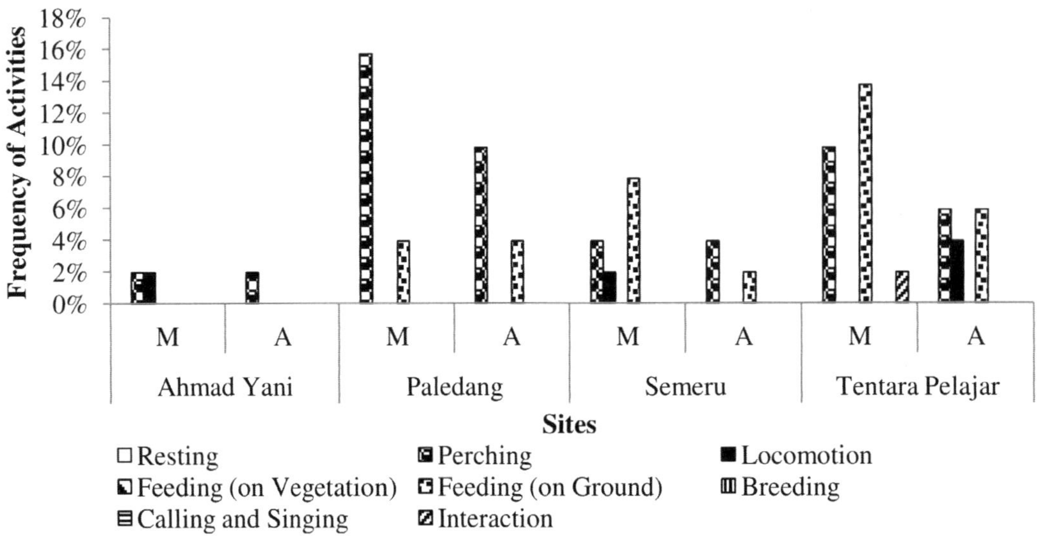

Figure 9. Frequency of activities of Eurasian Tree Sparrow; (M: morning; A: afternoon)

3.3. Relationship between noise and birds activity
Perching, feeding, calling and singing, and breeding activities were influenced by noise levels at the study sites, although the Chi-Square test showed that there was no correlation between noise and bird activities in this research (Chi-Square test: 0.33; $p>0.05$). Some species were not found in noisy location. Seven observed species can occur in all locations with different noise levels, but certain activities were not performed, or the frequency of the activity was decreased. Noise affected birds in the greenways because the highest noise levels occur when the birds do their activities [8]. This caused an overlap between bird's song frequency and traffic noise [18]. There was a pattern of bird activities in all research sites: the activity of most bird species were more varied at Jalan Ahmad Yani and Jalan Paledang, except for Javan Munia which activities were more varied in Jalan Tentara Pelajar. This may be due to several factors such as the suitability of the location of the bird habitat and the level of disturbance for birds [7]. Jalan Ahmad Yani and Jalan Semeru were dominated by similar vegetation types, but bird activity was more varied on Jalan Ahmad Yani. The noise on Jalan Semeru was suspected to be a prime disturbance that caused the birds to avoid this site.

4. Conclusions and Recommendations

4.1. Conclusions
Noise in this research sites were at the values of 71.9 to 83.3 dB (A). Traffic noise measured in heavy and light traffic lanes did not have significant differences. Although the Chi-Square test showed that there was no correlation between noise and bird activities in this research due to small sample of locations,birds occupied in the greenway with highest noise level did not perform certain activties such as breeding activities. Feeding, calling and singing and perching performed less than in the quieter sites. Certain activity which less performed on noisier site showed that a high noise level impacted the lives of birds at urban greenways in Bogor.

4.2. Recommendations
Developing the greenway on many roads in Bogor by increasing the structure of vegetation should be done. Planting more vegetation such as shrubs and herbs will add the composition of the vegetation structure of the greenway. The vegetation which composed of the ground to the upper canopy would

reduce more value of traffic noise. Attention to the wildlife in the greenway should be increased because its population can still be maintained. More green open space is also needed thus urban wildlife corridor can be increased inside the city.

5. Acknowledgement

We would like to thank Kantor Kesatuan Bangsa dan Politik Kota Bogor, Bogor Public Work Office and Bogor Sanitation and Parks Office for secondary data supports, also friends from KSHE 50 IPB as field observation company.

References

[1] Burger J 1994 *Estuaries*17 695-701
[2] Chng S C L, Eaton J A, Krishnasamy K, Shepherd C R and Nijman V 2015 *In The Market forExtinction: An Inventory of Jakarta's Bird Market* (Selangor: TRAFFIC)
[3] Dooling R J Popper A N 2007 *The Effects of Highway Noise on Birds* (Sacramento: California Department of Transportation Division of Environmental Analysis)
[4] Endah G P and Partasasmita R 2015 *Pros Sem Nas Masy Biodiv Indon*1 1289-94
[5] Fernández-Juricic E 2000 *Conservation Biology* 14 513-21
[6] Fernández-Juricic E, Telleria J 2000 *Bird Study* 4713-21
[7] Hernowo J B and Prasetyo L B 1989 *Media Konservasi.* 2 61-71
[8] Herrera-Montes M I and Aide M 2011 Impacts of traffic noise on anuran and birdcommunities *Urban Ecosyst*
[9] Decree of the State Minister for The Environment No.Kep-48/MENLH/11/1996 Concerning Noise Level Standards
[10] MacKinnon J, Philips K and van-Balen B 2010 *Burung-burung di Sumatera, Jawa, Bali, dan Kalimantan* ed Sumadipura S and Kartikasari A (Bogor: Burung Indonesia)
[11] Mardiastuti A 2016 *Orasi Ilmiah Guru Besar IPB* 30th Jan 2016 Bogor, Indonesia
[12] Martono H, Sukar and Sulistiyani N 2004 *Media Litbang Kesehatan* 14 44-50
[13] O'Donnell C F J and Dilks P J 1988 *Science and Research Series No. 4: A Method forQuantifying Habitat Use by Forest Birds* (Wellington: Department of Conservation)
[14] Ontario J, Hernowo J B, Haryanto and Ekarelawan 1991 *Pola Pembinaan Habitat Burung di Kawasan Pemukiman terutama di Perkotaan* (Bogor: Jurusan Konservasi Sumberdaya Hutan, Fakultas Kehutanan, Institut Pertanian Bogor)
[15] Putra G W, Harianto S P, and Nurcahyani N 2014 *Jurnal Sylva Lestari*2 93-100
[16] Reijnen R, Foppen R, Braak C T and Thissen 1995 *Journal of Applied Ecology*32 187-202
[17] Rheindt F E 2003 *Journal für Ornithologie* 144 295-306
[18] Slabbekoorn H and Ripmeester E A P 2008 *Molecular Ecology*17 72-83
[19] Sulistyadi E 2010 *Jurnal Biologi Indonesia* 6 237-253
[20] Wafiroh A H 2013 Pengukuran tingkat kebisingan di lingkungan SMPN 2 Jember [*undergraduate theses*] (Jember: Universitas Jember)
[21] Wiącek J, Polaka M, Kucharczyka M and Bohatkiewiczb J 2015 *Landscape and Urban Planning* 134 76-82

Reprofiling landscape of rainwater harvesting in supporting Semarang urban water resilience

Y Prihanto[1], R H Koestoer[1], D Sutjiningsih[2], and L Darmajanti[3]

[1] School of Environmental Science, Universitas Indonesia, Jakarta 10430, Indonesia
[2] Department of Civil Engineering, Faculty of Engineering, Universitas Indonesia, Depok 16424, Indonesia
[3] Department of Sociology, Faculty of Social and Political Science, Universitas Indonesia, Depok 16424, Indonesia

E-mail: putranusa212@yahoo.com

Abstrack. The uniqueness of the urban physiographic landscape could share advantage as well as constraint for the sake of urban water availability. Physiography landscape of urban area will slightly affect the ability of a city in providing water needs of urban dwellers. An alternative solution to resolve the problem is rainwater harvesting. Semarang is one of secondary cities, faces various problems such as ground water scarcity, land subsidence, flood, and drought. Water resilience problem of Semarang would become more severe when development activities do not pay attention to 2 kinds of landscapes: social and physiography dimension. This study aims to examine the distribution and characteristics of landscape physiography associated with social condition, considering appropriate form of rainwater harvesting application. The basic methods used are spatial analyses and combined with non-parametric statistical technique based on questionnaires. Landscape physiography variables embrace various aspects of altitude, slope, geology, soil, ground water potential, landslide, flood, and land subsidence. While social variables involve several aspects of community attitudes, gender, education and income. The results shows that the application of rainwater harvesting needs to consider physiographic landscape together with social profile to ensure and support water resilience of urban areas.

Keywords: Landscape, Rainwater harvesting, Urban, Water resilience, , Community

1. Introduction

Many countries around the world have been well aware of increasingly critical water resources. This condition encourages various circles to find ways to ensure the availability of water in a sustainable manner. Many cities are faced with critical water problems as a result of population growth and excessive groundwater exploitation. This requires many cities to work to build water resistance so as not to hit the water crisis. Resilience can be defined as "the ability of a system, community or society to be exposed to hazards to resist, absorb, accommodate and recover from the efforts of a hazard in a timely and efficient manner" [1]. The definition of resilience by UNISDR is in line with the water security formulated by UN-Water. Water security is defined as capacity of the population to sustainable livelihoods, human well-being, socio-economic development, as well as for ensuring protection against water-borne pollution and water-related disasters, and for preserving ecosystems in a climate of peace and political stability [2].

Semarang is an urban area that has a unique physiographic landscape. The uniqueness of physiographic landscape owned by Semarang City, facing this city on water resistance problem. Semarang is one of secondary cities, faces various problems such as ground water scarcity, land subsidence, flood, and drought. An alternative solution to resolve the problem is rainwater harvesting. Unfortunately the implementation of the rainwater harvesting program in Semarang City has not run as expected. In fact, if you see the potential of annual rainfall of Semarang city ranging from 2,000-2,500 mm/year, the successful application of rainwater harvesting can be a promising solution.

Many studies on rain water harvesting have been done in different parts of the world [3, 4, 5, 6, 7, 8, 9]. Studies of the application of special rain water harvesting in urban areas are also widely published [3,4,5,]. Various rainwater harvesting related themes, such as calculating rainwater catchment potential [3, 6, 7, 10], selecting rainwater storage locations [10, 11], calculating cost savings that can be done [9, 12, 13], to quality water [14] as well as utilization of rainwater harvesting [15]. Studies on the social aspects of rainwater harvesting [16, 17] have also been widely practiced, especially regarding the level of public acceptance [16], and the factors that influence [16, 17].

Studies conducted by Mankad et al. in Australia [16] sought to identify key factors affecting people's willingness to apply rainwater harvesting principles. The study states that public attitudes are generally very supportive, but perception change factors play a role in encouraging the realization of community attitudes [16]. Another study by Donohue and Biggs [8] tries to make multi-dimensional spatial approaches to assessing the social environment in Nepal. The results of their study suggest that there is a correlation between social conditions of society with landscape topography factors [8]. This is in line with the results of the study Barthwal et al. [11] which states that social factors of income level are the decisive factor in the application of rainwater harvesting, and the application of rainwater harvesting should adapt the characteristics of the region. The examples of the studies conducted, in part, lead to one thought that the application of rainwater harvesting should pay attention to environmental aspects either social or physical.

Based on the condition of unsuccessful application of rainwater harvesting in Semarang City and based on previous studies, this study is interested to know whether or not the linkage of social condition and landscape physiography with attitude of society in Semarang City, in applying rainwater harvesting. This study aims to examine the distribution and characteristics of landscape physiography associated with rainwater harvesting application.

2. Methods

2.1. Study area
This study was conducted in the administrative area of Semarang City. The city administration is divided into 16 districts and 177 urban villages. The administrative population covers all sub-district administrations and the calculations in this study are conducted according to the sub-district analysis unit. The districts are grouped into 3 physiographic landscape units based on the identification and zonation of the region. An overview of the location of the study and the spatial distribution of administrative regions, presented in Figure 1.

2.2. Type of data
This study uses primary data and secondary data. Primary data used is obtained through the distribution of questionnaires and field observations. Primary data obtained from authorized institutions in accordance with the theme. Secondary data used in this research there are several kinds. Data of population, number of head of household and annual average rainfall of Semarang City obtained from Central Bureau of Statistics (BPS) for publication data year 2012-2016. Data on the distribution of residential areas (2012), high-resolution image data of World View (2012), and 30 m resolution SRTM images obtained from the Geospatial Information Agency. Spatial data of roof cover (2011-2012) and urban household water requirements were obtained from the Ministry of Public Works. Geological map, soil, ground water potential, landslide, flood, and land subsidence are obtained from Spatial Planning of Semarang City 2005-2030 document, Regional Planning and Development Board (BAPPEDA).

2.3. Sampling method
The population of this study is the adult population of Semarang City that can determine the attitude in managing household water needs. The gender of the respondents can be male or female. This study uses combined sampling method. Combined sampling methods are area, and random methods. In the initial stage the number of samples is calculated according to the zone area, then the determination of

the respondents is determined randomly. Based on calculations, each zone of territory should be represented by a minimum of 100 respondent samples to meet 90% accuracy.

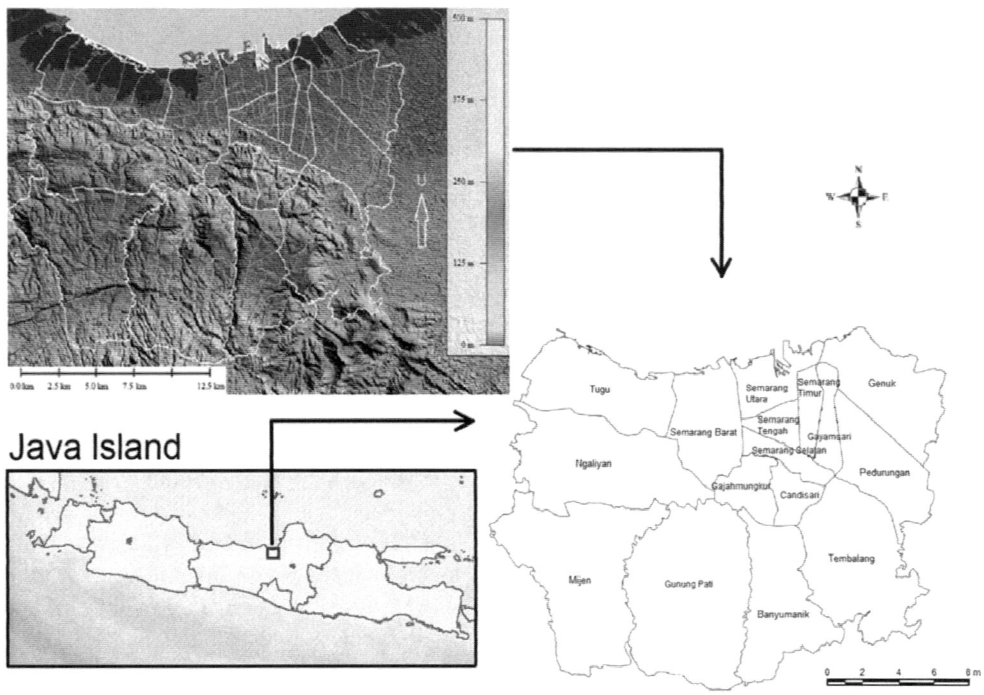

Figure 1. Location of study.

2.4. Variables
The variables in this study are differentiated into independent variables and dependent variables. The independent variables consist of: physiographic landscape zone, ground water potential, disaster vulnerability, gender, education, and income. The dependent variable of this study is the attitude of the community in applying rainwater harvesting. The independent variables tested for their correlation relationship with dependent variable are: physiographic landscape zone, gender, education, and income. The variables of ground water potential and disaster vulnerability are only used to look spatially related to the potential and vulnerability of rainwater harvesting application in Semarang City.

2.5. Questionnaire
The questionnaire used in this study consists of 2 parts. The first section contains information on respondents, namely: location of residence zone, gender, education level, and income level. The second part contains questions relating to the assessment of individual attitudes toward the application of rainwater harvesting. Questions submitted consist of 8 questions as presented in Table 1. Assessment of attitude done using Likert scale. Each question can be judged by the numbers 1 through 5, so the value of all the answers to the question can be totaled to the final value. The total value of all the answers to questions can be a minimum of 8 and a maximum of 40. The average value of individual attitudes is calculated based on the number of respondents in each analysis group.

2.6. Data analysis method
Some of the analytical methods used in this study are described in the following sections. The identification and zoning of landscape physiography in this study was done visually and mathematically using geographic information system (GIS). Determination of the limits of phonograph landscape zonation using the administrative boundary of the sub-district in Semarang City. Profiling geology, soil, and ground water potential are done based on thematic data obtained from the spatial plan document of Semarang City 2005-2030. Calculation and analysis using GIS

tools and software. The same method is used to process and analyze the profile of disaster vulnerability (landslide, flood, and land subsidence). For the analysis of the distribution of societal attitudes by gender, education and income, descriptive statistical methods are used. The respondents' attitudes were derived from the sum of the scores from each question (8 questions with a score of 1-5). The calculation is done to know the average value of respondent attitude on different group of variables. The method used to know the correlation of public attitudes with gender, education and income or correlation between attitude with landscape, is method of crosstab analysis. This method is used to know the value of Pearson Chi-Square and significance value of relationship between variables, so it can be known, there is or not relationship between these variables.

Table 1. Questions of attitude variable.

No.	Questions on attitude variable
1	Do you agree, if the central/regional government promotes rain harvesting in your area by developing a reservoir/tank to meet sufficiency water needs?
2	Do you agree, if the government promotes rain harvesting in your area through development of absorption wells and/or biopores to absorb water and reducing flooding risk?
3	Do you agree, if you are asked to get involved in rain harvesting activities by preparing your own tanks/tubs to collect rainwater in your home?
4	Do you agree, if you are asked to be involved in rain harvesting activities by making your own absorption well and/or biopores to absorb rainwater in the vicinity of your residential area?
5	Do you agree, if the government provides equipment and community assistance to provide means and infrastructure of rain harvesting, such as tank, well, and drill?
6	Do you agree, if the government provides incentives in the form of tax deductions or exemptions on earth and buildings if you apply rain harvesting through applying absorption wells, biopore and/or rainwater reservoirs?
7	Do you agree if - harvesting or collecting rainwater activities in a tub/water tank for sufficient water needs should be managed through community or urban village activities?
8	Do you agree, if the activities to create and to maintain absorption wells and biopore to absorb water, to reduce flooding and drought risks should become routine joint activities managed by community or urban village?

Source: setup by author

Analysis of the volume of rainwater and population water needs is done to assess the potential harvesting of rainwater by volume compared to the volume of household water needs. The volume of rainwater that is compared is the volume of rainwater obtained from the capture of the roof of the settlement with the daily household water needs. Calculation of rainfall potential (RWH) potential that can be harvested using average rainfall (P, in mm/year), the width of the roof or the catchment area (A, in m2), and the runoff coefficient (RC, nondimensional) [21] as shown in equation (1). The calculation of the population's water needs, using the reference set by the Ministry of Public Works Indonesian, 150 liters/day/person. The calculation of the community's water needs uses the formula (2) that is, the number of population (P, in person) multiplied by the daily water requirement (Wn, in liters/day/person) multiplied by the number of days (365, in day) in one year.

$$\text{RWH potential} = P.A.RC \quad (1)$$

$$\text{Annual water needs} = POP.Wn.365 \quad (2)$$

The calculation of the population for estimates of 2030 is calculated using population projection formula approach with geometry method. To estimate the 2030 rooftop area the 2030 population projection approximation method is divided by the average number of family members (2012) times the average roof per head of the family in 2012. Calculation of population in 2030 can be estimated by using equation (3):

$$P_n = P_o\,(1+r)^n \quad (3)$$

Remarks:

P_n = number of population in year n
P_o = number of population in year 0 or base year
n = number of years between 0 and n
r = population growth rate per year (in %)

3. Result and Discussion

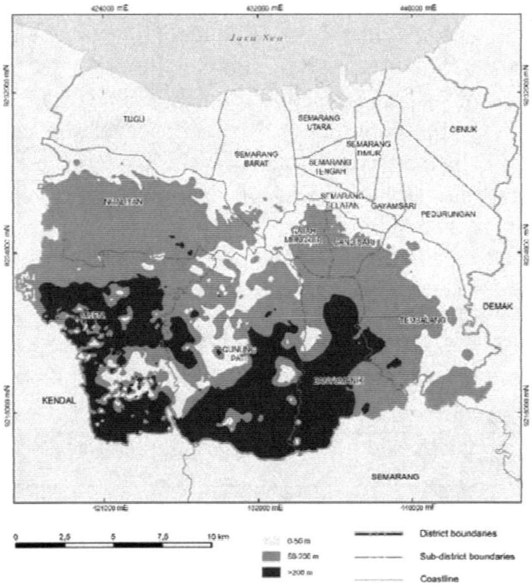

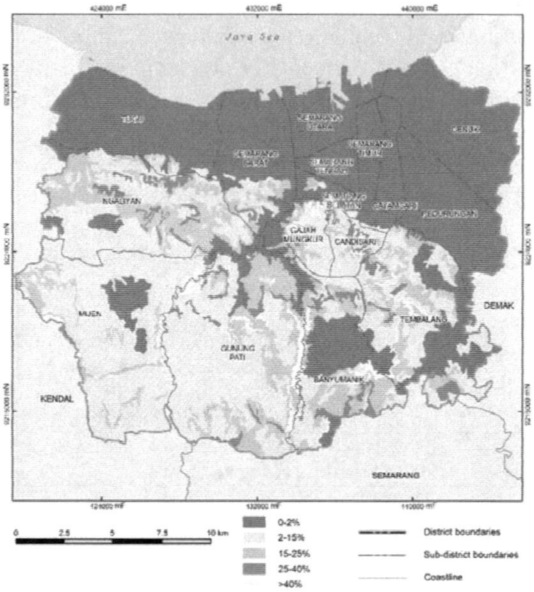

Figure 2. Land altitude. **Figure 3.** Slope of Semarang City.

Figure source: processing data from document of Semarang City Spatial Planning 2005-2030.

3.1. Identification and zoning of physiography landscape

Based on altitude and slope, the administrative area of Semarang City is divided into 3 areas according to landscape physiography. The division also refers to the ecosystem and region zones according to the hydrological cycle. The division of the 3 zones is: the lower Semarang zone, middle Semarang and upper Semarang. According to the calculation the distribution of land based on the altitude class, the lower Semarang region has a height of 0-50 meters from sea level. The lower Semarang area covers an area of 20,105 ha. Semarang central zone area has a height ranging from 50-200 meters from sea level with an area of 12,004 ha. The upper Semarang zone area has a height of more than 200 meters from sea level covering an area of 6,735 ha. Based on the slope, the area of Semarang City can be grouped into 5 classes of slopes. The wide distribution for each slope class by zone is presented in Table 2.

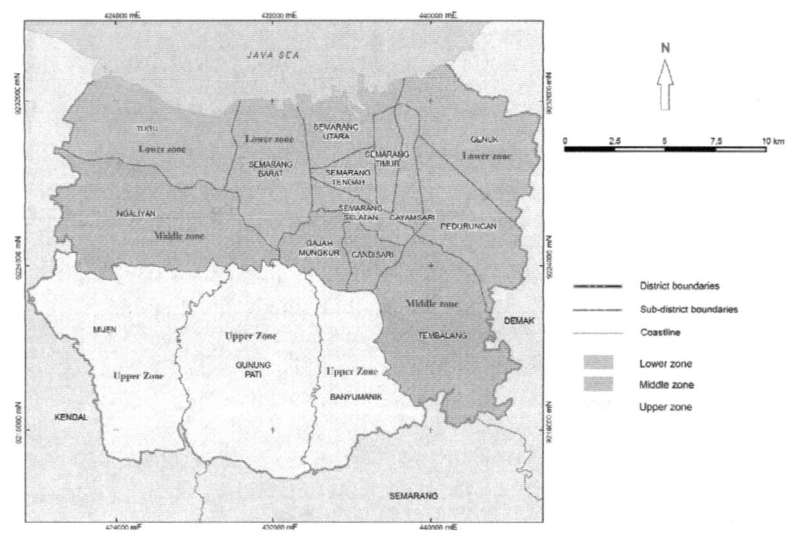

Figure 4. Zone division.
Figure source: processing data by author.

3.2. Profiling geology, soil, and ground water potential

Based on geological map, the lower zone area is dominated by alluvium surface sediment type, which is 12,304 ha. This formation is a coastal plain and river sediment plains. The dominant material of this formation is clay and sand, with a thickness of more than 50 meters. The middle zone of Semarang City is dominated by the basic sedimentary breccia formation, covering an area of 7,440 ha from the total area of 10,237 ha. The upper zone is dominated by geological formation of volcanic breccia sedimentary rocks of 8,803 ha from the total area of the upper zone of 14,626 ha. In general, Semarang City is dominated by surface sediment of 13,725 ha of alluvium and 11,408 ha of basic sedimentary bricks. Based on the geological map it is known that the formation of volcanic formation is more common in the upper zone. While alluvium and marine formations are more in the lower zone.

Table 2. Slope distribution by zone.

Zone	Area slope (ha)				
	0-2%	2-15%	15-25%	25-40%	> 40%
Lower	13,128	567	244	43	0
Middle	1,718	4,572	2,925	675	348
Upper	1,763	8,094	3,064	830	875
Total area of slope	16,609	13,232	6,233	1,548	1,223

Source: processing and calculation by author.

Soil type in Semarang city, dominated by type of reddish brown latosol of 12,931 ha. This type of soil is most often found in the upper zone, an area of 10,604 ha. Another dominant type of soil is the gray alluvial association, most commonly found in the lower zone, covering an area of 7,603 ha of the total area of 9,761 ha. Mediteran brown type most commonly found in the middle zone of 5,353 ha, of the total area of 7,681 ha.

Semarang City groundwater is generally available in limited quantities with moderate productivity (medium productivity, broad distribution or medium productivity, and local productivity) with an area of 22,135 ha. Areas with productive aquifers (productive and productive classes) cover an area of 7,002 ha. Areas with productive aquifers are more common in the lower Semarang zone, while the remainder can be found in the central and upper Semarang zone areas. Areas with rare earth water potentials cover a total area of 5,635 ha and areas with brackish water covering an area of 3,149 ha. Areas with high groundwater potential only cover a very small area of 922 ha only.

3.3. Profiling of disaster vulnerability (landslide, flood, and land subsidence)

Based on the three vulnerabilities of disaster, the most impacted disaster vulnerability in Semarang City area is landslide (medium and high risk) covering 9,479 ha. The second rank is a flood disaster that impacts an area of 7,893 ha. The disaster vulnerability due to the decreasing of the land surface has an impact on the area of 5.309 ha. Some areas of Semarang City are at very low landslide risk, which is 22,696 ha (58.43%) of the total area. The area of the vulnerable zone is the middle and upper zone. Total 3,483 ha area of Semarang City, located at a location with high landslide risk.

Table 3. Area of landslide risk.

Zone	Area of landslide risk level (ha)			
	Very low	Low	Medium	High
Lower	13,509	222	219	32
Middle	3,655	3,147	2,999	437
Upper	5,532	3,301	2,778	3,014
Total area	22,696	6,670	5,996	3,483

Source: processing and calculation by author.

Most of the flood-prone areas are located in the Semarang lower zone, which is 6,682 ha. This condition is related to the position of the area of the Semarang lower zone that is directly adjacent to the sea and potentially affected by rising tide. A small percentage of flood risk areas are found in the Semarang central zone (826 ha) and the Semarang upper zone (385 ha).

Semarang City area that experienced the largest decline in land surface is in the lower zone of Semarang. The zone region of central and upper Semarang, has not decreased ground level. This condition is understood, because the basic material landscape area of the lower Semarang zone area is alluvial sediment material of lower density than 2 other zone areas.

3.4. Distribution of community attitudes according to gender, education and income

Based on the distribution of the average value of respondents' attitudes in the application of rainwater harvesting presented in Table look some attitude patterns according to gender variables, education level and income. In the gender variables it appears that the attitude of men in the middle zone is much better than other zone areas. This condition is actually inversely proportional to the attitude of women in the middle zone who tend to have the lowest value compared to other zones. Female respondents had relatively uniform attitudinal values for each zone. While male respondents have more fluctuating attitude values in each zone.

Based on educational variables, it is indicated that the average value of respondents' attitudes in the application of rainwater harvesting increases in line with the rising level of education. If it is associated with the location of the zone of the respondents, then only on the group of highly educated respondents whose respondents' attitudes increase as the zone changes from bottom to top. For middle and low education groups tend to form cross pattern.

Based on average earnings in general it appears that the value of attitude increases with increasing levels of income. If it is associated with the zone area it appears that there is a difference in attitude values but the pattern of values formed to form a pattern of crossing/zig zag. To ascertain the relationship between sex variables, education level and income with respondents' attitude values in the application of rainwater harvesting it is necessary to test the correlation which will be discussed in section 3.5.

Table 4. Value distribution of attitude on variable classification.

Variable	Classification	Zone		
		Lower	Middle	Upper
Gender	Male	28.77	30.23	29.40
	Female	29.76	29.17	29.96
Education	Low	28.57	26.83	27.08
	Midle	28.56	29.28	29.06
	High	30.02	30.15	31.08
Income	Low	28.73	28.31	2909
	Midle	29.10	29.87	29.90
	High	32.27	32.29	31.56

Source: processing and calculation by author

3.5. Correlation of community attitudes with gender, education and income

The correlation test in this study used crosstab analysis. The calculation results are presented in Table 5. Based on the comparison of the calculated value with the value of the table and the comparison of its significance level at the 0.05 reference in Table 5, it can be concluded the correlation result between the variables. The sex variables were not related to attitudinal values, but the educational level and average income variables were related to community attitudes in applying rainwater harvesting. This is in line with the statement of Barthwal et al. [11] which states that social factors of income level are the decisive factor in the application of rainwater harvesting.

Table 5. Correlation of community attitudes with gender, education and income.

Respondent characteristics	Pearson Chi-Square attitude				
	Value	df	Table value	Asymp. Sig. (2-sided)	Sig. Reference
Gender	1.280	2	5.991	0.527	0.05
Education	46.001	10	18.307	0.000	0.05
Income	31.425	8	15.507	0.000	0.05

Source: processing and calculation by author using Crosstab analysis

3.6. Correlation of community attitudes with physiography landscape

The calculation of the correlation between community attitudes in applying rainwater harvesting with physiography landscape is presented in Table 6. Based on the calculation it can be concluded that there is a correlation relationship between the value of community attitudes in applying rainwater harvesting with physiography landscape where they live. This result reinforces Donohue and Biggs' [8] statement about the results of their study suggesting that there is a correlation between social conditions of society with landscape topography factors. To understand this connection, it is necessary to understand the theory of urban development patterns. The pattern of urban development can be determined by the pattern of transport network which is also influenced by the physiography landscape pattern. In case of Semarang City, the flood vulnerability and the land subsidence in lower zone are predicted to be the factors driving the community to find better settlements in the south zone area. People with higher levels of education and income tend to prefer residence in the middle and upper zone area.

Table 6. Correlation of community attitudes with physiography landscape.

Physiography landscape	Pearson Chi-Square attitude				
	Value	df	Table value	Asymp. Sig. (2-sided)	Sig. Reference
Zone 1,2 and 3	97.348	50	67.505	0.000	0.05

Source: processing and calculation by author using Crosstab analysis

3.7. Analysis volume of rainwater and water needs of the population.

The calculation of the potential volume of rainwater that can be captured and the estimated volume of household water demand is presented in Table 6. Based on the approach of the roof area and the number of population can be calculated the need and the possibility of supply for the prediction of 2030. Year 2030 is taken as a reference because that year the spatial plan the area of Semarang City must be rearranged. The calculation results show that the lower and middle zone areas have the potential to catch more rain water because the population density is higher than the upper zone area. Based on the water requirement, the lower zone area tends to require more water than twice that of the other zone. When compared directly between water demand and rainwater harvesting supply potential, the lower zone area is likely to have a water deviation of 20,884.64 x 10^6 liters per year, while the middle zone has a water surplus of 2,430.05 x 10^6 liters and an upper zone of 1,329 x 10^6 liters per year.

If the condition of supply and demand of water occurs as calculated in Table 6, it is necessary to consider the management of rainfall harvesting concerning physical condition (physiography landscape, geology, soil), social condition (education level, income), and disaster vulnerability comprehensively. The composition of the population according to the level of education, income and physiography landscape can be the basis of consideration in planning and design of implementation rainwater harvesting technology. By knowing the interrelationship between variables then it is possible to do better management of rainwater harvesting, so it can support water resilience of urban area.

Table 7. *Analysis volume of rainwater and water needs of the population.*

	Zone		
	Lower	Middle	Upper
Roof area estimation 2030	1,788.40 ha	1,758.55 ha	1,169.23 ha
Volume estimates rainwater harvest	34,337.36 (x10^6) litre	33,764.19 (x10^6) litre	22,449.30 (x10^6) litre
Projected population 2030	1,008,621	572,313	385,759
Volume estimates household water needs	55,222.00 (x10^6) litre	31,334.14 (x10^6) litre	21,120.31 (x10^6) litre

Source: processing and calculation by author

4. Conclusion

Based on the results obtained it can be concluded that the distribution and landscape characteristics have an impact or influence the attitude of the community in the application of rainwater harvesting. These results were obtained through a correlation analysis between community attitudes and their presence in different landscape physiography zones. Communities in higher physiography landscape areas are indicated to have better attitudes than people in lower and flat physiography landscape.

IOP Conf. Series: Earth and Environmental Science **179** (2018) 012043 doi:10.1088/1755-1315/179/1/012043

Differences in attitude responses are also related to the level of education and income community. In line with physiography landscape, public attitudes to apply rainwater harvesting are also indicated driven by education and income level factors. Referring to these conditions it can be concluded that the planning application of rainwater harvesting can not only consider the technical aspects only. There needs to be social considerations related to the physiography landscapes. Physiography landscape based social considerations is one of key to successful implementation of rainwater harvesting in urban areas.

Acknowledgments
The authors acknowledgee School of Environmental Sciences Universitas Indonesia (UI) publication support program. Acknowledgments are also conveyed to the Geospatial Information Agency (BIG) for data and software used in this study, and Universitas Diponegoro, especially Urban Planning Studies Program for supports in fieldwork process. Authors thanks to Sylvia Prisca Delima for her helpful comment on the first draft of this paper. Special thanks delivered to Sesa Wiguna, for his helps on processing and map making.

References

[1] United Nations International Strategy for Disaster Risk Reduction (UNISDR) 2009 *Terminology on disaster risk reduction.* UNISDR, Geneva, Switzerland.
[2] UN University 2013 *Water Security & the A Global Water Agenda* UN-Water Analytical Brief Canada.
[3] Akter A and Shoukat A 2015 Potentiality of rainwater harvesting for an urban community in Bangladesh *Journal of Hydrology* **528** 84-93
[4] Zhang D, Gersberg R M, Wilhelm C, Voigt M 2009 Decentralized water management rainwater harvesting and greywater reuse in an urban area of Beijing China *Urban Water Journal* **6** (5) 375–385
[5] Zeng B, Tan H, Wu L 2007 A New Approach to Urban Rainwater Management *Journal of China University of Mining & Technology* **17** (1) 82-84.
[6] Zaizen M, Urakawa T, Matsumoto Y, Takai H 2000 The collection of rainwater from dome stadiums in Japan *Urban Water Journal* **1** (4) 355–359
[7] Villarreal E L and Dixon A 2005 Analysis of a rainwater collection system for domestic water supply in Ringdansen Norrköping Sweden *Building and Environment Journal* **40** (9) 1174–1184
[8] Donohue C and Biggs E 2015 Monitoring socio-environmental change for sustainable development: Developing a Multidimensional Livelihoods Index (MLI) *Applied Geography* **62** 391-403
[9] Chiu Y R, Liaw C H, Chen L C 2009 Optimizing rainwater harvesting systems as an innovative approach to saving energy in hilly communities *Renewable Energy* **34** 492-498
[10] Mankad A, Walton A, Alexander K 2015 Key dimensions of public acceptance for managed aquifer recharge of urban stormwater *Journal of Cleaner Production* **89** 214-223
[11] Barthwal S, Barthwal S C, Goyal H, Nirmani B, Awasthi B 2013 Socio-economic acceptance of rooftop rainwater harvesting A case study *Urban Water Journal*
[12] Ghisi E 2006 Potential for potable water savings by using rainwater in the residential sector of Brazil *Building and Environment Journal* **41** (11) 1544–1550
[13] Nolde E 2007 Possibilities of rainwater utilisation in densely populated areas including precipitation runoffs from traffic surfaces *Desalination Journal* **215** (1–3) 1–11
[14] Abdulla F A and Al-Shareef A W 2009 Roof rainwater harvesting systems for household water supply in Jordan *Journal of Desalination* **243** (1-3) 195-207
[15] Jeremy G, Sam T and Diveshkumar M 2012 Mandatory urban rainwater harvesting: learning from experience *Water Science & Technology* **65** (7) 1200-1207
[16] Rida A, Abdullah D, Ghada S 2010 Combining GIS with multicriteria decision making for siting water harvesting ponds in Northern Jordan *Journal of Arid Environments* **74** 1471-1477

[17] Guebaili A, Djebbar Y, Guedri A, Boukhari S 2011 Rainwater harvesting in North Africa: A novel method for reservoir sizing *Journal of Materials and Environmental Science* **2** *(S1) 469-472*

[18] Farrenya R, Gabarrella X, Rieradevall J 2011 Cost-efficiency of rainwater harvesting strategies in dense Mediterranean neighbourhoods *Resources, Conservation and Recycling* **55** 686–694

[19] Jaemin S, Mooyoung H, Tschung-il K, Jee-eun S 2009 Rainwater harvesting as a sustainable water supply option in Banda Aceh *Desalination* **248** 233–240

[20] Winnaar G, Jewitt G P W, Horan M 2007 A GIS-based approach for identifying potential runoff harvesting sites in the Thukela River basin, South Africa *Physics and Chemistry of the Earth* **32** 1058–1067

[21] Ramon F, Tito M P, Albert G, Carlota T, Joan R, Xavier G 2011 Roof selection for rainwater harvesting: Quantity and quality assessments in Spain *Water Research* **45** 3245-3254